Complex Varia[illegible]
and their Applications

International Mathematics Series

Consulting Editor: A Jeffrey, University of Newcastle upon Tyne

Forthcoming titles in the Series

Introduction to Numerical Analysis
A Wood

Other titles in the Series

Discrete Mathematics: Numbers and Beyond
Stephen Barnett

Complex Variables and their Applications

ANTHONY D. OSBORNE

Department of Mathematics
Keele University

ADDISON WESLEY LONGMAN

HARLOW, ENGLAND ■ READING, MASSACHUSETTS ■ NEW YORK
MENLO PARK, CALIFORNIA ■ DON MILLS, ONTARIO ■ AMSTERDAM
BONN ■ SYDNEY ■ SINGAPORE ■ TOKYO ■ MADRID ■ SAN JUAN
MILAN ■ MEXICO CITY ■ SEOUL ■ TAIPEI

Addison Wesley Longman Limited
Edinburgh Gate
Harlow
Essex CM20 2JE
England

and Associated Companies throughout the world.

Typeset by 32
Printed and bound in Great Britain by Biddles Ltd., Guildford and King's Lynn.

First published 1999

ISBN 0-201-34290-1

British Library Cataloguing-in-Publication Data
A catalogue record for this book is available from the British Library

Contents

Preface

In writing this volume it has been my intention to produce a book which is fairly versatile. My hope is that not only is this book suitable for both a first and second course in complex variables for mathematicians, but it contains enough applications to be of use and interest to engineering and other science students. It is written specifically with today's undergraduates in mind and only assumes a knowledge of basic real analysis and calculus. Some of the material has evolved from a lecture course that I have given to second-year undergraduates over a number of years. The text includes the standard techniques and applications of complex variables, with plentiful examples. Generally, I have tried to give generic applications in the form of residue theory, differential equations and integral transforms, but specific applications are given in Chapter 9. The book also presents the important analytical concepts and techniques used in deriving the standard results in complex analysis. Readers who are more interested in applications may wish to leave these derivations and go straight to the calculations and examples.

My hope is that this book can be used at different levels, depending on the sections that are chosen. Any material which is normally covered in a first course is included within the first six chapters. The next three chapters deal with applications other than residue theory. Although some of the applications concerning differential equations are not often given in books on complex variables, it is my belief that this is the natural setting for the material I have included. Some of the material in Chapters 5, 6, 9, 10, and possibly 11, is suitable for inclusion in a second course. A short bibliography, which suggests further reading, but which is by no means intended to be exhaustive, is given at the end of the book.

Notation which is possibly non-standard is defined when used. When no ambiguity can occur, Theorem 2.5 is sometimes abbreviated to 2.5 and equation (2.5) is sometimes abbreviated to (2.5).

I have endeavoured to provide a large number of exercises at the end of each chapter, ranging from routine, through those that test understanding of the main concepts, to hopefully quite challenging questions. The most challenging exercises are marked with a star. A large number of answers have been included. A solutions manual, which provides the full solutions to all the questions, is available to lecturers from the publishers.

I am indebted to Dr David Bedford, who read some of the first draft of my manuscript and made helpful suggestions. I also thank Professor Jeffrey and the anonymous reviewers for their helpful advice and constructive criticisms. Finally, I am grateful to the editorial staff at Addison Wesley Longman for their support and advice, and all the staff in the production department who made this book possible.

Anthony D. Osborne
Keele, April 1998

1 Functions of a Complex Variable

This book is concerned with functions which map complex numbers to complex numbers. Although anyone wishing to study complex variables will have some knowledge of complex numbers, we begin with a review of such numbers and their properties. The chapter continues by introducing functions of a complex variable and the standard elementary functions.

Complex Numbers

It first becomes clear that we need to consider numbers other than real numbers and, in particular, roots of negative real numbers, when we are seeking solutions of quadratic equations. For example, consider the quadratic equation,

$$x(10 - x) = 40 \tag{1.1}$$

Then $x^2 - 10x + 40 = 0 \quad \Rightarrow \quad (x-5)^2 + 15 = 0 \quad \Rightarrow \quad x = 5 \pm \sqrt{-15}$ and the square of any real number is clearly non-negative. Hence x is not a real number and the two solutions are examples of complex numbers. Geometrically, the fact that there are no real solutions to (1.1) corresponds to the fact that the parabola with equation $y = x^2 - 10x + 40$ does not intersect the x-axis.

Definition

A **complex number** is a number of the form $x + iy$, where x and y are any real numbers and i represents $\sqrt{-1}$. Such a number is often written as the ordered pair (x, y).

Notation

The set of all complex numbers will be denoted by $\mathbb{C}$ and the set of all real numbers by $\mathbb{R}$.

The notation $x + iy$ for the ordered pair (x, y) can be formally justified by defining the sum and product of any two complex numbers as below. Note that $\mathbb{R}$ is a proper subset of $\mathbb{C}$.

Note

Engineers and scientists tend to use the symbol j for $\sqrt{-1}$ since i is reserved for current.

Historical Note

Roots of negative real numbers were first introduced by Cardano in his *Ars Magna*, on the solution of algebraic equations, in 1545. He gave a single example, the quadratic equation (1.1). Bombelli was the first mathematician to manipulate complex numbers, in his *Algebra*, in 1572. However, full acceptance of complex numbers only came in the nineteenth century. Gauss gave complex numbers their present name and used them in his proof of the fact that any polynomial equation of degree n has exactly n roots in $\mathbb{C}$. It was Euler who introduced the symbol i in a memoir in 1777.

Definitions

Let $z = x + iy = (x, y) \in \mathbb{C}$. Then x and y are called the **real part** and **imaginary part** of z respectively. We write $x = \text{Re } z$ and $y = \text{Im } z$.

Note that it is y, not iy, that is the imaginary part of z. The word *imaginary* is unfortunate, but for historical reasons this misnomer has stuck. Equality of complex numbers corresponds to the definition of equality of ordered pairs.

Definition

Two complex numbers, z_1 and z_2, are **equal**, written $z_1 = z_2$ in the usual way, if their real parts are equal and their imaginary parts are equal. This is often termed comparing real and imaginary parts.

Turning now to the arithmetic of complex numbers, the most fruitful way to define the usual operations is in such a way as to ensure that $i^2 = -1$ and that $\mathbb{C}$ has the same algebraic properties as $\mathbb{R}$.

Definitions

Consider complex numbers as ordered pairs of real numbers. Given $z_1 = (x_1, y_1)$ and $z_2 = (x_2, y_2) \in \mathbb{C}$, their **sum** and **product** are denoted and defined respectively by

$$z_1 + z_2 = (x_1 + x_2, y_1 + y_2)$$

$$z_1 z_2 = (x_1 x_2 - y_1 y_2, x_1 y_2 + y_1 x_2)$$

Notice that these definitions include the sum and product of real numbers as special cases and give the usual componentwise addition and multiplication by a scalar when $\mathbb{C}$ is treated as the vector space $\mathbb{R}^2$. It is easily checked that the second definition gives $(0, 1)(0, 1) = (-1, 0)$; that is, $i^2 = -1$. The first definition gives the formal justification for the notation $x + iy$ for (x, y). Written in a different way, these definitions read

$$(x_1 + iy_1) + (x_2 + iy_2) = (x_1 + x_2) + i(y_1 + y_2)$$

$$(x_1 + iy_1)(x_2 + iy_2) = x_1 x_2 + i(y_1 x_2 + x_1 y_2) + i^2 y_1 y_2$$

Hence complex numbers may be added and multiplied by manipulating them as real numbers and replacing i^2 by -1 whenever it occurs. It can be checked, using the definitions, that $\mathbb{C}$ has essentially the same algebraic properties as $\mathbb{R}$; that is, $\mathbb{C}$ is a **field**. This is left as an exercise.

Note The Field Properties of $\mathbb{C}$

Let z_1, z_2 and $z_3 \in \mathbb{C}$. Then

1. $z_1 + z_2 \in \mathbb{C}$ and $z_1 z_2 \in \mathbb{C}$
2. $z_1 + z_2 = z_2 + z_1$ and $z_1 z_2 = z_2 z_1$ (commutative laws)
3. $(z_1 + z_2) + z_3 = z_1 + (z_2 + z_3)$ and $(z_1 z_2) z_3 = z_1 (z_2 z_3)$ (associative laws)
4. $z_1(z_2 + z_3) = z_1 z_2 + z_1 z_3$ (distributive law)
5. There is a unique number $0 = (0, 0) \in \mathbb{C}$ such that $z + 0 = z$ for all $z \in \mathbb{C}$
6. Given $z \in \mathbb{C}$ there is unique number $-z \in \mathbb{C}$ such that $z + (-z) = 0$. Then $z_2 - z_1 = z_2 + (-z_1)$ (subtraction)
7. There is a unique number $1 = (1, 0) \in \mathbb{C}$ such that $1z = z$ for all $z \in \mathbb{C}$
8. Given $z \in \mathbb{C}$, $z \neq 0$ there is a unique number $z^{-1} \in \mathbb{C}$ such that $zz^{-1} = 1$. Then $\dfrac{z_2}{z_1} = z_2 z_1^{-1}$, $z_1 \neq 0$ (division)

Important Notes

(i) Using these eight rules, complex numbers may be added, subtracted, multiplied and divided by formally manipulating them as real numbers and replacing i^2 by -1 whenever it occurs.

(ii) The usual ordering in $\mathbb{R}$, which compares positions of real numbers on the real line, cannot be extended to $\mathbb{C}$ which, geometrically speaking, consists of points in the plane. Thus, a statement such as $z_1 < z_2$ has no meaning unless $z_1, z_2 \in \mathbb{R}$.

Example 1.1

(i) $(6 + 3i) - (4 - 2i) = 2 + 5i$

(ii) $(7 + 4i)(3 - 2i) = 21 + 12i - 14i - 8i^2 = 29 - 2i$

(iii) Suppose that a solution is sought to the equation

$$5x + i(3 - y) + (x + 1) + i(2y - 1) = (3 + 2i)(1 - i)$$

By using the usual algebraic rules and replacing i^2 when it occurs by -1, the equation simplifies to

$$(6x + 1) + i(y + 2) = 5 - i$$

It then follows by the definition of equality that

$$6x + 1 = 5 \quad \text{and} \quad y + 2 = -1 \text{ so that } x = 2/3 \quad \text{and} \quad y = -3$$

(iv) Let $(2+3i)/(1-2i) = x+iy$. Then

$$\begin{aligned} 2+3i &= (1-2i)(x+iy) = (x+2y)+i(y-2x) \\ \Rightarrow \quad & x+2y=2 \quad \text{and} \quad y-2x=3 \\ \Rightarrow \quad & x=-4/5 \quad \text{and} \quad y=7/5 \end{aligned}$$

In practice, the division process is straightforward, once the ideas of the modulus and complex conjugate of a complex number have been introduced.

Definitions

Let $z = x+iy \in \mathbb{C}$. The **modulus** of z is denoted and defined by $|z| = \sqrt{x^2+y^2}$. The **complex conjugate** of z is denoted and defined by $\overline{z} = x - iy$.

Notes

(i) We follow the usual convention that if a is any positive real number, then $\sqrt{a}$ denotes the positive root of a.

(ii) Unlike in the real case, $|z|^2 \neq z^2$ in general. Although $z \in \mathbb{C}$, $|z| \in \mathbb{R}$. You have been warned!

Note that $|z|$ gives a measure of the magnitude of z. In particular, $z = 0$ if and only if $|z| = 0$. Note also that z is real if and only if $\overline{z} = z$. The following two results give the important properties of $\overline{z}$ and $|z|$. The modulus has some of the properties of its real counterpart.

Lemma 1.1. Properties of the Complex Conjugate

(i) $\overline{z_1+z_2} = \overline{z_1}+\overline{z_2}$
(ii) $\overline{z_1 z_2} = \overline{z_1}\,\overline{z_2}$
$\Big\}$ for all $z_1, z_2 \in \mathbb{C}$

(iii) $\overline{\overline{z}} = z$
(iv) $z\overline{z} = |z|^2$
$\Big\}$ for all $z \in \mathbb{C}$ □

Proof

The results follow directly from the definitions.

(i) Let $z_1 = x_1+iy_1$ and $z_2 = x_2+iy_2$. Then

$$\begin{aligned} \overline{z_1+z_2} &= \overline{(x_1+x_2)+i(y_1+y_2)} = (x_1+x_2)-i(y_1+y_2) \\ \Rightarrow \quad \overline{z_1+z_2} &= (x_1-iy_1)+(x_2-iy_2) = \overline{z_1}+\overline{z_2} \end{aligned}$$

(ii) This is similar and is left as an exercise.

(iii) This is trivial.

(iv) Let $z = x + iy$. Then

$$z\bar{z} = (x + iy)(x - iy) = x^2 + y^2 = |z|^2$$ ■

Note that (iv) gives the factorisation of the sum of two squares. And from (ii) it follows that $\overline{-z_2} = -(\overline{z_2})$, hence (i) implies that $\overline{z_1 - z_2} = \overline{z_1} - \overline{z_2}$.

Lemma 1.2. Properties of the Modulus of a Complex Number

(i) $|z_1 z_2| = |z_1||z_2|$ for all $z_1, z_2 \in \mathbb{C}$

(ii) $|\bar{z}| = |z|$

(iii) $z + \bar{z} = 2\,\mathrm{Re}\, z \leqslant 2|z|$

(iv) $z - \bar{z} = 2\,\mathrm{Im}\, z \leqslant 2|z|$

(ii)–(iv) for all $z \in \mathbb{C}$

(v) $|z_1 + z_2| \leqslant |z_1| + |z_2|$ for all $z_1, z_2 \in \mathbb{C}$ (triangle inequality) □

Proof

(i)
$$\begin{aligned} |z_1 z_2|^2 &= (z_1 z_2)(\overline{z_1 z_2}) && \text{by 1.1(iv)} \\ &= (z_1 \overline{z_1})(z_2 \overline{z_2}) && \text{by 1.1(ii)} \\ &= |z_1|^2 |z_2|^2 && \text{by 1.1(iv)} \end{aligned}$$

The result follows on taking positive square roots since $|z| \geqslant 0$ for all $z \in \mathbb{C}$.

(ii) This is trivial.

(iii) Let $z = x + iy$. Then

$$z + \bar{z} = (x + iy) + (x - iy) = 2x \leqslant 2\sqrt{x^2 + y^2}$$

and similarly for (iv).

(v)
$$\begin{aligned} |z_1 + z_2|^2 &= (z_1 + z_2)(\overline{z_1 + z_2}) && \text{by 1.1(iv)} \\ &= z_1 \overline{z_1} + z_2 \overline{z_1} + z_1 \overline{z_2} + z_2 \overline{z_2} && \text{by 1.1(i)} \\ &= |z_1|^2 + |z_2|^2 + (z_1 \overline{z_2} + \overline{z_1 \overline{z_2}}) && \text{by 1.1(ii)–(iv)} \\ &\leqslant |z_1|^2 + |z_2|^2 + 2|z_1 \overline{z_2}| && \text{by (iii)} \\ &= (|z_1| + |z_2|)^2 && \text{by (i), (ii)} \end{aligned}$$

Then taking the positive square root gives the result. ■

Important Notes

(i) Let $z_1, z_2 \in \mathbb{C}$ with $z_2 \neq 0$. Then, by 1.1(iv),

$$\frac{z_1}{z_2} = \frac{z_1 \overline{z_2}}{z_2 \overline{z_2}} = \frac{z_1 \overline{z_2}}{|z_2|^2}$$

Since $|z_2|$ is real, this provides a practical method for division.

(ii) As in the real case, it follows by 1.2(v) that

$$|z_i| = |z_i \pm z_j \mp z_j| \leqslant |z_i \pm z_j| + |z_j| \quad \text{for } i, j = 1, 2\ (i \neq j)$$

Hence

$$||z_1| - |z_2|| \leqslant |z_1 \pm z_2|$$

Example 1.2

Using the above procedure for division,

$$\frac{2+3i}{1-2i} = \frac{(2+3i)(1+2i)}{(1-2i)(1+2i)} = \frac{-4+7i}{1+4} = -\frac{4}{5} + \frac{7i}{5}$$

(Compare this with Example 1.1(iv).)

The power of the triangle inequality is illustrated in the following example.

Example 1.3 The Enestrom–Kakeya Theorem

Let $a_i \in \mathbb{R}$, $i = 0, 1, \ldots, n$ with $a_0 > a_1 > a_2 \ldots > a_n > 0$. Prove that all the roots of the polynomial

$$P(z) = a_n z^n + a_{n-1} z^{n-1} + \ldots + a_1 z + a_0$$

satisfy $|z| > 1$.

Solution

Note that

$$a_0 = (1-z)P(z) + (a_0 - a_1)z + \ldots + (a_{n-1} - a_n)z^n + a_n z^{n+1}$$

Hence by the triangle inequality, which is clearly a strict inequality in this case, for $|z| \leqslant 1$ with $z \neq 1$,

$$|a_0| < |(1-z)P(z)| + (a_0 - a_1) + \ldots + (a_{n-1} - a_n) + a_n$$

$$\Rightarrow \quad a_0 < |(1-z)P(z)| + a_0 \quad \Rightarrow \quad |P(z)| > 0$$

and $P(1) \neq 0$, as required.

Note

For any $z \in \mathbb{C}$, z^n for $n \in \mathbb{Z}$ is defined exactly as for powers of real numbers, so the usual rules of indices apply.

Exercise 1.1.1 Let $z_1 = 2 + i$, $z_2 = 3 - 4i$ and $z_3 = 7 + 5i$. Find

(i) $z_1 - 2z_2$

(ii) $z_1 z_3 + z_2$

(iii) z_2^3

(iv) $\dfrac{z_1}{z_3}$

(v) $\dfrac{z_1 z_2}{z_1 + \overline{z}_3}$

Exercise 1.1.2 Give a simple example to show that if $z \in \mathbb{C}$, $|z|^2 \neq z^2$ in general.

Exercise 1.1.3 Find the values of

(i) $\displaystyle\sum_{n=0}^{11} i^n$

(ii) $\left|\dfrac{1}{1+3i} - \dfrac{1}{1-3i}\right|$

Exercise 1.1.4 Let

$$z = \frac{1 - \cos\theta + i\sin\theta}{1 + \cos\theta - i\sin\theta}$$

where θ is real. Show that $\operatorname{Re} z = 0$ and $\operatorname{Im} z = \tan(\theta/2)$.

Exercise 1.1.5 Find the real numbers a and b such that $z = 1 + 2i$ is a solution to the cubic equation

$$z^3 + az + b = 0$$

and find all other solutions in this case.

Exercise 1.1.6 Prove that $z_1 z_2 = 0$ if and only if $z_1 = 0$ or $z_2 = 0$.

Exercise 1.1.7 Find the roots of the equation $z^3 = 1$. If $z^3 = 1$ with $z \neq 1$, show that $z^2 + z^4 = -1$.

Exercise 1.1.8 Find the values of $z \in \mathbb{C}$ which satisfy $|z - i| \leqslant |z - 2|$.

Exercise 1.1.9 Show that for $\alpha,\ \beta \in \mathbb{R}$ with $0 \leqslant \alpha \leqslant 1$ and $0 \leqslant \beta \leqslant 1$, $\alpha^2 + \beta^2 \leqslant 1 + \alpha^2\beta^2$. Hence prove that for $a,\ z \in \mathbb{C}$, with $|a| \leqslant 1$ and $|z| \leqslant 1$,

$$\left|\frac{z - a}{1 - \overline{a}z}\right| \leqslant 1$$

Exercise 1.1.10

(i) Find all 2×2 matrices $\boldsymbol{A}$, with real entries, which satisfy $\boldsymbol{A}^2 = -\boldsymbol{I}_2$.

(ii) Prove that if n is odd, then there is no $n \times n$ matrix $\boldsymbol{A}$ with real entries such that $\boldsymbol{A}^2 = -\boldsymbol{I}_n$.

Exercise **1.1.11** Let

$$S = \left\{ \begin{bmatrix} x & y \\ -y & x \end{bmatrix}, x, y \in \mathbb{R} \right\}$$

and define $f : \mathbb{C} \to S$ by

$$f(z) = f(x+iy) = \begin{bmatrix} x & y \\ -y & x \end{bmatrix}$$

Prove that $f(z_1 + z_2) = f(z_1) + f(z_2)$ and $f(z_1 z_2) = f(z_1)f(z_2) = f(z_2)f(z_1)$ for all $z_1, z_2 \in \mathbb{C}$. (S is then algebraically identical to $\mathbb{C}$.)

Exercise **1.1.12** Prove that $\mathbb{C}$ satisfies the field axioms.

The Complex Plane

Recall that real numbers may be represented as points on a line and that a complex number z is an ordered pair of real numbers. Hence such a number is represented uniquely by the point P with Cartesian coordinates (x, y), as shown in Fig. 1.1. In this way, $\mathbb{C}$ is identified with $\mathbb{R}^2$. The x-axis is the **real axis** and the y-axis is the **imaginary axis**.

Note that the length of OP gives $|z|$ and the reflection of P in the x-axis represents $\bar{z}$. More generally, if a point P_1 represents z_1 and a point P_2 represents z_2, then $|z_1 - z_2|$ is the length of the line segment P_1P_2, by Pythagoras' theorem. The complex number z can be thought of as the position vector of the point $P(x, y)$. In this way, geometrically speaking, addition and subtraction of complex numbers is equivalent to addition and subtraction of the corresponding vectors in the complex plane. The triangle inequality for complex numbers, Lemma 1.2(v), corresponds to the triangle inequality for (two-dimensional) vectors. Lastly, notice that P can be located using the usual plane polar coordinates r and θ, (Fig. 1.1), so $x + iy = r(\cos\theta + i\sin\theta)$.

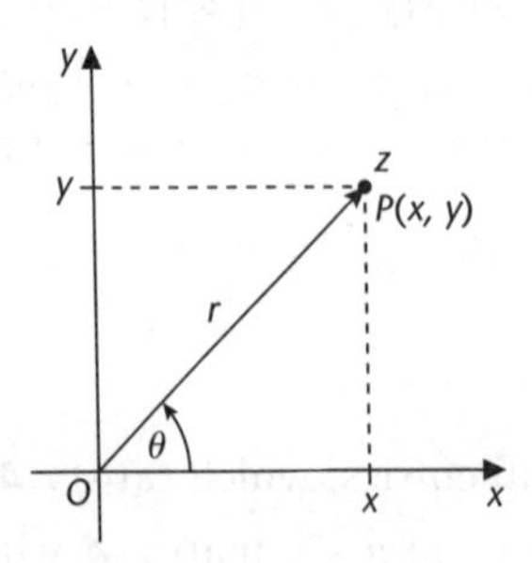

Figure 1.1

Historical Note

Complex numbers were first represented geometrically in John Wallis's *Algebra* of 1673. And in 1797 Casper Wessel published a paper describing the representation of a complex number as a point in the plane. This approach was investigated independently by Argand in 1806. In 1837 Hamilton gave the definition of a complex number as an ordered pair of real numbers. Together with some rules for their manipulation, Hamilton's definition placed complex numbers on a firm algebraic footing.

Since any point in the plane with Cartesian coordinates (x, y) can be identified with a complex number z, it is possible to translate Cartesian equations of curves in the plane to equations involving the single complex variable z, and vice versa. Perhaps the simplest and most important case is that of a circle. The following result is a consequence of the comments above.

Important Note

The equation of a circle, centred at $\alpha \in \mathbb{C}$, with radius r, is $|z - \alpha| = r$. This result can be verified algebraically by letting $z = x + iy$ and $\alpha = a + ib$.

Example 1.4

(i) Consider the equation

$$|z + 3i| = \operatorname{Im} z + 4$$

Letting $z = x + iy$ gives

$$x^2 + (y + 3)^2 = (y + 4)^2 \quad \Rightarrow \quad 2y = x^2 - 7$$

Hence the equation represents a parabola in the complex plane.

(ii) The equation $|z - 1| = 2$ is the equation of a circle centre $(1, 0)$ and radius 2. Hence the set of points satisfying $|z - 1| < 2$ is the set of points 'inside' this circle, excluding the circle itself. Thus the set of points satisfying $|z - 1| < 2$ and $|z + 1| < |z - 3|$ consists of those points (x, y) inside the given circle such that $x < 1$. This is shown as the shaded area in Fig. 1.2.

Certain curves in the complex plane are best described parametrically. This is particularly true of line segments. It follows easily using vector algebra that any point on the line passing through the endpoints of two fixed, non-parallel vectors $\boldsymbol{a}$ and $\boldsymbol{b}$ has position vector $\boldsymbol{r} = (\boldsymbol{b} - \boldsymbol{a})t + \boldsymbol{a}$, where t is a real parameter. Hence we have the following result.

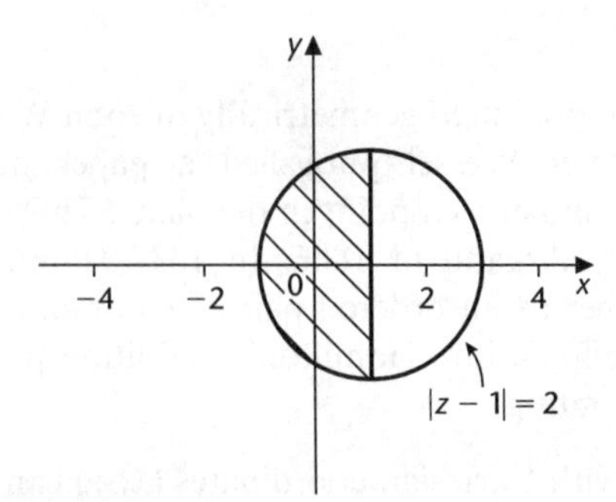

Figure 1.2

Important Note

The line segment with endpoints α and $\beta \in \mathbb{C}$ is the set of points $\{z = (1-t)\alpha + \beta t : 0 \leqslant t \leqslant 1\}$. For example, the line segment joining 1 to i is the set of points $\{z = (1-t) + it : 0 \leqslant t \leqslant 1\}$. This is equivalent to saying that the line segment has Cartesian equation $y = 1 - x$, $0 \leqslant x \leqslant 1$. In this case, if x is chosen as parameter, x runs between 1 and 0. More generally, a curve with Cartesian equation $y = f(x)$, $a \leqslant x \leqslant b$, is the set of points $\{z = t + if(t) : a \leqslant t \leqslant b\}$.

We require certain fundamental topological concepts in the complex plane for later work. Since $\mathbb{C}$ can be identified with $\mathbb{R}^2$, these are precisely the usual topological definitions in $\mathbb{R}^2$, which are natural generalisations of those in $\mathbb{R}$. Of particular importance are the ideas of open and closed neighbourhoods, which are the two-dimensional analogues of open and closed intervals of the real line. An open δ -neighbourhood of a point $\alpha \in \mathbb{C}$ is the set of all points whose distance is less than δ from α and so is the 'inside' of the circle, centre α and radius δ. A closed δ-neighbourhood includes all points on this circle as well. In a deleted δ-neighbourhood of α, the point α itself is excluded.

Definitions

Let $\alpha \in \mathbb{C}$ and $\delta \in \mathbb{R}^+$, the set of positive real numbers. An **open δ-neighbourhood (or open δ-disc) of α** is the set of points $\mathcal{N}(\alpha, \delta) = \{z : |z-\alpha| < \delta\}$. A **closed δ-neighbourhood (or closed δ-disc) of α** is the set of points $\overline{\mathcal{N}}(\alpha, \delta) = \{z : |z-\alpha| \leqslant \delta\}$. An **open deleted δ-neighbourhood (or punctured δ-disc) of α** is the set of points $\mathcal{N}'(\alpha, \delta) = \{z : 0 < |z-\alpha| < \delta\}$.

Two open neighbourhoods are pictured in Fig. 1.3. The ideas of interior, exterior and boundary points essentially correspond to our intuitive ideas.

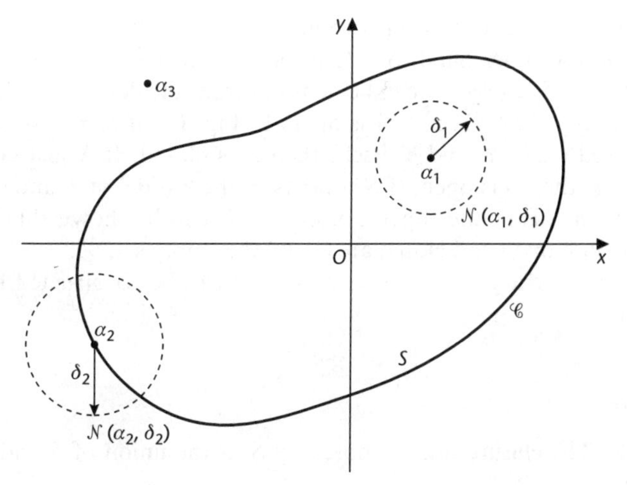

Figure 1.3

Definitions

Let $S \subseteq \mathbb{C}$. The complex number α is an **interior point** of S if there is some open δ-neighbourhood, $\mathcal{N}(\alpha, \delta)$, such that $\mathcal{N}(\alpha, \delta) \subset S$. The complex number α is a **boundary point** of S if every open neighbourhood of α contains points belonging to S and points not in S. If a point is not an interior or boundary point of S, then it is an **exterior** point of S.

For example, in Fig. 1.3, suppose that the set S consists of all points 'inside' or on the closed curve $\mathscr{C}$ shown. The point α_1 is an interior point of S, α_2 is a boundary point of S and α_3 is an exterior point of S. This is also true if S just consists of all points 'inside' $\mathscr{C}$.

In general, $S \subseteq \mathbb{C}$ need not have all three types of points. For example, a finite set has no interior points, and $\mathbb{C}$ has no exterior points.

Definition

The point $\alpha \in \mathbb{C}$ is a **limit point** of $S \subseteq \mathbb{C}$ if every deleted neighbourhood of α contains at least one point of S.

Thus, any limit point of S must have other points of S 'arbitrarily close' to it, but need not belong to S. For example, in Fig 1.3, α_1 and α_2 are limit points of S. If S excludes $\mathscr{C}$ itself, then $\alpha_2 \notin S$. It is clear that any limit point is an interior or boundary point.

Definitions

Let $S \subseteq \mathbb{C}$. Then S is **open** if every point of S is an interior point. The set S is **closed** if S contains all its limit points.

Intuitively, an open set is a set without a boundary. Any open neighbourhood is open. On the other hand, $S \subset \mathbb{C}$ is closed if it has a boundary. Any closed neighbourhood is closed. It can be deduced from the definitions that S is closed if the set $\mathbb{C} \backslash S = \{z \in \mathbb{C} : z \notin S\}$ is open. In Fig. 1.3, if S consists of the 'inside' of the closed curve $\mathscr{C}$ and $\mathscr{C}$ itself, then S is closed. If S just consists of the 'inside' of $\mathscr{C}$, then S is open. If S consists of the 'inside' of $\mathscr{C}$ and some, but not all of $\mathscr{C}$, then S is neither open nor closed. It can be shown that the only sets which are both open and closed are $\mathbb{C}$ and the empty set, $\varnothing$.

Given an arbitrary set $S \subset \mathbb{C}$, a closed set can be constructed by 'adding its boundary'.

Definition

Let $S \subseteq \mathbb{C}$. The **closure** of S, denoted by $\overline{S}$, is the union of S and the set of its limit points.

Clearly, the closure of any set is closed, and if S is closed then $\overline{S} = S$.

Definitions

A **(polygonally) connected set** $S \subseteq \mathbb{C}$ is a set for which any two points of S can be joined by a path consisting of line segments, all points of which lie in S. A **region** is a non-empty, open connected set.

Intuitively, a connected set consists of a single piece of the complex plane. The set S in Fig. 1.3 is connected. As we shall see later, it is often convenient to take the domain of a function of a complex variable to be a region of $\mathbb{C}$.

Definition

A connected set S is **simply connected** if $\mathbb{C} \backslash S$ is connected.

Basically speaking, $S \subseteq \mathbb{C}$ is simply connected if it has no 'holes'. For instance, the open neighbourhood $\{z : |z| < 2\}$ is a simply connected region, since $\{z : |z| \geqslant 2\}$ is connected, whereas the annular region $\{z : 1 < |z| < 2\}$ is not simply connected, since $A = \{z : |z| \leqslant 1 \text{ or } |z| \geqslant 2\}$ is not connected. This is demonstrated in Fig. 1.4, where there is clearly no path of line segments connecting α to β lying within A.

Definitions

A set $S \subseteq \mathbb{C}$ is **bounded** if there is a positive real constant M such that $|z| \leqslant M$ for all $z \in S$. An **unbounded** set is one which is not bounded. A set is **compact** if it is closed and bounded. Notice that any closed δ-neighbourhood is compact.

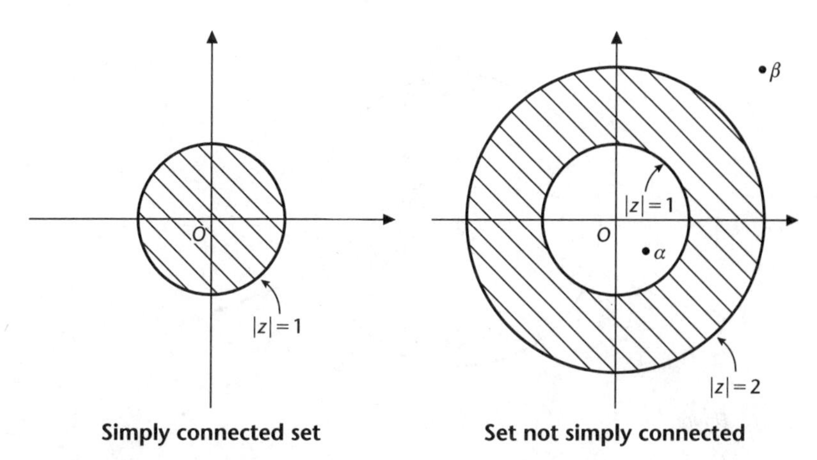

Figure 1.4

Important Note

Any closed curve in the plane which is not self-intersecting except at its end points divides the plane into two regions, one bounded, called the **inside**, and the other unbounded, called the **outside**, having the curve as a common boundary. This is the **Jordan curve theorem**.

The Riemann Sphere

Often it is useful to extend $\mathbb{C}$ by including a **point at infinity**. For example, this can be useful when considering the behaviour of complex functions as $|z| \to \infty$. This extended set, denoted by $\tilde{\mathbb{C}}$, is the **extended complex plane**. $\mathbb{C}$ can be regarded as being embedded in $\mathbb{R}^3$ by identifying $x + iy$ with $(x, y, 0)$. A geometric model for $\tilde{\mathbb{C}}$ is then the unit sphere with Cartesian equation $x^2 + y^2 + u^2 = 1$ in $\mathbb{R}^3$. A given point z in the plane may be associated with that point Z on the sphere such that the line passing through the north pole $N(0, 0, 1)$ and z intersects the sphere at Z, as shown in Fig. 1.5. Thus if $|z| < 1$, then Z lies in the southern hemisphere with the south pole $S(0, 0, -1)$ corresponding to 0; if $|z| = 1$, then Z coincides with z; if $|z| > 1$, then Z lies in the northern hemisphere, with N corresponding to the point at infinity. This geometric model for $\tilde{\mathbb{C}}$ is the **Riemann sphere** and the mapping $z \to Z$ is **stereographic projection**.

Important Note

There is only *one* point at infinity in $\tilde{\mathbb{C}}$, represented geometrically by the north pole on the Riemann sphere.

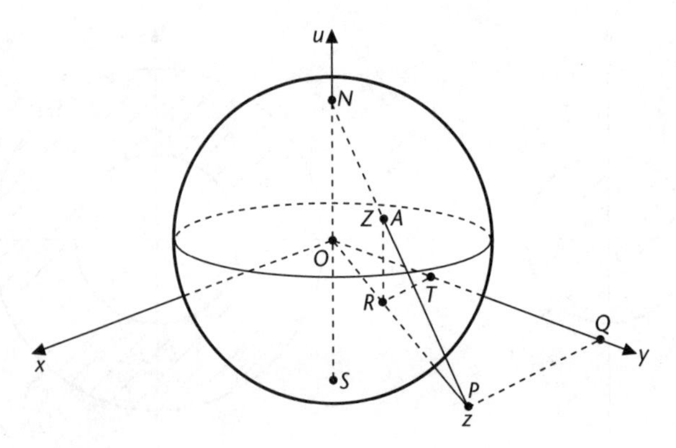

Figure 1.5

We can further investigate the correspondence between $z \in \mathbb{C}$ and its image Z on the Riemann sphere by using elementary geometry, as in the proof of the following lemma.

Lemma 1.3. Stereographic Projection

Let $Z(X, Y, U)$ be the image on the Riemann sphere of the point $z(x, y, 0)$. Then

$$\frac{x}{X} = \frac{y}{Y} = \frac{1}{1-U} \tag{1.2}$$

$$X = \frac{2x}{|z|^2+1} \qquad Y = \frac{2y}{|z|^2+1} \qquad U = \frac{|z|^2-1}{|z|^2+1} \tag{1.3}$$

□

Proof

Referring to Fig. 1.5, since triangles OPQ and ORT are similar,

$$\frac{x}{X} = \frac{y}{Y} = \frac{|z|}{\sqrt{X^2+Y^2}} \tag{1.4}$$

Since triangles ONP and RAP are also similar,

$$\frac{1}{|z|} = \frac{U}{|z| - \sqrt{X^2+Y^2}} \quad \Rightarrow \quad 1 - U = \frac{\sqrt{X^2+Y^2}}{|z|} \tag{1.5}$$

and (1.2) follows from (1.4) and (1.5). Finally, since $X^2 + Y^2 + U^2 = 1$, it follows from (1.2) that

$$|z|^2 + 1 = \frac{X^2}{(1-U)^2} + \frac{Y^2}{(1-U)^2} + 1 = \frac{2-2U}{(1-U)^2} = \frac{2}{1-U} \tag{1.6}$$

Then (1.3) follows from (1.2) and (1.6). ■

It can be shown, using Lemma 1.3, that under stereographic projection, the image of any circle in the complex plane is a circle on the Riemann sphere which does not pass through N, whereas the image of any line in the plane is a circle on the sphere which does pass through N. For example, the unit circle, centre 0, is clearly mapped to the equator on the sphere.

Example 1.5

(i) Consider the line with equation $x + y = 1$ in the complex plane. It follows from (1.2) that the image of this line on the Riemann sphere is given by

$$\frac{X}{1-U} + \frac{Y}{1-U} = 1 \quad \Rightarrow \quad X + Y + U = 1$$

where $X^2 + Y^2 + U^2 = 1$, i.e. the intersection of a plane and the sphere, which is a circle passing through (0, 0, 1).

(ii) The image of the circle with equation $x^2 + y^2 = 3$ in the complex plane, i.e. $|z|^2 = 3$, is given by $X = x/2$, $Y = y/2$ and $U = 1/2$ by (1.3). Hence the image of this circle has equations $4X^2 + 4Y^2 = 3$, $U = 1/2$, which is a circle on the sphere that does not pass through (0, 0, 1).

Exercise **1.2.1** Sketch the set of values of z in the complex plane for which

(i) $|z-2| < 2$ and $|z-i| > 2$

(ii) $\operatorname{Re} z - \operatorname{Im} z < 1$

(iii) $\left|\dfrac{z+2i}{z-2i}\right| \leqslant 3$

Exercise **1.2.2** Prove that

$$|z_1 + z_2|^2 + |z_1 - z_2|^2 = 2\left(|z_1|^2 + |z_2|^2\right)$$

for all $z_1, z_2 \in \mathbb{C}$. What is the geometrical interpretation of this result?

Exercise **1.2.3** Use a geometrical argument to support the fact that

$$|z - 3i| + |z + 3i| = 12$$

is the equation of an ellipse in the complex plane. Prove this fact using an algebraic approach.

Exercise **1.2.4** Show that the equation

$$(\bar{z} + z)^2 = 2(\alpha\bar{z} + \bar{\alpha}z) \qquad (\alpha \in \mathbb{C})$$

represents a parabola in the complex plane if α is not real. What curve(s) does it represent if α is real?

Exercise **1.2.5** Which of the following sets of complex numbers are (a) open, (b) closed, (c) connected, (d) simply connected, (e) bounded, (f) compact? Give brief reasons for your answers.

(i) $\{z : \text{Im } z < 0\}$

(ii) $\{z : |z - \alpha| \geqslant 4\}$

(iii) $\{1, i, -1, -i\}$

(iv) $\{z : |z| < 1\} \cup \{z : |z - 2i| < 1\}$

(v) $\{z : 1 \leqslant |\text{Re } z| + |\text{Im } z| < 4\}$

(vi) $\mathbb{C}$

Exercise **1.2.6** Prove that $S \subseteq \mathbb{C}$ is closed if $\mathbb{C} \backslash S$ is open.

Exercise **1.2.7** Find the image of the circle with equation

$$(x - a)^2 + (y - b)^2 = r^2 \qquad (a, b, r \in \mathbb{R})$$

under stereographic projection.

Exercise ***1.2.8** Let A be the image of $z_1 = x_1 + iy_1$ and B the image of $z_2 = x_2 + iy_2$ on the Riemann sphere. Let $d(z_1, z_2)$ be the distance from A to B. Show that

$$d(z_1, z_2) = \begin{cases} \dfrac{2|z_1 - z_2|}{\sqrt{\left(1 + |z_1|^2\right)\left(1 + |z_2|^2\right)}} & (z_2 \neq \infty) \\[2ex] \dfrac{2}{\sqrt{1 + |z_1|^2}} & (z_2 = \infty) \end{cases}$$

The distance function d is the **chordal metric** for the Riemann sphere and can be shown to be a metric in the usual sense. Note that $d(z_1, z_2) \leqslant 2$ and so $\tilde{\mathbb{C}}$ is bounded relative to this metric. It then follows that $\tilde{\mathbb{C}}$ is compact with respect to d.

The Polar Form of a Complex Number

Consider any non-zero $z \in \mathbb{C}$ represented by a point $P(x, y)$ in the plane, as in Fig. 1.1. It is usual to denote $|z|$ by $r \geqslant 0$. If OP makes an angle θ, measured anticlockwise in radians, with the (positive) x-axis, then $x = r\cos\theta$ and $y = r\sin\theta$. Hence $z = x + iy$ can be expressed as

$$z = r(\cos\theta + i\sin\theta) \tag{1.7}$$

where $r = |z|$ and $\tan\theta = y/x$. This is the **polar representation** of z. This form of any complex number is particularly useful when calculating powers and roots and also gives a practical application of complex numbers to trigonometry. This form is sometimes abbreviated to $z = r\,\text{cis}\,\theta$, and in engineering texts it often appears as $r \angle \theta$.

Definitions

Any angle θ defining z in (1.7) is an **argument** of z and is written $\theta = \arg z$. We shall use the convention that the unique value of θ such that $-\pi < \theta \leqslant \pi$ is the **principal argument** of z and is denoted by Arg z.

Note

The given range of values for Arg z is not a universal convention. The reason for choosing this particular convention will become clear later.

Lemma 1.4. Multiplication and Division in Polar Form

Let $z_1 = r_1 \operatorname{cis} \theta_1$ and $z_2 = r_2 \operatorname{cis} \theta_2$. Then

(i) $z_1 z_2 = r_1 r_2 \operatorname{cis} (\theta_1 + \theta_2)$

(ii) $\dfrac{z_1}{z_2} = \dfrac{r_1}{r_2} \operatorname{cis} (\theta_1 - \theta_2),\ z_2 \neq 0$ □

Proof

(i)
$$\begin{aligned} z_1 z_2 &= r_1 r_2 (\cos\theta_1 + i\sin\theta_1)(\cos\theta_2 + i\sin\theta_2) \\ &= r_1 r_2 (\cos\theta_1\cos\theta_2 - \sin\theta_1\sin\theta_2 + i(\sin\theta_1\cos\theta_2 + \cos\theta_1\sin\theta_2)) \\ &= r_2 r_2 (\cos\ (\theta_1 + \theta_2) + i\sin(\theta_1 + \theta_2)) \end{aligned}$$

(ii)
$$\begin{aligned} \frac{z_1}{z_2} &= \frac{r_1(\cos\theta_1 + i\sin\theta_1)(\cos\theta_2 - i\sin\theta_2)}{r_2(\cos\theta_2 + i\sin\theta_2)(\cos\theta_2 - i\sin\theta_2)} \\ &= \frac{r_1}{r_2}\,\frac{\cos\theta_1\cos\theta_2 + \sin\theta_1\sin\theta_2 + i(\sin\theta_1\cos\theta_2 - \cos\theta_1\sin\theta_2)}{\cos^2\theta_2 + \sin^2\theta_2} \\ &= \frac{r_1}{r_2}(\cos(\theta_1 - \theta_2) + i\sin(\theta_1 - \theta_2)) \end{aligned}$$ ■

The following result is an important and elegant use of polar representation.

Theorem 1.5. De Moivre's Theorem

Let $z = r \operatorname{cis} \theta$. Then $z^n = r^n \operatorname{cis} n\theta$, for all $n \in \mathbb{N}$. □

Proof

We use the principle of mathematical induction. The result is trivially true for $n = 1$, so suppose that the result is true for $n = k$. Then

$$z^{k+1} = zz^k = (r \operatorname{cis} \theta)(r^k \operatorname{cis} k\theta) \quad \text{by assumption}$$

$$\Rightarrow \quad z^{k+1} = r^{k+1} \operatorname{cis} (k+1)\theta \qquad \text{by 1.4(i)}$$

Thus, the result is true for $n = 1$ and if true for $n = k$, it is true for $n = k + 1$. Hence the result is true for all $n \in \mathbb{N}$ by the principle of induction. ■

Example 1.6

Simplify $\dfrac{\left(\sqrt{3} - i\right)^{30}}{(1 + i)^{20}}$

Solution

$|\sqrt{3} - i| = 2$, $\text{Arg}(\sqrt{3} - i) = -\pi/6$, $|1 + i| = \sqrt{2}$ and $\text{Arg}(1 + i) = \pi/4$. Then by De Moivre's theorem and Lemma 1.4,

$$\frac{(\sqrt{3} - i)^{30}}{(1 + i)^{20}} = \frac{2^{30}(\text{cis}\,(-\pi/6))^{30}}{2^{10}(\text{cis}\,(\pi/4))^{20}} = \frac{2^{20}\,\text{cis}\,(-5\pi)}{\text{cis}\,(5\pi)} = 2^{20}\,\text{cis}(-10\pi) = 2^{20}$$

The advantage of using polar form in this example is clear!

De Moivre's theorem can be applied to derive certain trigonometric identities. The method has the advantage of proving two identities at the same time. Note, however, that the proof of De Moivre's theorem itself depends explicitly on the addition formulae for sines and cosines.

Example 1.7

Use De Moivre's theorem to prove

$$\sin 4\theta = 4 \sin\theta \cos\theta - 8 \sin^3\theta \cos\theta \qquad (\text{for all } \theta \in \mathbb{R})$$

Solution

Let $s = \sin\theta$ and $c = \cos\theta$. By Theorem 1.5,

$$\text{cis}\,4\theta = \cos 4\theta + i \sin 4\theta = (\text{cis}\,\theta)^4 = (c + is)^4$$

$$\begin{aligned} \Rightarrow \quad \cos 4\theta + i \sin 4\theta &= c^4 + 4c^3(is) + 6c^2(is)^2 + 4c(is)^3 + (is)^4 \\ &= (c^4 - 6c^2s^2 + s^4) + i(4c^3s - 4cs^3) \end{aligned}$$

since the binomial theorem clearly holds for complex numbers. Then comparing imaginary parts gives

$$\begin{aligned} \sin 4\theta &= 4\cos^3\theta \sin\theta - 4\cos\theta \sin^3\theta \\ &= 4\cos\theta \sin\theta\,(1 - \sin^2\theta) - 4\cos\theta \sin^3\theta \\ &= 4\cos\theta \sin\theta - 8\cos\theta \sin^3\theta \end{aligned}$$

(Comparing real parts gives $\cos 4\theta = \cos^4\theta - 6\cos^2\theta \sin^2\theta + \sin^4\theta$, etc.)

Example 1.8

Prove, using De Moivre's theorem, that

$$\cos^5\theta = \frac{1}{16}\cos 5\theta + \frac{5}{16}\cos 3\theta + \frac{5}{8}\cos\theta \qquad (\text{for all } \theta \in \mathbb{R})$$

Solution

The given identity may be proved by employing the technique of the previous example to the right-hand side. However, a better way to proceed is as follows. Let $z = \text{cis}\,\theta$. Then by 1.5 and 1.4(ii),

$$z^n + z^{-n} = \text{cis}\,n\theta + \text{cis}(-n\theta) = 2\cos n\theta \qquad (\text{for all } n \in \mathbb{N})$$

Hence

$$\begin{aligned}(2\cos\theta)^5 &= (z + z^{-1})^5 \\ &= z^5 + 5z^4z^{-1} + 10z^3z^{-2} + 10z^2z^{-3} + 5zz^{-4} + z^{-5} \\ &= (z^5 + z^{-5}) + 5(z^3 + z^{-3}) + 10(z + z^{-1}) \\ &= 2\cos 5\theta + 10\cos 3\theta + 20\cos\theta\end{aligned}$$

using the binomial theorem. The result then follows easily.

De Moivre's theorem also provides a practical way of determining the nth roots of a complex number. Let $w = z^{1/n}$ for $n \in \mathbb{N}$ so that, as in the real case, $z = w^n$. Let $z = r\,\text{cis}\,\theta$ and $w = s\,\text{cis}\,\phi$. Then by 1.5,

$$r(\cos\theta + i\sin\theta) = s^n(\cos n\phi + i\sin n\phi)$$

Comparing real and imaginary parts then gives

$$r\cos\theta = s^n\cos n\phi \qquad r\sin\theta = s^n\sin n\phi$$

Squaring and adding these two equations then gives $s = \sqrt[n]{r}$ so that $\tan\theta = \tan n\phi$, hence $\phi = (\theta + 2k\pi)/n$ for any $k \in \mathbb{Z}$. Now

$$\text{cis}\left(\frac{\theta + 2(k+n)\pi}{n}\right) = \text{cis}\left(\frac{\theta + 2k\pi}{n}\right) \qquad (\text{for any } k \in \mathbb{Z})$$

so that only n distinct values of $\text{cis}\,\phi$ exist. For simplicity, we take $k = 0, 1, \ldots, n-1$ to produce these values. Hence, any complex number has exactly n nth roots.

Corollary 1.6 Roots of Complex Numbers

Let $z = r\,\text{cis}\,\theta$ where $\theta = \text{Arg}\,z$ without loss of generality. Then

$$z^{1/n} = \sqrt[n]{r}\,\text{cis}\left(\frac{\theta + 2k\pi}{n}\right) \qquad k = 0, 1, \ldots n-1, \text{ for all } n \in \mathbb{N}$$

□

Example 1.9

Find the three exact cube roots of $1 - i$ in Cartesian form.

Solution

$|1 - i| = \sqrt{2}$ and $\operatorname{Arg}(1 - i) = -\pi/4$. Then by 1.6,

$$(1 - i)^{1/3} = \sqrt[6]{2}\operatorname{cis}\left(\frac{-\pi/4 + 2k\pi}{3}\right) \qquad (k = 0, 1, 2)$$

Let the three cube roots of $1 - i$ be z_0, z_1 and z_2. Using the fact that $\pi/12 = \pi/3 - \pi/4$ and the standard trigonometric addition formulae, we obtain $\cos(\pi/12) = (\sqrt{6} + \sqrt{2})/4$ and $\sin(\pi/12) = (\sqrt{6} - \sqrt{2})/4$. Hence

$$z_0 = \sqrt[6]{2}(\cos(\pi/12) - i\sin(\pi/12)) = \frac{\sqrt[6]{2}}{4}\left((\sqrt{6} + \sqrt{2}) - i(\sqrt{6} - \sqrt{2})\right)$$

$$z_1 = \sqrt[6]{2}(\cos(7\pi/12) + i\sin(7\pi/12)) = \sqrt[6]{2}(-\sin(\pi/12) + i\cos(\pi/12))$$

$$= \frac{\sqrt[6]{2}}{4}\left((\sqrt{2} - \sqrt{6}) + i(\sqrt{2} + \sqrt{6})\right)$$

$$z_2 = \sqrt[6]{2}(\cos(5\pi/4) + i\sin(5\pi/4)) = -\sqrt[6]{2}(\cos(\pi/4) + i\sin(\pi/4))$$

$$= \frac{-(1 + i)}{\sqrt[3]{2}}$$

Example 1.10 The nth Roots of Unity

The nth roots of unity are the n roots of the equation $z^n = 1 = \operatorname{cis} 0$. Hence by 1.6, the n nth roots of unity are

$$\operatorname{cis}(2k\pi/n),\ k = 0, 1, \ldots, n - 1;\ \text{that is, } 1, \omega, \omega^2, \ldots, \omega^{n-1}$$

where $\omega = \operatorname{cis}(2\pi/n)$. In the complex plane, these all lie on the unit circle, $|z| = 1$, and form the vertices of a regular n-gon. Note that if $\omega \neq 1$ then

$$\omega^n = 1 \quad \Rightarrow \quad (\omega - 1)(1 + \omega + \omega^2 + \ldots + \omega^{n-1}) = 0$$
$$\Rightarrow \quad 1 + \omega + \omega^2 + \ldots + \omega^{n-1} = 0$$

Taking real parts, for instance, then gives

$$\cos(2\pi/n) + \cos(4\pi/n) + \ldots + \cos(2(n - 1)\pi/n) = -1 \qquad (\text{for all } n \in \mathbb{N})$$

Using polar representation, curves in the complex plane which are naturally defined in terms of polar coordinates can be rewritten in terms of a single complex variable. For example, the equation

$$|z|^3 = a\operatorname{Re}(z^2) \qquad (a \in \mathbb{R})$$

is equivalent to the polar equation

$$r = a\cos 2\theta$$

that is, the equation of a four-leaved rose.

Exercise **1.3.1** Find the polar representation of $\sin\theta - i\cos\theta, \theta \in \mathbb{R}$.

Exercise **1.3.2** Express the following in the form $a + ib$ by first finding the polar representation of each of the complex numbers involved:

(i) $(1+i)^{1000}$

(ii) $(1-i)^8(1+i\sqrt{3})^3$

(iii) $\dfrac{(\sqrt{3}-i)^3}{(-1+i\sqrt{3})^5}$

(iv) $27^{1/3}\, i^{-1/2}$

(v) $(-1)^{1/8}$

(vi) $(\sqrt{3}+i)^{1/4}$

Exercise **1.3.3** Find the two square roots of $3 + 4i$ in the form $a + ib$. Hence solve the equation

$$3z^2 + (2+7i)z + (2i-4) = 0$$

Exercise ***1.3.4** Use polar representation to prove that every complex number $z \neq -1$ of unit modulus can be expressed as

$$z = \frac{1+it}{1-it} \qquad \text{(for some } t \in \mathbb{R}\text{)}$$

Exercise **1.3.5** Use De Moivre's theorem to prove the following identities in real numbers. (You may assume that $\sin^2\theta + \cos^2\theta = 1$.)

(i) $\cos 5\theta = 16\cos^5\theta - 20\cos^3\theta + 5\cos\theta$

(ii) $\tan 4\theta = \dfrac{4\tan\theta - 4\tan^3\theta}{1 - 6\tan^2\theta + \tan^4\theta}$

(iii) $\sin^4\theta = \frac{3}{8} - \frac{1}{2}\cos 2\theta + \frac{1}{8}\cos 4\theta$

(iv) $\cos^8\theta + \sin^8\theta = \frac{1}{64}\cos 8\theta + \frac{7}{16}\cos 4\theta + \frac{35}{64}$

Exercise **1.3.6**

(i) Find the sixth roots of 1 and prove that they are the vertices of a regular hexagon, centre 0, in the complex plane.

*(ii) Prove that

$$\sin(\pi/n)\sin(2\pi/n)\ \ldots\ \sin((n-1)\pi/n) = n2^{1-n} \qquad (n \in \mathbb{N}, n \neq 1)$$

(*Hint*. Find the product of the non-zero roots of $(1-z)^n = 1$.)

Exercise **1.3.7** Let $k \in \mathbb{N}$. Prove that the non-negative integer powers of $\operatorname{cis}(3\pi/k)$ form a multiplicative cyclic group. Find the order of this group when (i) 3 divides k, (ii) 3 does not divide k.

Exercise **1.3.8** Show that the equation $|z+1||z-1| = 1$ is equivalent to the polar equation $r^2 = 2\cos 2\theta$, a lemniscate, if $z = r \operatorname{cis} \theta$. Sketch the curve in the complex plane.

Functions of a Complex Variable

This book is primarily concerned with functions which map A to B, where A and B are subsets of $\mathbb{C}$. The usual definitions associated with functions mapping sets to sets, which should be familiar to the reader, are given below.

Definitions

Let A and B be non-empty subsets of $\mathbb{C}$. A **function f mapping A to B** is a rule which associates with each $z \in A$, a unique $w \in B$. We write $f: A \to B$. We say w is the **image of z under f** or the **value of f at z** and write $w = f(z)$. A is the **domain** of f and B is the **codomain** of f. The **range** of f, denoted by $f(A)$, is the set of values of f. We say f is a **surjection** if $f(A) = B$; that is, the range of f is the whole of its codomain. The function f is an **injection** if $f(z_1) = f(z_2) \quad \Rightarrow \quad z_1 = z_2$, for all $z_1, z_2 \in A$; that is, distinct values of z give distinct values of w. Lastly, f is a **bijection** if it is an injection and a surjection.

Example 1.11

(i) $f: \mathbb{C} \to \mathbb{C}$ defined by $f(z) = z^2$ is a function with domain $\mathbb{C}$. It is not an injection since $f(-1) = f(1) = 1$. The range of f is $\mathbb{C}$, so it is a surjection. On the other hand, g defined by $g(z) = z^{1/2}$ is not a function, since for every value of $z \neq 0$ there are two values of g.

(ii) $f: \mathbb{C} \to \mathbb{C}$ given by $f(z) = \bar{z}$ is a function which is a bijection. Geometrically speaking, f reflects any point in the real axis.

(iii) $f: \mathbb{C}\backslash\{0\} \to (-\pi, \pi]$ given by $f(z) = \operatorname{Arg} z$ is a function which is clearly a surjection but not an injection. For example, $f(i) = f(2i) = \pi/2$.

(iv) $f: \mathbb{C} \to \mathbb{R}$ given by $f(z) = \operatorname{Im} z$ is clearly a function which is a surjection but not an injection.

Important Conventions

(i) For simplicity, we usually take $B = \mathbb{C}$.

(ii) Most of the definitions and results concerning functions of real variables will be taken for granted and quoted without proof.

(iii) The letters x, y, u and v will always denote real variables and $z = x + iy$ and $w = u + iv$ will always denote complex variables. Thus, if $f: A \to \mathbb{C}$ and $w = f(z)$, then $u + iv = f(x + iy)$.

(iv) In later work, it is sometimes convenient to identify a function f with its values $f(z)$. With this convention, we can talk about 'the function $f(z^2)$, for example, instead of the composition of two functions, $f \circ g$, given in this case by $(f \circ g)(z) = f(g(z)) = f(z^2)$.

From a geometric point of view, functions which map $A \subseteq \mathbb{C}$ to $\mathbb{C}$ map points in the plane to points in the plane, so they map curves and regions in the z-plane to curves and regions in the w-plane. Hence such functions can be used to distort complicated curves and regions into simple curves and regions. This fact is extremely useful in applications to physical problems.

Example 1.12

Show that $f: \mathbb{C}\backslash\{0\} \to \mathbb{C}$, given by $w = f(z) = 1/z$, maps the inside of the circle $x^2 + y^2 = a^2$ in the z-plane to the outside of the circle $u^2 + v^2 = 1/a^2$ in the w-plane.

Solution

The equation of the given circle in the z-plane is $|z| = a$. Thus, along this circle, $|w| = 1/|z| = 1/a$; that is, $u^2 + v^2 = 1/a^2$. Also, $|z| < a \quad \Rightarrow \quad |w| > 1/a$, so that any non-zero point, A say, inside the circle $x^2 + y^2 = a^2$, gets mapped to a point A', outside the circle $u^2 + v^2 = 1/a^2$. This is shown in Fig. 1.6, which illustrates the case $a > 1$.

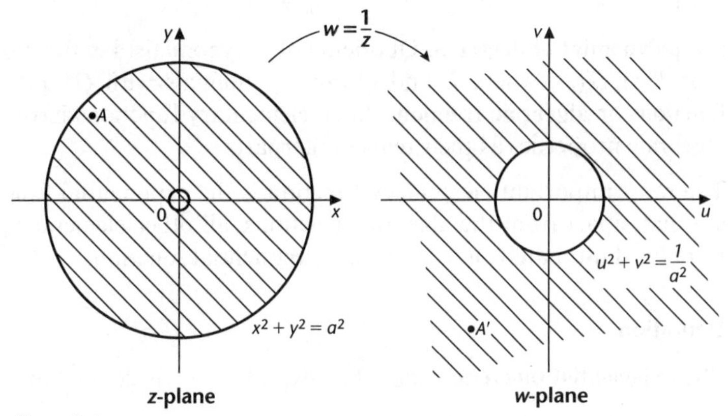

Figure 1.6

Convention

We shall adopt the usual convention that if a complex number is represented by a point A in the plane, then the image of A under some function f is denoted by A', etc.

Note that $f: A \subseteq \mathbb{C} \to \mathbb{C}$ is equivalent to a pair of real-valued functions $u, v : \mathbb{R}^2 \to \mathbb{R}$; that is, if $w = u + iv = f(z) = f(x + iy)$, it is always (theoretically) possible to find $u(x, y)$ and $v(x, y)$.

Example 1.13

Let $w = f(z) = z^3$ for all $z \in \mathbb{C}$. Then

$$w = u + iv = (x + iy)^3 = (x^3 - 3xy^2) + i(3yx^2 - y^3)$$

by the binomial theorem. Equating real and imaginary parts,

$$u(x, y) = x^3 - 3xy^2 \qquad v(x, y) = 3yx^2 - y^3$$

The Elementary Functions

We now investigate the elmentary functions that are the analogues of the usual elementary functions of real analysis. These functions have many properties in common with their real counterparts, although there are some fundamental differences.

Definitions

If $n \in \mathbb{Z}_{\geqslant 0}$, the set of non-negative integers, and $\alpha_i \in \mathbb{C}, i = 0, 1, \ldots, n$, then the function $P : \mathbb{C} \to \mathbb{C}$ defined by

$$P(z) = \alpha_0 + \alpha_1 z + \ldots + \alpha_n z^n \qquad (\alpha_n \neq 0)$$

is a **polynomial of degree *n*.** Quotients of polynomials, i.e. functions which map z to $P(z)/Q(z)$, where P and Q are polynomials with $Q(z) \not\equiv 0$, are **rational functions** or **algebraic fractions.** Such elementary functions have the same basic algebraic properties as their real counterparts.

The most important elementary function is the exponential function. We shall see that, apart from the algebraic fractions, all other elementary functions can be defined ultimately in terms of the exponential function.

Definition

The **exponential function**, denoted by $\exp : \mathbb{C} \to \mathbb{C}$, is defined for all $z \in \mathbb{C}$ by

$$\exp z = e^z = e^{x+iy} = e^x(\cos y + i \sin y)$$

Note

Although we take this as a definition, it is sometimes justified by examining the Maclaurin series expansions of the real functions involved. As a definition, it can be justified by noting that exp is the unique function that clearly includes the real exponential function as a special case, and as in the real case, is its own derivative. This will be demonstrated in the next chapter.

Historical Note

Euler stated that $e^{\sqrt{-1}x} = \cos x + \sqrt{-1}\sin x$ for real x in his *Introductio in Analysin Infinitorum* in 1748. The same statement, in a different form, was published by Cotes as early as 1714.

Important Note

Using the definition of the exponential function, the polar representation of any complex number z can be written simply as $z = re^{i\theta}$, where $r = |z|$ and $\theta = \operatorname{Arg} z$.

The exponential function obeys most of the usual rules for indices. However, unlike its real counterpart, it is periodic, which affects the rules for roots.

Lemma 1.7. Elementary Properties of the Exponential Function

(i) $e^{z+2\pi i} = e^z$ for all $z \in \mathbb{C}$

$$\left.\begin{aligned} &\text{(ii)}\quad e^{z_1+z_2} = e^{z_1}e^{z_2} \\ &\text{(iii)}\quad e^{z_1-z_2} = \frac{e^{z_1}}{e^{z_2}} \end{aligned}\right\} \text{for all } z_1, z_2 \in \mathbb{C}$$

$$\left.\begin{aligned} &\text{(iv)}\quad (e^z)^n = e^{nz} \\ &\text{(v)}\quad (e^z)^{1/n} = e^{(z+2k\pi i)/n} \\ &\qquad\quad (k = 0, 1, \ldots, n-1) \end{aligned}\right\} \text{for all } z \in \mathbb{C}, n \in \mathbb{N}$$

□

Proof

Let $z = x + iy$. Then

$$e^{z+2\pi i} = e^{x+i(y+2\pi)} = e^x(\cos(y+2\pi) + i\sin(y+2\pi))$$

$$\Rightarrow \quad e^{z+2\pi i} = e^x(\cos y + i\sin y) = e^z$$

This proves (i). Since $e^{iy} = \operatorname{cis} y$, (ii) and (iii) are essentially a restatement of Lemma 1.4, and (iv) and (v) are essentially restatements of 1.5 and 1.6 respectively. ■

Important Note

By definition, $|e^z| = e^x$, so the range of exp is $\mathbb{C}\backslash\{0\}$. Also,

$$|e^{iy}| = 1 \qquad \text{(for any } y \in \mathbb{R})$$

Hence the circle with equation $|z - \alpha| = r$ has the parametric polar equation

$$z(\theta) = \alpha + re^{i\theta} \qquad (-\pi < \theta \leqslant \pi)$$

The exponential function has a practical application in finding certain real trigonometric sums, as the following example demonstrates.

Example 1.14

Prove that

$$1 + z + z^2 + \ldots + z^n = \frac{1 - z^{n+1}}{1 - z} \qquad \text{(for all } z \in \mathbb{C}\backslash\{1\}, n \in \mathbb{N})$$

Hence simplify

$$1 + \cos\theta\cos\phi + \cos^2\theta\cos 2\phi + \ldots + \cos^n\theta\cos n\phi \qquad (\theta, \phi \in \mathbb{R})$$

Solution

Let $S = 1 + z + z^2 + \ldots + z^n$. Then, as in the real case,

$$S - zS = (1 - z)S = 1 - z^{n+1}$$

Now let $z = \cos\theta e^{i\phi}$, so that $z^k = \cos^k\theta e^{ki\phi}$ for all $k \in \mathbb{N}$ by 1.7(iv), hence the given sum consists of powers of z when real parts are taken. Then, by the above,

$$1 + \cos\theta e^{i\phi} + \cos^2\theta e^{2i\phi} + \ldots + \cos^n\theta e^{ni\phi} = \frac{1 - \cos^{n+1}\theta e^{(n+1)i\phi}}{1 - \cos\theta e^{i\phi}}$$

Using the definition of $e^{i\phi}$ and the standard technique for complex division then gives

$$\begin{aligned}
&1 + \cos\theta\cos\phi + \cos^2\theta\cos 2\phi + \ldots + \cos^n\theta\cos n\phi \\
&= \operatorname{Re}(1 + \cos\theta e^{i\phi} + \cos^2\theta e^{2i\phi} + \ldots + \cos^n\theta e^{ni\phi}) \\
&= \operatorname{Re}\left(\frac{(1 - \cos^{n+1}\theta e^{(n+1)i\phi})(1 - \cos\theta e^{-i\phi})}{(1 - \cos\theta e^{i\phi})(1 - \cos\theta e^{-i\phi})}\right) \\
&= \frac{1 - \cos\theta\cos\phi - \cos^{n+1}\theta\cos(n+1)\phi + \cos^{n+2}\theta\cos n\phi}{1 - 2\cos\theta\cos\phi + \cos^2\theta}, \text{ etc.}
\end{aligned}$$

The hyperbolic functions are defined as in the case of real variables.

Definitions

For all $z \in \mathbb{C}$,

$$\sinh z = \tfrac{1}{2}(e^z - e^{-z}) \qquad \cosh z = \tfrac{1}{2}(e^z + e^{-z})$$

Then

$$\tanh z = \frac{\sinh z}{\cosh z} \quad (\cosh z \neq 0) \qquad \coth z = \frac{1}{\tanh z} \quad (\sinh z \neq 0)$$

$$\operatorname{sech} z = \frac{1}{\cosh z} \quad (\cosh z \neq 0) \qquad \operatorname{csch} z = \frac{1}{\sinh z} \quad (\sinh z \neq 0)$$

Since these definitions are algebraically identical to their real counterparts, the identities involving the hyperbolic functions are the same as those involving real variables. In particular, we have the following results.

Lemma 1.8. Elementary Properties of sinh and cosh

Let z, z_1 and z_2 denote any complex numbers. Then

(i) $\cosh^2 z - \sinh^2 z = 1$

(ii) $\sinh(z_1 + z_2) = \sinh z_1 \cosh z_2 + \cosh z_1 \sinh z_2$

(iii) $\cosh(z_1 + z_2) = \cosh z_1 \cosh z_2 + \sinh z_1 \sinh z_2$

(iv) $\sinh(-z) = -\sinh z$ and $\cosh(-z) = \cosh z$ □

Proof

The results follow from the definitions and the properties of e^z. We prove (i) as an example and the rest are left as an exercise.

$$\begin{aligned}\cosh^2 z - \sinh^2 z &= \tfrac{1}{4}(e^z + e^{-z})^2 - \tfrac{1}{4}(e^z - e^{-z})^2 \\ &= \tfrac{1}{4}(e^{2z} + e^{-2z} + 2 - e^{2z} - e^{-2z} + 2) \qquad \text{by 1.7} \\ &= 1\end{aligned}$$

■

It follows from the definition of $\exp : \mathbb{C} \to \mathbb{C}$ that

$$e^{iy} = \cos y + i \sin y \qquad e^{-iy} = \cos y - i \sin y \qquad (\text{for all } y \in \mathbb{R})$$

$$\Rightarrow \quad \sin y = \frac{1}{2i}(e^{iy} - e^{-iy}) \qquad \cos y = \frac{1}{2}(e^{iy} + e^{-iy}) \qquad (\text{for all } y \in \mathbb{R})$$

It is therefore natural to make the following definitions.

Definitions

For all $z \in \mathbb{C}$,

$$\sin z = \frac{1}{2i}(e^{iz} - e^{-iz}) \qquad \cos z = \frac{1}{2}(e^{iz} + e^{-iz})$$

Then

$$\tan z = \frac{\sin z}{\cos z} \quad (\cos z \neq 0) \qquad \cot z = \frac{1}{\tan z} \quad (\sin z \neq 0)$$

$$\sec z = \frac{1}{\cos z} \quad (\cos z \neq 0) \qquad \csc z = \frac{1}{\sin z} \quad (\sin z \neq 0)$$

These definitions include the corresponding trigonometric functions of real variables as special cases. All the usual algebraic identities hold, but note that it does not make sense to talk about 'the angle z' when finding $\sin z$, etc., if z is not real.

Lemma 1.9. Elementary Properties of sin and cos

Let z, z_1 and z_2 denote any complex numbers. Then

(i) $\sin^2 z + \cos^2 z = 1$

(ii) $\sin(z_1 + z_2) = \sin z_1 \cos z_2 + \cos z_1 \sin z_2$

(iii) $\cos(z_1 + z_2) = \cos z_1 \cos z_2 - \sin z_1 \sin z_2$

(iv) $\sin(-z) = -\sin z$ and $\cos(-z) = \cos z$ □

Proof

Once again, these identities follow directly from the definitions and the properties of e^z. We prove (ii) as an example.

$$\begin{aligned}
&\sin z_1 \cos z_2 + \cos z_1 \sin z_2 \\
&= \frac{1}{4i}(e^{iz_1} - e^{-iz_1})(e^{iz_2} + e^{-iz_2}) + \frac{1}{4i}(e^{iz_1} + e^{-iz_1})(e^{iz_2} - e^{-iz_2}) \\
&= \frac{1}{4i}(e^{i(z_1+z_2)} + e^{i(z_1-z_2)} - e^{i(z_2-z_1)} - e^{-i(z_1+z_2)} \\
&\quad + e^{i(z_1+z_2)} - e^{i(z_1-z_2)} + e^{i(z_2-z_1)} - e^{-i(z_1+z_2)}) \qquad \text{by 1.7} \\
&= \frac{1}{2i}(e^{i(z_1+z_2)} - e^{-i(z_1+z_2)}) = \sin(z_1 + z_2)
\end{aligned}$$

■

The following correspondence follows directly from the definitions. In complex analysis, the trigonometric functions and the hyperbolic functions are not independent concepts; they are intimately related.

Lemma 1.10. Correspondence Between Trigonometric and Hyperbolic Functions

$$\sinh(iz) = i\sin z \qquad \cosh(iz) = \cos z$$

$$\sin(iz) = i\sinh z \qquad \cos(iz) = \cosh z$$

□

Hence Lemma 1.9 follows immediately from Lemma 1.8, or vice versa, using Lemma 1.10.

Word of Warning

Although functions of complex variables share many properties with their counterparts for real variables, this is not always the case. It is all too tempting to generalise familiar properties of functions of a real variable to the complex case, where they may not be true. For example, there is no real number x such that $\cosh x = 0$. On the other hand, for $z \in \mathbb{C}$, $\cosh z = 0$ if and only if $\cos(iz) = 0$ by Lemma 1.10, and it is easily shown that $\cos : \mathbb{C} \to \mathbb{C}$ has the same zeros as its real counterpart. Hence $\cosh z = 0$ if and only if $iz = (2n+1)\pi/2, n \in \mathbb{N}$.

The following result is a special case of 1.9(ii) and (iii), with $z_1 = x$ and $z_2 = iy$, and follows by 1.10. Corresponding results for sinh and cosh follow from 1.8 and 1.10.

Lemma 1.11. Real and Imaginary Parts of $\sin z$ and $\cos z$

Let $z = x + iy \in \mathbb{C}$. Then

(i) $\sin z = \sin x \cosh y + i\cos x \sinh y$

(ii) $\cos z = \cos x \cosh y - i\sin x \sinh y$ □

It follows immediately from 1.11 that the trigonometric functions are periodic with the same periods as their real counterparts. It also shows that, unlike in the real case, there exists $z \in \mathbb{C}$ such that $|\sin z| > 1$ and similarly for cos. For example, $\sin i = i\sinh 1 \Rightarrow |\sin i| = \sinh 1 > 1$. Lemma 1.11 can also be used to show that the trigonometric functions have the same zeros as their real counterparts.

Example 1.15

Show that $\cos z = 0$ if and only if $z = (n + 1/2)\pi$, for any $n \in \mathbb{Z}$, as in the case of a real variable.

Solution

Let $z = x + iy$. Then by 1.11,

$$\begin{aligned} |\cos z|^2 &= \cos^2 x \cosh^2 y + \sin^2 x \sinh^2 y \\ &= \cos^2 x(1 + \sinh^2 y) + \sin^2 x \sinh^2 y \qquad \text{by 1.8} \\ &= \cos^2 x + \sinh^2 y \qquad \text{by 1.9} \end{aligned}$$

Then $\cos z = 0 \Leftrightarrow |\cos z| = 0 \Leftrightarrow \cos^2 x = -\sinh^2 y, x, y \in \mathbb{R} \Leftrightarrow \cos x = \sinh y = 0 \Leftrightarrow x = (n + 1/2)\pi$ and $y = 0$, as required.

Example 1.16

Show that $\sin : \mathbb{C} \to \mathbb{C}$ maps the given points, boundary and shaded region in the z-plane, to the given points, boundary and the upper half w-plane, as shown in Fig. 1.7.

Solution

Let $w = \sin z$ with $z = x + iy$ and $w = u + iv$. By 1.11,

$$u(x, y) = \sin x \cosh y \qquad v(x, y) = \cos x \sinh y$$

Hence along the boundary line given by $x = -\pi/2$ and $y \geqslant 0$, $u = -\cosh y \leqslant -1$ and $v = 0$, with $A(-\pi/2, 0)$ mapped to $A'(-1, 0)$.

Along the boundary line given by $-\pi/2 \leqslant x \leqslant \pi/2$ and $y = 0$, $u = \sin x$ and $v = 0$, so that $-1 \leqslant u \leqslant 1$ and $v = 0$, with $O(0, 0)$ mapped to $O'(0, 0)$ and $B(\pi/2, 0)$ mapped to $B'(1, 0)$. Along the boundary line given by $x = \pi/2$ and $y \geqslant 0$, $u = \cosh y \geqslant 1$ and $v = 0$.

Finally, let $C(x, y)$ be any point in the shaded region in the z-plane, so that $-\pi/2 < x < \pi/2$ and $y > 0$. Then $v > 0$ but there is no restriction on u. Hence the shaded region in the z-plane is mapped to the upper half w-plane.

To complete our list of elementary functions, we need the complex analogue of the logarithmic function for real variables. Recall that the logarithmic function, $\text{Log} : \mathbb{R}^+ \to \mathbb{R}$, is defined as the inverse of the exponential function, $\exp : \mathbb{R} \to \mathbb{R}^+$, i.e.

$$\text{if} \quad x = e^y \quad \text{then} \quad y = \text{Log}\, x \qquad (\text{for all } x \in \mathbb{R}^+)$$

Clearly, $\text{Log} : \mathbb{R}^+ \to \mathbb{R}$ is a function since $\exp : \mathbb{R} \to \mathbb{R}^+$ is a bijection.

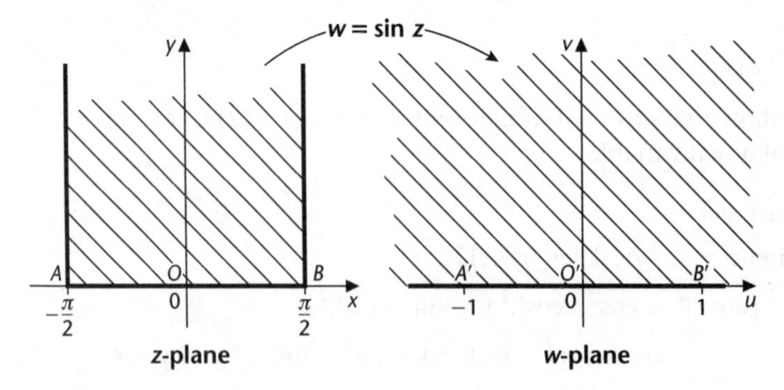

Figure 1.7

Notation

We use $\text{Log} : \mathbb{R}^+ \to \mathbb{R}$ to denote $\ln : \mathbb{R}^+ \to \mathbb{R}$ in this setting. The reason for this notation will become clear.

This situation is altered in the complex case since $\exp : \mathbb{C} \to \mathbb{C}\backslash\{0\}$ is not an injection by Lemma 1.7. Nevertheless, we try to mirror the real definition as far as possible.

Definition

The relation $\log : \mathbb{C}\backslash\{0\} \to \mathbb{C}$ is defined by

$$w = \log z \quad \text{if} \quad z = e^w \qquad (\text{for all } z \in \mathbb{C}\backslash\{0\})$$

Note the use of lower case 'l' in this definition. Then 'log' is not a function since $\log z$ takes on an infinite number of values.

Lemma 1.12. Values of log

$$\log z = \text{Log}|z| + i(\text{Arg}\, z + 2k\pi) \text{ where } k \text{ is any integer} \qquad \square$$

Proof

Let $r = |z|$ and $\theta = \text{Arg}\, z$, so that $z = r\,\text{cis}\,\theta$. Let $z = e^w$ with $w = u + iv$. Then

$$z = e^w \quad \Rightarrow \quad r(\cos\theta + i\sin\theta) = e^u(\cos v + i\sin v)$$

Comapring real and imaginary parts and simplifying gives

$$r = e^u \text{ and } \sin\theta = \sin v \quad \Rightarrow \quad u = \text{Log}\, r \text{ and } v = \theta + 2k\pi \qquad (k \in \mathbb{Z}) \quad \blacksquare$$

Note that by 1.12, $\log(re^{i\theta}) = \text{Log}\, r + i(\theta + 2k\pi)$.

Since $\log z$ has an infinite number of values, when we write $\log z$, we mean the set of values taken by $\log z$. In order to obtain a function from log, we need to restrict k to a particular value. Conventionally, this is $k = 0$ but any other fixed value of k will provide a (different) function.

Definitions

As z varies throughout $\mathbb{C}\backslash\{0\}$, each set of values of $\log z$ for a particular value of k is called a **branch** of log. That branch corresponding to $k = 0$ is called the **principal branch** of log. The **logarithmic function**, denoted by $\text{Log} : \mathbb{C}\backslash\{0\} \to C$ is defined by

$$\text{Log}\, z = \text{Log}|z| + i\,\text{Arg}\, z \qquad (\text{for all } z \in \mathbb{C}\backslash\{0\})$$

Thus, Log is the principal branch of log and includes $\text{Log} : \mathbb{R}^+ \to \mathbb{R}$ as a special case. Since the principal branch of log is always taken when finding $\text{Log}\, z$,

$\text{Log}: \mathbb{C}\backslash\{0\} \to \mathbb{C}$ does not generally obey the same properties as its real counterpart. In particular, note that $\text{Log}(e^z) \neq z$ in general. For example, $\text{Log}(e^{-i\pi}) = \text{Log}(-1) = i\pi$.

Lemma 1.13. Properties of Log

Let z, z_1 and z_2 denote any complex numbers. Then

(i) $\text{Log}(z_1 z_2) = \text{Log}\, z_1 + \text{Log}\, z_2 + 2k(z_1, z_2)\pi i$

(ii) $\text{Log}(z_1/z_2) = \text{Log}\, z_1 - \text{Log}\, z_2 + 2k(z_1, z_2)\pi i$

(iii) $\text{Log}(z^n) = n\,\text{Log}\, z + 2k(z)\pi i$ for all $n \in \mathbb{N}$

where in each case, k is a particular integer depending on z_1 and z_2 or z. □

Proof

(i) Let $|z_i| = r_i$ and $\text{Arg}\, z_i = \theta_i$, $i = 1, 2$. Then

$$z_1 z_2 = r_1 r_2 \,\text{cis}\,(\theta_1 + \theta_2) \qquad \text{by 1.4}$$

$$\Rightarrow \quad \text{Log}(z_1 z_2) = \text{Log}(r_1 r_2) + i(\theta_1 + \theta_2) + 2k(z_1, z_2)\pi i$$

$$\Rightarrow \quad \text{Log}(z_1 z_2) = (\text{Log}\, r_1 + i\theta_1) + (\text{Log}\, r_2 + i\theta_2) + 2k(z_1, z_2)\pi i$$

where $k(z_1, z_2)$ is an integer chosen so that $\theta_1 + \theta_2 + 2k\pi$ is the principal argument of $z_1 z_2$, as required.

(ii) This is similar and is left as an exercise.

(iii) Let $|z| = r$ and $\theta = \text{Arg}\, z$. Then $z^n = r^n \,\text{cis}\, n\theta$ by 1.5. Hence

$$\text{Log}(z^n) = \text{Log}(r^n) + in\theta + 2k(z)\pi i = n(\text{Log}\, r + i\theta) + 2k(z)\pi i$$

where $k(z)$ is an integer chosen so that $n\theta + 2k\pi$ is the principal argument of z^n. ■

Note

In Lemma 1.13(i) and (ii), $k(z_1, z_2)$ must take one of the values -1, 0 or 1.

Example 1.17

(i) $\text{Log}(i(i-1)) = \text{Log}(-1-i) = \text{Log}\,\sqrt{2} - 3\pi i/4$. Also, $\text{Log}\, i = i\pi/2$ and $\text{Log}(i-1) = \text{Log}\,\sqrt{2} + 3\pi i/4$. Hence

$$\text{Log}(i(i-1)) = \text{Log}\, i + \text{Log}(i-1) - 2\pi i$$

(ii) $\text{Log}(-1+i)^{10} = \text{Log}(2^5 \,\text{cis}\,(15\pi/2)) = 5\,\text{Log}\, 2 - i\pi/2$. Hence

$$\text{Log}(-1+i)^{10} = 10\,\text{Log}(-1+i) - 8\pi i$$

Having defined the complex exponential and logarithmic functions, it is easy to define complex exponents. From 1.12,

$$e^{p \log z} = \exp\,(p \operatorname{Log} r + ip\theta + 2k\pi ip) \qquad (\text{for all } p \in \mathbb{Q})$$

where $r = |z|$ and $\theta = \operatorname{Arg} z$. Hence

$$e^{p \log z} = r^p(\cos\,(p\theta + 2k\pi p) + i \sin\,(p\theta + 2k\pi p)) = z^p$$

by De Moivre's theorem and Corollary 1.6. Also, for any $x, a \in \mathbb{R}$, $x^a = e^{a \operatorname{Log} x}$. In order to be consistent with these results, we define z^α, where z and $\alpha \in \mathbb{C}$ as follows.

Definitions

$z^\alpha = e^{\alpha \log z}$ for all z, $\alpha \in \mathbb{C}, z \neq 0$. Such powers are clearly multivalued in general. The **principal branch** of z^α is defined by

$$z^\alpha = e^{\alpha \operatorname{Log} z} \qquad (\text{for all } z, \alpha \in \mathbb{C}, z \neq 0)$$

Convention

Unless otherwise stated, z^α will always be taken to mean the principal branch of z^α. For example,

$$i^{-i} = e^{-i \operatorname{Log} i} = e^{-i(\operatorname{Log} 1 + i\pi/2)} = e^{\pi/2}$$

and so is real!

Historical Note

After explaining that $i^{-i} = e^{\pi/2}$ to one of his classes, Benjamin Peirce, a professor at Harvard in the nineteenth century, stated, 'Gentlemen, this is surely true, it is absolutely paradoxical, we can't understand it, and we haven't the slightest idea what the equation means, but we may be sure that it means something very important.'

Notice that if $n \in \mathbb{N}$, then the principal branch of $z^{1/n}$ is given by $\exp((\operatorname{Log} r + i\theta)/n) = \sqrt[n]{r} \operatorname{cis}(\theta/n)$, i.e., by $k = 0$ in 1.6. In particular, the principal value of $(-1)^{1/2} = \operatorname{cis}(\pi/2) = i$, with our convention of $-\pi < \operatorname{Arg} z \leqslant \pi$.

Note

Complex exponents obey some of the usual rules of indices, such as

$$z^{\alpha_1 + \alpha_2} = z^{\alpha_1} z^{\alpha_2} \qquad z^{\alpha_1 - \alpha_2} = z^{\alpha_1} / z^{\alpha_2}$$

but not all, since the exponential function does not (see Lemma 1.7 and the exercises).

The inverses of the trigonometric and hyperbolic functions are defined as for real variables. In general, if $f: A \to B$ where $A, B \subseteq \mathbb{C}$ and B is chosen to be the range of f, so that f is a surjection, then $f^{-1}: B \to A$ is defined by $f^{-1}(z) = w$ if $z = f(w)$. Since f may not be an injection, f^{-1} need not be a function. Since the trigonometric and hyperbolic functions are defined in terms of the exponential function, their inverses are expressible in terms of the relation log.

Example 1.18

The inverse of $\sin : \mathbb{C} \to \mathbb{C}$ is denoted by $\sin^{-1} : \mathbb{C} \to \mathbb{C}$ and is defined by

$$w = \sin^{-1} z \quad \text{if} \quad z = \sin w \qquad (\text{for all } z \in \mathbb{C})$$

Now if $z = \sin w$,

$$z = \frac{1}{2i}(e^{iw} - e^{-iw})$$

by definition and multiplying by $2ie^{iw}$ then gives

$$e^{2iw} - 2ize^{iw} - 1 = 0 \quad \Rightarrow \quad e^{iw} = iz + (1 - z^2)^{1/2}$$

solving the quadratic equation in e^{iw}. Hence

$$w = \sin^{-1} z = -i \log (iz + (1 - z^2)^{1/2})$$

To obtain the inverse **function**, the principal branch of log and the root are taken. This ensures that $\sin^{-1} 0 = 0$ as in the real case. That is, the **inverse sine function** is given by

$$\sin^{-1} z = -i \operatorname{Log} (iz + (1 - z^2)^{1/2})$$

Exercise 1.4.1 For each of the following functions, express $f(z)$ in the form $u(x, y) + iv(x, y)$. Decide which of the functions are (a) an injection and (b) a surjection. Give reasons for your answers.

(i) $f: \mathbb{C} \to \mathbb{C}$ given by $f(z) = 3z + 1$

(ii) $f: \mathbb{C} \to \mathbb{C}$ given by $f(z) = z(2z - 1)$

(iii) $f: \mathbb{C}\backslash\{2/3\} \to \mathbb{C}$ given by $f(z) = \dfrac{\bar{z}}{3z + 2}$

(iv) $f: \mathbb{C}\backslash\mathbb{R}_{\geqslant 0} \to \mathbb{C}$ given by $f(z) = 1/\operatorname{Arg} \bar{z}$

(v) $f: \mathbb{C} \to \mathbb{R}$ given by $f(z) = \operatorname{Re} z$

Exercise 1.4.2 Let $w = u + iv$ and $z = x + iy = r \operatorname{cis} \theta$. Show that the transformation $w = z + 1/z$ maps the circle $x^2 + y^2 = a^2$, $a \neq 1$ in the z-plane, onto the ellipse with parametric equations $u = (a + 1/a) \cos \theta$, $v = (a - 1/a) \sin \theta$ in the w-plane.

Exercise **1.4.3** Let $w = z^{1/2}$, where $w = u + iv$ and $z = x + iy$. Find u and v as explicit functions of x and y.

Exercise **1.4.4** Let $z = x + iy$. Show that

$$\left| e^{ie^{\bar{z}}} \right| = e^{e^x \sin y}$$

Exercise **1.4.5** Use the elementary properties of the exponential function to evaluate the following sums:

(i) $\displaystyle\sum_{k=0}^{n} \sin k\theta$

(ii) $\displaystyle\sum_{k=0}^{n} \cos(\theta + k\phi)$

where $\theta, \phi \in \mathbb{R}$.

Exercise **1.4.6** Using only the definitions of the hyperbolic sine and cosine functions and properties of the exponential function, prove that

$$\cosh(z_1 + z_2) = \cosh z_1 \cosh z_2 + \sinh z_1 \sinh z_2 \qquad (\text{for all } z_1, z_2 \in \mathbb{C})$$

(i) Use this result to prove that if $z = x + iy$, then

$$|\cosh z|^2 = \sinh^2 x + \cos^2 y$$

(ii) Find all the roots of the equation $\cosh z = 1/2$.

Exercise **1.4.7** Let $z = x + iy \in \mathbb{C}$. Find the real and imaginary parts of $\tan z$ in terms of x and y. Hence find all the roots of the equation $\tan z = 2i$.

Exercise **1.4.8** Find the images of the points $A(0, a)$, $B(\pi/2, a)$, $C(\pi, a)$, $D(3\pi/2, a)$ and $E(2\pi, a)$ in the z-plane, where $a \in \mathbb{R}^+$, under $w = \sin z$. Show that sin maps the line segment AE to an ellipse, with centre 0, in the w-plane.

Exercise **1.4.9** Find the images of the points, boundary and shaded region in the z-plane (Fig. 1.8) under $w = \operatorname{Log} z$.

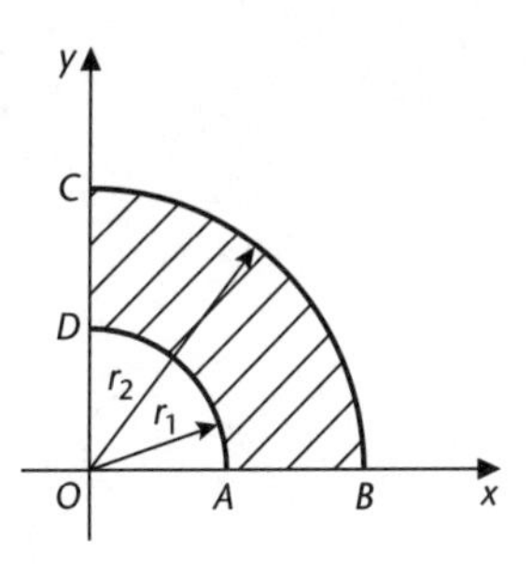

Figure 1.8

Exercise **1.4.10** Prove Lemma 1.13(ii).

Exercise **1.4.11**

(i) Prove that $z^\alpha z^\beta = z^{\alpha+\beta}$, for all z, α, $\beta \in \mathbb{C}$, $z \neq 0$. Take principal branches.

(ii) Prove that $(z_1 z_2)^\alpha = z_1{}^\alpha z_2{}^\alpha e^{2\pi i k \alpha}$, for all z_1, z_2, $\alpha \in \mathbb{C}$, $z \neq 0$, for some integer k depending on z_1 and z_2. Take principal branches.

(iii) Give a simple example to show that $(z^\alpha)^\beta \neq z^{\alpha\beta}$ in general.

Exercise **1.4.12** Find all the values of $\tanh^{-1} 0$.

Exercise **1.4.13** Prove that the principal branch of $\operatorname{csch}^{-1} : \mathbb{C}\backslash\{0\} \to \mathbb{C}$ is given by

$$\operatorname{csch}^{-1} z = \operatorname{Log}(1/z + (1/z^2 + 1)^{1/2})$$

Exercise **1.4.14** Functions of time occur in the analysis of many types of physical system. These functions may describe exponential growth or decay, oscillatory behaviour, or behaviour which oscillates with an amplitude that grows or decays exponentially with time. Many such functions f can be defined by

$$f(t) = \operatorname{Re}(Fe^{st})$$

where $s = \sigma + i\omega \in \mathbb{C}$, is called the **complex frequency** of oscillation and $F \in \mathbb{C}$. The number F is the **phasor** associated with f and is independent of t.

(i) Show that if $|F| = F_0$, $\operatorname{Arg} F = \theta$ and $\phi = \theta + \pi/2$, then

$$f(t) = F_0 e^{\sigma t} \cos(\theta + \omega t) = F_0 e^{\sigma t} \sin(\phi + \omega t)$$

For example, if s and $F \in \mathbb{R}$, so that $\omega = \theta = 0$, f represents exponential growth or decay. If θ, $\omega \neq 0$ and $\sigma < 0$, then f represents sinusoidal motion with decaying amplitude, and so on.

(ii) If f is given by $f(t) = \operatorname{Re}(Fe^{st})$, as above, prove that the phasor F, of f, is unique provided that $\omega \neq 0$. What can we say when $\omega = 0$?

For an account of phasors, with applications, see the appendix to Chapter 3 in A. D. Wunsch, *Complex Variables with Applications*, 2nd edn, Addison-Wesley, 1994.

2 Differentiation and the Cauchy–Riemann Equations

This chapter introduces the limit of a function. It is a straightforward generalisation of the corresponding concept for a function of a real variable. This leads on to the ideas of continuity and derivatives, as for functions of a real variable. It is assumed that the reader is already familiar with these ideas from real analysis. The consideration of continuity also leads to the idea of branch points in the complex plane. We also give a famous characterisation of differentiable functions of a complex variable, in the form of the Cauchy–Riemann equations, which has some unexpected applications. The Cauchy–Riemann equations are of fundamental importance in complex analysis. Finally, we introduce the ideas of singular points and zeros of functions.

Limits of Functions

As in the case of real variables, to say $f(z)$ has limit ℓ as z tends to α essentially means that $f(z)$ can be made 'arbitrarily close' to ℓ by making z 'close enough' to α but distinct from it. This idea is formalised in the following definition.

Definition

Let $A \subseteq \mathbb{C}$ be an open set, $f: A \to \mathbb{C}$ and $\alpha \in \overline{A}$. Then **$f(z)$ has limit ℓ as z tends to α** if and only if, given any real $\varepsilon > 0$, there exists a real $\delta > 0$ (depending on ε) such that

$$0 < |z - \alpha| < \delta \quad \Rightarrow \quad |f(z) - \ell| < \varepsilon$$

If this definition is satisfied, we write

$$\lim_{z \to \alpha} f(z) = \ell$$

Geometrically speaking, the definition states that any open ε-neighbourhood of ℓ contains all the values of f, for z in some open δ-neighbourhood of α, except possibly for the value $f(\alpha)$. This is indicated in Fig. 2.1. In general, the smaller the given value of ε, the smaller δ will have to be.

Note

$\lim_{z \to \alpha} f(z)$ does not depend on $f(\alpha)$, since $f(\alpha)$ does not appear in the definition. Notice that α need not belong to the domain of f.

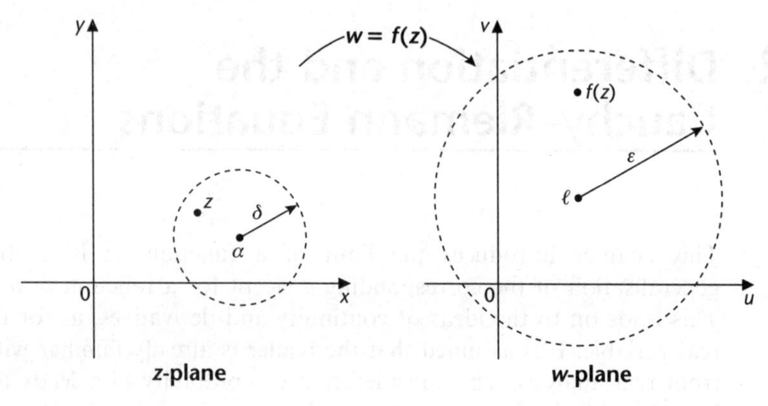

Figure 2.1

Example 2.1

(i) It follows immediately from the definition that $\lim_{z\to\alpha} z = \alpha$. Also $|\bar{z} - \bar{\alpha}| = |\overline{z - \alpha}| = |z - \alpha|$ so that it follows from the definition that $\lim_{z\to\alpha} \bar{z} = \bar{\alpha}$.

(ii) Suppose we wish to use the definition to prove that $\lim_{z\to i}(z^2 + 1) = 0$. It follows by the triangle inequality that

$$|(z^2 + 1) - 0| = |(z - i)(z - i + 2i)| \leqslant |z - i|(|z - i| + 2)$$

then $0 < |z - i| < \delta \quad \Rightarrow \quad |(z^2 + 1) - 0| < \delta(\delta + 2) < 3\delta$

as long as $\delta \leqslant 1$, so that $\delta^2 \leqslant \delta$. Then given $\varepsilon > 0$, we choose $\delta = \min(1, \varepsilon/3)$, so that for this choice of δ

$$0 < |z - i| < \delta \quad \Rightarrow \quad |(z^2 + 1) - 0| < \varepsilon$$

as required.

Important Notes

(i) With real analyis, when taking limits, there are only two directions in which a real variable can approach a real number a. But with complex analysis, in the definition above, z can approach α from any direction, along any curve.

(ii) Most of the results concerning limits of functions of a complex variable are essentially the same as the corresponding results for functions of a real variable. Since the definition is formally the same as the definition for a real variable and the triangle inequality holds for complex numbers (see Lemma 1.2(v)), the proofs of the following standard results are essentially the same as for a real variable. For this reason, these proofs are omitted.

Theorem 2.1. Properties of Limits of Functions

(i) If $\lim_{z \to \alpha} f(z)$ exists then it is unique.

Providing the following limits exist:

(ii) $\lim_{z \to \alpha} (kf(z)) = k \lim_{z \to \alpha} f(z)$ for any $k \in \mathbb{C}$

(iii) $\lim_{z \to \alpha} (f(z) + g(z)) = \lim_{z \to \alpha} f(z) + \lim_{z \to \alpha} g(z)$

(iv) $\lim_{z \to \alpha} (f(z)g(z)) = (\lim_{z \to \alpha} f(z))(\lim_{z \to \alpha} g(z))$

(v) $\lim_{z \to \alpha} \left(\dfrac{f(z)}{g(z)} \right) = \dfrac{\lim_{z \to \alpha} f(z)}{\lim_{z \to \alpha} g(z)}$, as long as $\lim_{z \to \alpha} g(z) \neq 0$ □

Note that in (v), it follows by definition that $\lim_{z \to \alpha} g(z) \neq 0$ ensures that $g(z) \neq 0$ in some deleted δ-neighbourhood of α.

Example 2.2

(i) It can be shown that $\lim_{z \to 0} (\sin z)/z$ exists. Recall that $\lim_{x \to 0} (\sin x)/x = 1$ for $x \in \mathbb{R}$. Hence letting $z \to 0$ along the positive x-axis and using 2.1(i) gives $\lim_{z \to 0} (\sin z)/z = \lim (\sin x)/x = 1$.

(ii) Using the same technique, $\lim_{z \to 0} |z|/z$ does not exist, since if $z \to 0$ along the positive x-axis, $\lim_{z \to 0} |z|/z = \lim_{x \to 0^+} x/x = 1$, whereas if $z \to 0$ along the positive y-axis, $\lim_{z \to 0} |z|/z = \lim_{y \to 0^+} y/iy = -i$, contradicting 2.1(i).

(iii) $$\lim_{z \to i} \frac{z^2 + iz + 2}{z^2 - 3iz - 2} = \lim_{z \to i} \frac{(z - i)(z + 2i)}{(z - i)(z - 2i)} = \lim_{z \to i} \frac{z + 2i}{z - 2i} = -3$$
using 2.1(iii), (v).

Continuity

The concept of continuity at a point in complex analysis is again a straightforward generalisation of the corresponding concept in real analysis.

Definitions

Let $A \subseteq \mathbb{C}$ be open. A function $f: A \to \mathbb{C}$ is **continuous at** $\alpha \in A$ if and only if $\lim_{z \to \alpha} f(z) = f(\alpha)$. Function f is **continuous on a region** $\mathscr{R}$ if and only if it is continuous at every point of $\mathscr{R}$. Naively speaking, continuous functions map continuous curves in the z-plane to continuous curves in the w-plane. Once again, results concerning continuity at a point are essentially the same as those for real variables.

Theorem 2.2. Elementary Properties of Continuous Functions

(i) If $\lim_{z \to \alpha} f(z) = \ell$ and g is continuous in some open neighbourhood of ℓ, then

$$\lim_{z \to \alpha} g(f(z)) = g(\ell) = g(\lim_{z \to \alpha} f(z)) \qquad \text{(composite rule)}$$

(ii) Let $k \in \mathbb{C}$ and suppose that f and g are continuous at $\alpha \in \mathbb{C}$. Then $kf, f + g$ and fg are continuous at α, and f/g is continuous at α as long as $g(\alpha) \neq 0$.

(iii) If f and g are continuous at α and $f(\alpha)$ respectively, then the composition $g \circ f$ is continuous at α. □

It follows by Theorem 2.2 that every polynomial is continuous everywhere and, more generally, every algebraic fraction is continuous at those points at which the denominator is non-zero.

Note

As in real analysis, if a function f is continuous on a compact set, then f is bounded. This result is proved in Chapter 4 and appears as Theorem 4.5.

Example 2.3

(i) Let $f: \mathbb{C} \to \mathbb{C}$ be defined by $f(z) = \bar{z}$. From Example 2.1(i), it follows that $\lim_{z \to \alpha} f(z) = \bar{\alpha} = f(\alpha)$ for any $\alpha \in \mathbb{C}$, so that f is continuous everywhere.

(ii) Let $f: \mathbb{C} \to \mathbb{C}$ be defined by

$$f(z) = \begin{cases} 1 \text{ if } |z| \text{ is rational} \\ 0 \text{ if } |z| \text{ is irrational} \end{cases}$$

It follows by definition that f is continuous at $\alpha \in \mathbb{C}$ if and only if, given any $\varepsilon > 0$, there exists a $\delta > 0$ such that

$$|z - \alpha| < \delta \quad \Rightarrow \quad |f(z) - f(\alpha)| < \varepsilon$$

Suppose that $|\alpha|$ is rational. No matter how small $\delta > 0$ is chosen, there exists z_1 with $|z_1 - \alpha| < \delta$, such that $|z_1|$ is irrational. Then

$$|f(z_1) - f(\alpha)| = |0 - 1| = 1 \not< \varepsilon \qquad (\varepsilon \leqslant 1)$$

Hence f is not continuous at α. Similarly, f is not continuous at any α for which $|\alpha|$ is irrational. Thus, f is discontinuous everywhere.

(iii) $f: \mathbb{C} \to \mathbb{C}$ defined by $f(z) = \sec z = 1/\cos z$ (with f defined arbitrarily at values of z for which $\cos z = 0$, i.e. at $z = \alpha_n = (n + 1/2)\pi$, $n \in \mathbb{Z}$)) is continuous everywhere except at α_n, by 2.2(ii) and the fact that $\lim_{z \to \alpha_n} f(z)$ does not exist in $\mathbb{C}$.

(iv) The exponential function is easily shown to be continuous everywhere, so that by 2.2(i),

$$\lim_{z \to i} e^{\sin z} = e^{\lim_{z \to i} \sin z} = e^{i \sinh 1} = \cos(\sinh 1) + i \sin(\sinh 1)$$

Branch Points and Riemann Surfaces

As seen in Chapter 1, many elementary functions of a complex variable occur as specific branches of many-valued relations defined in terms of the relation arg. Such functions will be discontinuous at points along any line acting as a 'boundary' between different branches, corresponding to a discontinuity of the associated principal arguments. These somewhat arbitrary and artificial lines of discontinuity are known as branch cuts.

For example, because of our convention for Arg z, each branch of arg z is discontinuous along the line segment given by $z = x,\ x \leqslant 0$. Choosing a different convention for Arg z will produce a different line of discontinuity of each branch of arg z, emanating from the origin. Notice that 0 is the only point common to all branch cuts. Any complete continuous circuit of a closed curve, passing through $\alpha \in \mathbb{C}$ and not enclosing 0, i.e. one for which $|z|$ and arg z vary continuously, does not alter the initial value of arg α. On the other hand, upon completion of a continuous circuit of any closed curve passing through α and enclosing 0, any argument of α increases or decreases by 2π, so that arg α does not take the same value; the value of arg α now lies in a different branch since a branch cut has been crossed. This is indicated in Fig. 2.2.

This type of behaviour can be undesirable and needs to be identified when it occurs. These ideas are formalised in the following definitions.

Definitions

A many-valued relation which associates a subset of $\mathbb{C}$ with each point in some subset of $\mathbb{C}$ will be called a **multifunction**. A point $\alpha \in \mathbb{C}$ is a **branch point** of a multifunction g, defined on some open subset A of $\mathbb{C}$, if there exists at least one

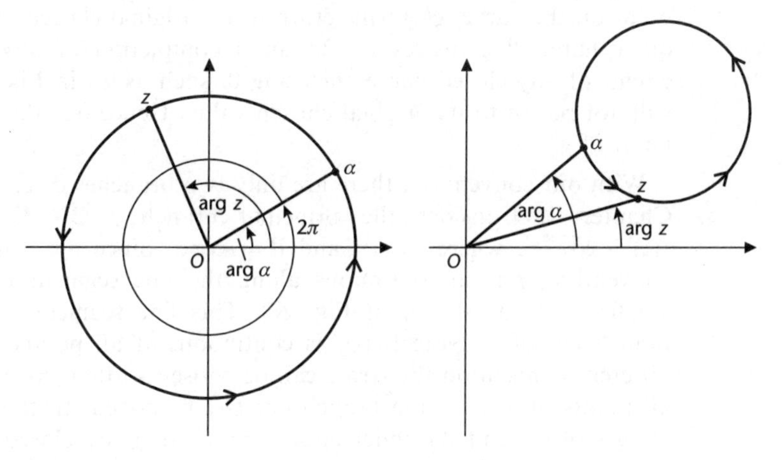

Figure 2.2

closed curve $\mathscr{C}$ enclosing α and lying in A, such that upon completion of a circuit of $\mathscr{C}$ for which any argument of $z-\alpha$ increases continuously by 2π, $g(z)$ does not return to its orginal chosen value. A **branch cut** of g is then a chosen line segment with initial point at a branch point α of g, such that there is one and only one branch cut in any neighbourhood of α containing no other branch point of g. A **branch** of g is any function with the same domain as g and obtained from g by making it single-valued, which is continuous at all points of its domain except along any chosen branch cuts of g.

Notes

When no ambiguity can occur, we shall use for multifunctions the same notation as we used for functions. The choice of branch cuts for a multifunction with branch points, set up to separate distinct branches, depends very much on a chosen convention. Crossing a branch cut has the effect of changing to a different branch of the multifunction. Notice that the image of any closed curve enclosing just one branch point of a multifunction g, under any branch of g, will not be closed, since any branch will be discontinuous along a branch cut. On the other hand, given a closed curve $\mathscr{C}$ not enclosing a branch point of g, there is a choice of convention which ensures that $\mathscr{C}$ does not cross any branch cut of g, so that the image of $\mathscr{C}$ under any branch of g is closed.

Example 2.4

(i) Consider the multifunction g defined on $\mathbb{C}$ by $g(z)=z^{1/2}$, i.e. $g(z)=\sqrt{r}e^{i\theta/2}$, where $r=|z|$ and $\theta=\arg z$. Any value of θ is not altered after a complete continuous circuit of any closed curve not enclosing 0, such as $\mathscr{C}_1$ or $\mathscr{C}_2$ shown in Fig. 2.3, so that if z is any point on the curve, $g(z)$ will return to its original chosen value. On the other hand, θ increases by 2π upon completion of any continuous circuit of any closed curve enclosing 0, such as $\mathscr{C}_3$ in Fig. 2.3, so $g(z)$ will not return to its original chosen value. Hence 0 is the only branch point of g.

With our convention, there are only two branches of g, as defined in Chapter 1. Consider the principal branch $f:\mathbb{C}\to\mathbb{C}$ defined by $f(z)=\sqrt{r}e^{i\theta/2}$, where $r=|z|$ and $\theta=\operatorname{Arg} z$. Since $-\pi<\operatorname{Arg} z\leqslant\pi$ by convention, f is discontinuous along the line segment OA given by $x\leqslant 0$, $y=0$, as shown in Fig. 2.3. This line segment is our chosen branch cut of g. Note that f is continuous at all points of $\mathscr{C}_1$ and a different convention for Arg z can be chosen so that f is continuous at all points of $\mathscr{C}_2$, i.e. the branch cut OA is moved. In these cases, the images of $\mathscr{C}_1$ and $\mathscr{C}_2$ under either branch of g are closed curves. The image of $\mathscr{C}_1$ under f is shown in Fig. 2.3. Starting at the point P, the

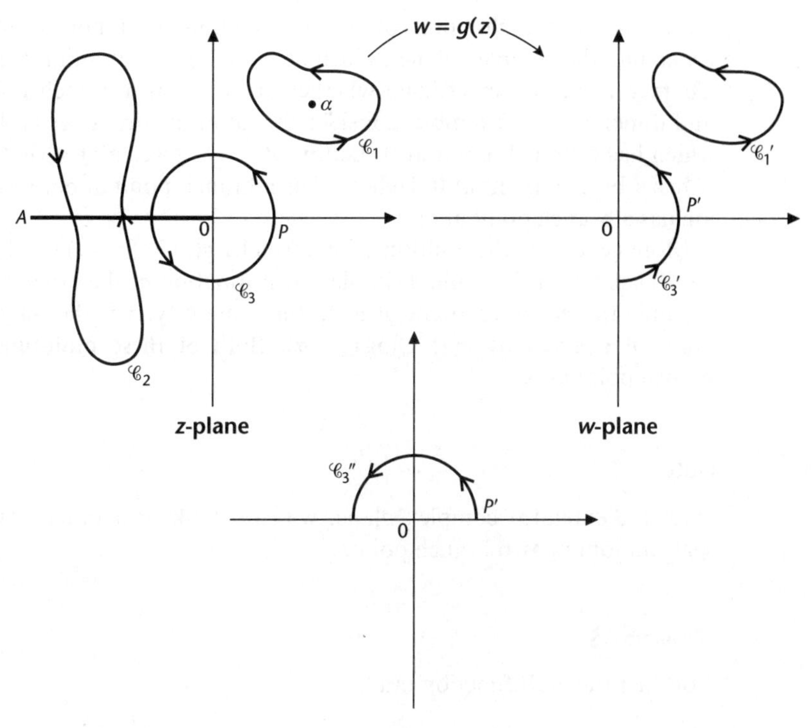

Figure 2.3

image of $\mathscr{C}_3$ under f, shown as $\mathscr{C}_3'$, is not a closed curve. A continuous image of $\mathscr{C}_3$ can be obtained by swopping branches of g when the branch cut OA is crossed; this is shown as $\mathscr{C}_3''$ in Fig. 2.3. Note that such a curve will still not be closed. Continuously traversing $\mathscr{C}_3$ twice gets us back to the orginal branch of g, so the image in this case, using both branches, will be closed.

(ii) Consider the multifunction log defined by

$$\log z = \operatorname{Log}|z| + i \arg z \qquad (z \neq 0)$$

(see Lemma 1.12). Once again 0 is the one and only branch point of log, and because of our convention chosen for Arg z, the line segment $x \leqslant 0$, $y = 0$ is the branch cut for log. Each branch of log, as defined in Chapter 1, is discontinous along the branch cut, so the image of any closed curve enclosing 0, under any branch of log, is not closed. In this case, since log has an infinite number of branches, no matter how many times a closed curve enclosing 0 is continuously traversed, its image can never be a closed curve, even if a different branch is employed when the branch cut is crossed.

Note that a branch point of a multifunction need not be in its domain. Extending the complex plane to include ∞, the point at infinity, as described in Chapter 1, we can investigate whether or not ∞ is a branch point of a given multifunction. For example, consider the multifunction g, given by $g(z) = z^{1/2}$, which has a branch point at 0. Letting $Z = 1/z$ gives $g(z) = G(Z) = Z^{-1/2}$ and G has a branch point at 0. Hence g has a branch point at ∞. In the same way, log has a branch point at ∞.

More generally, the multifunction given by $g(z) = (z - \alpha)^{1/n}$ where $\alpha \in \mathbb{C}$ and $n \in \mathbb{N}$ has n branches, and a simple transformation of the origin shows that α is the only finite branch point of g. In the same way, α is the only finite branch point of g given by $g(z) = \log(z - \alpha)$. Both of these multifunctions have a branch point at ∞.

Note

With the extended complex plane, we can think of a branch cut as any line segment joining two branch points.

Example 2.5

Consider the multifunction $\tan^{-1}$.

$$w = \tan^{-1} z \quad \Rightarrow \quad z = \frac{e^{2iw} - 1}{i(e^{2iw} + 1)} \quad \Rightarrow \quad e^{2iw} = \frac{1 + iz}{1 - iz}$$

$$\Rightarrow \quad w = \frac{1}{2i}\log\left(\frac{1 + iz}{1 - iz}\right)$$

$$\Rightarrow \quad 2i\tan^{-1} z = \log(i(z - i)) - \log(-i(z + i)) \qquad (z \neq \pm i)$$

(see Lemma 1.13). Hence $\tan^{-1}$ has branch points at $\pm i$ (but not at ∞ in $\tilde{\mathbb{C}}$).

Example 2.6

Define the multifunction g by $g(z) = (z^2 + 4)^{1/2}$. Show that $\pm 2i$ are branch points of g. Show also that a complete continuous circuit around any simple closed curve enclosing both points produces no change in an original chosen value of $g(z)$, on a particular branch of g, where z is any point on the closed curve. (A curve is **simple** if it has no self-intersections.) Indicate three possible branch cuts for g.

Solution

Note that $g(z) = (z + 2i)^{1/2}(z - 2i)^{1/2}$. Let $\mathscr{C}_1$ be a simple closed curve enclosing $2i$, and $\mathscr{C}_2$ be a simple closed curve enclosing $-2i$, as shown in Fig. 2.4.

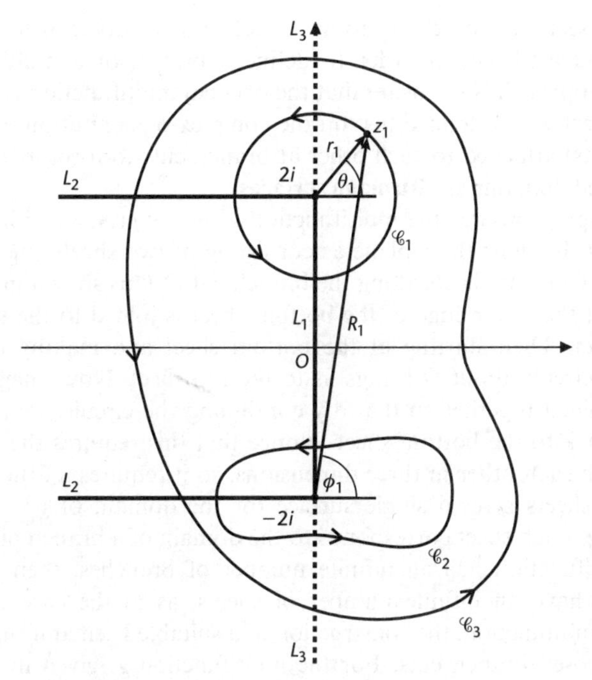

Figure 2.4

Let $z - 2i = re^{i\theta}$ and $z + 2i = Re^{i\phi}$. Choose a particular point z_1 on $\mathscr{C}_1$ and let $z_1 = r_1 e^{i\theta_1} + 2i = R_1 e^{i\phi_1} - 2i$. ($\theta_1$ can be any chosen value of arg $(z_1 - 2i)$ and ϕ_1 can be any chosen value of arg $(z_1 + 2i)$.) Let

$$w_1 = g(z_1) = \sqrt{r_1 R_1} e^{i(\theta_1/2 + \phi_1/2)}$$

As $\mathscr{C}_1$ is traversed anticlockwise once, so that θ and ϕ vary continuously, θ_1 increases by 2π whereas ϕ_1 is unchanged, so upon returning to z_1,

$$g(z_1) = \sqrt{r_1 R_1} e^{i(\theta_1/2 + \pi + \phi_1/2)} = -\sqrt{r_1 R_1} e^{i(\theta_1/2 + \phi_1/2)} = -w_1$$

Hence g does not return to its original chosen value, so $2i$ is a branch point of g. Similarly, $-2i$ is a branch point of g. Now suppose that $\mathscr{C}_3$ is a simple closed curve enclosing $2i$ and $-2i$ (Fig. 2.4). Then, using the same technique as above, if z is any point on $\mathscr{C}_3$, after one continuous circuit of $\mathscr{C}_3$, the change in arg $(z - 2i) = 2\pi$ and the change in arg$(z + 2i) = 2\pi$, so that the change in arg $(g(z)) = 2\pi$, and $g(z)$ returns to its original chosen value, as required.

Clearly, $\pm 2i$ are the only finite branch points of g. Letting $Z = 1/z$ gives $g(z) = G(Z) = (4Z^2 + 1)^{1/2}/Z$, so that $Z = 0$ is not a branch point of G and ∞ is not branch point of g. Three possible branch cuts for g are labelled L_1, L_2 and L_3 in Fig. 2.4. In the figure, L_1 joins $-2i$ and $2i$ and passes through 0, whereas L_3 is the rest of the imaginary axis.

As we have seen, it is necessary to introduce branch cuts, which are somewhat arbitrary and artificial, in order to define a branch of a multifunction. An alternative approach is to ensure that the original multifunction is a continuous function directly, not defined just on the complex plane, but on many complex planes (sheets) attached to each other at branch cuts to form its domain. Such many-sheeted domains are **Riemann surfaces**.

For example, consider the multifunction given by $g(z) = z^{1/2}$ investigated in Example 2.4. Imagine the z-plane as consisting of two sheets superimposed on each other. 'Cut' the sheets along the branch cut OA, as shown in Fig. 2.3, and imagine that the lower edge of the bottom sheet is joined to the upper edge of the top sheet. Then starting at the bottom sheet and making one complete continuous circuit about O brings us to the top sheet. Now imagine the other cut edges joined together so that, by continuing the circuit, we pass from the top sheet back to the bottom sheet. Notice that this requires the two sheets to pass through each other in three dimensions, so it requires a little imagination! The joined sheets gives a single surface for the domain of g – the Riemann surface for g. Each sheet corresponds to the domain of a branch of g.

If a multifunction has an infinite number of branches, then the Riemann surface will have an infinite number of sheets, as in the case of log. For a particular multifunction, the construction of a suitable Riemann surface depends upon the chosen branch cuts. For the multifunction g, given in Example 2.6, choosing the branch cut L_1 shown in Fig 2.4, a possible Riemann surface for g consists of two sheets, with opposite edges joined along L_1, as described above.

Note

It is possible to give a formal topological definition of a Riemann surface but this is beyond the scope of this book. An extensive topological theory of Riemann surfaces exists and this more sophisticated approach gives a much deeper insight into multifunctions.

Historical Note

Riemann introduced the notion of a Riemann surface during the 1850s. Riemann surfaces were basic to the progress of analysis and topology in the twentieth century. For a comprehensive account of Riemann surfaces, see for example, G. Springer, *Introduction to Riemann Surfaces,* Addison-Wesley, 1957.

Exercise

2.1.1 Use the definition of a limit to prove the following results:

(i) $\lim_{z\to 0} (2z^2 + iz + i) = i$

(ii) $\lim_{z\to 1+i} z + \bar{z} = 2$

(iii) $\lim_{z\to -i} \dfrac{z+4}{z-3i} = i + 1/4$

Exercise **2.1.2** Use the definition of a limit to prove the following theorems:

(i) If $\lim_{z\to\alpha} f(z)$ exists then it is unique.

(ii) Providing the limits exist,

$$\lim_{z\to\alpha}(f(z)+g(z)) = \lim_{z\to\alpha} f(z) + \lim_{z\to\alpha} g(z)$$

Exercise **2.1.3** Find the following limits, quoting any results that you use:

(i) $\lim_{z\to 0}\dfrac{(z+i)^2+1}{z}$

(ii) $\lim_{z\to 2i}\dfrac{z^2-4iz-4}{z^2-3iz-2}$

Exercise **2.1.4** Show that the following limits do not exist, by letting $z\to 0$ along distinct paths:

(i) $\lim_{z\to 0}\dfrac{z}{\bar{z}}$

(ii) $\lim_{z\to 0}\left(\dfrac{z}{\bar{z}}\right)^2$

Exercise **2.1.5** Let $f:\mathbb{C}\to\mathbb{C}$ be defined by $f(z)=\operatorname{Re} z$. Prove, using the definition, that $\lim_{z\to\alpha} f(z) = \operatorname{Re}\alpha$, so that f is continuous everywhere.

Exercise **2.1.6** Use the $\varepsilon-\delta$ definition to prove that $f:\mathbb{C}\to\mathbb{C}$ defined by $f(z)=|z|^3/z, z\neq 0, f(0)=0$, is continuous at 0.

Exercise **2.1.7** Use the properties of limits of functions to prove that if f and g are continuous at α with $g(\alpha)\neq 0$, then f/g is continuous at α.

Exercise **2.1.8** Find the image of the circle $|z|=a$ under $w=\operatorname{Log} z$. Starting at $z=a$, with the principal branch of log, find the image of $|z|=a$, traversed three times anticlockwise so that $\arg z$ increases continuously, under $w=\log z$.

Exercise **2.1.9** Locate the finite branch points of the following multifunctions:

(i) $g(z)=\dfrac{1}{(z^4-16)^{1/2}}$

(ii) $g(z)=\dfrac{\log(z^2-1)}{\cos(z^{1/2})}$

(iii) $g(z)=\coth^{-1} z$

Exercise **2.1.10**

(a) Show that $\pm i$ are branch points of the multifunction g, given by $g(z)=(z^2+1)^{1/n}$, $n\in\mathbb{N}$, $n>1$, and find four different branch cuts for g.

(b) Show that 1, -2 and i are branch points of the multifunction g, given by $g(z)=(z^3+z^2-iz^2-2z-iz+2i)^{1/2}$.

Exercise **2.1.11** Show that 0 is a branch point of g given by $g(z) = z^{1/3} - z^{1/4}$. After how many complete continuous circuits of a closed curve enclosing 0 will $g(z)$ resume its original value?

Exercise **2.1.12** Indicate three closed curves in the z-plane which cannot be deformed into each other without crossing a branch point of the multifunction g, given by $g(z) = (z^2 - 1)^{1/2}$, and which are such that when traversed continuously once, $g(z)$ resumes its initial value.

Derivatives

The definition of the derivative of a function of a complex variable is formally the same as the derivative of a function of a real variable, although in this case it does not have the same simple geometrical interpretation.

Definition

Let $A \subseteq \mathbb{C}$ be open, $f: A \to \mathbb{C}$ and $\alpha \in A$. Then the **derivative** of f at α (if it exists) is denoted and defined by

$$f'(\alpha) = \lim_{h \to 0} \frac{f(\alpha + h) - f(\alpha)}{h}$$

Function f is **differentiable** at α if and only if $f'(\alpha)$ exists.

Notes

(i) Remember that $h \in \mathbb{C}$ and can approach 0 in any direction along any path in the complex plane.

(ii) Letting $h = z - \alpha$ gives an alternative form of the definition as

$$f'(\alpha) = \lim_{z \to \alpha} \frac{f(z) - f(\alpha)}{z - \alpha}$$

(iii) As in the real case, if $w = f(z)$ we sometimes write $dw/dz = f'(z)$. Higher derivatives are defined in the usual way.

(iv) As in the case of real variables, if f is differentiable at α then it is continuous there (see Exercises 2.2).

(v) As the definition is identical in form to the definition for real variables, the following standard results can be derived from the definition by essentially the same steps used in real analysis, bearing in mind the results of Chapter 1 and the results of this chapter on limits of functions.

Theorem 2.3. Elementary Properties of Derivatives

(a) $f(z) = z^n$, $n \in \mathbb{N}$ $\Rightarrow$ $f'(z) = nz^{n-1}$ for all $z \in \mathbb{C}$.
(b) At each point where the derivatives exist:
 (i) If $h(z) = f(z) + g(z)$ then $h'(z) = f'(z) + g'(z)$.
 (ii) If $h(z) = f(z)g(z)$ then $h'(z) = f'(z)g(z) + f(z)g'(z)$.
 (iii) If $h(z) = (g \circ f)(z)$ and $w = f(z)$ then $h'(z) = g'(w)f'(z)$.
 (iv) If $h(z) = 1/f(z)$ then $h'(z) = -f'(z)/(f(z))^2$. □

Proof

The proof of each statement is identical in form to the proof of the corresponding result for a real variable. As examples, we prove (a) and (b)(ii).
(a) Using the binomial theorem gives

$$f'(z) = \lim_{h \to 0} \frac{(z+h)^n - z^n}{h}$$

$$= \lim_{h \to 0} \frac{(z^n + nz^{n-1}h + n(n-1)z^{n-2}h^2/2 + \ldots + h^n) - z^n}{h} = nz^{n-1}$$

(b) (ii) It follows from the definition that

$$h'(z) = \lim_{h \to 0} \frac{f(z+h)(g(z+h) - g(z)) + g(z)(f(z+h) - f(z))}{h}$$

$$= \lim_{h \to 0} (f(z+h)g'(z) + g(z)f'(z)) = f(z)g'(z) + g(z)f'(z)$$

by Theorem 2.1. ■

Notes

(i) It follows by 2.3(b)(ii) that if $h(z) = kf(z)$, then $h'(z) = kf'(z)$ for any $k \in \mathbb{C}$.
(ii) Letting $W = g(w)$ in 2.3(b)(iii), the result can be written as

$$\frac{dW}{dz} = \frac{dW}{dw} \cdot \frac{dw}{dz}$$

and is known as the **chain rule**, as in real calculus.
(iii) It follows by 2.3(b)(ii) and (iv) that

$$h(z) = \frac{f(z)}{g(z)} \quad \Rightarrow \quad h'(z) = \frac{g(z)f'(z) - f(z)g'(z)}{(g(z))^2}$$

at points where the derivatives exist, as in the real case. Hence by Theorem 2.3 polynomials and, more generally, algebraic fractions can be differentiated as in the real case.

Example 2.7

Use the definition of the derivative of a function to show that $f : \mathbb{C} \to \mathbb{R}$ defined by $f(z) = \operatorname{Re} z$ is nowhere differentiable.

Solution

Take any $z \in \mathbb{C}$. Since $f'(z)$ is a limit, if it exists it is unique, no matter how $h \to 0$, by 2.1(i). Let $z = x + iy$ and $h = \alpha + i\beta$, where x, y, α and β are real. Let $h \to 0$ along the real axis (in either direction), so that $\beta = 0$. Then

$$f'(z) = \lim_{h \to 0} \frac{\operatorname{Re}(z+h) - \operatorname{Re} z}{h} = \lim_{\alpha \to 0} \frac{(x+\alpha) - x}{\alpha} = 1$$

Now let $h \to 0$ along the imaginary axis, so that $\alpha = 0$. Then

$$f'(z) = \lim_{i\beta \to 0} \frac{x - x}{i\beta} = \lim_{\beta \to 0} \frac{0}{i\beta} = 0$$

This contradicts 2.1(i), so that f is nowhere differentiable.

It is easily shown that f is continuous everywhere (see Exercises 2.1). In contrast to functions mapping $\mathbb{R}$ to $\mathbb{R}$, functions mapping $\mathbb{C}$ to $\mathbb{C}$ which are continuous everywhere and differentiable nowhere are easy to find.

Analytic Functions and the Cauchy–Riemann Equations

Of great importance in complex analysis is the following result, which gives a necessary condition for differentiability. The proof uses the technique of Example 2.7.

Definitions

A function f is **analytic**, or **regular** or **holomorphic**, on a region $\mathscr{R}$ of $\mathbb{C}$ if and only if it is differentiable at each point of $\mathscr{R}$. A function f is **entire** if and only if it is analytic on $\mathbb{C}$.

Notation

Let $u : \mathbb{R}^2 \to \mathbb{R}$ and $w = u(x, y)$. We will usually write the first partial derivative of u with respect to x at (x, y) as $u_x(x, y)$ rather than as $\partial w / \partial x$. When no ambiguity can occur, we will write $u_x(x, y)$ as u_x, and similarly for y. We shall usually denote

$$\frac{\partial}{\partial x}\left(\frac{\partial w}{\partial y}\right)$$

by $u_{yx}(x, y)$ or simply by u_{yx}, and so on.

Example 2.10 The Cauchy-Riemann Equations in Polar Form

If $w = f(z) = f(re^{i\theta}) = u(r, \theta) + iv(r, \theta)$ (2.5)

show that the Cauchy–Riemann equations reduce to

$$u_r(r, \theta) = \frac{1}{r} v_\theta(r, \theta), \; v_r(r, \theta) = \frac{-1}{r} u_\theta(r, \theta) \qquad (r \neq 0) \tag{2.6}$$

Hence find $f(z)$ if f is entire and $u(r, \theta) = r^2 \cos^2 \theta$.

Solution

Note that $r = \sqrt{x^2 + y^2}$ and $\tan \theta = y/x$. Hence by the chain rule for functions of two real variables

$$\frac{\partial u}{\partial x} = \frac{\partial u}{\partial r}\frac{\partial r}{\partial x} + \frac{\partial u}{\partial \theta}\frac{\partial \theta}{\partial x} = \frac{x}{r}\frac{\partial u}{\partial r} - \frac{y}{r^2}\frac{\partial u}{\partial \theta}$$

$$\frac{\partial u}{\partial y} = \frac{\partial u}{\partial r}\frac{\partial r}{\partial y} + \frac{\partial u}{\partial \theta}\frac{\partial \theta}{\partial y} = \frac{y}{r}\frac{\partial u}{\partial r} + \frac{x}{r^2}\frac{\partial u}{\partial \theta}$$

Similarly $\dfrac{\partial v}{\partial x} = \dfrac{x}{r}\dfrac{\partial v}{\partial r} - \dfrac{y}{r^2}\dfrac{\partial v}{\partial \theta}, \; \dfrac{\partial v}{\partial y} = \dfrac{y}{r}\dfrac{\partial v}{\partial r} + \dfrac{x}{r^2}\dfrac{\partial v}{\partial \theta}$

Hence the Cauchy–Riemann equations give

$$\cos\theta u_r - \frac{\sin\theta}{r} u_\theta = \sin\theta v_r + \frac{\cos\theta}{r} v_\theta \tag{2.7}$$

$$\sin\theta u_r + \frac{\cos\theta}{r} u_\theta = -\cos\theta v_r + \frac{\sin\theta}{r} v_\theta \tag{2.8}$$

Mutiplying (2.7) by $\cos\theta$ and (2.8) by $\sin\theta$, and adding gives the first equation of (2.6). Multiplying (2.7) by $\sin\theta$ and (2.8) by $\cos\theta$ and subtracting gives the second equation of (2.6) as required.

If f is entire then the Cauchy–Riemann equations (2.6) are satisfied everywhere ($r \neq 0$), so $u(r, \theta) = r^2 \cos 2\theta$ gives

$$\begin{aligned}
& u_r = 2r\cos 2\theta \qquad u_\theta = -2r^2 \sin 2\theta \\
\Rightarrow \quad & v_\theta = 2r^2 \cos 2\theta \qquad v_r = 2r \sin 2\theta \\
\rightarrow \quad & v(r, \theta) = r^2 \sin 2\theta + F(r) = r^2 \sin 2\theta + G(\theta)
\end{aligned}$$

where F and G are arbitrary functions. Hence

$$v(r, \theta) = r^2 \sin 2\theta + k \quad \Rightarrow \quad f(z) = f(re^{i\theta}) = r^2(\cos 2\theta + i \sin 2\theta) + ik$$

where k is an arbitrary real constant. The form of $f(z)$ is suggested by letting $\theta = 0$ in $f(re^{i\theta})$. This gives $f(r) = r^2 + ik$. In general,

$$f(z) = r^2(\cos\theta + i\sin\theta)^2 + ik = z^2 + ik$$

by De Moivre's theorem.

The following result is the partial converse of Theorem 2.4.

Theorem 2.6. Sufficient Conditions for Differentiability

Let a function f be given by (2.1) and suppose that the Cauchy–Riemann equations (2.2) are satisfied on a region $\mathscr{R}$, with the first-order partial derivatives of $u(x, y)$ and $v(x, y)$, with respect to x and y, continuous on $\mathscr{R}$. Then f is analytic on $\mathscr{R}$ and its derivative at any point $z \in \mathscr{R}$ is given by Corollary 2.5. □

Proof

Let $z = x + iy \in \mathscr{R}$ and $h = \alpha + i\beta$, where x, y, α and β are real. Then

$$f(z+h) - f(z) = [u(x+\alpha, y+\beta) - u(x, y)] + i[v(x+\alpha, y+\beta) - v(x, y)] \quad (2.9)$$

using (2.1). It follows by the mean value theorem for functions of one real variable that

$$\begin{aligned} u(x+\alpha, y+\beta) - u(x, y) &= [u(x+\alpha, y+\beta) - u(x, y+\beta)] \\ &\quad + [u(x, y+\beta) - u(x, y)] \\ &= \alpha u_x(x+\theta\alpha, y+\beta) + \beta u_y(x, y+\beta\psi) \end{aligned} \quad (2.10)$$

where $0 < \theta < 1$ and $0 < \psi < 1$. Now let

$$u_x(x+\theta\alpha, y+\beta) = u_x(x, y) + \varepsilon_1 \qquad u_y(x, y+\beta\psi) = u_y(x, y) + \varepsilon_2 \quad (2.11)$$

Then since u_x and u_y are continuous in $\mathscr{R}$,

$$\varepsilon_1,\ \varepsilon_2 \to 0 \quad \text{as} \quad (\alpha, \beta) \to (0, 0)$$

independent of the chosen path. Substituting (2.11) into (2.10),

$$u(x+\alpha, y+\beta) - u(x, y) = \alpha u_x(x, y) + \beta u_y(x, y) + \alpha\varepsilon_1 + \beta\varepsilon_2 \quad (2.12)$$

Similarly,

$$v(x+\alpha, y+\beta) - v(x, y) = \alpha v_x(x, y) + \beta v_y(x, y) + \alpha\eta_1 + \beta\eta_2 \quad (2.13)$$

where $\eta_1,\ \eta_2 \to 0$ as $(\alpha, \beta) \to (0, 0)$, independent of the chosen path. Hence from (2.9), (2.12) and (2.13), since the Cauchy–Riemann equations are satisfied in $\mathscr{R}$, it follows that

$$\begin{aligned} f(z+h) - f(z) &= (\alpha u_x - \beta v_x + \alpha\varepsilon_1 + \beta\varepsilon_2) + i(\alpha v_x + \beta u_x + \alpha\eta_1 + \beta\eta_2) \\ &= h(u_x + iv_x) + \alpha\sigma_1 + \beta\sigma_2 \text{ say} \end{aligned} \quad (2.14)$$

where $\lim_{h\to 0}\sigma_1 \lim_{h\to 0}\sigma_2 = 0$, independent of the chosen path in the complex plane. Hence, since

$$\left|\frac{\alpha\sigma_1+\beta\sigma_2}{h}\right| \leqslant \frac{\max(|\alpha|,|\beta|)}{\sqrt{\alpha^2+\beta^2}}|\sigma_1+\sigma_2| \leqslant |\sigma_1+\sigma_2| \leqslant |\sigma_1|+|\sigma_2|$$

it follows from (2.14) that

$$f'(z) = \lim_{h\to 0}\frac{f(z+h)-f(z)}{h} = u_x + iv_x \qquad (\text{for all } z \in \mathscr{R})$$

as required. The rest follows by 2.5. ■

Example 2.11

(i) Let $f:\mathbb{C}\to\mathbb{C}$ be defined by $f(z)=e^z$. It follows by definition that if f is given by (2.1) then

$$u(x,y) = e^x\cos y, \qquad v(x,y) = e^x \sin y$$
$$\Rightarrow u_x = e^x\cos y = v_y \qquad u_y = -e^x\sin y = -v_x$$

so that the Cauchy–Riemann equations are satisfied and the first-order partial derivatives are continuous everywhere. Hence f is entire by Theorem 2.6, and by Corollary 2.5

$$f'(z) = e^x(\cos y + i\sin y) = e^z = f(z)$$

as in the real case.

It is easy to show that exp is the unique entire function f with this property, such that $f(0)=1$. Suppose a function f is entire and is given by (2.1) with $f(z)=f'(z)$. Then by 2.4, $u(x,y)=u_x(x,y)$ and $v(x,y)=v_x(x,y)$, so that

$$u(x,y) = e^xF(y) \quad \text{and} \quad v(x,y) = e^xG(y)$$

where $F(y)=G'(y)$ and $F'(y)=-G(y)$ by the Cauchy–Riemann equations. Hence $G''(y)=-G(y) \quad\Rightarrow\quad G(y)=\alpha\sin y+\beta\cos y$ where α and β are real constants. Then $f(z)=(\alpha+i\beta)e^z$. Using the initial condition $f(0)=1$ gives the result.

(ii) Let $f:\mathbb{C}\to\mathbb{C}$ be defined by $f(z)=\sin z$. It follows by Lemma 1.11 that if f is given by (2.1) then

$$u(x,y) = \sin x\cosh y \qquad v(x,y) = \cos x \sinh y$$
$$\Rightarrow \quad u_x = \cos x\cosh y = v_y \qquad u_y = \sin x\sinh y = -v_x$$

so that the Cauchy–Riemann equations are satisfied and the first-order partial derivatives are continuous at all points of $\mathbb{C}$. Hence by 2.6, f is differentiable for all $z\in\mathbb{C}$ and by 2.5 and 1.11,

$$f'(z) = \cos x\cosh y - i\sin x\sinh y = \cos z$$

as expected.

(iii) When investigating certain functions using Theorem 2.6, it is sometimes convenient to use alternative forms of the Cauchy–Riemann equations, such as the polar form given in (2.6). This form is particularly useful when investigating Log and associated functions.

For example, consider $f: \mathbb{C}\backslash\{0\} \to \mathbb{C}$ given by $f(z) = \operatorname{Log} z$. Let $z = re^{i\theta}$ where $r = |z|$ and $\theta = \operatorname{Arg} z$, and let f be given by (2.5). Then

$$u(r, \theta) = \operatorname{Log} r \qquad v(r, \theta) = \theta$$

$$\Rightarrow \quad u_r = \frac{1}{r} = \frac{1}{r} v_\theta \qquad v_r = 0 = \frac{-1}{r} u_\theta$$

Hence (2.6) is satisfied and the first-order partial derivatives are continuous at all points, except those given by $\operatorname{Arg} z = \pi$ and $z = 0$. Hence, by 2.6 and 2.4, f is differentiable for all $z \in \mathbb{C}$ except $z = 0$ and those points given by $\operatorname{Arg} z = \pi$, i.e. except at the branch point and along the branch cut of the associated multifunction. By 2.5, at points where $f'(z)$ exists,

$$f'(z) = u_x + iv_x = \left(\frac{x}{r} u_r - \frac{y}{r^2} u_\theta\right) + i\left(\frac{x}{r} v_r - \frac{y}{r^2} v_\theta\right)$$

using the results from Example 2.10. Then

$$f'(z) = \frac{x}{r^2} - \frac{iy}{r^2} = \frac{\cos\theta - i\sin\theta}{r} = \frac{1}{r(\cos\theta + i\sin\theta)} = \frac{1}{z}$$

Note

Theorem 2.6 is not true in general without the extra condition that the first-order partial derivatives are continuous, as the following example shows.

Example 2.12

Consider the function $f: \mathbb{C} \to \mathbb{C}$ defined by

$$f(z) = \frac{x^3(1 - i) + y^3(1 + i)}{x^2 + y^2} \qquad (z \neq 0 \quad \text{and} \quad f(0) = 0)$$

If f is given by (2.1), then

$$u(x, 0) = x \qquad u(0, y) = y \qquad v(x, 0) = -x \qquad v(0, y) = y$$

$$\Rightarrow \quad u_x(0, 0) = v_y(0, 0) = 1 \qquad u_y(0, 0) = -v_x(0, 0) = 1$$

so that the Cauchy–Riemann equations are satisfied at $(0, 0)$. However, letting $h \to 0$ along the real axis in the definition of $f'(0)$, where $h = \alpha + i\beta$, gives

$$f'(0) = \lim_{h \to 0} \frac{f(h)}{h} = \lim_{\alpha \to 0} \frac{\alpha(1 - i)}{\alpha} = 1 - i$$

Letting $h \to 0$ along the line $\alpha = \beta$ in the definition gives

$$f'(0) = \lim_{\alpha \to 0} \frac{\alpha((1-i)+(1+i))}{2\alpha(1+i)} = \frac{(1-i)}{2}$$

Hence f is not differentiable at 0. Notice that, in general,

$$u(x,y) = \frac{x^3+y^3}{x^2+y^2} \quad \text{and} \quad v(x,y) = \frac{y^3-x^3}{x^2+y^2}$$

and the first-order partial derivatives of u and v are clearly not continuous at $(0,0)$.

Important Notes

(i) It follows from Examples 2.11 and some of the previous comments, together with Theorem 2.3 and the definitions of Chapter 1, that the derivatives of the elementary functions and their inverses are formally the same as the corresponding results for functions of a real variable, at points where the functions are differentiable.

(ii) Since $z = x + iy \quad \Rightarrow \quad x = (z+\bar{z})/2,\ y = (z-\bar{z})/2i,\ u(x,y)$ and $v(x,y)$ in (2.1) may be formally regarded as functions of two independent variables, z and $\bar{z}$. Let $w = f(z)$. If u and v have continuous first-order partial derivatives with respect to x and y, then a necessary and sufficient condition that w is independent of $\bar{z}$ is $w_{\bar{z}} = 0$. This condition reduces to

$$w_{\bar{z}} = u_{\bar{z}} + iv_{\bar{z}} = 0 \quad \Leftrightarrow \quad u_x x_{\bar{z}} + u_y y_{\bar{z}} + i(v_x x_{\bar{z}} + v_y y_{\bar{z}}) = 0$$

$$\Leftrightarrow \quad \frac{1}{2}(u_x - v_y) + \frac{i}{2}(u_y + v_x) = 0$$

Hence this condition is satisfied if and only if the Cauchy–Riemann equations are satisfied. In other words, in any rule of association defining an analytic function, x and y can only occur in the combination $x + iy = z$.

Example 2.13

(i) Since $f(z) = e^z \quad \Rightarrow \quad f'(z) = e^z$ for all $z \in \mathbb{C}$, it follows by 2.3(b)(iii) that $f(z) = e^{kz} \quad \Rightarrow \quad f'(z) = ke^{kz}$ for all $z \in \mathbb{C}$ and any complex constant k. Now let

$$g(z) = \cos z = \tfrac{1}{2}(e^{iz} + e^{-iz}) \qquad (\text{for all } z \in \mathbb{C})$$

Then using the above result and 2.3(b)(i), it follows that

$$g'(z) = \frac{1}{2}(ie^{iz} - ie^{-iz}) = \frac{-1}{2i}(e^{iz} - e^{-iz}) = -\sin z \qquad (\text{for all } z \in \mathbb{C})$$

Similarly $\dfrac{d}{dz}(\sin z) = \cos z$

Then if $h(z) = \tan z = \dfrac{\sin z}{\cos z}$

it follows by 2.3(b)(ii) and (iv) that

$$h'(z) = \frac{\cos^2 z + \sin^2 z}{\cos^2 z} = \sec^2 z \qquad (z \neq (2n+1)\pi/2, n \in \mathbb{Z})$$

and so on.

(ii) Let $f : \mathbb{C} \to \mathbb{C}$ be defined by $f(z) = \sinh^{-1} z$. Using the process demonstrated in Chapter 1, $f(z) = \operatorname{Log}(z + (z^2+1)^{1/2})$. Then

$$f'(z) = \frac{1 + z(z^2+1)^{-1/2}}{z + (z^2+1)^{1/2}} = \frac{((z^2+1)^{1/2} + z)((z^2+1)^{1/2} - z)}{(z(z^2+1)^{1/2} + z^2 + 1)((z^2+1)^{1/2} - z)}$$

$$\Rightarrow \quad f'(z) = \frac{1}{(z^2+1)^{3/2} - z^2(z^2+1)^{1/2}} = \frac{1}{(z^2+1)^{1/2}} \qquad (z \neq \pm i)$$

using the result of Example 2.11(iii) and Theorem 2.3.

Alternatively, let $w = f(z) = \sinh^{-1} z$ so that $z = g(w) = \sinh w$. Then by (i) above, Theorem 2.3 and the results of Chapter 1,

$$g(w) = \frac{1}{2}(e^w - e^{-w}) \quad \Rightarrow \quad g'(w) = \frac{dz}{dw} = \frac{1}{2}(e^w + e^{-w}) = \cosh w$$

$$\Rightarrow \quad f'(z) = \frac{dw}{dz} = \left(\frac{dz}{dw}\right)^{-1} = \frac{1}{\cosh w} = \frac{1}{(\sinh^2 w + 1)^{1/2}} = \frac{1}{(z^2+1)^{1/2}}$$

Harmonic Functions

Let a function $f : A \subseteq \mathbb{C} \to \mathbb{C}$ be given by (2.1) and suppose that f is analytic in a region $\mathcal{R} \subseteq \mathbb{C}$, with the first- and second-order partial derivatives of $u(x, y)$ and $v(x, y)$ continuous in $\mathcal{R}$. Then, by Theorem 2.4,

$$u_{xx} + u_{yy} = v_{yx} - v_{xy} = 0 \qquad v_{xx} + v_{yy} = -u_{yx} + u_{xy} = 0$$

that is, $u(x, y)$ and $v(x, y)$ are both solutions of **Laplace's equation**,

$$\nabla^2 \phi = \phi_{xx} + \phi_{yy} = 0 \tag{2.15}$$

in the region $\mathscr{R}$. (In fact, it will be shown in the next chapter that if f is analytic in $\mathscr{R}$, then the partial derivatives of $u(x, y)$ and $v(x, y)$, of all orders, exist and are continuous in $\mathscr{R}$.)

For example, $f: \mathbb{C} \to \mathbb{C}$ given by $f(z) = z^3$ is entire and

$$\begin{aligned} z = x + iy \quad &\Rightarrow \quad f(z) = (x^3 - 3xy^2) + i(3x^2y - y^3) \\ \Rightarrow \quad u(x, y) = x^3 - 3xy^2 \quad &\quad v(x, y) = 3x^2y - y^3 \end{aligned}$$

It is easily checked that u and v satisfy Laplace's equation throughout the plane.

Laplace's equation is of the greatest importance in practical applications in many different areas of physics, such as in finding steady temperatures, electrostatic potentials, gravitational potentials in Newtonian theory, and in steady fluid flow.

Definitions

A function $\phi(x, y)$ is **harmonic** in a region $\mathscr{R}$ if it has continuous second-order partial derivatives and satisfies Laplace's equation (2.15) in $\mathscr{R}$. A function $v(x, y)$ is a **harmonic conjugate** of a harmonic function $u(x, y)$ if u and v satisfy the Cauchy–Riemann equations (2.2).

It follows that any number of harmonic functions can be generated simply by writing down any number of analytic functions and finding their real and imaginary parts. Note that any linear combination of harmonic functions is harmonic and that a constant is harmonic.

Example 2.14

Verify that $u(x, y) = \exp(x^2 - y^2)\cos 2xy$ is harmonic and find its harmonic conjugate. If $f: \mathbb{C} \to \mathbb{C}$ is entire and is given by (2.1), with $f(0) = 1$, find $f(z)$.

Solution

$$\begin{aligned} &u_x = 2e^{(x^2 - y^2)}(x\cos 2xy - y\sin 2xy) \\ &\Rightarrow \quad u_{xx} = 2e^{(x^2 - y^2)}((2x^2 - 2y^2 + 1)\cos 2xy - 4xy\sin 2xy) \\ &u_y = -2e^{(x^2 - y^2)}(y\cos 2xy + x\sin 2xy) \quad \Rightarrow \quad u_{yy} = -u_{xx} \end{aligned}$$

as required. Let $v(x, y)$ be a harmonic conjugate of $u(x, y)$. Then

$$\begin{aligned} &v_x = -u_y \quad \text{and} \quad v_y = u_x \\ &\Rightarrow \quad v_x = 2e^{(x^2 - y^2)}(y\cos 2xy + x\sin 2xy) \\ &\qquad\; v_y = 2e^{(x^2 - y^2)}(x\cos 2xy - y\sin 2xy) \\ &\Rightarrow \quad v_x = \left(e^{(x^2 - y^2)}\right)_x \sin 2xy + e^{(x^2 - y^2)}(\sin 2xy)_x \\ &\Rightarrow \quad v(x, y) = e^{(x^2 - y^2)}\sin 2xy + F(y) \end{aligned}$$

where $F(y)$ is an arbitrary function of y. Note that this solution for $v(x,y)$ is suggested by the form of $u(x,y)$. Then from the expression for v_y, it follows that $F'(y) = 0$ and so

$$f(x+iy) = u + iv = e^{(x^2-y^2)}(\cos 2xy + i(\sin 2xy + k))$$

where k is a real constant. Since $f(0) = 1, k = 0$. The form of $f(z)$ is suggested by taking $y = 0$, say, in the above expression. When $y = 0$, $f(x) = \exp x^2$ and this suggests that $f(z) = \exp z^2$ in general, as is indeed the case. From the expression above,

$$f(z) = e^{(x^2-y^2)}e^{2ixy} = e^{(x+iy)^2} = e^{z^2}$$

Example 2.15

From Example 2.11(iii), recall that $f\colon \mathbb{C}\backslash\{0\} \to \mathbb{C}$ given by $f(z) = \text{Log}\, z$ is analytic for all z such that $\text{Arg}\, z \neq \pi$ and $z \neq 0$. Hence $u(x,y) = \text{Re}(f(z)) = \text{Log}|z| = \text{Log}\, r$, where $r^2 = x^2 + y^2$ is harmonic except at $r = 0$, as is easily checked. Then, more generally, $u(x,y) = a\text{Log}\, r + b$, where a and b are real constants, is harmonic, except at $r = 0$.

Suppose we wish to find the steady temperature of water bounded by two concentric cylinders of radii 1 and 2. The inner cylinder is packed in ice and kept at 0 °C, while the outer cylinder is heated to K °C. The temperature does not depend on the length of the cylinders, so we can take a cross-section of the problem (Fig. 2.5). The steady temperature T is unchanging with time, so $T = T(x,y)$. It can be shown that T satisfies Laplace's equation (2.15) and that there is a unique solution to the problem. We thus require a harmonic function, $T(x,y)$, which satisfies the boundary conditions, $T = 0$ when $r = 1$ and $T = K$ when $r = 2$, as shown in Fig. 2.5.

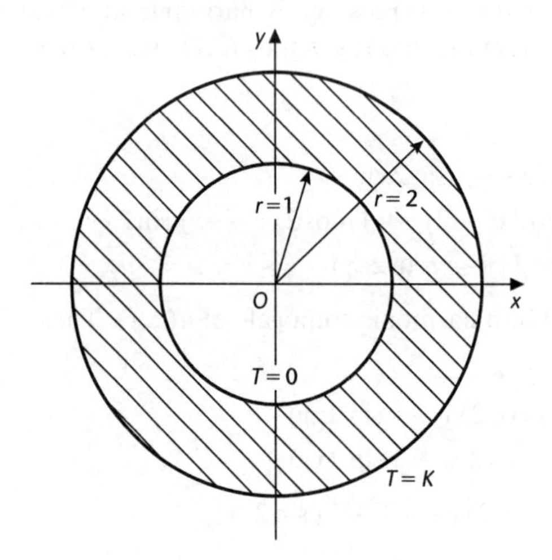

Figure 2.5

Since the boundaries are circles and the region excludes $r = 0$, the obvious choice for the solution is $T(x, y) = T(r) = a \operatorname{Log} r + b$ where a and b are constants. $T(r = 1) = 0 \Rightarrow b = 0$ and $T(r = 2) = K \Rightarrow a = K/\operatorname{Log} 2$. Hence, the required solution is

$$T(x, y) = T(r) = \frac{K \operatorname{Log} r}{\operatorname{Log} 2}$$

Harmonic functions are extremely important in applications and a large proportion of Chapter 9 is devoted to their properties and applications.

Exercise **2.2.1** Use the definition of the derivative of a function to prove that, at each point where the derivatives exist,

(i) if $h(z) = f(z) + g(z)$ then $h'(z) = f'(z) + g'(z)$

(ii) if $h(z) = 1/f(z)$ then $h'(z) = -f'(z)/(f(z))^2$

Exercise **2.2.2** Prove, using the definitions, that the existence of the derivative of a function of a complex variable at a point implies the continuity of the function there (as in real analysis). Show also, by considering $f: \mathbb{C} \to \mathbb{C}$ defined by $f(z) = |z|^2$, the continuity of a function of a complex variable at a point does not imply the existence of a derivative there.

Exercise **2.2.3** Show, directly from the definition, that the function $f: \mathbb{C} \to \mathbb{C}$ given by $f(z) = \bar{z}$ is nowhere differentiable.

Exercise **2.2.4** Use the Cauchy–Riemann equations and Theorems 2.4 to 2.6 to determine at which points, if any, the following functions are differentiable and find their derivatives at such points.

(i) $f: \mathbb{C} \to \mathbb{C}, f(z) = z^3$

(ii) $f: \mathbb{C} \to \mathbb{C}, f(z) = e^{\bar{z}}$

(iii) $f: \mathbb{C} \to \mathbb{C}, f(z) = \cosh z$

(iv) $f: \mathbb{C} \to \mathbb{C}, f(z) = z \operatorname{Im} z$

(v) $f: \mathbb{C} \setminus \{0\} \to \mathbb{C}, f(z) = 1/z$

(vi) $f: \mathbb{C} \to \mathbb{C}, f(z) = \sin \bar{z}$

Exercise **2.2.5** Use the Cauchy–Riemann equations to find the most general entire function f such that $\operatorname{Re}(f'(z)) = 0$.

Exercise **2.2.6** Use the Cauchy–Riemann equations to prove the following:

(i) If f is analytic on a region $\mathscr{R}$ and $|f(z)|$ is constant on $\mathscr{R}$, then $f(z)$ is constant on $\mathscr{R}$.

(ii) There is no analytic function f such that $\operatorname{Im}(f(z)) = x^2 - 2y$.

Exercise **2.2.7** Show that the family of curves given by $u(x, y) = k_1$, where k_1 is a constant, have gradient $-u_x/u_y$ at each point. Deduce that if f is entire and $f(z) = u(x, y) + iv(x, y)$, then the curves of this family intersect the curves of the family $v(x, y) = k_2$ (k_2 constant) orthogonally.

Exercise **2.2.8** If f is entire and $z = x + iy$, prove that, for all $z \in \mathbb{C}$,

$$\left(\frac{\partial^2}{\partial x^2} + \frac{\partial^2}{\partial y^2}\right)|f(z)|^2 = 4|f'(z)|^2$$

Exercise **2.2.9** Let $f: \mathbb{C} \to \mathbb{C}$ and let $w = Re^{i\phi} = f(z) = f(x + iy)$. Use the chain rule for partial derivatives of functions of two real variables to show that the Cauchy–Riemann equations become

$$\frac{1}{R}R_x(x, y) = \phi_y(x, y) \quad \frac{1}{R}R_y(x, y) = -\phi_x(x, y)$$

Suppose that f is entire and $\phi(x, y) = 2xy$. Find $R(x, y)$ and hence $w = f(z)$.

Exercise **2.2.10** Let $f: \mathbb{C} \to \mathbb{C}$ be defined by $f(z) = z^5/|z|^4, z \neq 0, f(0) = 0$. Prove that f is not differentiable at 0, but that the Cauchy–Riemann equations are satisfied there.

Exercise **2.2.11** Using the polar form of the Cauchy–Riemann equations (2.6) and Theorems 2.4 to 2.6, determine at which points $f: \mathbb{C} \to \mathbb{C}$ given by $f(z) = z^{1/2}$ is differentiable and find its derivative there. Use the fact that $f(z) = \exp\left(\frac{1}{2}\operatorname{Log} z\right)$ to verify your answers; assume the derivatives of exp and Log and use Theorem 2.3.

Exercise **2.2.12** Find the derivative of $\tanh^{-1} z$ at points where it exists

(i) by expressing $\tanh^{-1}$ in terms of Log

(ii) by letting $z = \tanh w$

Exercise **2.2.13** Verify that the following functions of two real variables are harmonic and find their harmonic conjugates. If f given by $f(z) = u(x, y) + iv(x, y)$ is entire, find $f(z)$ in each case.

(i) $u(x, y) = y^3 - 3x^2y$

(ii) $u(x, y) = \sinh x \sin y$

Exercise **2.2.14** By considering the entire function f given by $f(z) = z^2$, show that if $v(x, y)$ is a harmonic conjugate of $u(x, y)$ in some region, it is not in general true that u is a harmonic conjugate of v in that region.

Exercise **2.2.15** Show that $v(x, y) = \theta = \tan^{-1}(y/x), 0 \leqslant \theta < \pi$, is harmonic. Hence find the steady temperature $T(x, y)$, in the upper half-plane, given that $T(x, 0) = 0, x > 0$ and $T(x, 0) = K, x < 0$.

Singular Points and Zeros

We end this chapter by introducing the idea of singular points of functions, which are of fundamental importance when it comes to integration. Closely related is the idea of zeros of functions.

Definitions

(i) A function f is **analytic at a point** $\alpha \in \mathbb{C}$ if it is differentiable at α and at each point in some open neighbourhood of α.

(ii) A point $\alpha \in \mathbb{C}$ is a **singular point** of a function f if f fails to be analytic at α but is analytic at some point in every open neighbourhood of α.

(iii) A singular point of a function f is **isolated** if there some open neighbourhood of the point throughout which f is analytic except at the point itself.

Notes

(i) Definition (i) ensures that α is contained in a region on which f is analytic. Definition (ii) excludes the possibility of a function having a domain consisting entirely of singular points.

(ii) Singular points of a function need not belong to the domain of the function.

(iii) If f is not continuous at $\alpha \in \mathbb{C}$, then f is not analytic at α, so that α may be a singular point of f. It is easiest to look for this type of singular point first.

Example 2.16

(i) Let $f: \mathbb{C}\backslash\{0\} \to \mathbb{C}$ be defined by $f(z) = 1/z$. Then 0 is the only singular point of f (f is not continuous there), hence it is isolated.

(ii) It is easy to check using Theorem 2.4 that $f: \mathbb{C}\backslash\{0\} \to \mathbb{C}$ defined by $f(z) = 1/\bar{z}$ is differentiable nowhere and so is analytic nowhere. Hence, in contrast to (i), 0 is not a singular point of f.

(iii) Since $\sin(1/z) = 0$ for $z = 1/n\pi$ for all $n \in \mathbb{Z}$, it follows that 0 is a non-isolated singular point of f given by $f(z) = 1/\sin(1/z)$.

Note

It follows from the earlier section on branch points that points on a branch cut of a multifunction and the branch point itself are non-isolated singular points of any branch of that multifunction. However, as Example 2.16(iii) shows, not all non-isolated singular points fall into this category.

Apart from singular points arising from branch points and branch cuts of multifunctions, singular points can be classified in the following way.

Definitions

Let $\alpha \in \mathbb{C}$ be a singular point of a function f.

(i) The point α is a **removable singular point** of f if and only if $\lim_{z \to \alpha} f(z)$ exists.

(ii) The point α is a **pole of order** n ($n \in \mathbb{N}$) if and only if $\lim_{z \to \alpha} (z - \alpha)^n f(z) = k \neq 0$. A pole of order 1 is a **simple pole.**

(iii) A singular point of a function f which is not a removable singularity, a pole, or associated with a branch point or branch cut of a multifunction, is an **essential singular point**.

Notes

(i) Suppose that α is a removable singular point of f, so that $f'(\alpha)$ does not exist for some reason but $\lim_{z \to \alpha} f(z) = k$ say. Then redefining $f(\alpha) = k$ gives $f'(\alpha) = \lim_{z \to \alpha} f'(z)$. Hence the term 'removable' singular point.

(ii) If α is a pole of order n of f, then f can be expressed as $f(z) = g(z)/(z - \alpha)^n$, where $\lim_{z \to \alpha} g(z)$ exists and is non-zero. Hence either g is analytic at α or has a removable singular point at α and so can be made analytic at α by redefining $g(\alpha)$, by (i).

Hence if α is a pole of order n of f, then f can be expressed in the form $f(z) = g(z)/(z - \alpha)^n$, where g is analytic at α, with $g(\alpha) \neq 0$, without loss of generality. Notice that $\lim_{z \to \alpha} 1/f(z) = 0$.

Example 2.17

(i) Let $f : \mathbb{C}\backslash\{0\} \to \mathbb{C}$ be defined by $f(z) = (\sin z)/z$, $z \neq 0$. Then 0 is a removable singular point of f since $\lim_{z \to 0} f(z) = 1$, and the domain of f can be extended so that $f(0) = 1$.

(ii) Function f defined by

$$f(z) = \frac{z + 2}{(z^2 + 1)(z - 4)^2 (z + 2i)^7}$$

has simple poles at $\pm i$, a pole of order 2 at 4 and a pole of order 7 at $-2i$.

(iii) Let f be defined by $f(z) = \csc z = 1/\sin z$. Then f has isolated singular points at $z = n\pi$, $n \in \mathbb{Z}$, since f is not continuous there. Note that $\lim_{z \to 0} z f(z) = 1$, so that 0 is a simple pole of f. Similarly, $n\pi$ is a simple pole of f for any $n \in \mathbb{Z}$.

(iv) Function f defined by $f(z) = 1/\sin(1/z)$ has a non-isolated singular point at 0. Clearly, this singular point is not removable and is not a pole. Also, f is not a branch of a multifunction, so that 0 is an essential singular point of f.

The function g defined by $g(z) = e^{-1/z}$ also has an essential singularity at 0, but in this case, the singular point is clearly isolated.

The following result is often useful when investigating singular points and is formally the same as its real counterpart.

Theorem 2.7. L'Hôpital's Rule

Let f and g be analytic in a region containing the point $\alpha \in \mathbb{C}$ and suppose that $f(\alpha) = g(\alpha) = 0$ with $g'(\alpha) \neq 0$. Then

$$\lim_{z \to \alpha} \frac{f(z)}{g(z)} = \frac{f'(\alpha)}{g'(\alpha)}$$ □

Proof

By hypothesis, since $g'(\alpha) \neq 0$,

$$\lim_{z \to \alpha} \frac{f(z)}{g(z)} = \lim_{z \to \alpha} \left(\frac{f(z) - f(\alpha)}{z - \alpha} \cdot \frac{z - \alpha}{g(z) - g(\alpha)} \right) = \frac{f'(\alpha)}{g'(\alpha)}$$

using Theorem 2.1 as required. ■

Notes

(i) As in the case of real variables, limits of other so-called indeterminate forms can often be evaluated by making the appropriate modifications to L'Hôpital's rule.

(ii) The rule can clearly be applied again if $f'(\alpha)$ and $g'(\alpha)$ are also zero, but $g''(\alpha) \neq 0$, etc.

Example 2.18

(i) $$\lim_{z \to i} \frac{z^6 + 1}{z^2 + 1} = \lim_{z \to i} \frac{6z^5}{2z} = \lim_{z \to i} 3z^4 = 3$$

by L'Hôpital's rule

(ii) $$\lim_{z \to 0} \frac{\sin z - \tan z}{z^2} = \lim_{z \to 0} \frac{\cos z - \sec^2 z}{2z} = \lim_{z \to 0} \frac{-\sin z - 2\sec^2 z \tan z}{2} = 0$$

by two applications of L'Hôpital's rule

(iii) $$\lim_{z\to 0}\frac{\operatorname{Log}(\cos z)}{z^2} = \lim_{z\to 0}\frac{(-\sin z)/\cos z}{2z}$$
$$= \left(\lim_{z\to 0}\frac{\sin z}{z}\right)\left(\lim_{z\to 0}\frac{-1}{2\cos z}\right) = -\frac{1}{2}$$

by one application of L'Hôpital's rule and Theorem 2.1. Now let $w = (\cos z)^{1/z^2}$. Then $\operatorname{Log} w = (\operatorname{Log}(\cos z))/z^2 \Rightarrow \lim_{z\to 0}\operatorname{Log} w = -1/2$ by above. Then by Theorem 2.2,

$$\operatorname{Log}(\lim_{z\to 0} w) = -1/2 \quad\Rightarrow\quad \lim_{z\to 0} w = \lim_{z\to 0}(\cos z)^{1/z^2} = e^{-1/2}$$

Example 2.19

(i) Returning to Example 2.17(i), $\lim_{z\to 0}(\sin z)/z = \lim_{z\to 0}\cos z = 1$, by 2.7. Redefining $f(0) = 1$ gives

$$f'(0) = \lim_{h\to 0}\frac{f(h)-1}{h} = \lim_{h\to 0}\frac{\sin h - h}{h^2} = \lim_{h\to 0} f'(h)$$

by 2.7 again. Then by a further two applications of 2.7,

$$f'(0) = \lim_{h\to 0}\frac{\cos h - 1}{2h} = \lim_{h\to 0}\frac{-\sin h}{2} = 0$$

(ii) Returning to Example 2.17(iii), by 2.7 it follows that

$$\lim_{z\to n\pi}(z-n\pi)f(z) = \lim_{z\to n\pi}\frac{z-n\pi}{\sin z} = \lim_{z\to n\pi}\frac{1}{\cos z} = (-1)^n$$

so that $n\pi$, $n \in \mathbb{Z}$, is a simple pole of f.

We turn now to the concept of the order of a zero of a given function.

Definition

A point $\alpha \in \mathbb{C}$ is a **zero of order *n*** ($n \in \mathbb{N}$) of a function f if and only if f can be expressed in the form $f(z) = (z-\alpha)^n g(z)$ for some function g with $\lim_{z\to\alpha} g(z) \neq 0$.

Note

If α is a zero of f, then $f(\alpha) = 0$. Once again, we can assume that g is analytic at α without loss of generality, with $g(\alpha) \neq 0$. Then α is a zero of order n of f if and only if it is a pole of order n of $1/f$.

Example 2.20

(i) Function f given by $f(z) = (z - 2)^2(z^2 + 1)$ for all $z \in \mathbb{C}$ has a zero of order 2 at 2 and two zeros of order 1 (**simple zeros**) at $\pm i$.

(ii) Function f given by $f(z) = \tan z$ $(z \neq n\pi + \pi/2, n \in \mathbb{Z})$ clearly has zeros given by $\sin z = 0$ and so has zeros at $z = n\pi, n \in \mathbb{Z}$. Now

$$\lim_{z \to n\pi} \frac{f(z)}{z - n\pi} = \lim_{z \to n\pi} \frac{f'(z)}{1} = f'(n\pi) = \sec^2(n\pi) = 1 \neq 0$$

by L'Hôpital's rule, so that f has simple zeros at $z = n\pi$.

It is straightforward to show that the zeros of a function are isolated from each other in the plane.

Theorem 2.8. Zeros are Isolated

Let α be a zero of order n of f. Then this zero is **isolated**; that is, there exists an open neighbourhood of α which contains no other zero of f. □

Proof

By hypothesis, it follows that $f(z) = (z - \alpha)^n g(z)$, where g is analytic at α without loss of generality, and $g(\alpha) \neq 0$. Let $g(\alpha) = 2\beta$ say. Since g is continuous at α, there exists a real $\delta > 0$ such that

$$|z - \alpha| < \delta \quad \Rightarrow \quad |g(z) - g(\alpha)| < |\beta|$$

Hence, by the triangle inequality,

$$|z - \alpha| < \delta \quad \Rightarrow \quad |g(z)| \geqslant \Big| |g(\alpha)| - |g(z) - g(\alpha)| \Big| > |\beta|$$

so that $g(z)$ is not zero in the neighbourhood $|z - \alpha| < \delta$. ■

Note

Since f has a pole of order n at α if and only if $1/f$ has a zero of order n at α, it follows that poles of functions are isolated.

Exercise

2.3.1 Find the following limits by using L'Hôpital's rule:

(i) $\displaystyle\lim_{z \to -i} \frac{z^{11} - i}{z^7 - i}$

(ii) $\displaystyle\lim_{z \to 1} \frac{1 + \cos \pi z}{z^2 - 2z + 1}$

(iii) $\displaystyle\lim_{z \to i} (z + 1 - i)^{1/(z - i)}$

Exercise **2.3.2** Locate and classify the singular points of the following functions. Give reasons for your answers.

(i) $f(z) = \dfrac{z+1}{z(z^2+1)^3}$

(ii) $f(z) = \dfrac{\mathrm{Log}(z^2+9)}{(z+1)^2}$

(iii) $f(z) = \tan z$

(iv) $f(z) = \mathrm{sech}\, z$

(v) $f(z) = \dfrac{e^z - 1}{z}$

(vi) $f(z) = \sin(z + 1/z)$

Exercise **2.3.3** Locate the zeros of the following functions and find their orders:

(i) $f(z) = (z^2-1)^2(z^4+1)(z-i)^7$

(ii) $f(z) = \sinh^2 z$

(iii) $f(z) = \dfrac{e^z - 1}{z - 1}$

(iv) $f(z) = \dfrac{\mathrm{Log}\, z}{z^2}$

Exercise **2.3.4** Prove that α is a pole of order n of f if and only if it is a zero of order n of $1/f$.

3 Integration, Cauchy's Theorems and Related Results

We introduce the idea of a complex definite integral in this chapter and give the famous results of Cauchy, together with related results. Some of the rich applications of these results will be demonstrated. For example, these techniques may be used to evaluate certain real definite integrals very simply, which would otherwise be difficult to evaluate. We shall also see how these results help us to locate the position of zeros of polynomials in the complex plane.

Definite Integrals

Initially, it seems logical to suppose that a definite integral of a complex function, evaluated between two points in the complex plane, will depend on the curve joining the two points. Hence the definition of a complex definite integral is very similar to that of a real line integral in the plane. We shall assume that the notion of a real definite integral (at least in the Riemannian sense) is already familiar to the reader. We first need to consider the case in which the complex-valued integrand is a function of one real variable only.

Preliminary Definition

Let $[a, b]$ be a closed interval of $\mathbb{R}$ and let $p, q : [a, b] \to \mathbb{R}$. Suppose that p and q are piecewise continuous on $[a, b]$. (In other words, p and q are continuous on $[a, b]$ except possibly for a finite number of jump discontinuities.) Let $g : [a, b] \to \mathbb{C}$ be defined by $g(t) = p(t) + iq(t)$ for $t \in [a, b]$. Then $\int_a^b g(t)\,dt$ is defined by

$$\int_a^b g(t)\,dt = \int_a^b p(t)\,dt + i\int_a^b q(t)\,dt \tag{3.1}$$

Notes

The conditions given are sufficient to ensure that the real integrals on the right of (3.1) exist (in the Riemannian sense). This is the obvious definition to make – the integral is evaluated by integrating the real and imaginary parts of the integrand. Clearly, all the usual properties of real (Riemann) integrals hold for $\int_a^b g(t)\,dt$.

We now define a sufficiently 'well-behaved' curve $\mathscr{C}$ joining two points in the complex plane, so that $\int_{\mathscr{C}} f(z)\,dz$ can be defined.

Definitions

(i) An **arc** $\mathscr{C}$ is a set of points $\{(x(t), y(t)) : t \in [a, b]\}$ in the complex plane where x and y are continuous functions of the real parameter t. It is convenient to describe the points on $\mathscr{C}$ by

$$z = z(t) = x(t) + iy(t) \qquad (a \leqslant t \leqslant b) \tag{3.2}$$

$\mathscr{C}$ is a **simple** or **Jordan** arc if it does not cross itself; that is, $z(t_1) = z(t_2) \quad \Rightarrow \quad t_1 = t_2$ for all $t_1, t_2 \in [a, b]$.

(ii) An arc $\mathscr{C}$ is **smooth** if $z'(t)$ exists and is non-zero for $t \in [a, b]$. (This implies that $\mathscr{C}$ has a continuously turning tangent.)

(iii) A (simple) **contour** is an arc consisting of a finite number of (simple) smooth arcs joined end to end. When only the initial and final values of $z(t)$ are the same, the contour is a (simple) **closed contour**.

Notice that, in the sense we have defined them, any arc and hence any contour joins one point to another point.

Recall the Jordan curve theorem (see Chapter 1). This result implies that any simple closed contour divides the complex plane into two regions, the bounded **inside** and the unbounded **outside**, having the curve as a common boundary. We shall not prove this topological result, which is intuitively obvious but very difficult to prove.

Example 3.1

In Fig. 3.1, $\mathscr{C}_1$ is an arc, $\mathscr{C}_2$ is a simple arc, $\mathscr{C}_3$ is a smooth arc, $\mathscr{C}_4$ is a simple smooth arc, $\mathscr{C}_5$ is a contour and $\mathscr{C}_6$ is a simple closed contour.

The easiest contours to describe by means of a real parameter, as in (3.2), consist of line segments and arcs of circles (see Chapter 1). It turns out that such contours are the most useful in applications.

Important Convention

With all closed contours, the direction of **increasing** parameter will be taken as **anticlockwise**. (Think of using the angular polar coordinate as a parameter for a circle.)

Main Definition

Let $\mathscr{C}$ be any contour in the complex plane represented by (3.2), extending from $\alpha = z(a)$ to $\beta = z(b)$. Let the domain of $f : A \subseteq \mathbb{C} \to \mathbb{C}$ contain the contour $\mathscr{C}$ and let f be piecewise continuous on $\mathscr{C}$; that is, $f(z(t))$ is piecewise continuous on

Figure 3.1

$[a, b]$. Then the **complex definite line integral of f along $\mathscr{C}$** is denoted and defined by

$$\int_{\mathscr{C}} f(z)\, dz = \int_a^b f(z(t))z'(t)\, dt \tag{3.3}$$

Notes

(i) The definition is reminiscent of integration by substitution for real integrals. It is independent of the choice of the parametrisation of $\mathscr{C}$. To show this, let $\mathscr{C}$ be a smooth simple arc without loss of generality and let $\mathscr{C}$ be also represented by $z = z(u)$ where $u \in \mathbb{R}$ with $u \in [c, d]$ and $\alpha = z(c)$ and $\beta = z(d)$, say. Then

$$\int_c^d f(z(u))z'(u)du = \int_a^b f(z(t))\left(\frac{dz}{dt}\frac{dt}{du}\right)\frac{du}{dt}\,dt = \int_a^b f(z(t))z'(t)\, dt$$

by the chain rule (Theorem 2.3(b) (iii)).

(ii) Let f be given by $f(z) = u(x, y) + iv(x, y)$ where $z = x + iy$ for all points in its domain. Then, from the definition of a derivative, it follows that

$$f(z(t))z'(t) = (u(t) + iv(t))(x'(t) + iy'(t))$$
$$\Rightarrow \quad f(z(t))z'(t) = [u(t)x'(t) - v(t)y'(t)] + i[v(t)x'(t) + u(t)y'(t)]$$

Thus, by (3.1) of the preliminary definition,

$$\int_{\mathscr{C}} f(z)\,dz = \int_a^b (u(t)x'(t) - v(t)y'(t))\,dt + i\int_a^b (v(t)x'(t) + u(t)y'(t))\,dt \quad (3.4)$$

and so is well defined. The real integrals exist since $\mathscr{C}$ is a contour and f is piecewise continuous, so that the integrands are piecewise continuous.

Example 3.2

Evaluate $\int_{\mathscr{C}} z^2\,dz$ where $\mathscr{C}$ is the circle, centre 0 and radius 1, represented by $z(\theta) = e^{i\theta}$ where $-\pi \leqslant \theta \leqslant \pi$ without loss of generality (see Chapter 1).

Solution

By definition

$$\int_{\mathscr{C}} z^2 dz = \int_{-\pi}^{\pi} (e^{i\theta})^2 z'(\theta) d\theta = \int_{-\pi}^{\pi} e^{2i\theta} ie^{i\theta}\,d\theta = i\int_{-\pi}^{\pi} e^{3i\theta}\,d\theta$$

$$\Rightarrow \quad \int_{\mathscr{C}} z^2 dz = -\int_{-\pi}^{\pi} \sin 3\theta\,d\theta + i\int_{-\pi}^{\pi} \cos 3\theta\,d\theta = 0$$

It is no coincidence that the integrand is analytic inside and on $\mathscr{C}$ and the result is zero. The connection will be revealed later.

Note

As the following examples show, in general given two points α and β in the complex plane, $\int_\alpha^\beta f(z)\,dz$ will depend on the choice of contour joining α to β.

Example 3.3

(i) Evaluate $\int_{\mathscr{C}_i} \bar{z}\,dz, i = 1, 2$, where $\mathscr{C}_1$ and $\mathscr{C}_2$ are the contours shown in Fig. 3.2.

(ii) Let $f: \mathbb{C} \to \mathbb{C}$ be defined by $f(z) = x + iky$ for some $k \in \mathbb{Z}$. Let $\mathscr{C}_n$ be the contour joining 0 to $1 + i$ given by $y = x^n$, $n \in \mathbb{N}$. Show that $\int_{\mathscr{C}_n} f(z)\,dz$ is independent of n if and only if $k = 1$.

Solution

(i) It is advisable to try to find the simplest representation of a given contour, since the integral does not depend on the choice of parametrisation. For example, a suitable parametrisation of $\mathscr{C}_1$ is $z(t) = t$ for $0 \leqslant t \leqslant 1$ and $z(t) = 1 + i(t-1)$ for $1 \leqslant t \leqslant 2$, but this is not the simplest choice. This is equivalent to splitting up $\mathscr{C}_1$ into two line

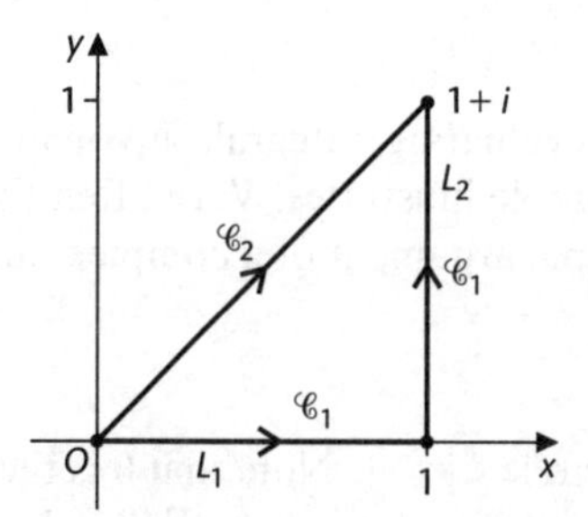

Figure 3.2

segments L_1 and L_2 as shown, where along L_1, $z(t) = t, 0 \leqslant t \leqslant 1$, and along L_2, $z(t) = 1 + it$, $0 \leqslant t \leqslant 1$. Then

$$\int_{\mathscr{C}_1} \bar{z}\, dz = \underset{(L_1)}{\int_0^1 t\, dt} + \underset{(L_2)}{\int_0^1 (1 - it)i\, dt} = 2\int_0^1 t\, dt + i\int_0^1 dt = 1 + i$$

The obvious representation of $\mathscr{C}_2$ is $z(t) = t + it, 0 \leqslant t \leqslant 1$, since $\mathscr{C}$ is part of the line $y = x$. Then by definition

$$\int_{\mathscr{C}_2} \bar{z}\, dz = \int_0^1 (t - it)(1 + i)\, dt = 2\int_0^1 t\, dt = 1$$

This shows that $\int_0^{1+i} \bar{z}\, dz$ depends on the choice of contour joining 0 to $1 + i$. Note that the integral taken round the whole triangle $\mathscr{C}$, in an anticlockwise sense, is given by

$$\int_{\mathscr{C}} \bar{z}\, dz = \int_{\mathscr{C}_1} \bar{z}\, dz - \int_{\mathscr{C}_2} \bar{z}\, dz = i$$

Note also that, in this case, the integrand is not analytic inside or on the closed contour $\mathscr{C}$.

(ii) The simplest parametrisation of each $\mathscr{C}_n$ is $z(t) = t + it^n, 0 \leqslant t \leqslant 1$. Then by definition,

$$\int_{\mathscr{C}_n} f(z)\, dz = \int_0^1 (t + ikt^n)(1 + int^{n-1})\, dt$$

$$= \int_0^1 (t - nkt^{2n-1})\, dt + i\int_0^1 (k + n)t^n dt = \frac{1}{2}(1 - k) + i\left(\frac{n + k}{n + 1}\right)$$

as required. It follows by the Cauchy–Riemann equations and 2.6 that f is analytic if and only if $k = 1$. In this special case,

$$\int_{\mathscr{C}_n} z\, dz = i = \left[\frac{z^2}{2}\right]_0^{1+i}$$

as in the real case. Once again, this is no accident.

Note

Some care has to be taken when evaluating integrals involving branches of multifunctions, as the following example illustrates. Very often they depend on the chosen convention for the principal argument of a complex number.

Example 3.4

Consider $\int_{\mathscr{C}} z^{-1/2}dz$, where $\mathscr{C}$ is the circle $|z| = 1$. Note that 0 is the branch point of the integrand, treated as a multifunction (see Example 2.4(i)). Now according to our convention, $z^{1/2} = \sqrt{r}e^{i\theta/2}$, where $r = |z|$, $\theta = \operatorname{Arg} z$, $-\pi < \theta \leqslant \pi$. Hence the simplest parametrisation of $\mathscr{C}$ to choose is $z(\theta) = e^{i\theta}$, $-\pi \leqslant \theta \leqslant \pi$, as in Example 3.2. Then by definition,

$$\int_{\mathscr{C}} \frac{dz}{z^{1/2}} = \int_{-\pi}^{\pi} ie^{i\theta}e^{-i\theta/2}d\theta = i\int_{-\pi}^{\pi} \cos(\theta/2)d\theta - \int_{-\pi}^{\pi} \sin(\theta/2)d\theta = 4i$$

Another parametrisation of $\mathscr{C}$ is $z(\phi) = e^{i\phi}$, $0 \leqslant \phi \leqslant 2\pi$. Then

$$\int_{\mathscr{C}} \frac{dz}{z^{1/2}} = \int_0^{\pi} ie^{i\phi}e^{-i\phi/2}d\phi + \int_{\pi}^{2\pi} ie^{i\phi}e^{-i(\phi-2\pi)/2}d\phi = 4i$$

as above. However, if the chosen convention is $0 \leqslant \operatorname{Arg} z < 2\pi$,

$$\int_{\mathscr{C}} \frac{dz}{z^{1/2}} = \int_0^{2\pi} ie^{i\phi}e^{-i\phi/2}d\phi = -4$$

The following elementary properties of definite integrals follow directly from the definition.

Lemma 3.1. Properties of Definite Integrals

Let f and g be piecewise continuous on $\mathscr{C}$.

(i) $$\int_{\mathscr{C}} (\alpha f(z) + \beta g(z))\, dz = \alpha\int_{\mathscr{C}} f(z)\, dz + \beta\int_{\mathscr{C}} g(z)\, dz$$

for any constants $\alpha, \beta \in \mathbb{C}$.

(ii) If $\mathscr{C}$ consists of a contour $\mathscr{C}_1$ joining α to β and a contour $\mathscr{C}_2$ joining β to γ, then

$$\int_{\mathscr{C}} f(z)\, dz = \int_{\mathscr{C}_1} f(z)\, dz + \int_{\mathscr{C}_2} f(z)\, dz$$

(iii) Let a contour $\mathscr{C}$ be described by $z = z(t)$, $a \leqslant t \leqslant b$ and let $-\mathscr{C}$ be the same set of points, taken in reverse order, so that $-\mathscr{C}$ is described by $z = z(-t)$, $-b \leqslant t \leqslant -a$. Then

$$\int_{-\mathscr{C}} f(z)\, dz = -\int_{\mathscr{C}} f(z)\, dz$$

□

Proof

All three results follow directly from the definition. We shall prove (i) as an example and leave the other two as exercises. Let $f(z) = u(x, y) + iv(x, y)$ and $g(z) = U(x, y) + iV(x, y)$ at all points of their domains, and let a parametrisation of $\mathscr{C}$ be given by (3.2). It follows by (3.4) that

$$\int_{\mathscr{C}} (f(z) + g(z))\, dz = \int_a^b ((u(t) + U(t))x'(t) - (v(t) + V(t))y'(t))\, dt$$

$$+i \int_a^b ((v(t) + V(t))x'(t) + (u(t) + U(t))y'(t))\, dt$$

$$\Rightarrow \quad \int_{\mathscr{C}} (f(z) + g(z))\, dz = \int_{\mathscr{C}} f(z)\, dz + \int_{\mathscr{C}} g(z)\, dz \tag{3.5}$$

using the linear property of real integrals and (3.4) again. Now let $\gamma = r + is, r, s \in \mathbb{R}$, be any complex constant. Then by (3.4) again,

$$\int_{\mathscr{C}} \gamma f(z)\, dz = \int_a^b ((ru(t) - sv(t))x'(t) - (rv(t) + su(t))y'(t))\, dt$$

$$+ i \int_a^b ((rv(t) + su(t))x'(t) + (ru(t) - sv(t))y'(t))\, dt$$

$$= (r + is)\left(\int_a^b (u(t)x'(t) - v(t)y'(t))\, dt + i \int_a^b (v(t)x'(t) + u(t)y'(t))\, dt\right)$$

$$\Rightarrow \quad \int_{\mathscr{C}} \gamma f(z)\, dz = \gamma \int_{\mathscr{C}} f(z)\, dz \tag{3.6}$$

Then the result follows by (3.5) and (3.6). ■

Notes

It follows from 3.1(ii) that if $\mathscr{C}$ is a closed contour then $\int_{\mathscr{C}} f(z)\, dz$ is independent of the choice of initial point of $\mathscr{C}$. Lemma 3.1(ii) is analogous to the result

$$\int_a^c f(t)\, dt = \int_a^b f(t)\, dt + \int_b^c f(t)\, dt$$

for real definite integrals, and 3.1(iii) is analogous to

$$\int_b^a f(t)\, dt = -\int_a^b f(t)\, dt$$

The following result, which gives an upper bound for the modulus of any complex definite integral, is of fundamental importance for future results and techniques.

Lemma 3.2. The 'ML Lemma'

Let $\mathscr{C}$ be any contour and let the function f be piecewise continuous on $\mathscr{C}$. Let L denote the length of $\mathscr{C}$; that is, if $\mathscr{C}$ is represented by $z = z(t)$, $t \in [a, b]$, then $L = \int_a^b |z'(t)|\, dt$. Suppose that $|f(z)| \leqslant M$ on $\mathscr{C}$ for some positive real constant M. Then

$$\left| \int_{\mathscr{C}} f(z)\, dz \right| \leqslant ML \qquad \square$$

Proof

We begin by considering $|\int_a^b g(t)\, dt|$ where $g : [a, b] \to \mathbb{C}$ is piecewise continuous, as in the preliminary definition. Suppose that $\int_a^b g(t)\, dt \neq 0$ and that the modulus of the integral is r and its principal argument is θ . Then $r = \int_a^b e^{-i\theta} g(t)\, dt$ since $e^{i\theta}$ is a constant. Then since r is real, $e^{-i\theta} g(t)$ is also real, so that

$$r = \int_a^b e^{-i\theta} g(t) \leqslant \int_a^b |e^{-i\theta} g(t)|\, dt = \int_a^b |g(t)|\, dt$$

$$\Rightarrow \quad r = \left| \int_a^b g(t)\, dt \right| \leqslant \int_a^b |g(t)|\, dt \tag{3.7}$$

as in the case of real definite integrals.

Now let the contour $\mathscr{C}$ have parametric equation (3.2). Then by definition and (3.7),

$$\left| \int_{\mathscr{C}} f(z)\, dz \right| = \left| \int_a^b f(z(t)) z'(t)\, dt \right| \leqslant \int_a^b |f(z(t))||z'(t)|\, dt$$

$$\Rightarrow \quad \left| \int_{\mathscr{C}} f(z)\, dz \right| \leqslant M \int_a^b |z'(t)| dt = ML \qquad \blacksquare$$

Notes

(i) Since f is piecewise continuous on the closed set $\mathscr{C}$, f is bounded on $\mathscr{C}$, so that the constant M exists.

(ii) If the contour $\mathscr{C}$ is a segment of the real axis joining a to b, then Lemma 3.2 and the definition give

$$\left| \int_a^b f(t)\, dt \right| \leqslant M(b - a)$$

a standard result from real calculus.

(iii) In contrast to the real case, it is not in general true that $|\int_{\mathscr{C}} f(z)\, dz| \leqslant \int_{\mathscr{C}} |f(z)| dz$ for complex integrals. Try $f(z) = 1/z$ with $\mathscr{C}$ the circle $|z| = 1$, for instance.

Example 3.5

Let $\mathscr{C}$ be the circle $|z| = R$ with $R > 1$ and consider

$$\left| \int_{\mathscr{C}} \frac{\operatorname{Log} z}{z^n} \, dz \right|$$

where $n \in \mathbb{N}$. The length of $\mathscr{C}$ is $2\pi R$ and on $\mathscr{C}$,

$$\left| \frac{\operatorname{Log} z}{z^n} \right| = \frac{|\operatorname{Log} z|}{R^n}$$

Also, by definition,

$$\operatorname{Log} z = \operatorname{Log}|z| + i \operatorname{Arg} z \qquad (-\pi < \operatorname{Arg} z \leqslant \pi)$$

Hence on $\mathscr{C}$,

$$|\operatorname{Log} z| \leqslant ((\operatorname{Log} R)^2 + \pi^2)^{1/2} < \operatorname{Log} R + \pi \, (\text{since } \operatorname{Log} R, \pi > 0)$$

Then by lemma 3.2 it follows that

$$\left| \int_{\mathscr{C}} \frac{\operatorname{Log} z}{z^n} \, dz \right| \leqslant 2\pi R \left(\frac{\operatorname{Log} R + \pi}{R^n} \right) = 2\pi R^{1-n} (\operatorname{Log} R + \pi)$$

As in real variable calculus, if a function f has a known antiderivative, it is relatively simple to evaluate $\int_{\mathscr{C}} f(z)\, dz$.

Theorem 3.3. Fundamental Theorem of Calculus

Let $F : A \subseteq \mathbb{C} \to \mathbb{C}$ be an analytic function with continuous derivative f in a region $\mathscr{R}$ containing the contour $\mathscr{C}$ parametrised by $z = z(t)$, $t \in [a, b]$. Then

$$\int_{\mathscr{C}} f(z) \, dz = F(\beta) - F(\alpha) = [F(z)]_{\alpha}^{\beta}$$

where $\alpha = z(a)$ and $\beta = z(b)$. □

Proof

Suppose first of all that $\mathscr{C}$ is a smooth arc. Then by hypothesis and the definition,

$$\int_{\mathscr{C}} f(z) \, dz = \int_{\mathscr{C}} F'(z) \, dz = \int_a^b F'(z(t)) z'(t) \, dt = \int_a^b \frac{dF}{dz} \frac{dz}{dt} \, dt = \int_a^b \frac{dF}{dt} \, dt$$

by the chain rule. Let $F(z(t)) = U(t) + iV(t)$. Then, using the preliminary definition and the above result,

$$\int_{\mathscr{C}} f(z) \, dz = \int_a^b U'(t) \, dt + i \int_a^b V'(t) \, dt = [F(z(t))]_a^b = F(\beta) - F(\alpha),$$

as required. More generally, if $\mathscr{C}$ is a contour consisting of a finite number of smooth arcs $\mathscr{C}_k$, connecting α_k to β_k, then $\int_{\mathscr{C}_k} f(z) \, dz = F(\alpha_k) - F(\beta_k)$ for each k, by above. The result then follows by 3.1(ii). ■

Notes

(i) If f is the derivative of F, we write $F(z) = \int f(z)\,dz$ and call F an **indefinite integral** of f, as in real calculus. Since all the standard derivatives of the elementary functions are the same as in real variable calculus, the same applies to indefinite integrals. So, for example, $\int e^{iz} dz = e^{iz}/i$, and so on.

(ii) In contrast to the real case, the fundamental theorem of calculus is not the ultimate result concerning evaluation of definite integrals, as will soon be shown.

Example 3.6

Since sinh is the continuous derivative of the entire function cosh, it follows by Theorem 3.3 that

$$\int_{\text{Log}\,2 - i\pi}^{i\pi/2} \sinh z\,dz = \left[\cosh z\right]_{\text{Log}\,2 - i\pi}^{i\pi/2}$$

independent of the contour joining $\text{Log}\,2 - i\pi$ to $i\pi/2$. Hence

$$\int_{\text{Log}\,2 - i\pi}^{i\pi/2} \sinh z\,dz$$

$$= \cos(\pi/2) - \cosh(\text{Log}\,2)\cosh(i\pi) + \sinh(\text{Log}\,2)\sinh(i\pi)$$

$$= \cosh(\text{Log}\,2) = \tfrac{1}{2}\left(e^{\text{Log}\,2} + e^{\text{Log}\,1/2}\right) = \tfrac{5}{4}$$

using the results of Chapter 1.

The next result follows immediately from Theorem 3.3.

Corollary 3.4. Definite Integrals Independent of the Contour

If f is the continuous derivative of an analytic function in a region $\mathscr{R}$ containing α and β, then $\int_\alpha^\beta f(z)\,dz$ is independent of the contour joining α to β lying within $\mathscr{R}$, so that if $\mathscr{C}$ is any simple closed contour in $\mathscr{R}$, then $\int_{\mathscr{C}} f(z)\,dz = 0$. □

Functions f for which $\int_\alpha^\beta f(z)\,dz$ is independent of the choice of contour are sometimes called **integrable**.

Example 3.7

Consider $\int_{\mathscr{C}} z^{-2} dz$ where $\mathscr{C}$ is the circle $|z| = 1$. The integrand is the continuous derivative of the analytic function F given by $F(z) = -1/z, z \neq 0$, throughout any annular region containing $|z| = 1$ but excluding the origin. (Note that the region $\mathscr{R}$ in 3.4 does not have to be simply connected.) Hence by 3.4, $\int_{\mathscr{C}} z^{-2} dz = 0$, as can be checked using the definition.

Now consider $\int_{\mathscr{C}} z^{-2} dz$ where $\mathscr{C}$ is the circle $|z| = 1$ again. It is easy to check using the definition that $\int_{\mathscr{C}} z^{-2} dz = 2\pi i$. The fact that $\int_{\mathscr{C}} z^{-1} dz \neq 0$ is due to the fact that there is no region containing $\mathscr{C}$ throughout which the integrand is the continuous derivative of an analytic function. This is so because the derivative of $\operatorname{Log} z$ is $1/z$ and Log fails to be analytic along the branch cut given by $z = x$, $x \leqslant 0$, according to our convention. Under any convention, any region containing $\mathscr{C}$ will contain points on a branch cut of log, since any branch cut of log will pass through $\mathscr{C}$. Hence 3.4 does not apply to this integral. However, we can still use 3.3 to evaluate the integral, as long as we take some care. With our convention,

$$\int_{\mathscr{C}} \frac{dz}{z} = \lim_{\theta \to \pi-} [\operatorname{Log} z]_{z=e^{-i\theta}}^{z=e^{i\theta}} = \lim_{\theta \to \pi-} [\operatorname{Log}|z| + i \operatorname{Arg} z]_{z=e^{-i\theta}}^{z=e^{i\theta}}$$

(where $0 \leqslant \theta < \pi$), and so

$$\int_{\mathscr{C}} \frac{dz}{z} = \lim_{\theta \to \pi-} (2i\theta) = 2\pi i$$

The converse of Corollary 3.4 is also true.

Theorem 3.5. Integrable Functions are Derivatives

Let f be continuous in a region $\mathscr{R}$ and suppose that $\int_\alpha^\beta f(z)\, dz$ is independent of any contour joining α and β, contained wholly within $\mathscr{R}$. Then f is the derivative of an analytic function in $\mathscr{R}$. □

Proof

Let γ be any point in $\mathscr{R}$ and define F by $F(z) = \int_\gamma^z f(w)\, dw$, $z \in \mathscr{R}$. Then by hypothesis, F is well defined since the integral is independent of any contour joining γ to z, contained wholly within $\mathscr{R}$. Now,

$$\frac{F(z+h) - F(z)}{h} = \frac{1}{h}\left(\int_\gamma^{z+h} f(w)\, dw - \int_\gamma^z f(w)\, dw\right) = \frac{1}{h}\int_z^{z+h} f(w)\, dw$$

$$\Rightarrow \quad \left|\frac{F(z+h) - F(z)}{h} - f(z)\right| = \left|\frac{1}{h}\int_z^{z+h} (f(w) - f(z))\, dw\right| \tag{3.8}$$

where $z + h \in \mathscr{R}$ and $h \neq 0$, using 3.1 and 3.3. By hypothesis, the integration can be taken over the line segment joining z to $z + h$.

Since f is continuous in $\mathscr{R}$, given any $\varepsilon > 0$, there is a $\delta > 0$ such that

$$|h| < \delta \quad \Rightarrow \quad |f(z+h) - f(z)| < \varepsilon$$

It then follows by equation (3.8) and Lemma 3.2 that

$$0 < |h| < \delta \quad \Rightarrow \left|\frac{F(z+h) - F(z)}{h} - f(z)\right| < \frac{1}{|h|}(\varepsilon|h|) = \varepsilon$$

$$\Rightarrow \quad f(z) = \lim_{h \to 0} \frac{F(z+h) - F(z)}{h} = F'(z) \quad \text{as required.}$$

■

Exercise 3.1.1 Evaluate $\int_{\mathscr{C}} z|z|dz$ where $\mathscr{C}$ is the contour given by $z(t) = 1 - t$ for $0 \leqslant t \leqslant 1, z(t) = i(t-1)$ for $1 \leqslant t \leqslant 2$.

Exercise 3.1.2 Evaluate

(i) $\int_{\mathscr{C}_k} z\, dz$

(ii) $\int_{\mathscr{C}_k} \bar{z}\, dz$

$k = 1, 2$, where $\mathscr{C}_1$ and $\mathscr{C}_2$ are the contours shown in Fig. 3.3(a). ($\mathscr{C}_2$ is an arc of a circle, centre 1 and radius 1.)

Verify that

$$\int_{\mathscr{C}_k} z\, dz = \left[\frac{z^2}{2}\right]_0^{1+i} \qquad (k = 1 \text{ or } 2)$$

and that in (ii) the result depends on the choice of contour.

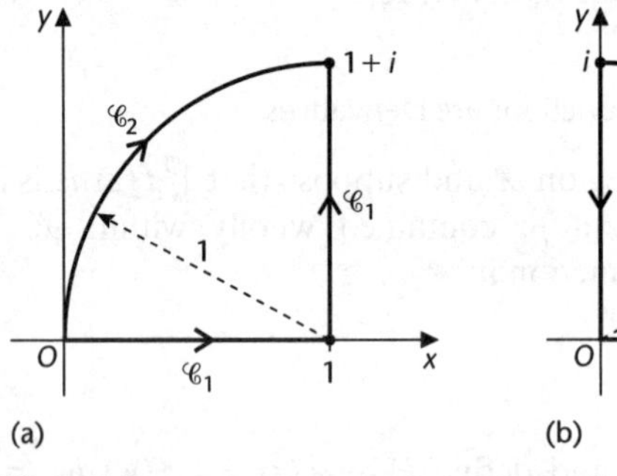

Figure 3.3

Exercise 3.1.3 Let $\mathscr{C}$ be the closed contour shown in Fig. 3.3(b), where the arc is part of a circle of unit radius, centre the origin. Show, by direct evaluation, that $\int_{\mathscr{C}} z^2 dz = 0$, but $\int_{\mathscr{C}} |z|^2 dz \neq 0$.

Exercise 3.1.4 Let $\mathscr{C}$ denote the circle $|z| = R$. Show, by direct evaluation, that $\int_{\mathscr{C}} \operatorname{Log} z\, dz$ and $\int_{\mathscr{C}} ((\operatorname{Log} z)/z) dz$ depend on the radius of $\mathscr{C}$, whereas $\int_{\mathscr{C}} z^n dz$ for any $n \in \mathbb{Z}$ does not.

Exercise 3.1.5 Evaluate

(i) $\int_{\mathscr{C}} z \operatorname{Im} z\, dz$

(ii) $\int_{\mathscr{C}} |z|^2 dz$

where $\mathscr{C}$ is the triangle with vertices at 0, 3 and $3 + 4i$.

Exercise **3.1.6** Evaluate $\int_{\mathscr{C}_k} \bar{z}\, dz$, $k = 1, 2$, where $\mathscr{C}_1$ is the arc of the ellipse with Cartesian equation $x^2/a^2 + y^2/b^2 = 1$, $x \geqslant 0$, and $\mathscr{C}_2$ has Cartesian equation $y = e^x$, $0 \leqslant x \leqslant 1$.

Exercise **3.1.7** Evaluate $\int_{\mathscr{C}} (1 - \operatorname{Im} z)\, dz$ where $\mathscr{C}$ is one arch of the cycloid with parametric equations $x = \theta - \sin\theta$, $y = 1 - \cos\theta$, $0 \leqslant \theta \leqslant 2\pi$.

Exercise **3.1.8** Evaluate $\int_{\mathscr{C}} z^{-1} dz$ where $\mathscr{C}$ is the rectangle with vertices at $\pm i \pm 2$.

Exercise **3.1.9** (a) Provided that the integrals exist, prove that if a contour $\mathscr{C}$ consists of a contour $\mathscr{C}_1$ joining α to β and a contour $\mathscr{C}_2$ joining β to γ, then $\int_{\mathscr{C}} f(z)\, dz = \int_{\mathscr{C}_1} f(z)\, dz + \int_{\mathscr{C}_2} f(z)\, dz$.

(b) Let a contour $\mathscr{C}$ be described by $z = z(t)$, $a \leqslant t \leqslant b$ and let $-\mathscr{C}$ be described by $z = z(-t)$, $-b \leqslant t \leqslant -a$. Prove that $\int_{-\mathscr{C}} f(z)\, dz = -\int_{\mathscr{C}} f(z)\, dz$, provided that the integrals exist.

Exercise **3.1.10** Without actually evaluating the integrals, use the *ML* lemma to show that

(i) $\left| \int_{\mathscr{C}} \frac{dz}{1 + z^2} \right| \leqslant \frac{3\pi}{16}$, where $\mathscr{C}$ is the arc of the circle $|z| = 3$, joining 3 to $3i$ in the first quadrant

(ii) $|\int_{\mathscr{C}} e^{\bar{z} \operatorname{Im} z} dz| \leqslant 2\pi\sqrt{e}$, where $\mathscr{C}$ is the circle $|z| = 1$

(iii) $|\int_{\mathscr{C}} e^{\bar{z} \operatorname{Im} z} dz| \leqslant (2 + \sqrt{2})e$, where $\mathscr{C}$ is the triangle with vertices 0, 1 and $1 + i$

Exercise **3.1.11** Use the fundamental theorem of calculus to evaluate $\int_0^{\pi + i} \sin 2z\, dz$.

Exercise **3.1.12** Use the fundamental theorem of calculus to evaluate

(i) $\int_{\mathscr{C}_k} z^{1/2} dz$

(ii) $\int_{\mathscr{C}_k} \operatorname{Log} z\, dz$

$k = 1, 2$, where $\mathscr{C}_1$ is a simple closed contour not enclosing the origin and $\mathscr{C}_2$ is the circle $|z| = R$.

Cauchy's Theorem

It is often the case that a given complex definite integral can be evaluated without knowing the antiderivative of the integrand. In particular, the following famous result enables us to evaluate $\int_{\mathscr{C}} f(z)\, dz$, where $\mathscr{C}$ is a simple closed contour, without having to resort to the definition or the fundamental theorem of calculus, in a large number of cases. It is one of the foundations of the theory of analytic functions.

Theorem 3.6. Cauchy's Theorem

Let $f: A \subseteq \mathbb{C} \to \mathbb{C}$ be analytic inside and on a simple closed contour $\mathscr{C}$, with f' continuous inside and on $\mathscr{C}$. Then $\int_{\mathscr{C}} f(z)\,dz = 0$. □

Notes

(i) The result states that $\int_{\mathscr{C}} f(z)\,dz$ is independent of the contour and the integrand as long as f is analytic and f' is continuous inside and on $\mathscr{C}$. The result is easily extended to the case where $\mathscr{C}$ is expressible as a finite number of simple closed contours. For example,

$$\int_{\mathscr{C}} \frac{e^z}{z^2+9}\,dz = 0 \quad \text{where } \mathscr{C} \text{ is the circle } |z| = 1$$

$$\int_{\mathscr{C}} e^{z^2}\,dz = \int_{\mathscr{C}} \sin(z^3)\,dz = 0 \quad \text{where } \mathscr{C} \text{ is any closed contour}$$

(ii) It may be that for a particular contour $\mathscr{C}$, the conditions of 3.6 are not necessary for $\int_{\mathscr{C}} f(z)\,dz = 0$. For example, using the definition, $\int_{|z|=1} ((\operatorname{Log} z)/z)\,dz = 0$, although the conditions of 3.6 are not met.

The following revised form of 3.6 shows that the condition that f' is continuous inside and on $\mathscr{C}$ can be dropped.

Theorem 3.7. The Cauchy–Goursat Theorem

Let f be analytic in a simply connected region $\mathscr{R}$. Then if $\mathscr{C}$ is any simple closed contour lying totally within $\mathscr{R}$, $\int_{\mathscr{C}} f(z)\,dz = 0$. □

Historical Note

Cauchy first proved his theorem in 1814. Goursat proved his surprising amended version at the turn of the twentieth century. Almost all of the applications of the result were found by Cauchy.

The proof of Cauchy's theorem is staightforward and just uses Green's theorem from real calculus. It is assumed that the reader is familiar with this result. Without the additional assumption of f' continuous inside and on $\mathscr{C}$, the proof is much harder and we defer the proof of 3.7 until the end of the chapter.

Theorem 3.8. Green's Theorem

Let S be a closed set consisting of a simple closed contour $\mathscr{C}$ and its inside. If P, $Q: \mathbb{R}^2 \to \mathbb{R}$ are continuous with continuous first-order partial derivatives on S, then

$$\int_{\mathscr{C}} (P(x,y)\,dx + Q(x,y)dy) = \iint_S (Q_x(x,y) - P_y(x,y))\,dx\,dy$$ □

Proof of 3.6

Recall that if $z(t) = x(t) + iy(t)$, $a \leqslant t \leqslant b$, is any parametrisation of $\mathscr{C}$ and $f(z) = u(x, y) + iv(x, y)$, then

$$\int_{\mathscr{C}} f(z)\, dz = \int_a^b (u(t)x'(t) - v(t)y'(t))\, dt + i \int_a^b (v(t)x'(t) + u(t)y'(t))\, dt \tag{3.4}$$

This can be expressed in terms of real line integrals as

$$\int_{\mathscr{C}} f(z)\, dz = \int_{\mathscr{C}} (u(x, y)\, dx - v(x, y)dy) + i \int_{\mathscr{C}} (v(x, y)\, dx + u(x, y)dy \tag{3.9}$$

Since f' is continuous on the closed set S, u_x, u_y, v_x and v_y are continuous in S by Theorem 2.5, and so by Green's theorem and (3.9) it follows that

$$\int_{\mathscr{C}} f(z)\, dz = -\iint_S (v_x + u_y)\, dx\, dy + i \iint_S (u_x - v_y)\, dx\, dy$$

Since f is analytic on S, the Cauchy–Riemann equations are satisfied on S by Theorem 2.4; that is, $u_y = -v_x$ and $u_x = v_y$ on S. Hence $\int_{\mathscr{C}} f(z)\, dz = 0$ as required. ■

The following example demonstrates how Cauchy's theorem may be used to evaluate certain real definite integrals.

Example 3.8

Let $\mathscr{C}$ be any simple closed contour not enclosing the origin. Then by Cauchy's theorem, $\int_{\mathscr{C}} (e^{iz}/z)dz = 0$. In particular, let $\mathscr{C}$ be the contour shown in Fig. 3.4. The semicircular arc Γ is parametrised by $z(\theta) = Re^{i\theta}, 0 \leqslant \theta \leqslant \pi$ so that by definition and (3.7),

$$\left| \int_\Gamma \frac{e^{iz}}{z} dz \right| = \left| \int_0^\pi ie^{iR\cos\theta - R\sin\theta} d\theta \right| \leqslant \int_0^\pi |ie^{iR\cos\theta}||e^{-R\sin\theta}| d\theta$$

$$\Rightarrow \quad \left| \int_\Gamma \frac{e^{iz}}{z} dz \right| \leqslant \int_0^\pi e^{-R\sin\theta} d\theta = 2\int_0^{\pi/2} e^{-R\sin\theta} d\theta \leqslant 2\int_0^{\pi/2} e^{-2R\theta/\pi} d\theta$$

since $2\theta/\pi \leqslant \sin\theta$ for $0 \leqslant \theta \leqslant \pi/2$. Then

$$\left| \int_\Gamma \frac{e^{iz}}{z} dz \right| \leqslant \frac{\pi}{R}\left(1 - e^{-R}\right) < \frac{\pi}{R} \quad \text{so that} \quad \int_\Gamma \frac{e^{iz}}{z} dz \to 0 \quad \text{as} \quad R \to \infty$$

Also, $z(\theta) = \varepsilon e^{i\theta}$, $\pi \leqslant \theta \leqslant 0$ on the semicircular arc γ. Then

$$\int_\gamma \frac{e^{iz}}{z} dz = -\int_0^\pi ie^{i\varepsilon e^{i\theta}} d\theta \quad \Rightarrow \quad \lim_{\varepsilon \to 0} \int_\gamma \frac{e^{iz}}{z} dz = -\int_0^\pi i\, d\theta = -i\pi$$

since the integral is a continuous function of ε.

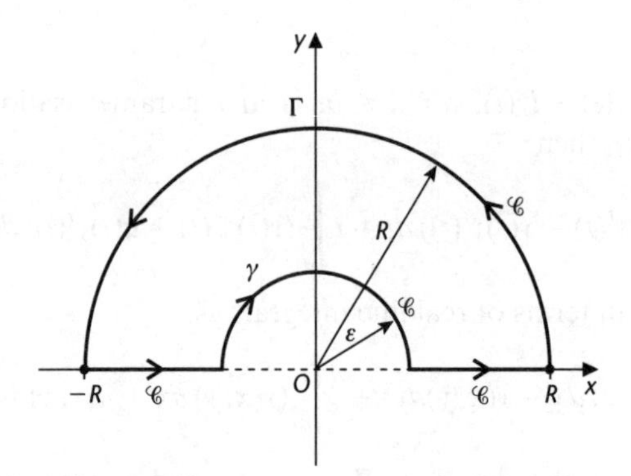

Figure 3.4

Hence, letting $R \to \infty$ and $\varepsilon \to 0$ gives

$$0 = \int_{\mathscr{C}} \frac{e^{iz}}{z}\,dz = \lim_{R\to\infty} \int_{-R}^{R} \frac{e^{ix}}{x}\,dx - i\pi \quad \Rightarrow \quad \int_{-\infty}^{\infty} \frac{\sin x}{x}\,dx = \pi$$

$$\Rightarrow \quad \int_{0}^{\infty} \frac{\sin x}{x}\,dx = \frac{\pi}{2}$$

by comparing imaginary parts and noting that the last integrand is even. This real definite integral cannot be obtained directly by using the fundamental theorem of real variable calculus.

Corollary 3.9. Definite Integrals Independent of the Contour

Let α and β be any two points in a simply connected region $\mathscr{R}$. If $f : A \subseteq \mathbb{C} \to \mathbb{C}$ is analytic throughout $\mathscr{R}$, then $\int_{\alpha}^{\beta} f(z)\,dz$ is independent of the choice of contour contained wholly within $\mathscr{R}$ and joining α to β. □

Proof

Let $\mathscr{C}_1$ and $\mathscr{C}_2$ be any two non-intersecting contours (without loss of generality) lying entirely within $\mathscr{R}$ and joining α to β, as shown in Fig 3.5. (Arrows denote the direction of increasing parameter.) $\mathscr{C}_1$ and $\mathscr{C}_2$ form a simple closed contour $\mathscr{C}$, and since $\mathscr{R}$ is simply connected, it follows by 3.7 and 3.1 that

$$\int_{\mathscr{C}} f(z)\,dz = \int_{\mathscr{C}_2} f(z)\,dz - \int_{\mathscr{C}_1} f(z)\,dz = 0$$

as required. ■

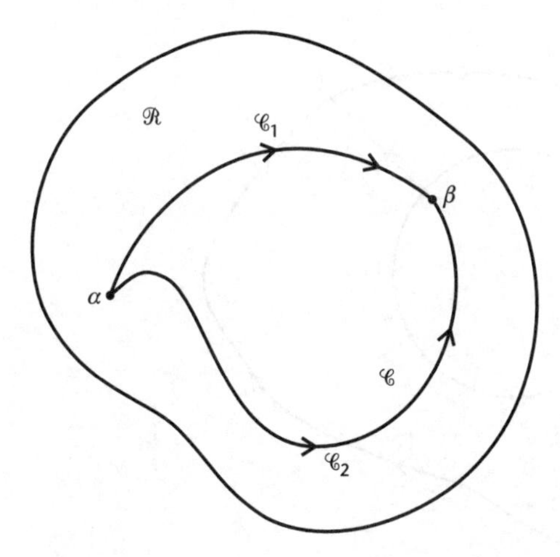

Figure 3.5

Note

Compare Corollaries 3.4 and 3.9. It follows by 3.9 and Theorem 3.5 that if f is analytic in a simply connected region $\mathcal{R}$ then there exists an analytic function F in $\mathcal{R}$ such that $F' = f$.

Cauchy's Integral Formula

Cauchy's integral formula concerns definite integrals around closed contours, in which the integrand has just one singularity, a simple pole, inside the closed contour. We require, first of all, the following result, which is important in its own right.

Lemma 3.10. A Deformation Result

Let $f: A \subseteq \mathbb{C} \to \mathbb{C}$ be analytic in a simply connected region containing a simple closed contour $\mathcal{C}$, except at a point α lying inside $\mathcal{C}$. If $\mathcal{C}'$ is a circle lying totally inside $\mathcal{C}$, centred at α, then

$$\int_{\mathcal{C}} f(z)\,dz = \int_{\mathcal{C}'} f(z)\,dz \qquad \square$$

Note how this result says that any simple closed contour enclosing just one singular point of the integrand can be conveniently replaced by a circle. The resulting integral is then easily evaluated using the definition.

Proof

Let β and γ be any two points on $\mathcal{C}$ and β' and γ' be any two points on $\mathcal{C}'$. Create contours Γ_1 and Γ_2 as shown in Fig. 3.6. (Note the direction of

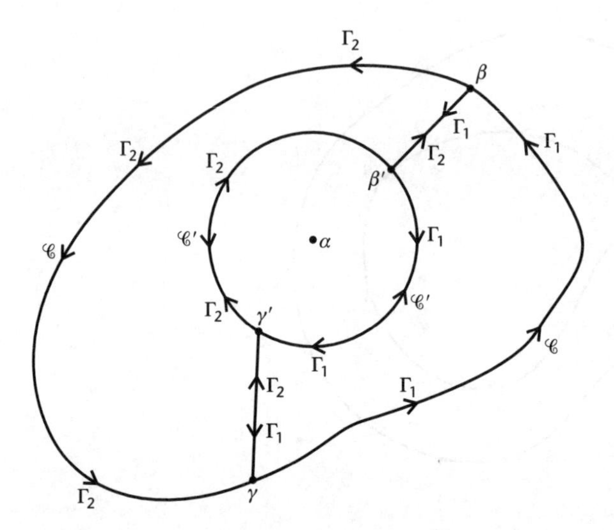

Figure 3.6

increasing parameters.) By 3.1, $\int_{\beta\,\Gamma_1}^{\beta'} f(z)\,dz = -\int_{\beta\,\Gamma_2}^{\beta'} f(z)\,dz$ and similarly for the other line segment,

$$\int_{\mathscr{C}} f(z)\,dz - \int_{\mathscr{C}'} f(z)\,dz = \int_{\Gamma_1} f(z)\,dz + \int_{\Gamma_2} f(z)\,dz$$

by 3.1 again. But f is analytic in a simply connected region containing both Γ_1 and Γ_2, so that by 3.7,

$$\int_{\mathscr{C}} f(z)\,dz - \int_{\mathscr{C}'} f(z)\,dz = 0$$

■

Notes

(i) Lemma 3.10 still holds if $\mathscr{C}'$ is replaced by any simple closed contour lying totally inside $\mathscr{C}$ and enclosing α.

(ii) Lemma 3.10 does not hold for integrands with a single branch point since branch points are not isolated singular points. For example, from Exercise 3.1.12(i), if $\mathscr{C}$ is the circle $|z| = R$ then $\int_{\mathscr{C}} z^{1/2} dz = -4iR^{3/2}/3$, so the result depends on the radius of the circle.

Example 3.9

Evaluate $\int_{\mathscr{C}} z^n dz$ for any $n \in \mathbb{Z}$ where $\mathscr{C}$ is any simple closed contour (i) enclosing 0, (ii) not enclosing 0.

Solution

Note that if $n \geqslant 0$ then the integrand is entire, so in either case $\int_{\mathscr{C}} z^n dz = 0$, by 3.6 or 3.7. Also note that in case (ii), if $n < 0$ then the integrand is analytic inside and on $\mathscr{C}$, so $\int_{\mathscr{C}} z^n dz = 0$, also by 3.6 or 3.7. Now consider (i) with $n < 0$, so Cauchy's theorem is not applicable. It follows by 3.10 that $\mathscr{C}$ may be replaced by the unit circle, centred at the origin. Hence let $z(\theta) = e^{i\theta}$ where $-\pi \leqslant \theta \leqslant \pi$ on $\mathscr{C}$ without loss of generality. Then by definition,

$$\int_{\mathscr{C}} z^n dz = \int_{-\pi}^{\pi} e^{ni\theta} ie^{i\theta} d\theta = 0 \qquad (n \neq -1) \quad \text{and} \quad \int_{\mathscr{C}} z^{-1} dz = 2\pi i$$

Alternatively, Theorem 3.3 may be used as in Example 3.7.

An extension of this result with $n = -1$ is the following lemma.

Lemma 3.11. An Integral Around a Circle

Let $\mathscr{C}'$ be the circle of radius ρ, centred at α. Then

$$\int_{\mathscr{C}'} \frac{dz}{z - \alpha} = 2\pi i$$

□

Proof

If z is any point on $\mathscr{C}'$, then $|z - \alpha| = \rho$, so that a suitable parametrisation of $\mathscr{C}'$ is $z(\theta) = \rho e^{i\theta} + \alpha, \; -\pi \leqslant \theta \leqslant \pi$ say. Then by definition,

$$\int_{\mathscr{C}'} \frac{dz}{z - \alpha} = \int_{-\pi}^{\pi} \frac{i\rho e^{i\theta}}{\rho e^{i\theta}} d\theta = 2\pi i$$

■

The last two results lead to the following remarkable result.

Theorem 3.12. Cauchy's Integral Formula

Let $f : A \subseteq \mathbb{C} \to \mathbb{C}$ be analytic in a simply connected region $\mathscr{R}$. If $\mathscr{C}$ is any simple closed contour in $\mathscr{R}$ and α is any point inside $\mathscr{C}$, then

$$f(\alpha) = \frac{1}{2\pi i} \int_{\mathscr{C}} \frac{f(z)}{z - \alpha} dz$$

□

Note

The result says that the given integral is independent of $\mathscr{C}$ and only depends on the value of f at the simple pole α of the integrand.

Proof

The proof can be conveniently split into two parts.

Step 1

$$I = \int_{\mathscr{C}} \frac{f(z)}{z-\alpha}\,dz = \int_{\mathscr{C}'} \frac{f(\alpha)}{z-\alpha}\,dz + \int_{\mathscr{C}'} \frac{f(z)-f(\alpha)}{z-\alpha}\,dz$$

where $\mathscr{C}'$ is any circle, centred at α, lying within $\mathscr{C}$ by 3.10.

$$\text{Hence} \quad I = 2\pi i f(\alpha) + I_1 \quad \text{where} \quad I_1 = \int_{\mathscr{C}'} \frac{f(z)-f(\alpha)}{z-\alpha}\,dz$$

by 3.11. Notice by 3.2, taking the radius of $\mathscr{C}'$ to be ρ, $|I_1| \leqslant 2\pi\rho M$, where

$$\left|\frac{f(z)-f(\alpha)}{z-\alpha}\right| = \frac{|f(z)-f(\alpha)|}{\rho} \leqslant M \quad \text{on } \mathscr{C}'$$

It remains to show that $I_1 = 0$.

Step 2

Since f is analytic at α, it is continuous at α, so that by definition, given any $\varepsilon > 0$, there exists a $\delta > 0$ such that

$$|z-\alpha| < \delta \quad \Rightarrow \quad |f(z)-f(\alpha)| < \varepsilon$$

Then from step 1,

$$|I_1| < 2\pi\varepsilon \quad \text{whenever} \quad |z-\alpha| < \delta$$

But on $\mathscr{C}'$, $|z-\alpha| = \rho$, so that choosing $\rho < \delta$ gives $|I_1| < 2\pi\varepsilon$. Hence $\lim_{\rho\to 0} I_1 = 0$, as required. ■

For example, it follows directly from 3.12 that if $\mathscr{C}$ is the rectangle with vertices at $\pm i$ and ± 2, then $\int_{\mathscr{C}} z^{-1}\,dz = 2\pi i$. Compare this method to using the definition – see the solution to Exercise 3.1.8!

Example 3.10

Let $\mathscr{C}$ be the circle $|z| = 3$. Using partial fractions,

$$\int_{\mathscr{C}} \frac{\cosh(z^2)}{z(z^2+4)}\,dz = \int_{\mathscr{C}} \frac{\cosh(z^2)}{4z}\,dz - \int_{\mathscr{C}} \frac{\cosh(z^2)}{8(z+2i)}\,dz - \int_{\mathscr{C}} \frac{\cosh(z^2)}{8(z-2i)}\,dz$$

The numerator in the integrand is entire and the points $0, \pm 2i$ lie inside $\mathscr{C}$. Hence by Cauchy's integral formula,

$$\int_{\mathscr{C}} \frac{\cosh(z^2)}{z(z^2+4)}\,dz = \frac{2\pi i}{8}(2\cosh 0 - 2\cosh(-4)) = \frac{\pi i}{2}(1 - \cosh 4)$$

Example 3.11

Let $\mathscr{C}$ be the unit circle, centred at the origin, parametrised by $z(\theta) = e^{i\theta}$, $-\pi \leqslant \theta \leqslant \pi$. Evaluate $\int_{\mathscr{C}} (e^{az}/z)\, dz$ where a is a non-zero real constant, hence evaluate the real integral $\int_0^{\pi} e^{a\cos\theta} \cos(a \sin\theta) d\theta$.

Solution

Note that the given integrands have no indefinite integrals expressible in terms of elementary functions. However, $f(z) = e^{az}$ is entire and 0 is inside $\mathscr{C}$. Then by Cauchy's integral formula,

$$\int_{\mathscr{C}} \frac{e^{az}}{z} dz = 2\pi i e^0 = 2\pi i$$

Then by definition, it follows that

$$\int_{\mathscr{C}} \frac{e^{az}}{z} dz = \int_{-\pi}^{\pi} \frac{(\exp(ae^{i\theta}))(ie^{i\theta})}{e^{i\theta}} d\theta = i\int_{-\pi}^{\pi} e^{a(\cos\theta + i\sin\theta)} d\theta = 2\pi i$$

$$\Rightarrow \quad \int_{-\pi}^{\pi} e^{a\cos\theta}(\cos(a\sin\theta) + i\sin(a\sin\theta))\, d\theta = 2\pi$$

Equating real parts gives

$$\int_{-\pi}^{\pi} e^{a\cos\theta} \cos(a\sin\theta)\, d\theta = 2\pi \quad \Rightarrow \quad \int_0^{\pi} e^{a\cos\theta} \cos(a\sin\theta)\, d\theta = \pi$$

since the integrand is even.

A very important corollary of Cauchy's integral formula is the following result.

Theorem 3.13. Cauchy's Integral Formula for Derivatives

Let f be analytic in a simply connected region $\mathscr{R}$ and let $\mathscr{C}$ be a simple closed contour in $\mathscr{R}$. If α is any point inside $\mathscr{C}$ then

$$f^{(n)}(\alpha) = \frac{n!}{2\pi i}\int_{\mathscr{C}} \frac{f(z)}{(z-\alpha)^{n+1}} dz \qquad n \in \mathbb{N}$$ □

Proof

Once again, we split the proof up into two parts.

Step 1

To illustrate the technique of the proof, we first show that the result is true for $n = 1$. By hypothesis and 3.12, if $\alpha + h$ lies inside $\mathscr{C}$ then

$$\frac{f(\alpha+h) - f(\alpha)}{h} = \frac{1}{2\pi i}\int_{\mathscr{C}} \frac{f(z)}{h}\left(\frac{1}{z-\alpha-h} - \frac{1}{z-\alpha}\right) dz = \frac{1}{2\pi i} I_1$$

where $I_1 = \int_{\mathscr{C}} \frac{f(z)}{(z-\alpha-h)(z-\alpha)} dz$

Now let $I_2 = \int_{\mathscr{C}} \frac{f(z)}{(z-\alpha)^2} dz$

It certainly seems reasonable that $I_1 \to I_2$ as $h \to 0$. We now show rigorously that this is indeed the case.

Let $|f(z)| \leqslant M$ on $\mathscr{C}$, L be the length of $\mathscr{C}$ and d the shortest distance from α to any point z on $\mathscr{C}$. Then $|z-\alpha| \geqslant d$ and it follows by the triangle inequality that

$$|z-\alpha| \leqslant |z-\alpha-h| + |h| \quad \Rightarrow \quad |z-\alpha-h| \geqslant d - |h|$$

Hence, altogether,

$$|z-\alpha| \geqslant d \quad \text{and} \quad 2|h| \leqslant d \quad \Rightarrow \quad |z-\alpha-h| \geqslant d/2 \tag{3.10}$$

and it follows by 3.2 that for $2|h| \leqslant d$,

$$|I_1 - I_2| = \left| h \int_{\mathscr{C}} \frac{f(z)}{(z-\alpha)^2(z-\alpha-h)} dz \right| \leqslant \frac{2|h|ML}{d^3}$$

so that $I_1 - I_2 \to 0$ as $h \to 0$. Then

$$f'(\alpha) = \lim_{h\to 0} \frac{f(\alpha+h) - f(\alpha)}{h} = \frac{1}{2\pi i} I_2 = \frac{1}{2\pi i} \int_{\mathscr{C}} \frac{f(z)}{(z-\alpha)^2} dz$$

Step 2

The given result then follows by induction. In practice the inductive step is a little tedious, but uses the same technique as step 1. By above, the result is true for $n = 1$. Suppose that

$$f^{(k)}(\alpha) = \frac{k!}{2\pi i} \int_{\mathscr{C}} \frac{f(z)}{(z-\alpha)^{k+1}} dz$$

for some $k \in \mathbb{N}$.

Then $\frac{f^{(k)}(\alpha+h) - f^{(k)}(\alpha)}{h} = \frac{k!}{2\pi i} I_1$

where $I_1 = \int_{\mathscr{C}} \frac{f(z)(w^{k+1} - (w-h)^{k+1})}{hw^{k+1}(w-h)^{k+1}} dz$

and where $w = z - \alpha$.

Let $I_2 = (k+1) \int_{\mathscr{C}} \frac{f(z)}{w^{k+2}} dz$ then $|I_1 - I_2| = \left| \int_{\mathscr{C}} \frac{f(z)P(w,h)}{hw^{k+2}(w-h)^{k+1}} dz \right|$

where,

$$P(w,h) = w(w^{k+1} - (w-h)^{k+1}) - (k+1)h(w-h)^{k+1}$$

$$\Rightarrow \quad P(w,h) = w^{k+2} - (w+(k+1)h)(w-h)^{k+1}$$

$$\Rightarrow \quad P(w,h) = w^{k+2} - (w+(k+1)h)(w^{k+1} - (k+1)w^k h + \frac{k(k+1)}{2} w^{k-1}h^2$$
$$+ \text{ terms in higher powers of } h$$

$$\Rightarrow \quad P(w,h) = \tfrac{1}{2}(k+1)(k+2)h^2 w^k + \text{terms in higher powers of } h$$

using the binomial theorem. Then using Lemma 3.2 and (3.10),

$$|I_1 - I_2| \leqslant \frac{ML|P(w,h)|}{|h|d^{k+2}(d/2)^{k+1}}$$
$$\leqslant \frac{ML2^{k+1}}{d^{2k+3}}\left(\tfrac{1}{2}(k+1)(k+2)D^k + \text{terms in powers of } |h|\right)|h|$$

where $|z-\alpha| \leqslant D$ on $\mathscr{C}$. Then $I_1 \to I_2$ as $h \to 0$ and

$$f^{(k+1)}(\alpha) = \lim_{h\to 0} \frac{f^{(k)}(\alpha+h) - f^{(k)}(\alpha)}{h} = \frac{(k+1)k!}{2\pi i}\int_{\mathscr{C}} \frac{f(z)}{(z-\alpha)^{k+2}}\,dz$$

as required. ■

Note

What the bulk of this proof is really doing is showing that differentiation under the integral sign with respect to α is valid. Once this has been shown, the inductive proof of 3.13 is very straightforward (see Exercises 3.2).

Example 3.12

Let $\mathscr{C}$ be any simple closed contour enclosing 0. Then, since $f(z) = \sinh z$ is entire, it follows by 3.13 that

$$\int_{\mathscr{C}} \frac{\sinh z}{z^3}\,dz = \frac{2\pi i}{2!} f''(0) = \pi i \sinh 0 = 0$$

If a function f is analytic at a point α, there is an open neighbourhood of α in which f is analytic, so 3.13 holds in this neighbourhood. This gives the following immediate corollary.

Corollary 3.14. Derivatives of Analytic Functions

If a function f is analytic at a point, then its derivatives of all orders are also analytic at that point. □

Note

This result was used in Chapter 2 when discussing harmonic functions. But it does not hold for functions mapping $\mathbb{R}$ to $\mathbb{R}$. It is easy to construct examples of such functions which are differentiable only once at a point. For instance, consider $f: \mathbb{R} \to \mathbb{R}$ defined by $f(x) = x^{3/2}$, $x \geqslant 0$ and $f(x) = 0$, $x < 0$. This function is differentiable only once at 0.

The following result, which is the converse of the Cauchy–Goursat theorem, can be deduced easily from 3.5 and 3.14. This result is often useful in determining whether or not a given function is analytic in some region. For instance, it is a useful tool in the study of differentiability of functions defined by infinite series.

Theorem 3.15. Morera's Theorem

Let f be piecewise continuous in a simply connected region $\mathscr{R}$ and suppose that $\int_{\mathscr{C}} f(z)\,dz = 0$ for all closed contours contained within $\mathscr{R}$. Then f is analytic in $\mathscr{R}$. □

Proof

It follows by hypothesis that if α and β are any two points in $\mathscr{R}$ then, using the argument of the proof of 3.9, $\int_{\alpha}^{\beta} f(z)\,dz$ is independent of any contour joining α to β and contained within $\mathscr{R}$. Then by 3.5, f is the derivative of an analytic function in $\mathscr{R}$. Hence f is also analytic in $\mathscr{R}$ by 3.14. ■

Exercise **3.2.1** Show, using the definition, that if $\mathscr{C}$ is the circle $|z| = 1$, then

$$\int_{\mathscr{C}} \frac{\operatorname{Log}(z^{1/2})}{z}\,dz = 0$$

even though the integrand does not meet the conditions of Cauchy's theorem.

Exercise **3.2.2** Use Cauchy's theorem or Cauchy's integral formula to evaluate

$$\int_{\mathscr{C}} \frac{z^2}{z^2+9}\,dz$$

(i) where $\mathscr{C}$ is the circle $|z| = 2$

(ii) where $\mathscr{C}$ is the circle $|z - 2i| = 2$

(iii) where $\mathscr{C}$ is the triangle with vertices at 0, $1 - 4i$ and $-2 - 5i$

Exercise **3.2.3** Using Cauchy's theorem, evaluate $\int_{\mathscr{C}} e^z\,dz$, where $\mathscr{C}$ is the semicircle comprising the semicircular arc $|z| = a$, $-\pi/2 \leqslant \operatorname{Arg} z \leqslant \pi/2$ and the line segment $|\operatorname{Im} z| \leqslant a, \operatorname{Re} z = 0$. Verify the result by using the fundamental theorem of calculus. Use the result to evaluate the real integral

$$\int_0^{\pi/2} e^{a\cos\theta}\cos(a\sin\theta + \theta)\,d\theta$$

Exercise **3.2.4**

(i) Evaluate $\int_{\mathscr{C}} e^{-z^2}\,dz$ where $\mathscr{C}$ is any simple closed contour.

(ii) Given that $\int_0^\infty e^{-x^2}\,dx = \sqrt{\pi}/2$, use the result of (i) with $\mathscr{C}$ as the rectangle with vertices $0, a, a+ib$ and ib, to evaluate $\int_0^\infty e^{-x^2}\cos 2bx\,dx$, and to show that

$$\int_0^\infty e^{-x^2}\sin 2bx\,dx = e^{-b^2}\int_0^b e^{x^2}\,dx$$

(iii) Use the result of (i) with $\mathscr{C}$ as the boundary of the sector given by $0 \leqslant |z| \leqslant R$, $0 \leqslant \operatorname{Arg} z \leqslant \pi/4$, to deduce that

$$\int_0^\infty \cos(x^2)\,dx = \int_0^\infty \sin(x^2)\,dx = \frac{1}{2}\sqrt{\frac{\pi}{2}} \qquad \text{(Fresnel's integrals)}$$

You may assume that the given integrals converge and that

$$\lim_{R\to\infty}\int_0^{\pi/4} Re^{-R^2\cos 2\theta}\,d\theta = 0$$

Exercise **3.2.5** Evaluate $\int_{\mathscr{C}} z^{-1/2}dz$, where $\mathscr{C}$ is the circle $|z| = R$, using the fundamental theorem of calculus. Hence show that the deformation result, Lemma 3.10, does not hold in this case.

Exercise **3.2.6** Let a simple closed contour $\mathscr{C}$ enclose just two singular points of a function f at α and β. Let $\mathscr{C}'$ and $\mathscr{C}''$ be disjoint circles lying totally inside $\mathscr{C}$, centred at α and β respectively. Prove that $\int_{\mathscr{C}} f(z)\,dz = \int_{\mathscr{C}'} f(z)\,dz + \int_{\mathscr{C}''} f(z)\,dz$.

Exercise **3.2.7** Let $\mathscr{C}$ be a simple closed contour enclosing 0 and 1. Use the result of Exercise 3.2.6, Cauchy's theorem and the definition only, to evaluate $\int_{\mathscr{C}}((2z-1)/z^2 - z))dz$.

Exercise **3.2.8** Use Cauchy's integral formula to prove that

$$\int_{\mathscr{C}} \frac{e^{zt}}{z^2+4}\,dz = \pi i \sin 2t \qquad (t > 0)$$

where $\mathscr{C}$ is the circle $|z| = 3$.

Exercise **3.2.9** Evaluate $\int_{\mathscr{C}}((\cosh nz)/z)dz$ where $n \in \mathbb{N}$ and $\mathscr{C}$ is the circle $|z| = 1$, using Cauchy's integral formula. Hence evaluate $\int_0^{\pi/2}\cosh(n\cos\theta)\cos(n\sin\theta)\,d\theta$.

Exercise **3.2.10** Evaluate $\int_{\mathscr{C}}(e^{iz^2}/z)dz$ where $\mathscr{C}$ is the circle $|z| = 1$. Hence evaluate $\int_0^{2\pi} e^{-\sin 2\theta}\cos(\cos 2\theta)\,d\theta$.

Exercise **3.2.11** Use Cauchy's integral formula for derivatives to evaluate $\int_{\mathscr{C}}(e^{2z}/z^2)\,dz$, where $\mathscr{C}$ is the circle $|z| = 1$. Hence evaluate $\int_0^{\pi} e^{2\cos\theta}\cos(2\sin\theta - \theta)\,d\theta$.

Exercise **3.2.12** Let f be analytic in a simply connected region containing a simple closed contour $\mathscr{C}$. Assuming Cauchy's integral formula and that differentiation under the integral sign with respect to α is valid, prove Cauchy's integral formula for derivatives by induction.

Consequences of Cauchy's Integral Formulae

We now prove several important results which can be derived from Cauchy's integral formulae. The first result states that the only bounded entire functions are constant.

Theorem 3.16. Liouville's Theorem

If f is entire and $|f(z)| \leqslant M$ for some positive real constant M, for all $z \in \mathbb{C}$, then f is constant. □

Proof

Let $\mathscr{C}$ be a circle of radius r, centred at α. Since f is entire and bounded by M, it follows by 3.13 and 3.2 that

$$|f'(\alpha)| = \frac{1}{2\pi}\left|\int_{\mathscr{C}} \frac{f(z)}{(z-\alpha)^2}\,dz\right| \leqslant \frac{1}{2\pi}\left(\frac{M}{r^2}\right)2\pi r = \frac{M}{r}$$

for any chosen r. Hence $|f'(\alpha)| = 0$, so $f'(\alpha) = 0$ for any $\alpha \in \mathbb{C}$. It follows by the results of Chapter 2 that f is a constant. ■

Note

Liouville's theorem does not hold for functions mapping $\mathbb{R}$ to $\mathbb{R}$. For instance, $f\colon \mathbb{R} \to \mathbb{R}$ given by $f(x) = \sin x$ is entire and $|\sin x| \leqslant 1$ for all $x \in \mathbb{R}$. On the other hand, $|\sin z|$ is not bounded for all $z \in \mathbb{C}$.

Liouville's theorem can be used to give an elegant proof of the following famous result.

Theorem 3.17. The Fundamental Theorem of Algebra

Every polynomial equation

$$P(z) = a_0 + a_1 z + \ldots + a_n z^n = 0$$

with degree n, $n \in \mathbb{N}$, has at least one root. It follows that $P(z) = 0$ has exactly n roots. □

Proof

Suppose that $P(z) = 0$ has no roots. Then $1/P(z)$ is entire and $|1/P(z)| \to 0$ as $|z| \to \infty$ so that $1/P(z)$ is bounded for all z. Hence $1/P(z)$ is a constant by 3.16, and this is a contradiction. Thus $P(z) = 0$ has at least one root. Suppose this root is $z = \alpha$. If $n > 1$ then $P(z) = P(z) - P(\alpha) = (z - \alpha)Q(z)$ say, where Q is a polynomial of degree $n - 1$, so $Q(z) = 0$ has at least one root by the above. Continuing this process shows that $P(z) = 0$ has exactly n roots, as required. ■

Another two very useful consequences of Cauchy's integral formula are the so-called maximum and minimum modulus theorems.

Theorem 3.18. The Maximum Modulus Theorem

If f is analytic inside a simple closed contour $\mathscr{C}$ and continuous on $\mathscr{C}$, and is not identically equal to a constant, then the maximum value of $|f(z)|$, inside and on $\mathscr{C}$, occurs on $\mathscr{C}$. □

Proof

Step 1

By hypothesis, f and hence $|f|$ is continuous inside and on $\mathscr{C}$. Suppose that the maximum value of $|f(z)|$ is attained at a point α inside $\mathscr{C}$ and $|f(\alpha)| = M$. Let $\mathscr{C}'$ be a circle lying inside $\mathscr{C}$, centred at α. Then, by hypothesis, there exists a point β inside $\mathscr{C}'$ such that $|f(\beta)| < M$, so that $|f(\beta)| = M - \varepsilon$ for some $\varepsilon > 0$, with $|f|$ continuous at β. Then given $\varepsilon/2 > 0$ there exists $\delta > 0$ such that

$$|z - \beta| < \delta \quad \Rightarrow \quad \big||f(z)| - |f(\beta)|\big| < \varepsilon/2$$

$$\Rightarrow \quad |f(z)| < \varepsilon/2 + |f(\beta)| = M - \varepsilon/2$$

Step 2

Suppose that $\mathscr{C}_1$ is a circle centred at β with radius $\rho < \delta$, chosen so that $\mathscr{C}_1$ is contained within $\mathscr{C}'$ as shown in Fig. 3.7. It then follows by step 1 that

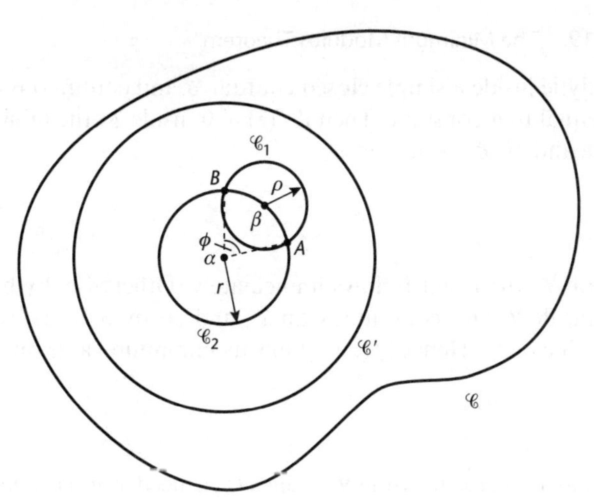

Figure 3.7

$|f(z)| < M - \varepsilon/2$ for all z inside $\mathscr{C}_1$. Now consider the circle $\mathscr{C}_2$ of radius r, centred at α, and passing through β as shown in Fig. 3.7. On the arc AB of $\mathscr{C}_2$, which is inside $\mathscr{C}_1$, $|f(z)| < M - \varepsilon/2$, and on the arc BA of $\mathscr{C}_2$, $|f(z)| \leqslant M$. Let ϕ be the angle subtended at α by the arc AB, as shown in Fig. 3.7.

Step 3

The circle $\mathscr{C}_2$ is parametrised by $z = \alpha + re^{i\theta}$, $0 \leqslant \theta \leqslant 2\pi$ say, where the point A is given by $\theta = 0$, without loss of generality. Then, since f is analytic at α, it follows by Cauchy's integral formula and the definition that

$$f(\alpha) = \frac{1}{2\pi i}\int_{\mathscr{C}_2} \frac{f(z)}{z-\alpha}\,dz = \frac{1}{2\pi}\int_0^{2\pi} f(\alpha + re^{i\theta})\,d\theta$$

$$\Rightarrow \quad f(\alpha) = \frac{1}{2\pi}\int_0^{\phi} f(\alpha + re^{i\theta})\,d\theta + \frac{1}{2\pi}\int_{\phi}^{2\pi} f(\alpha + re^{i\theta})\,d\theta$$

$$\Rightarrow \quad |f(\alpha)| \leqslant \frac{1}{2\pi}\int_0^{\phi} (M - \varepsilon/2)\,d\theta + \frac{1}{2\pi}\int_{\phi}^{2\pi} M\,d\theta = M - \frac{\phi\varepsilon}{4\pi}$$

Hence $|f(\alpha)| = M \leqslant M - \phi\varepsilon/4\pi$, which is a contradiction, so $|f(z)|$ must attain its maximum value on $\mathscr{C}$. ■

Note

Since f is continuous on the closed set consisting of $\mathscr{C}$ and its inside, it does have a maximum value on this domain, from the comments in Chapter 2 (see Theorem 4.5).

Theorem 3.19. The Minimum Modulus Theorem

Let f be analytic inside a simple closed contour $\mathscr{C}$ and continuous on $\mathscr{C}$, and not identically equal to a constant. Then if $f(z) \neq 0$ inside $\mathscr{C}$, the minimum value of $|f(z)|$, inside and on $\mathscr{C}$, occurs on $\mathscr{C}$. □

Proof

If $f(z) = 0$ on $\mathscr{C}$, the result follows immediately. Otherwise, by hypothesis, $1/f$ is analytic inside $\mathscr{C}$ and continuous on $\mathscr{C}$, and so by 3.18, $|1/f(z)|$ attains its maximum value on $\mathscr{C}$. Hence $|f(z)|$ attains its minimum value on $\mathscr{C}$. ■

Note

It is clear that if $f(z) = 0$ inside $\mathscr{C}$, then $|f(z)|$ need not assume its minimum value, i.e. 0, on $\mathscr{C}$. Consider $f(z) = z$ and let $\mathscr{C}$ be the circle $|z| = 1$ for instance.

The Location of Roots of Equations

Another important application of the preceding results is a result concerning the number of poles and zeros of a particular type of function inside a simple closed contour, and the related Rouché's theorem, both very useful in locating the position of roots of equations in the complex plane. We first need the following generalisation of Lemma 3.10.

Lemma 3.20. A Generalised Deformation Result

Let $f: A \subseteq \mathbb{C} \to \mathbb{C}$ be analytic inside and on a simple closed contour $\mathscr{C}$, except at a finite number of singular points $z_1, z_2, \ldots, z_m$, lying inside $\mathscr{C}$. Then if $\mathscr{C}_k$ is a circle centred at z_k lying totally inside $\mathscr{C}$, for each $k = 1, \ldots, m$, such that none of the circles intersect, then

$$\int_{\mathscr{C}} f(z)\,dz = \sum_{k=1}^{m} \int_{\mathscr{C}_k} f(z)\,dz \qquad \square$$

Proof

Note that, since there are only a finite number of singular points of f inside $\mathscr{C}$, it is possible to construct the circles $\mathscr{C}_k$ with radii chosen so that none of them intersect. Connect each circle to $\mathscr{C}$ by two lines and create simple closed contours Γ and Γ_k for each k, which do not enclose $z_1, \ldots, z_m$, as shown in Fig. 3.8.

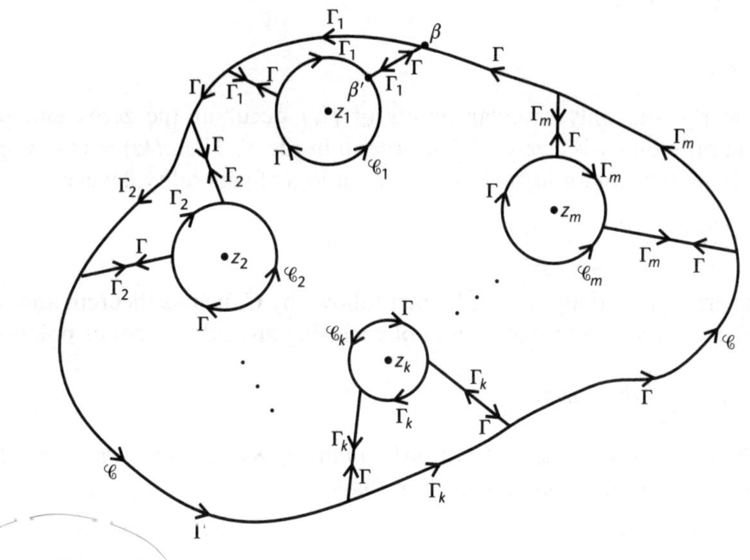

Figure 3.8

It follows by hypothesis and Cauchy's theorem that

$$\int_{\Gamma} f(z)\,dz = \int_{\Gamma_k} f(z)\,dz = 0 \qquad (k = 1, \ldots, m)$$

Also, by 3.1, $\int_{\beta\,\Gamma_1}^{\beta'} f(z)\,dz = -\int_{\beta\,\Gamma}^{\beta'} f(z)\,dz$ and similarly for the other line segments, so that, by 3.1 again,

$$\int_{\mathscr{C}} f(z)\,dz - \sum_{k=1}^{m} \int_{\mathscr{C}_k} f(z)\,dz = \int_{\Gamma} f(z)\,dz + \sum_{k=1}^{m} \int_{\Gamma_k} f(z)\,dz = 0$$

■

Definition

A function which is analytic on a region $\mathscr{R}$, except possibly for poles, is **meromorphic** on $\mathscr{R}$.

Any function which has no branch points or essential singular points is meromorphic on $\mathbb{C}$. For example, algebraic fractions, the hyperbolic and trigonometric functions are all meromorphic on $\mathbb{C}$. An important result which concerns meromorphic functions is the following theorem.

Theorem 3.21. Poles and Zeros of Meromorphic Functions

Let f be analytic inside and on a simple closed contour $\mathscr{C}$, except possibly for P poles inside $\mathscr{C}$. Let $f(z) \neq 0$ on $\mathscr{C}$ and let f have Z zeros inside $\mathscr{C}$. (A pole or zero of order n is counted n times.) Then

$$\frac{1}{2\pi i} \int_{\mathscr{C}} \frac{f'(z)}{f(z)}\,dz = Z - P$$

□

Proof

Clearly, the only singular points of f'/f occur at the zeros and poles of f. Suppose that α is a zero of f of order n inside $\mathscr{C}$. Then $f(z) = (z - \alpha)^n g(z)$ where g is analytic and non-zero at α, without loss of generality. Hence

$$\frac{f'(z)}{f(z)} = \frac{n}{z - \alpha} + \frac{g'(z)}{g(z)}$$

where g'/g is analytic at α. It then follows by Cauchy's theorem and 3.11 that if $\mathscr{C}_\alpha$ is a circle, centred at α and not enclosing any other zeros or poles of f,

$$\int_{\mathscr{C}_\alpha} \frac{f'(z)}{f(z)}\,dz = 2n\pi i \qquad (3.11)$$

Similarly, if β is a pole of f of order m inside $\mathscr{C}$ then $f(z) = h(z)/(z - \beta)^m$, where h is analytic and non-zero at β, so that

$$\frac{f'(z)}{f(z)} = \frac{h'(z)}{h(z)} - \frac{m}{z - \beta}$$

where h'/h is analytic at β. Then if $\mathscr{C}_\beta$ is a circle, centred at β and not enclosing any other zeros or poles of f,

$$\int_{\mathscr{C}_\beta} \frac{f'(z)}{f(z)}\,dz = -2m\pi i \tag{3.12}$$

The result then follows directly from (3.11), (3.12) and Lemma 3.20. ■

Example 3.13

Let $f(z) = (z^2+1)/z^3$. Then f has simple zeros at $\pm i$ and a pole of order 3 at 0. The number of zeros Z of f inside the circle $|z| = 2$, say, is 2 and the number of poles P of f inside the same circle is 3. Hence, by 3.21,

$$\int_{\mathscr{C}} \frac{f'(z)}{f(z)}\,dz = 2\pi i(Z-P) \quad \Rightarrow \quad \int_{\mathscr{C}} \frac{z^2+3}{z(z^2+1)}\,dz = 2\pi i$$

where $\mathscr{C}$ is the circle $|z| = 2$. This result can be checked by using partial fractions and Cauchy's integral formula:

$$\int_{\mathscr{C}} \frac{z^2+3}{z(z^2+1)}\,dz = \int_{\mathscr{C}} \frac{3}{z}\,dz - \int_{\mathscr{C}} \frac{dz}{z+i} - \int_{\mathscr{C}} \frac{dz}{z-i} = 6\pi i - 4\pi i = 2\pi i$$

Theorem 3.21 can be used to locate roots of equations in each of the four quadrants of the complex plane, as the following example demonstrates.

Example 3.14 The Principle of the Argument

It follows by 3.21 that if f is analytic inside and on a simple closed contour $\mathscr{C}$ and $f(z) \neq 0$ on $\mathscr{C}$, then the number of zeros of f inside $\mathscr{C}$ is

$$Z = \frac{1}{2\pi i}\int_{\mathscr{C}} \frac{f'(z)}{f(z)}\,dz$$

Also, for any particular branch of log, that is for a specific $k \in \mathbb{Z}$,

$$\frac{d}{dz}(\log(f(z))) = \frac{d}{dz}(\mathrm{Log}|f(z)| + i(\mathrm{Arg}(f(z)) + 2k\pi))) = \frac{f'(z)}{f(z)}$$

by Theorem 2.3(b)(iii), except along the associated branch cut. It follows by Theorem 3.3 that

$$Z = \frac{1}{2\pi}\Delta_{\mathscr{C}} \arg(f(z)) \tag{3.13}$$

where $\Delta_{\mathscr{C}} \arg(f(z))$ is the total change in any continuously varying argument of f around the contour $\mathscr{C}$. The formal proof of (3.13), usually known as the **principle of the argument**, is omitted.

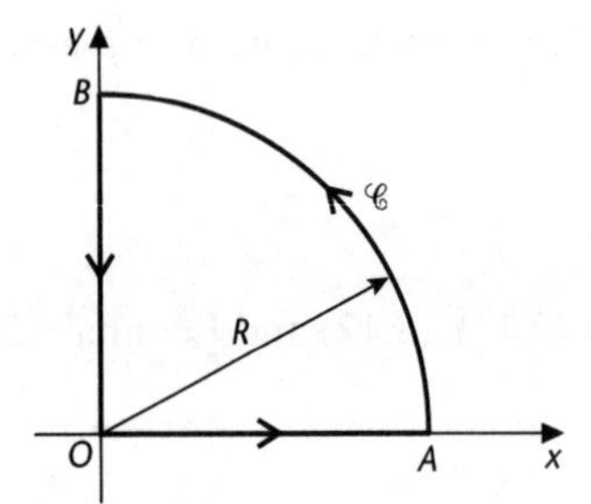

Figure 3.9

Now consider the equation $f(z) = z^5 - z + 16 = 0$. It is easily checked that this equation has no purely imaginary roots. Also for $x \in \mathbb{R}$, $f(x) = x^5 - x + 16 > 0$ for $x \geqslant 0$, (since $x < x^5 + 16$ for $x < 16$ and $x(x^4 - 1) + 16 > 0$ for $x \geqslant 1$). Function f has a local maximum when $x < 0$ with $5x^4 = 1$ and since $F(-1) > 0$ and $F(-2) < 0$, it follows (by the intermediate value theorem) that the given equation has exactly one root on the negative real axis.

We now use the principle of the argument to show that the given equation has exactly one root in the first quadrant. Note that f is entire. Consider the closed contour OAB, consisting of two line segments and the arc of a circle of radius R, shown in Fig. 3.9. Choose $\arg(f(z)) = 0$ on OA. Then on the arc AB, $z = Re^{i\theta}$, $0 \leqslant \theta \leqslant \pi/2$, so that $f(z) = R^5e^{5i\theta} - Re^{i\theta} + 16$. Then as $R \to \infty$, $\Delta_{AB}\arg(f(z)) \to 5\pi/2$. On the line segment BO, $f(z) = i(y^5 - y) + 16$ $\Rightarrow \arg(f(z)) = \tan^{-1}(y^4(y-1)/16)$. Note that $g(y) = y^4(y-1)/16 = 0$ when $y = 0$ or 1 and that $g(y) \to \infty$ as $y \to \infty$. Hence as $R \to \infty$, $\Delta_{BO}\arg(f(z)) \to -\pi/2$. Then $\Delta_{\mathscr{C}}\arg(f(z)) \to 2\pi$ as $R \to \infty$ and it follows by (3.13) that f has exactly one zero in the first quadrant. Since f must have five zeros altogether (by the fundamental theorem of algebra) and its complex zeros occur in conjugate pairs, f has exactly one zero in each quadrant.

Theorem 3.21 also leads to the following theorem, which gives another method of locating roots of equations.

Theorem 3.22. Rouché's Theorem

If f and g are analytic inside and on a simple closed contour $\mathscr{C}$ and $|g(z)| < |f(z)|$ on $\mathscr{C}$, then f and $f + g$ have the same number of zeros inside $\mathscr{C}$. □

Proof

Step 1

Let the function F be defined inside and on $\mathscr{C}$ by $F(z, t) = f(z) + tg(z)$ where t is real with $t \in [0, 1]$. Since f and g are analytic inside and on $\mathscr{C}$, F has no poles inside or on $\mathscr{C}$, for fixed t. Also, since $|g(z)| < |f(z)|$ on $\mathscr{C}$, F does not have a zero at any point on $\mathscr{C}$. (Note that at a possible zero of f on $\mathscr{C}$, $f(z) = -tg(z) \Rightarrow |f(z)| \leqslant |g(z)|$ and this is a contradiction.) It follows by Theorem 3.21 that if

$$Z(t) = \frac{1}{2\pi i}\int_{\mathscr{C}} \frac{f'(z) + tg'(z)}{f(z) + tg(z)}\,dz$$

then $Z(t)$ is the number of zeros of F inside $\mathscr{C}$.

Step 2

We now show that Z is a continuous function of t. For $s \in [0, 1]$,

$$|Z(t) - Z(s)| = \frac{|t-s|}{2\pi}\left|\int_{\mathscr{C}} \frac{(f(z)g'(z) - f'(z)g(z))}{F(z,t)F(z,s)}\,dz\right| \tag{3.14}$$

It follows by the triangle inequality that

$$|F(z,t)| \geqslant |f(z)| - t|g(z)| \geqslant |f(z)| - |g(z)| > 0 \quad \text{on } \mathscr{C}$$

$$\text{and} \quad |F(z,s)| \geqslant |f(z)| - |g(z)| > 0 \quad \text{on } \mathscr{C}$$

Since f and g and their derivatives are continuous on $\mathscr{C}$, and $\mathscr{C}$ is a compact subset of $\mathbb{C}$, there exists a constant M such that

$$\left|\frac{f(z)g'(z) - f'(z)g(z)}{(|f(z)| - |g(z)|)^2}\right| \leqslant M \quad \text{on } \mathscr{C}$$

It then follows by the above, Lemma 3.2 and (3.14) that

$$|Z(t) - Z(s)| \leqslant \frac{ML}{2\pi}|t-s|$$

where L is the length of $\mathscr{C}$, so that Z is a continuous function of t, as required. Since Z is integer-valued it follows (by the intermediate value theorem) that $Z(t)$ is constant for $t \in [0, 1]$. Hence $Z(0) = Z(1)$, where $Z(0)$ is the number of zeros of f and $Z(1)$ is the number of zeros of $f + g$. ■

Rouché's theorem is particularly useful for locating roots of equations in circular and annular regions, as the following example demonstrates.

Example 3.15

Use Rouché's theorem to show that the four roots of $z^4 + 4(1+i)z + 1 = 0$ lie inside the circle $|z| = 2$, three of which lie in the annular region $1 \leqslant |z| < 2$.

Solution

Take $f(z) = z^4$ and $g(z) = 4(1+i)z + 1$, so that f and g are entire. On the circle $\mathscr{C}$ with equation $|z| = 2$,

$$|f(z)| = |z^4| = 16 \text{ and } |g(z)| \leqslant 4|1+i||z| + 1 = 4\sqrt{2}\cdot 2 + 1 < 16$$

so that $|g(z)| < |f(z)|$ on $\mathscr{C}$. Also f has one zero, a zero of order 4 at 0, and this point lies inside $\mathscr{C}$. Hence $f+g$ has exactly four zeros inside the same curve.

Now let $F(z) = 4(1+i)z$ and $G(z) = z^4 + 1$, so that, once again, F and G are entire. On the circle $\mathscr{C}'$ with equation $|z| = 1$

$$|F(z)| = 4\sqrt{2}|z| = 4\sqrt{2} \quad \text{and} \quad |G(z)| \leqslant |z|^4 + 1 = 2 < 4\sqrt{2}$$

so that $|G(z)| < |F(z)|$ on $\mathscr{C}'$. Clearly F has a simple zero at 0, which lies inside $\mathscr{C}'$, so $F+G$ has exactly one zero inside $\mathscr{C}'$, as required.

Exercise **3.3.1** Use Cauchy's integral formula for derivatives to prove the following results:

(i) If f is analytic in a simply connected region containing a circle with centre α and radius r, then

$$f^{(n)}(\alpha) = \frac{n!}{2\pi r^n}\int_0^{2\pi} e^{-ni\theta} f(\alpha + re^{i\theta})d\theta$$

(ii) If f is analytic with $|f(z)| \leqslant M$ in a simply connected region containing the circle $|z| = r$, then

$$|\alpha| < r \quad \Rightarrow \quad |f^{(n)}(\alpha)| \leqslant \frac{Mrn!}{(r - |\alpha|)^{n+1}}$$

Exercise **3.3.2** Use Cauchy's theorem and Cauchy's integral formula to prove that if f is analytic in a simply connected region containing the circle $\mathscr{C}$ with equation $|z| = r$, and α is any non-zero point inside $\mathscr{C}$, then

$$f(\alpha) = \frac{1}{2\pi i}\int_{\mathscr{C}}\left(\frac{1}{z-\alpha} - \frac{1}{z - r^2/\overline{\alpha}}\right) f(z)\,dz$$

Deduce that if $0 < \rho < r$, then

$$f(\rho e^{i\phi}) = \frac{1}{2\pi}\int_0^{2\pi} \frac{(r^2 - \rho^2)}{r^2 - 2r\rho\cos(\theta - \phi) + \rho^2} f(re^{i\theta})\,d\theta$$

(This result is known as Poisson's integral formula for a circle.)

Exercise **3.3.3** Let f be entire and $|f(z)| \geqslant m > 0$ for all $z \in \mathbb{C}$, where m is a constant. Use Liouville's theorem to prove that f is a constant.

Exercise **3.3.4** Use Liouville's theorem to prove that if

$$f(z + 2\pi) = f(z + 2\pi i) = f(z) \qquad \text{(for all } z) \qquad (*)$$

and f is entire, then f is a constant. (A function satisfying condition (*) is a **doubly periodic function**.)

Exercise **3.3.5*** Let f be an entire function which is not identically zero. Suppose there exist two positive real numbers r and M and a non-negative integer n such that

$$|f(z)| < M|z|^n \qquad (|z| > r)$$

Use Liouville's theorem and induction to prove that f is a polynomial of at most degree n.

Exercise **3.3.6*** Let f be a function that is analytic inside the unit circle $|z| = 1$ and which satisfies

$$f(0) = 0 \quad \text{and} \quad |f(z)| \leqslant 1 \qquad (0 < |z| < 1)$$

Use the maximum modulus theorem to prove **Schwarz's lemma**:

$$|f'(0)| \leqslant 1 \quad \text{and} \quad |f(z)| \leqslant |z| \qquad (0 < |z| < 1)$$

(*Hint*: Consider the function g defined by $g(z) = f(z)/z$ for $0 < |z| < 1$, with $g(0) = f'(0)$. Then g is analytic inside the unit circle.)

Exercise **3.3.7** Use Theorem 3.21 with $f(z) = (z-4)^3/(1+z^2)$ to evaluate

$$\int_{\mathscr{C}} \frac{z^2 + 8z + 3}{(z-4)(1+z^2)}\,dz$$

where $\mathscr{C}$ is the circle $|z| = 5$.

Exercise **3.3.8** Use the principle of the argument to show that $z^8 + 4z^3 + 5z + 3 = 0$ has exactly two roots in the first quadrant.

Exercise **3.3.9** Use the principle of the argument to investigate the position of the roots of the equation $z^7 + 3z + 1 = 0$, determining how many roots lie on each axis and in each quadrant of the complex plane.

Exercise **3.3.10** Use Rouché's theorem to prove that the equation $e^{2iz} = 12z^n$, $n \in \mathbb{N}$, has n roots inside the circle $|z| = 1$.

Exercise **3.3.11** Use Rouché's theorem to show that the equation $z^6 + 7z + 1 = 0$ has six roots inside the circle $|z| = 2$, five of which lie in the annular region $1 \leqslant |z| < 2$.

Exercise **3.3.12** Determine the number of roots of the equation $8z^3 + 4z^2 + 2z - 3 = 0$ inside the circle $|z| = 1$. (*Hint*: Multiply the given equation by $2z - 1$ and apply Rouché's theorem.)

Exercise **3.3.13** Determine all the roots of $z^4 + 4 = 0$. Prove that the equation

$$(z^2 - 1)^2 + 2(z-1)^2 + 1 = 0$$

has exactly one root in each quadrant, by simplifying the equation and using Rouché's theorem.

Exercise **3.3.14** Use Rouché's theorem to show that $z^5 - z + 16 = 0$ has exactly one root in the first quadrant, which lies inside the circle $|z| = 2$. (Compare with Example 3.14.)

The Cauchy–Goursat Theorem

We end this chapter with the promised proof of Theorem 3.7. The easiest way to proceed with the proof is to prove that the result holds for a closed polygon first of all. The simplest closed polygon is a triangle, so that will be our starting point.

Theorem 3.23. Cauchy's Theorem for a Triangle

Let f be analytic in a simply connected region containing a triangle $\triangle$. Then

$$\int_{\triangle} f(z)\,dz = 0 \qquad \square$$

Proof

The basic idea is to approximate the integrand in a small enough region by an integrand which has an indefinite integral, so that the result follows by the fundamental theorem of calculus.

Step 1

Given the triangle $\triangle$, form four new triangles T_1, T_2, T_3 and T_4 by connecting the midpoints of the sides of $\triangle$ as shown in Fig. 3.10. Let $I = \int_{\triangle} f(z)\,dz$ and $J_k = \int_{T_k} f(z)\,dz$, $k = 1, \ldots, 4$. Then

$$I = J_1 + J_2 + J_3 + J_4 \quad \Rightarrow \quad |I| \leqslant |J_1| + |J_2| + |J_3| + |J_4|$$

by 3.1 (since there are equal and opposite contributions to the integrals J_k along the line segments M_1M_2, M_1M_3 and M_2M_3). Let I_1 be the the integral among the J_k with maximum modulus. It then follows by the above inequality that $|I| \leqslant 4|I_1|$. Relabel the associated triangle as 'new' triangle $\triangle_1$. By the same process, $\triangle_1$ can be divided into four congruent triangles such that the contour integral I_2 of f along one of these triangles, $\triangle_2$ say, satisfies $|I_1| \leqslant 4|I_2|$.

Continuing this process, after the nth stage, we obtain a sequence of triangles $\triangle$, $\triangle_1$, $\triangle_2, \ldots, \triangle_n$ for which the corresponding contour integrals I, I_1, $I_2, \ldots, I_n$ of f along these triangles satisfy

$$|I| \leqslant 4|I_1| \leqslant 4^2|I_2| \leqslant \ldots \leqslant 4^n|I_n| \qquad (3.15)$$

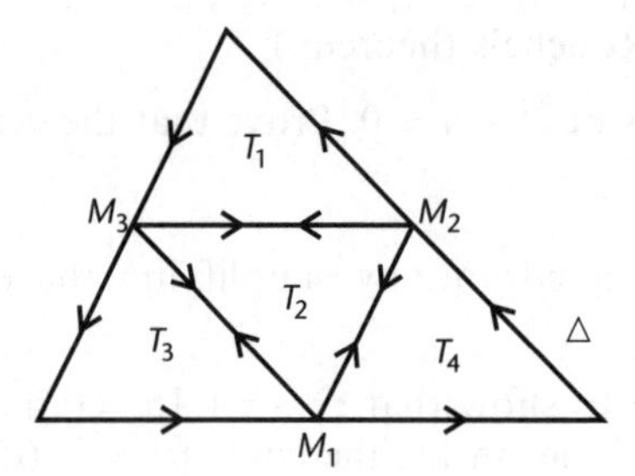

Figure 3.10

for any natural number n. The process also yields a sequence of nested triangular closed sets

$$\triangle^* \supseteq \triangle_1^* \supseteq \triangle_2^* \supseteq \ldots \supseteq \triangle_n^*$$

where $\triangle^*$ consists of $\triangle$ and its inside and similarly for $\triangle_k^*$.

Let L be the length of $\triangle$ and L_n be the length of $\triangle_n$ for each n. It follows by the way in which the triangles are constructed that $L_n = L/2^n$, so that $L_n \to 0$ as $n \to \infty$. Hence there is exactly one point, α say, which belongs to every $\triangle_n^*$.

Step 2

By hypothesis, f is differentiable at this point α, so that

$$f(z) = f(\alpha) + (z - \alpha)f'(\alpha) + \varepsilon(z, \alpha)(z - \alpha)$$

$$\text{where} \quad \varepsilon(z, \alpha) = \frac{f(z) - f(\alpha)}{z - \alpha} - f'(\alpha) \quad \Rightarrow \quad \lim_{z \to \alpha} \varepsilon(z, \alpha) = 0$$

It then follows that

$$I_n = \int_{\triangle_n} (f(\alpha) + (z - \alpha)f'(\alpha))\, dz + \int_{\triangle_n} \varepsilon(z, \alpha)(z - \alpha)\, dz$$

The integrand of the first integral is a linear function, so it has an indefinite integral. Hence, by Corollary 3.4, the integral is zero, so

$$I_n = \int_{\triangle_n} \varepsilon(z, \alpha)(z - \alpha)\, dz$$

Since $\lim_{z \to \alpha} \varepsilon(z, \alpha) = 0$, it follows that, given any $\varepsilon_1 > 0$, there is a $\delta > 0$ such that $|z - \alpha| < \delta \Rightarrow |\varepsilon(z, \alpha)| < \varepsilon_1$. Also, $\triangle_n$ lies within the circular region $|z - \alpha| < L_n$ and since $\lim_{n \to \infty} L_n = 0$, there is an $N \in \mathbb{N}$ such that $L_n < \delta$ for all $n > N$. Then by Lemma 3.2,

$$|I_n| \leqslant (\varepsilon_1 L_n) L_n \quad \Rightarrow \quad |I_n| \leqslant \varepsilon_1 \frac{L^2}{4^n} \qquad (n > N)$$

Hence from the chain of inequalities, (3.15), it follows that $|I| \leqslant L^2 \varepsilon_1$ for any $\varepsilon_1 > 0$. Hence $I = 0$ as required. ■

Note

Cauchy's theorem for a rectangle follows in almost the same way; only very minor modifications are needed in the proof.

The next step in proving the Cauchy–Goursat theorem is to show that it holds for any closed polygon, not just a triangle. Essentially this is true because any polygon can be split up into a finite number of triangles.

Theorem 3.24. Cauchy's Theorem for a Polygon

Let f be analytic in a simply connected region containing a simple closed polygon $\mathscr{P}$. Then $\int_{\mathscr{P}} f(z)\,dz = 0$. □

Part Proof

It can be shown that any simple closed polygon can be **triangulated**; that is the boundary and inside of the polygon is the union of triangles $\triangle_1, \ldots, \triangle_m$ say, and their insides, any two of which have either a vertex, a whole side or no point in common, such that each side of a triangle is either a side of one other triangle or a side of the polygon. Such a triangulation is shown in Fig. 3.11.

As in the case of the Jordan curve theorem, we shall not prove this intuitive result; the formal proof is quite lengthy. (For the details, see for example, J. W. Dettman, *Applied Complex Variables*, Macmillan, 1965.)

Since the net contribution to the integral along the line segments forming the triangles in the triangulation of $\mathscr{P}$ is zero by 3.1,

$$\int_{\mathscr{P}} f(z)\,dz = \sum_{k=1}^{m} \int_{\triangle_k} f(z)\,dz = 0 \qquad \text{(by 3.23)}$$

as required. ■

Note that the above result is easily extended to any closed polygon. The last step in the proof of the Cauchy–Goursat theorem is to show that the integral of a continuous function along any simple smooth arc can be approximated by an integral along a simple polygon.

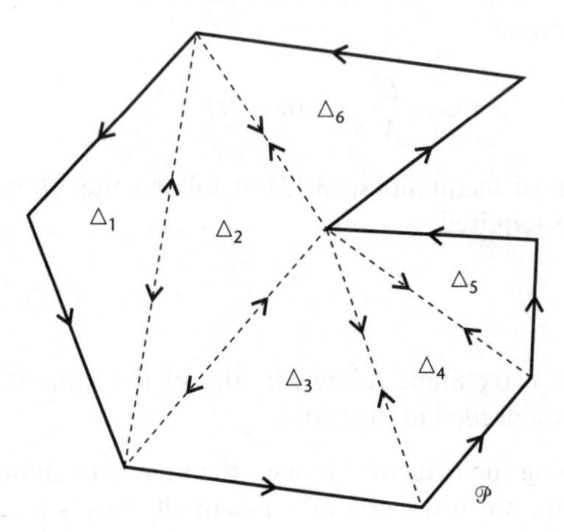

Figure 3.11

Theorem 3.25. Approximation by Polygons

Let $\mathscr{C}$ be a simple smooth arc and f a continuous function in some simply connected region $\mathscr{R}$ containing $\mathscr{C}$. Then given any $\varepsilon > 0$, there exists a simple polygon $\mathscr{P}$ contained in $\mathscr{R}$ such that

$$\left| \int_{\mathscr{C}} f(z)\,dz - \int_{\mathscr{P}} f(z)\,dz \right| < \varepsilon \qquad \square$$

Proof

Step 1

Note that $\mathscr{C}$ may be covered by a finite number of circles, centred at points on $\mathscr{C}$, which lie totally within $\mathscr{R}$, as indicated in Fig. 3.12. Let L denote the length of $\mathscr{C}$. Since f is continuous in $\mathscr{R}$, by choosing the largest radius of all the circles small enough, it is possible to construct such a system of circles which, together with their insides, form a closed set S with the property that given any $\varepsilon > 0$ there exists $\delta > 0$ such that for any points z and w in S,

$$|z - w| < \delta \quad \Rightarrow \quad |f(z) - f(w)| < \varepsilon/2L \tag{3.16}$$

Step 2

Let $\mathscr{C}$ be parametrised by $z = z(t),\ t \in [a, b]$. Recall that $\int_{\mathscr{C}} f(z)\,dz$ can be expressed in terms of real line integrals as in (3.9). For any particular subdivision $a = t_0, t_1, \ldots, t_n = b$ of $[a, b]$, let $S_n = \sum_{k=1}^{n} f(z_{k-1})(z_k - z_{k-1})$ where $z_k = z(t_k)$. It then follows from the definition of a Riemann line integral

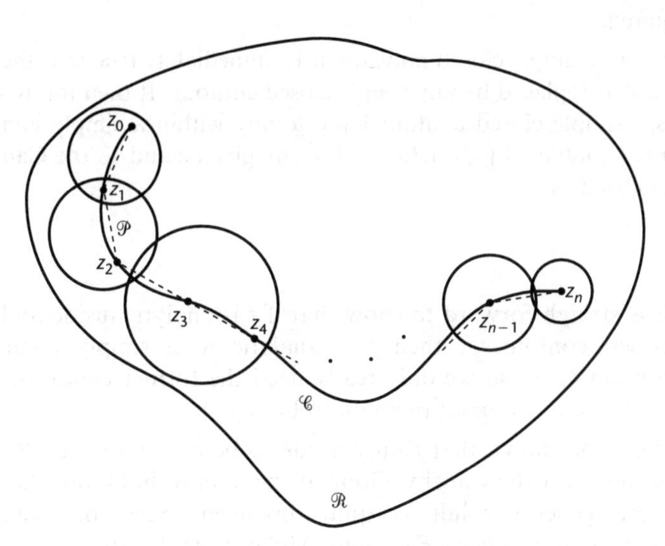

Figure 3.12

that $\int_{\mathscr{C}} f(z)\,dz = \lim_{\eta \to 0} S_n$ where $|z_k - z_{k-1}| \leqslant \eta$ for all k (so that n depends on η). Then, by (3.16), given $\varepsilon > 0$, there exists $N \in \mathbb{N}$ such that for all $n > N$,

$$\left| \int_{\mathscr{C}} f(z)\,dz - S_n \right| < \varepsilon/2 \tag{3.17}$$

and the polygon $\mathscr{P}$ with vertices $z_0, \ldots, z_n$ lies within the set S, with the property that $|f(z) - f(z_k)| < \varepsilon/2L$ along each line segment of $\mathscr{P}$ (Fig. 3.12).

Step 3

For the polygon $\mathscr{P}$ constructed in step 2,

$$\begin{aligned}\int_{\mathscr{P}} f(z)\,dz &= \int_{z_0}^{z_1} (f(z) - f(z_1) + f(z_1))\,dz + \ldots + \int_{z_{n-1}}^{z_n} (f(z) - f(z_n) + f(z_n))\,dz \\ &= \int_{z_0}^{z_1} (f(z) - f(z_1))\,dz + \ldots + \int_{z_{n-1}}^{z_n} (f(z) - f(z_n))\,dz + S_n\end{aligned}$$

It follows by the construction of step 2, (3.17) and 3.2 that

$$\left| S_n - \int_{\mathscr{P}} f(z)\,dz \right| < \frac{\varepsilon}{2L}(|z_1 - z_0| + \ldots + |z_n - z_{n-1}|) < \frac{\varepsilon}{2} \tag{3.18}$$

since $\sum_{k=1}^{n} |z_k - z_{k-1}|$, the length of $\mathscr{P}$, is necessarily less than L, the length of $\mathscr{C}$. Finally, by (3.17) and (3.18) we have

$$\begin{aligned}\left| \int_{\mathscr{C}} f(z)\,dz - \int_{\mathscr{P}} f(z)\,dz \right| &\leqslant \left| \int_{\mathscr{C}} f(z)\,dz - S_n \right| + \left| S_n - \int_{\mathscr{P}} f(z)\,dz \right| \\ &< \varepsilon/2 + \varepsilon/2 = \varepsilon\end{aligned}$$

as required. ■

When $\mathscr{P}$ is a simple closed polygon, it is immediately true that the above result holds if $\mathscr{C}$ is replaced by any simple closed contour. It then follows by 3.24 that if $\mathscr{C}$ is a simple closed contour lying totally within a simply connected set in which f is analytic, $|\int_{\mathscr{C}} f(z)\,dz| < \varepsilon$ for any given ε and so the Cauchy–Goursat theorem follows.

Notes

(i) It is straightforward to show that if f is analytic inside and on a simple closed contour $\mathscr{C}$, then f is analytic in a simply connected region containing $\mathscr{C}$, so we only really need the former condition in order for the Cauchy–Goursat theorem to hold.

(ii) It can be shown that f only needs to be continuous on $\mathscr{C}$ and analytic inside $\mathscr{C}$ for the Cauchy–Goursat theorem to hold, but the proof of this more general result is quite involved. See, for example, R. L. Goodstein, *Complex Functions*, McGraw-Hill, 1965.

4 Infinite Series, Taylor Series and Laurent Series

The main aim of this chapter is the study of complex power series, leading to Taylor series and their generalisation to Laurent series. Such series have many important applications. For instance, we need results concerning Laurent series in order to proceed further with complex integration. It turns out that the treatment of Taylor series in the complex case is actually easier and more satisfactory that in the real case. Power series expansions are also of fundamental use for solving linear differential equations, as described in Chapter 7. In order to study these special infinite series we begin, as in the real case, with sequences.

Sequences

Analogous to the idea of a sequence of real numbers, a sequence of complex numbers is a list of complex numbers derived from a definite rule.

Definition

A **sequence** of complex numbers, denoted by (z_n), is determined by a function $f: \mathbb{N} \to \mathbb{C}$ defined by $f(n) = z_n$, where z_n is the ***n*th term** in the sequence.

Example 4.1

Certain sequences of complex numbers defined by recurrence relations have attracted a lot of interest in recent years because of their connection with fractals and chaos.

Consider the sequence (z_n) defined by

$$z_{n+1} = z_n^2 + \alpha \quad \text{with} \quad z_1, \alpha \in \mathbb{C} \text{ chosen}$$

For a particular choice of z_1, the set of values of α for which this sequence is bounded is a **Mandlebrot set**, $M_2(z_1)$. Mandlebrot sets are **fractals**; that is, they are geometrical figures in the complex plane that consist of an identical pattern repeating itself on an ever reduced scale. The set $M_2(z_1)$, pictured as a geometrical figure, dramatically changes shape for small variations in the initial value z_1. Such Mandlebrot sets have surprisingly complicated structures. See, for example, B. B. Mandlebrot, *Fractals: Form, Chance and Dimension*, W. H. Freeman, 1977.

In many cases, definitions and results concerning sequences of complex numbers are analogous to definitions and results for sequences of real numbers. Also, in many cases the proof of a result is essentially the same as the proof of the corresponding result for sequences of real numbers, and usually we shall omit it.

If (z_n) is a sequence of complex numbers and z_n becomes 'arbitrarily close' to $\alpha \in \mathbb{C}$ as n becomes 'arbitrarily large', then α is the limit of the sequence. This intuitive idea is formalised in the following definition.

Definition

A sequence (z_n) **converges to limit** $\boldsymbol{\alpha}$, where $\alpha \in \mathbb{C}$, if given any real $\varepsilon > 0$, a natural number N (depending on ε) can be found such that

$$|z_n - \alpha| < \varepsilon \quad \text{whenever} \quad n > N$$

If (z_n) has limit α, we write $z_n \to \alpha$ as $n \to \infty$, or $\lim_{n\to\infty} z_n = \alpha$.

Example 4.2

Intuitively, $n/(1+in) = 1/(1/n + i) \to 1/i = -i$ as $n \to \infty$. Prove that $n/(1+in) \to -i$ as $n \to \infty$, using the definition.

Solution

$$\left|\frac{n}{1+in} + i\right| = \left|\frac{i}{1+in}\right| = \frac{1}{(1+n^2)^{1/2}} < \frac{1}{n} \qquad (n \in \mathbb{N})$$

Hence $$\left|\frac{n}{1+in} + i\right| < \frac{1}{n} < \frac{1}{N} \quad \text{whenever} \quad n > N$$

Then given any $\varepsilon > 0$, choose $N \in \mathbb{N}$ such that $N > 1/\varepsilon$. Then for this choice of N,

$$\left|\frac{n}{1+in} + i\right| < \varepsilon \quad \text{whenever} \quad n > N$$

as required.

The following theorem is a list of results which are analogous to results for real sequences and can be proved in exactly the same way, since $|z|$ for $z \in \mathbb{C}$ obeys the same properties as $|x|$ for $x \in \mathbb{R}$ and, in particular, the triangle inequality.

Theorem 4.1. Elementary Properties of Limits

(a) If the limit of a sequence exists, then it is unique.

(b) Any convergent sequence (z_n) is **bounded**; that is, there exists a positive real constant M such that $|z_n| \leqslant M$ for all $n \in \mathbb{N}$.

(c) Let $z_n \to \alpha$ and $w_n \to \beta$ as $n \to \infty$. Then
   (i) $\lambda z_n \to \lambda\alpha$ as $n \to \infty$ for any $\lambda \in \mathbb{C}$
   (ii) $z_n + w_n \to \alpha + \beta$ as $n \to \infty$
   (iii) $z_n w_n \to \alpha\beta$ as $n \to \infty$
   (iv) $z_n/w_n \to \alpha/\beta$ as $n \to \infty$ provided that $\beta \neq 0$

(d) Let $z_n \to \alpha$ as $n \to \infty$ and let $f : A \subseteq \mathbb{C} \to \mathbb{C}$ be continuous at each point of (z_n). Then $f(z_n) \to f(\alpha)$ as $n \to \infty$. □

Example 4.3

$n/(1+in) = 1/(1/n+i)$ and $1/n \to 0$ as $n \to \infty$. Hence by 4.1(c), $1/(1/n+i) \to 1/i$ as $n \to \infty$, so that $n/(1+in) \to -i$ as $n \to \infty$. (Compare with Example 4.2.)

Note

Results concerning real sequences involving inequalities of real numbers will not in general carry over to the complex case. For example, it makes no sense to talk of monotonic sequences of complex numbers, so there is no monotonic-bounded principle.

The following result implies that sequences of complex numbers can be investigated by examining their real and imaginary parts and using results for sequences of real numbers.

Theorem 4.2. Real and Imaginary Parts of Sequences

$z_n \to \alpha$ as $n \to \infty$ if and only if $\operatorname{Re} z_n \to \operatorname{Re} \alpha$ and $\operatorname{Im} z_n \to \operatorname{Im} \alpha$. □

Proof

Let $z_n = x_n + iy_n$ for all $n \in \mathbb{N}$ and $\alpha = a + ib$ where a and b are real. Suppose that $z_n \to \alpha$ as $n \to \infty$, so that given any $\varepsilon > 0$, there is a natural number N such that

$$|z_n - \alpha| = |(x_n - a) + i(y_n - b)| < \varepsilon \quad \text{whenever} \quad n > N$$

Now

$$|x_n - a| \leqslant |(x_n - a) + i(y_n - b)|$$
$$|y_n - b| \leqslant |(x_n - a) + i(y_n - b)|$$

for all $n \in \mathbb{N}$. Hence, given any $\varepsilon > 0$, there is a natural number N such that

$$|x_n - a| < \varepsilon \quad \text{and} \quad |y_n - b| < \varepsilon \quad \text{whenever} \quad n > N$$

Hence, by definition, $x_n \to a$ and $y_n \to b$ as $n \to \infty$.

Now suppose that these conditions hold. Then given any $\varepsilon > 0$, there exist natural numbers N_1 and N_2 such that $|x_n - a| < \varepsilon/2$ whenever $n > N_1$ and $|y_n - b| < \varepsilon/2$ whenever $n > N_2$. Then by the triangle inequality,

$$|z_n - \alpha| \leqslant |x_n - a| + |y_n - b| < \varepsilon \quad \text{whenever} \quad n > \max(N_1, N_2)$$

and so $z_n \to \alpha$ as $n \to \infty$, as required. ■

The standard results from real analysis concerning subsequences carry over to the complex case.

Definition

The sequence (w_n) is a **subsequence** of the sequence (z_n) if there exist natural numbers $n_1 < n_2 < \ldots$ such that $w_k = z_{n_k}, k \in \mathbb{N}$.

Theorem 4.3. Subsequences of Bounded Sequences

Any bounded sequence in $\mathbb{C}$ has a convergent subsequence. □

Proof

Let (z_n) be bounded so that there exists a positive real number M such that $|z_n| \leqslant M$ for all $n \in \mathbb{N}$. Then $|\text{Re}\, z_n| \leqslant M$, so that $(\text{Re}\, z_n)$ is a bounded sequence in $\mathbb{R}$ and hence has a convergent subsequence, $(\text{Re}\, z_{n_k}) = (\text{Re}\, w_k)$ say, in $\mathbb{R}$. By hypothesis, $(\text{Im}\, w_k)$ is a bounded real sequence and so has a convergent subsequence, $(\text{Im}\, w_{k_m})$ say, in $\mathbb{R}$. As a subsequence of $(\text{Re}\, w_k)$, the sequence $(\text{Re}\, w_{k_m})$ must converge too. Hence (w_{k_m}) is a convergent subsequence of (z_n), as required. ■

Recall the definition of a limit point given in Chapter 1. The following is a generalisation of a famous result from real analysis.

Theorem 4.4. The Bolzano–Weierstrass Theorem

Any bounded infinite subset S of $\mathbb{C}$ has a limit point. □

Proof

Since S is infinite, we may select a sequence (z_n) with the points z_n distinct and belonging to S. Since S is bounded, (z_n) has a convergent subsequence by 4.3. If

α is the limit of such a subsequence, by definition, every deleted neighbourhood of α contains points of (z_n) and hence S. Then α is a limit point of S, as required. ■

Using Theorem 4.4, we can proceed as in the real case to prove the following result, which was stated and used in the previous two chapters.

Theorem 4.5. Continuous Functions on Compact Sets

Let S be a compact subset of $\mathbb{C}$ and let $f: A \to \mathbb{C}$ be continuous at each point of S. Then f is bounded on S; that is, there exists a real constant M such that $|f(z)| \leqslant M$ for all $z \in S$. □

Proof

Suppose that f is unbounded on S. Then there exists distinct $z_n \in S$ such that $|f(z_n)| > n$ for each $n \in \mathbb{N}$. This defines a sequence (z_n) of points in S, which has a limit point α by 4.4. Note that $\alpha \in S$ since S is closed. Hence the function f is continuous at α, so $f(\alpha) = \lim_{z_n \to \alpha} f(z_n)$ by 4.1(d). But this is contradicted by the fact that $|f(z_n)| > n$ for each n. Hence f must be bounded in S. ■

In defining limits of sequences, we have defined $\lim_{n \to \infty} f(n)$ where $f: N \to \mathbb{C}$. In general, if $f: A \subseteq \mathbb{C} \to \mathbb{C}$, then $\lim_{z \to \infty} f(z)$ can be defined in terms of limits already defined in Chapter 2.

Definition

Let $f: A \subseteq \mathbb{C} \to \mathbb{C}$. Then ***f(z)* has limit ℓ as *z* tends to infinity**, written $\lim\limits_{z \to \infty} f(z) = \ell$, if given any $\varepsilon > 0$ there is a $\delta > 0$ (depending on ε) such that

$$\left|\frac{1}{z}\right| < \delta \quad \Rightarrow \quad |f(z) - \ell| < \varepsilon$$

This definition makes sense since there is only one point at infinity in the extended complex plane. The definition formalises the idea that, as $|z|$ becomes 'arbitrarily large', i.e. $1/z$ becomes 'arbitrarily close' to 0, $f(z)$ becomes 'arbitrarily close' to ℓ. This concept will prove useful in Chapter 5.

Notes

(i) Theorem 4.1 applies to $\lim_{z \to \infty} f(z)$; in particular, if the limit exists, it is unique, no matter how $z \to \infty$.

(ii) It is easy to show that $\lim_{z \to \infty} f(z) - \ell$ if and only if $\lim_{z \to 0} g(z) = \ell$, where $g(z) = f(1/z)$.

Example 4.4

(i) Intuitively, $\lim_{z\to\infty} 1/z^k = 0$, for any $k \in \mathbb{R}^+$. This can be proved using the definition as follows.

$$\left|\frac{1}{z}\right| < \delta \quad \Rightarrow \quad \left|\frac{1}{z^k} - 0\right| = \frac{1}{|z|^k} < \delta^k \quad \text{since} \quad k > 0$$

Then given any $\varepsilon > 0$, we choose $\delta = \varepsilon^{1/k}$. For this choice of δ,

$$\left|\frac{1}{z}\right| < \delta \quad \Rightarrow \quad \left|\frac{1}{z^k} - 0\right| < \varepsilon$$

(ii) Although $\lim_{x\to+\infty} e^{-x} = 0$, $\lim_{z\to\infty} e^{-z}$ does not exist. For example, letting $z = -x, x \in \mathbb{R}^+$, $\lim_{z\to\infty} e^{-z}$ does not exist. Alternatively, letting $z = iy, e^{-z} = e^{-iy} = \cos y - i\sin y$ and $\lim_{y\to\infty} e^{-iy}$ does not exist.

Sequences of Functions

Let $(f_n(z))$ denote a sequence of values of functions $f_n : A \subseteq \mathbb{C} \to \mathbb{C}$, $n \in \mathbb{N}$. Each value of z in the domain A gives a different sequence and if each such sequence converges, it is possible to construct a limit function with values $f(z)$, called the **pointwise limit function** of the sequence $(f_n(z))$. In general, such a limit function will not be continuous on A, even if each f_n is continuous.

Example 4.5

Let $f_n(z) = 1 - z^n$ for $|z| \leqslant 1$. Then $f_n(z) \to 1$ as $n \to \infty$ for $|z| < 1$ but $f_n(1) = 0 \to 0$ as $n \to \infty$.

Clearly, it is important to find conditions which ensure that $(f_n(z))$ converges to a continuous pointwise limit function, and more important, conditions under which termwise integration and differentiation of such a sequence of functions is valid. This leads to the concept of uniform convergence, the definition of which is essentially the same as in the real case.

Note

Readers who are primarily interested in applications of Taylor and Laurent series and who are willing to accept that power series may be differentiated and integrated termwise within their domain of convergence, may wish to omit references to uniform convergence.

Definition

Let $f_n : A \subseteq \mathbb{C} \to \mathbb{C}$ for each $n \in \mathbb{N}$. A sequence of functions $(f_n(z))$ **converges uniformly** to its pointwise limit function $f(z)$ in the domain A, if given any $\varepsilon > 0$ there exists $N \in \mathbb{N}$ (depending on ε but **independent of z**), such that

$$|f_n(z) - f(z)| < \varepsilon \quad \text{whenever } n > N \quad \text{for all } z \in A$$

Note

The crucial point to notice about uniform convergence is that the choice of N does not depend on the value of z; that is, the same N will satisfy the definition for all $z \in A$.

Example 4.6

The sequence (z^n) converges uniformly to the zero function for $|z| \leqslant 1/2$. To show this, notice first of all that the pointwise limit function is given by $f(z) = 0$, for $|z| \leqslant 1/2$. Now consider any value of z in the given domain. Then

$$|z| \leqslant 1/2 \quad \Rightarrow \quad |z^n - 0| = |z|^n \leqslant \frac{1}{2^n}$$

It follows by the binomial theorem that $2^n = (1+1)^n > n$ for all $n \in \mathbb{N}$. Hence

$$|z| \leqslant 1/2 \quad \Rightarrow \quad |z^n - 0| \leqslant \frac{1}{2^n} < \frac{1}{n} < \frac{1}{N} \quad \text{whenever} \quad n > N$$

Given any $\varepsilon > 0$, choose $N \in \mathbb{N}$ such that $N > 1/\varepsilon$. Note that N is independent of z. Then for this choice of N,

$$|z| \leqslant 1/2 \quad \Rightarrow \quad |z^n - 0| < \varepsilon \quad \text{whenever} \quad n > N$$

as required.

Note

In general, it is only possible to establish uniform convergence of a sequence of functions on a set A if A is compact, as in the above example.

Theorem 4.6. Uniform Convergence of Continuous Functions

If the sequence $(f_n(z))$ converges uniformly to $f(z)$ in a set A and if each f_n is continuous at each point of A, then f is continuous at each point of A. □

Proof

The proof is essentially the same as in the real case. For those readers unfamiliar with uniform convergence, we give the proof here. Let $\alpha \in A$ and z be any other point in A. By the triangle inequality it follows that

$$|f(z) - f(\alpha)| \leqslant |f(z) - f_n(z)| + |f_n(z) - f_n(\alpha)| + |f_n(\alpha) - f(\alpha)| \tag{4.1}$$

for any $n \in \mathbb{N}$. Since $(f_n(z))$ converges uniformly to $f(z)$ in A, given any $\varepsilon > 0$, there exists $N \in \mathbb{N}$ such that

$$|f(z) - f_n(z)| < \varepsilon/3 \quad \text{and} \quad |f_n(\alpha) - f(\alpha)| < \varepsilon/3 \quad \text{whenever} \quad n > N$$

Given the same $\varepsilon > 0$, there exists $\delta > 0$ such that

$$|z - \alpha| < \delta \quad \Rightarrow \quad |f_n(z) - f_n(\alpha)| < \varepsilon/3 \qquad (\text{for all } n \in \mathbb{N})$$

since each f_n is continuous at α. It follows from (4.1) that, given any $\varepsilon > 0$, there exists $\delta > 0$ such that

$$|z - \alpha| < \delta \quad \Rightarrow \quad |f(z) - f(\alpha)| < \varepsilon/3 + \varepsilon/3 + \varepsilon/3 = \varepsilon$$

Hence f is continuous at $\alpha \in A$ as required. ■

It follows by Example 4.5 that the sequence $(1 - z^n)$ for $|z| \leqslant 1$ has a discontinuous pointwise limit function, so that by 4.6 the convergence is not uniform on $|z| \leqslant 1$.

Notes

(i) It is not true in general that if $(f_n(z))$ converges pointwise to $f(z)$ in a set A and if each f_n and f are continuous in A, then the convergence is uniform on A.

(ii) To test whether or not a pointwise convergent sequence of functions, $(f_n(z))$, converges uniformly on A, first of all find the pointwise limit function $f(z)$. Then calculate $U_n = \sup_{z \in A} |f_n(z) - f(z)|$. It follows from the definition that if $U_n \to 0$ as $n \to \infty$, then the convergence is uniform. In practice, it may be difficult to find U_n for each n.

Theorem 4.7 Integration of Uniformly Convergent Sequences

If $(f_n(z))$ converges uniformly to $f(z)$ on a simple contour $\mathscr{C}$ and if each f_n is continuous at each point of $\mathscr{C}$, then

$$\lim_{n \to \infty} \int_{\mathscr{C}} f_n(z) dz = \int_{\mathscr{C}} \lim_{n \to \infty} f_n(z) dz = \int_{\mathscr{C}} f(z) dz$$ □

Proof

Since each f_n is continuous on $\mathscr{C}$, f is continuous on $\mathscr{C}$ by 4.6, so the integrals exist. Let L be the length of $\mathscr{C}$. Then, by hypothesis, given any $\varepsilon > 0$, there exists N such that

$$|f_n(z) - f(z)| < \varepsilon/L \quad \text{whenever} \quad n > N \quad \text{for all } z \in \mathscr{C}$$

Then

$$\left|\int_{\mathscr{C}} f_n(z)dz - \int_{\mathscr{C}} f(z)dz\right| < \varepsilon \qquad (n > N)$$

by Theorem 3.2, as required. ■

Note

Clearly, the continuity condition in 4.7 can be weakened.

Theorem 4.8. Differentiation of Uniformly Convergent Sequences

For each $n \in \mathbb{N}$, let f_n be analytic on a simply connected region $\mathscr{R}$ and let $(f_n(z))$ be uniformly convergent to $f(z)$ on each compact subset of $\mathscr{R}$. Then f is analytic on $\mathscr{R}$ and $(f'_n(z))$ converges uniformly to $f'(z)$ on each compact subset of $\mathscr{R}$. □

Proof

Step 1

We need to show first of all that f is analytic in $\mathscr{R}$. Since, by hypothesis, each f_n is continuous in $\mathscr{R}$, it follows that f is continuous in $\mathscr{R}$ by 4.6. Then if $\mathscr{C}$ is any simple closed contour of length L in $\mathscr{R}$,

$$\int_{\mathscr{C}} f(z)dz = \int_{\mathscr{C}} (f(z) - f_n(z))dz + \int_{\mathscr{C}} f_n(z)dz \qquad (\text{for any } n \in \mathbb{N})$$

But $\int_{\mathscr{C}} f_n(z)dz = 0$ for any $n \in \mathbb{N}$ by Cauchy's theorem since f_n is analytic inside and on $\mathscr{C}$. Since $(f_n(z))$ is uniformly convergent on any compact subset of $\mathscr{R}$, given any $\varepsilon > 0$ there exists $N \in \mathbb{N}$ such that for any $z \in \mathscr{C}$,

$$|f(z) - f_n(z)| < \varepsilon/L \quad \text{whenever} \quad n > N$$

Hence by Theorem 3.2,

$$\left|\int_{\mathscr{C}} f(z)dz\right| = \left|\int_{\mathscr{C}} f(z) - f_n(z)dz\right| < \varepsilon \quad \text{whenever} \quad n > N$$

so that $\int_{\mathscr{C}} f(z)dz = 0$. Hence f is analytic in $\mathscr{R}$ by Morera's theorem, 3.15.

Step 2

Now let z be any point in $\mathscr{R}$ and $\mathscr{C}$ be a circle with centre z and radius r, lying entirely within $\mathscr{R}$. By hypothesis, given any $\varepsilon > 0$, there exists $N \in \mathbb{N}$ such that for any $w \in \mathscr{C}$,

$$|f(w) - f_n(w)| < r\varepsilon \quad \text{whenever} \quad n > N$$

Then, by Cauchy's theorem for derivatives and Theorem 3.2,

$$|f'(z) - f_n'(z)| = \left| \frac{1}{2\pi i} \int_{\mathscr{C}} \frac{f(w) - f_n(w)}{(w - z)^2} \, dw \right| < \frac{2\pi r . r\varepsilon}{2\pi r^2} = \varepsilon$$

whenever $n > N$. Hence $(f_n'(z))$ converges uniformly to $f'(z)$ in any compact subset of $\mathscr{R}$.

■

Exercise **4.1.1** Use the definition of the limit of a sequence to prove that if α is any complex number satisfying $|\alpha| < 1$, then

$$1 + \frac{(i\alpha)^n}{n^2} \to 1 \quad \text{as } n \to \infty$$

Exercise **4.1.2** Use the definition to prove that if $z_n \to \alpha$ as $n \to \infty$, then

(i) $\overline{z}_n \to \overline{\alpha}$ as $n \to \infty$

(ii) $|z_n| \to |\alpha|$ as $n \to \infty$

(iii) $z_n^{1/2} \to \alpha^{1/2}$ as $n \to \infty$

Give a simple example to show that $|z_n| \to |\alpha|$ as $n \to \infty \quad \not\Rightarrow \quad z_n \to \alpha$ as $n \to \infty$.

Exercise **4.1.3** Let $z_n \to \alpha$ and $w_n \to \beta$ as $n \to \infty$. Use the definition to prove that (i) $z_n - w_n \to \alpha - \beta$ as $n \to \infty$, (ii) $z_n w_n \to \alpha\beta$ as $n \to \infty$. (Use the fact that any convergent sequence is bounded in (ii).)

Exercise **4.1.4** Use Theorem 4.1 and, in the second case, Exercise 4.1.2(iii) to find the following limits:

(i) $\displaystyle\lim_{n \to \infty} \frac{2n - in^2}{(1 + i)n^2 - 1}$

(ii) $\displaystyle\lim_{n \to \infty} \sqrt{n}\left((n + i)^{1/2} - (n - i)^{1/2}\right)$

Exercise **4.1.5** Prove that $\lim_{z \to \infty} f(z) = \ell$ if and only if $\lim_{z \to 0} g(z) = \ell$, where $g(z) = f(1/z)$.

Exercise **4.1.6** Prove that the sequence (n^{-z}) converges uniformly for $\operatorname{Re} z \geqslant a > 0$. Prove also that (n^{-z}) converges for $z \in A$, where $A = \{z : \operatorname{Re} z > 0 \text{ or } z = 0\}$ but that the convergence is not uniform on A.

Exercise **4.1.7** Prove that $(1/(1 + 2n^2 z))$ converges uniformly to 0 for all z satisfying $|z| \geqslant 1$.

Infinite Series

The concept of an infinite series of complex numbers is identical to that of an infinite series of real numbers. We begin with the usual definitions.

Definitions

Let (z_n) be a sequence of complex numbers. Then the sequence (S_k), defined by $S_k = \sum_{n=1}^{k} z_n = z_1 + z_2 + \ldots + z_k$, is the sequence of **partial sums** of the **infinite series** $z_1 + z_2 + z_3 + \ldots$. The infinite series **converges to sum** S if $S_k \to S$ as $k \to \infty$, and in this case we write $\sum_{n=1}^{\infty} z_n = S$. The infinite series **diverges** if (S_k) diverges.

Important Note

Let $R_k = \sum_{n=k+1}^{\infty} z_n$ for each k. Then R_k is called the **remainder** of the series after k terms. Since $|S_k - S| = |R_k - 0|$, it follows by the definition of the limit of a sequence that the series $\sum_{n=1}^{\infty} z_n$ converges to sum S if and only if $R_k \to 0$ as $k \to \infty$.

Once again, a lot of the properties of real series carry over to the complex case. It follows by the definition and 4.1(i) that if $\sum_{n=1}^{\infty} z_n$ converges then its sum is unique. Also the geometric series result is easily generalised to the complex case. The following standard results also follow from Theorem 4.1.

Theorem 4.9. Elementary Properties of Series

Suppose that the series $\sum_{n=1}^{\infty} z_n$ and $\sum_{n=1}^{\infty} w_n$ converge. Then

(i) $\displaystyle\sum_{n=1}^{\infty} (\lambda z_n) = \lambda \sum_{n=1}^{\infty} z_n$ for all $\lambda \in \mathbb{C}$

(ii) $\displaystyle\sum_{n=1}^{\infty} (z_n + w_n) = \sum_{n=1}^{\infty} z_n + \sum_{n=1}^{\infty} w_n$

(iii) $z_n \to 0$ as $n \to \infty$ □

Recall that condition (iii) is necessary for convergence but it is not sufficient; for example, $\sum_{n=1}^{\infty} 1/n$ satisfies condition (iii) but it still diverges. The following theorem shows that series of complex numbers can be investigated by examining their real and imaginary parts.

Theorem 4.10. Real and Imaginary Parts of Series

The series $\sum_{n=1}^{\infty} z_n$ converges to sum S if and only if $\sum_{n=1}^{\infty} \operatorname{Re} z_n$ converges to $\operatorname{Re} S$ and $\sum_{n=1}^{\infty} \operatorname{Im} z_n$ converges to $\operatorname{Im} S$. □

Proof

Let $z_n = x_n + iy_n$ for each $n \in \mathbb{N}$ and $S = A + iB$, where A and B are real. Then

$$S_k = \sum_{n=1}^{k} z_n = \sum_{n=1}^{k} x_n + i \sum_{n=1}^{k} y_n = A_k + iB_k \quad \text{say, for all } k$$

It follows by Theorem 4.2 that $S_k \to S$ as $k \to \infty$ if and only if $A_k \to A$ and $B_k \to B$ as $k \to \infty$, and the result follows by definition. ■

Instead of using the above result, just as for series of real numbers, if we want to test the convergence of $\sum_{n=1}^{\infty} z_n$, we can test for absolute convergence.

Definition

The series $\sum_{n=1}^{\infty} z_n$ **converges absolutely** if $\sum_{n=1}^{\infty} |z_n|$ converges. Note that $\sum_{n=1}^{\infty} |z_n|$ is a series of non-negative real numbers and all the standard convergence tests, such as the comparison, ratio and nth root test, apply.

Theorem 4.11. Absolute Convergence Implies Convergence

Any absolutely convergent series converges. □

Proof

Let $z_n = x_n + iy_n$. By hypothesis, $\sum_{n=1}^{\infty} (x_n^2 + y_n^2)^{1/2}$ converges, and since $|x_n| \leqslant (x_n^2 + y_n^2)^{1/2}$, $|y_n| \leqslant (x_n^2 + y_n^2)^{1/2}$ for all $n \in \mathbb{N}$, it follows by the comparison test that $\sum_{n=1}^{\infty} |x_n|$ and $\sum_{n=1}^{\infty} |y_n|$ converge. Hence $\sum_{n=1}^{\infty} x_n$ and $\sum_{n=1}^{\infty} y_n$ converge since any real absolutely convergent series converges. The result then follows by Theorem 4.10. ■

Notes

(i) It follows from results concerning series of real numbers and Theorems 4.10 and 4.11 that two absolutely convergent series can be multiplied to give another absolutely convergent series which converges to the product of the separate sums.

(ii) Recall that if a series is absolutely divergent, it is not necessarily divergent.

We now turn to the concept of uniform convergence of series of functions, which is the condition required for termwise differentiation and integration of such series.

Let $(f_n(z))$ be a sequence of values of functions $f_n : A \subseteq \mathbb{C} \to \mathbb{C}$. Then for each $z \in A$, $\sum_{n=1}^{\infty} f_n(z)$ is a series of complex numbers, which may or may not converge.

Definitions

A **series of functions** $\sum_{n=1}^{\infty} f_n(z)$ **converges pointwise** to **sum function** $F(z)$ if the sequence of **partial sum functions** $\left(\sum_{n=1}^{k} f_n(z)\right)$ converges pointwise to $F(z)$; that is, $\sum_{n=1}^{k} f_n(z) \to F(z)$ as $k \to \infty$ for each $z \in A$. The series of functions $\sum_{n=1}^{\infty} f_n(z)$ **converges uniformly** to $F(z)$ on A if the sequence $\left(\sum_{n=1}^{k} f_n(z)\right)$ converges uniformly to $F(z)$ on A.

The following result follows directly from the definition and Theorems 4.6, 4.7 and 4.8.

Theorem 4.12. Continuity, Differentiation and Integration of Uniformly Convergent Series

(i) If the series $\sum_{n=1}^{\infty} f_n(z)$ converges uniformly to $F(z)$ on a set A and each f_n is continuous at each point of A, then F is continuous at each point of A.

(ii) If the series $\sum_{n=1}^{\infty} f_n(z)$ converges uniformly to $F(z)$ on a simple contour $\mathscr{C}$ and if each f_n is continuous at each point of $\mathscr{C}$, then

$$\int_{\mathscr{C}} F(z)dz = \int_{\mathscr{C}} \sum_{n=1}^{\infty} f_n(z)dz = \sum_{n=1}^{\infty} \int_{\mathscr{C}} f_n(z)dz$$

(iii) If the series $\sum_{n=1}^{\infty} f_n(z)$ converges uniformly to $F(z)$ in any compact subset of a simply connected region $\mathscr{R}$ and each f_n is analytic in $\mathscr{R}$, then F is analytic in $\mathscr{R}$ with $F'(z) = \sum_{n=1}^{\infty} f'_n(z)$, the convergence being uniform on any compact subset of $\mathscr{R}$. □

Proof

(i) By definition, $\left(\sum_{n=1}^{k} f_n(z)\right)$ converges uniformly to $F(z)$ in A, and since each f_n is continuous in A, $\sum_{n=1}^{k} f_n$ is continuous in A. Then the result follows by 4.6.

(ii) By hypothesis, $\left(\sum_{n=1}^{k} f_n(z)\right)$ converges uniformly to $F(z)$ on $\mathscr{C}$ where each f_n is continuous on $\mathscr{C}$ and so by (i), F is continuous on $\mathscr{C}$ and the integrals exist. It follows by 4.7 that

$$\int_{\mathscr{C}} \sum_{n=1}^{k} f_n(z)dz = \sum_{n=1}^{k} \int_{\mathscr{C}} f_n(z)dz \to \int_{\mathscr{C}} F(z)dz \quad \text{as } k \to \infty$$

as required.

(iii) This is similar and uses 4.8. It is left as an exercise. ■

A very useful test for uniform convergence of series of functions, which carries over from the real case, is the following result.

Theorem 4.13. Weierstrass's *M* test

Suppose that for each $n \in \mathbb{N}$, there exists a positive real number M_n such that $|f_n(z)| \leqslant M_n$ for all $z \in A$ and that $\sum_{n=1}^{\infty} M_n$ converges. Then $\sum_{n=1}^{\infty} f_n(z)$ is absolutely and uniformly convergent on A. □

Proof

Note that $\sum_{n=1}^{\infty} f_n(z)$ is absolutely convergent on A by hypothesis and the comparison test. Let the remainder of $\sum_{n=1}^{\infty} f_n(z)$ after k terms be $R_k(z)$ for each $z \in A$, and let the remainder of $\sum_{n=1}^{\infty} M_n$ after k terms be R_k^*. Then by hypothesis, $R_k^* \to 0$ as $k \to \infty$, so that given any $\varepsilon > 0$, there exists $K^* \in \mathbb{N}$ such that $|R_k^*| < \varepsilon$ whenever $k > K^*$. Also, since $|f_n(z)| \leqslant M_n$ for each $n \in \mathbb{N}$, it follows by Exercise 4.2.4 and the triangle inequality that

$$|R_k(z)| = \left| \sum_{n=k+1}^{\infty} f_n(z) \right| \leqslant \sum_{n=k+1}^{\infty} |f_n(z)| \leqslant \sum_{n=k+1}^{\infty} M_n = R_k^*$$

for each k. Hence, given any $\varepsilon > 0$, there exists K^* such that $|R_k(z)| < \varepsilon$ whenever $k > K^*$ for all values of z in the domain of convergence. Hence $R_k(z) \to 0$ uniformly on A as $k \to \infty$, as required. ■

Example 4.7

Prove that $\zeta(z) = \sum_{n=1}^{\infty} n^{-z}$ is an analytic function for $\operatorname{Re} z > 1$. This is the **Riemann zeta function**, of fundamental importance in number theory.

Solution

Let $\operatorname{Re} z = x \geqslant \rho > 1$. Then $|n^{-z}| = n^{-x} \leqslant n^{-\rho}$. Let $M_n = n^{-\rho}$ in 4.13. Now $\sum_{n=1}^{\infty} n^{-\rho}$ converges by hypothesis and the real hyperharmonic series result, so the given series converges uniformly for $\operatorname{Re} z \geqslant \rho > 1$ by 4.13. Hence the given series converges uniformly on any compact subset of $\operatorname{Re} z > 1$, and so by 4.12, ζ is analytic on $\operatorname{Re} z > 1$ and $\zeta'(z) = \sum_{n=1}^{\infty} (\operatorname{Log} n) n^{-z}$ in this region.

Note

As in the real case, absolute convergence alone does not guarantee uniform convergence.

Power Series

The most important series of functions of complex variables consist of integer powers of a complex variable. These series have very wide applications.

Definition

A **power series** is a series of the form $\sum_{n=0}^{\infty} a_n(z-\alpha)^n$ where α and $a_n \in \mathbb{C}$ for any $n \in \mathbb{Z}_{\geqslant 0}$, and z may be any complex number in a stated domain. A trivial change of variable, $w = z - \alpha$, reduces such a series to the form $\sum_{n=0}^{\infty} a_n w^n$ and so we take $\alpha = 0$ without loss of generality. Note the slight change in convention for the values taken by n, which is for convenience.

Power series generally converge for certain values of z and diverge for others. Standard convergence tests for series of real numbers can be used to find the set of values of $|z|$ for which the series converges absolutely. As indicated earlier, it is generally much easier to test for absolute convergence than for convergence by other means.

Definition

The **radius of convergence** of the power series $\sum_{n=0}^{\infty} a_n z^n$ is denoted and defined by $R = \sup\{|z| : \sum_{n=0}^{\infty} |a_n z^n| \text{ converges}\}$. Then the series is absolutely convergent for all z inside the circle $\mathscr{C}$ given by $|z| = R$ and divergent for all z outside $\mathscr{C}$. $\mathscr{C}$ is the **circle of convergence** of the series.

Note

If a power series converges pointwise for all $z \in \mathbb{C}$, then the radius of convergence is taken to be infinite.

For convenience, we give below two standard convergence tests for real series of non-negative terms, which are the most useful for calculating the radius of convergence of a given power series.

Theorem 4.14. The Ratio and *n*th Root Tests

Let $\sum_{n=0}^{\infty} \alpha_n$ be a given series of complex numbers.

(i) **The ratio test**. Let $\lim_{n\to\infty} |\alpha_{n+1}/\alpha_n| = \lambda$, if it exists. Then $\sum_{n=0}^{\infty} |\alpha_n|$ converges if $\lambda < 1$ and diverges if $\lambda > 1$.

(ii) **The *n*th root test**. Let $\lim_{n\to\infty} |\alpha_n|^{1/n} = \lambda$, if it exists. Then $\sum_{n=0}^{\infty} |\alpha_n|$ converges if $\lambda < 1$ and diverges if $\lambda > 1$. □

Important Note

More generally, in the nth root test, if $\lim_{n\to\infty} |\alpha_n|^{1/n}$ does not exist, it can be replaced by $\limsup_{n\to\infty} |\alpha_n|^{1/n}$, the largest number to which any subsequence of $(|\alpha_n|^{1/n})$ converges, which always exists (allowing infinity as a possible value).

Note

The ratio and nth root tests give no information about the behaviour of a power series on the circle of convergence $\mathscr{C}$. For $|z| = R$, $\sum_{n=0}^{\infty} |a_n z^n| = \sum_{n=0}^{\infty} |a_n| R^n$ and the series can be tested for absolute convergence by other standard tests for real series of non-negative terms. More generally, it can be tested for convergence using Theorem 4.10.

Example 4.8

Find the radius of convergence of the following power series:

(i) $\displaystyle\sum_{n=0}^{\infty} \frac{z^n}{n!}$

(ii) $\displaystyle\sum_{n=1}^{\infty} n^2 z^n$

(iii) $3 + z + 3z^2 + z^3 + 3z^4 + \ldots$

Solution

(i) Let $\alpha_n = z^n/n!$. Then

$$\left|\frac{\alpha_{n+1}}{\alpha_n}\right| = \left|\frac{z^{n+1} n!}{(n+1)! z^n}\right| = \frac{|z|}{n+1} \quad \Rightarrow \quad \lim_{n \to \infty} \left|\frac{\alpha_{n+1}}{\alpha_n}\right| = 0 \quad \text{(for all } z \in \mathbb{C}\text{)}$$

Hence, by the ratio test, the given series is absolutely convergent for all z. It follows by this example that $\sum_{n=0}^{\infty} n! z^n$ converges only at $z = 0$.

(ii) Let $\alpha_n = n^2 z^n$. Then

$$\left|\frac{\alpha_{n+1}}{\alpha_n}\right| = \frac{(n+1)^2}{n^2}|z| = \left(1 + \frac{1}{n}\right)^2 |z| \quad \Rightarrow \quad \lim_{n \to \infty} \left|\frac{\alpha_{n+1}}{\alpha_n}\right| = |z|$$

Hence, by the ratio test, the given series is absolutely convergent for $|z| < 1$ and absolutely divergent for $|z| > 1$. In other words, the radius of convergence is 1. Note that when $|z| = 1$, then $|\alpha_n| = n^2$ and the series is absolutely divergent; it is in fact divergent, since in this case $\alpha_n = n^2 e^{ni \operatorname{Arg} z} \not\to 0$ as $n \to \infty$.

(iii) Let $\alpha_{2m} = 3z^{2m}$ and $\alpha_{2m+1} = z^{2m+1}$. In this case the ratio test is not appropriate since the required limit does not exist. On the other hand, $\lim_{m \to \infty} |\alpha_{2m}|^{1/2m} = \lim_{m \to \infty} 3^{1/2m}|z| = |z|$ and $\lim_{m \to \infty} |\alpha_{2m+1}|^{1/(2m+1)} = |z|$, so that $\lim_{n \to \infty} |\alpha_n|^{1/n} = |z|$. Hence, by the nth root test, the radius of convergence of the given series is 1.

Power series are particularly simple to manipulate since any power series is uniformly convergent within its domain of convergence.

Theorem 4.15. Uniform Convergence of Power Series

If the power series $\sum_{n=0}^{\infty} a_n z^n$ has radius of convergence R, then the series is uniformly convergent to its pointwise sum function in the closed set $|z| \leqslant \rho$ where $0 < \rho < R$. □

Proof

$|a_n z^n| \leqslant |a_n|\rho^n$ for $|z| \leqslant \rho$ and for each n. Also, $\sum_{n=0}^{\infty} |a_n|\rho^n$ is convergent by hypothesis. Hence $\sum_{n=0}^{\infty} a_n z^n$ is uniformly convergent for $|z| \leqslant \rho$ by Weierstrass's M test. ■

Since a power series is uniformly convergent within its circle of convergence, termwise integration is valid within this domain by 4.12(ii), and the next theorem follows immediately.

Theorem 4.16. Integration of Power Series

Suppose that $\sum_{n=0}^{\infty} a_n z^n$ converges to $f(z)$, inside the circle $|z| = R$ and that $\mathscr{C}$ is any contour lying inside this circle. Then

$$\int_{\mathscr{C}} f(z)dz = \int_{\mathscr{C}} \sum_{n=0}^{\infty} a_n z^n = \sum_{n=0}^{\infty} \int_{\mathscr{C}} a_n z^n dz$$ □

We also have the following special case of Theorem 4.12(iii), which says that a power series can be differentiated termwise within its domain of convergence.

Theorem 4.17. Differentiation of Power Series

Suppose that $\sum_{n=0}^{\infty} a_n z^n$ converges to $f(z)$ and that its radius of convergence is R. Then $\sum_{n=1}^{\infty} n a_n z^{n-1}$ converges to $f'(z)$, with radius of convergence R. □

Proof

The fact that the power series can be differentiated termwise is a consequence of 4.12(iii) and 4.15. It remains to show that $\sum_{n=1}^{\infty} n a_n z^{n-1}$ has radius of convergence R.

Step 1

We first show that $\lim_{n\to\infty} n^{1/n} = 1$, by using the binomial theorem. Let $n^{1/n} = 1 + \alpha_n$, so that

$$n = (1+\alpha_n)^n > 1 + \frac{n(n-1)}{2}\alpha_n^2 \quad \Rightarrow \quad \alpha_n^2 < \frac{2n-2}{n(n-1)} < \frac{2n}{n(n/2)} = \frac{4}{n}$$

for $n > 2$. Hence $\alpha_n < 2/\sqrt{n}, n > 2$, and so $\alpha_n \to 0$ as $n \to \infty$.

Step 2

We next show that $\limsup_{n\to\infty}|na_n|^{1/n} = \limsup_{n\to\infty}|a_n|^{1/n}$. From step 1, $\limsup_{n\to\infty}|a_n|^{1/n} \leqslant \limsup_{n\to\infty}|na_n|^{1/n} \leqslant \lim_{n\to\infty} n^{1/n}.\limsup_{n\to\infty}|a_n|^{1/n} = \limsup_{n\to\infty}|a_n|^{1/n}$, as required.

Step 3

Note that $\lim_{n\to\infty}|z|^{1/n} = 1$ for any finite $|z|$. Hence from step 2, $\limsup_{n\to\infty}|na_n z^{n-1}|^{1/n} = \limsup_{n\to\infty}|a_n z^n|^{1/n}$ and so $\sum_{n=1}^{\infty} na_n z^{n-1}$ and $\sum_{n=0}^{\infty} a_n z^n$ have the same radius of convergence by Theorem 4.14(ii). ■

Important Notes

(i) Theorem 4.17 states that a power series defines an analytic function within its domain of convergence. Letting $f(z) = \sum_{n=0}^{\infty} a_n z^n$ for $|z| < R$, where the given power series has radius of convergence R, it follows by 4.17 and 3.14 that the derivatives of f of all orders exist at any point satisfying $|z| < R$, and that $a_n = f^{(n)}(0)/n!$ for each $n \in \mathbb{N}$. More generally, if $f(z) = \sum_{n=0}^{\infty} a_n(z-\alpha)^n$ for $|z| < R$, then $a_n = f^{(n)}(\alpha)/n!$.

(ii) Because of (i), power series are sometimes used to define standard analytic functions. For instance, the power series $\sum_{n=0}^{\infty} z^n/n!$ is absolutely convergent for all z and can be used to define the exponential function, which is entire. We shall show shortly that the definition $e^z = \sum_{n=0}^{\infty} z^n/n!$ is equivalent to the definition given in Chapter 2.

(iii) It follows by Theorem 4.9 that two power series may be added or subtracted termwise within the intersection of their domains of convergence. It also follows from the note after Theorem 4.11 that two power series may also be multiplied within the intersection of their domains of absolute convergence. We give a proof of this fact in the next section.

Exercise **4.2.1** Use the definition of a convergent series to prove that $\sum_{n=1}^{\infty} z^{n-1} = 1/(1-z)$ if $|z| < 1$, and that the series diverges for $|z| \geqslant 1$. (You may assume that $\lim_{n\to\infty} z^n = 0$ for $|z| < 1$ and diverges otherwise.) Hence find $\sum_{n=1}^{\infty}(1/4i)^n$.

Exercise **4.2.2** Let $\sum_{n=1}^{\infty} z_n$ be a convergent series. Prove that

(i) if $\lambda \in \mathbb{C}$ then $\sum_{n=1}^{\infty}(\lambda z_n) = \lambda\sum_{n=1}^{\infty} z_n$

(ii) $z_n \to 0$ as $n \to \infty$

Exercise **4.2.3** Use standard results concerning real infinite series to prove the following:

(i) $\sum_{n=1}^{\infty} 1/n^{\alpha}$, where $\alpha \in \mathbb{C}$, converges for $\operatorname{Re}\alpha > 1$ and diverges for $\operatorname{Re}\alpha \leqslant 0$

(ii) $\sum_{n=1}^{\infty} e^{in}/n^2$ and $\sum_{n=1}^{\infty} (1/n^2)\sin(n+i)$ both converge

Exercise **4.2.4** Prove that if $\sum_{n=1}^{\infty} z_n$ is absolutely convergent, then

$$\left|\sum_{n=1}^{\infty} z_n\right| \leqslant \sum_{n=1}^{\infty} |z_n|$$

Exercise **4.2.5** Show that

$$\sum_{n=1}^{\infty} \frac{z^4}{(1+z^4)^{n-1}} = 1 + z^4 \quad \text{for} \quad |1+z^4| > 1$$

By considering the remainder of the series after k terms, prove that the series is uniformly convergent in the region given by $|1+z^4| > 1$, but not uniformly convergent for $|1+z^4| \geqslant 1$.

Exercise **4.2.6** Let $\sum_{n=1}^{\infty} f_n(z)$ be uniformly convergent to $F(z)$ in any compact subset of a simply connected region $\mathscr{R}$ and suppose that each f_n is analytic in $\mathscr{R}$. Use Theorem 4.8 to prove that F is analytic in $\mathscr{R}$ and that $F'(z) = \sum_{n=1}^{\infty} f'_n(z)$ for all z in $\mathscr{R}$.

Exercise **4.2.7** Use Weierstrass's M test to prove that

(i) $\sum_{n=1}^{\infty} z^n/n^2(n+1)$ is uniformly convergent for $|z| \leqslant 1$

(ii) $\sum_{n=1}^{\infty} 1/(z^2+n^2)$ is uniformly convergent in any annular region $m-1 < |z| < m$ where $m \in \mathbb{N}$

(iii) $\sum_{n=1}^{\infty} (\sin nz)/n^3$ is not uniformly convergent on $|z| \leqslant a$ but is uniformly convergent on $|x| \leqslant a$ where $x \in \mathbb{R}$.

Exercise **4.2.8** Find the radius of convergence of the following power series:

(i) $\displaystyle\sum_{n=1}^{\infty} \frac{n^2 z^{2n}}{(3n)!}$

(ii) $\displaystyle\sum_{n=1}^{\infty} \frac{(2z)^n}{\sqrt{n}}$

(iii) $\displaystyle\sum_{n=1}^{\infty} (-1)^n 3^{n+1} n z^{2n+1}$

(iv) $\displaystyle\sum_{n=1}^{\infty} n^n z^n$

(v) $\displaystyle\sum_{n=0}^{\infty} 3^n z^{(2n)!}$

(vi) $1 + 2z + z^2 + (2z)^3 + z^4 + (2z)^5 + \ldots$

Exercise **4.2.9** Show that, although $\sum_{n=1}^{\infty} f_n(z)$ and $\sum_{n=1}^{\infty} f'_n(z)$ have the same radius of convergence, where $f_n(z) = z^n/n^2$, they do not converge for the same set of values of z.

Exercise **4.2.10** Suppose that $\sum_{n=0}^{\infty} a_n z^n$ converges absolutely for $|z| < R$. Let z be any fixed point satisfying $|z| < R$ and let ρ satisfy $|z| < \rho < R$. Show that the series $\sum_{n=1}^{\infty} n|z|^n/\rho^n$ converges by the ratio test. Hence, by comparing $\sum_{n=1}^{\infty} |na_n z^{n-1}|$ with $\sum_{n=0}^{\infty} |a_n \rho^n|$, show that $\sum_{n=1}^{\infty} na_n z^{n-1}$ also converges absolutely for $|z| < R$.

Exercise **4.2.11** The exponential function may be defined by $e^z = \sum_{n=0}^{\infty} z^n/n!$ for all $z \in \mathbb{C}$.

(i) Use this definition to find the derivative of $f: \mathbb{C} \to \mathbb{C}$, defined by $f(z) = e^{-z} \cdot e^z$, for any $z \in \mathbb{C}$. Deduce that $e^{-z} = 1/e^z$ for all $z \in \mathbb{C}$.

(ii) Use the power series definition to find the derivative of $g : \mathbb{C} \to \mathbb{C}$, defined by $g(z) = e^\alpha \cdot e^z / e^{\alpha+z}$, α constant, for any $z \in \mathbb{C}$. Deduce that $e^{\alpha+z} = e^\alpha \cdot e^z$ for any z and $\alpha \in \mathbb{C}$.

Taylor Series

Suppose that the power series $\sum_{n=0}^{\infty} a_n(z-\alpha)^n$ converges absolutely to $f(z)$ within some circle of convergence, centred at α. We have shown in the previous section how it then follows by Theorem 4.17 that f is an analytic function and $a_n = f^{(n)}(\alpha)/n!$. We now show that the coefficients a_n can also be expressed in integral form, unlike in the case of real variables.

Let $f(z) = \sum_{n=0}^{\infty} a_n(z-\alpha)^n$ within some circle of convergence, centred at α. Then multiplying by $(z-\alpha)^{-1-m}$, where $m \in \mathbb{N}$, and integrating termwise, which is valid by Theorem 4.16, gives

$$\int_{\mathscr{C}} \frac{f(z)}{(z-\alpha)^{m+1}} dz = \sum_{n=0}^{\infty} a_n \int_{\mathscr{C}} (z-\alpha)^{n-m-1} dz \tag{4.2}$$

where $\mathscr{C}$ is any simple closed contour lying within the circle of convergence. It follows by the fundamental theorem of calculus that $\int_{\mathscr{C}} (z-\alpha)^{n-m-1} dz = 0$ if $n \neq m$, and it follows by Cauchy's integral formula that $\int_{\mathscr{C}} (z-\alpha)^{-1} dz = 2\pi i$. Hence, from (4.2) and Theorem 3.13, it follows that

$$a_m = \frac{1}{2\pi i} \int_{\mathscr{C}} \frac{f(z)}{(z-\alpha)^{m+1}} dz = \frac{f^{(m)}(\alpha)}{m!} \tag{4.3}$$

Essentially the converse of Theorem 4.17 is also true. In other words, any function which is analytic at a point $\alpha \in \mathbb{C}$ can be expanded in a power series, $\sum_{n=0}^{\infty} a_n(z-\alpha)^n$, about α, where the coefficients are given by (4.3). This power series is the Taylor series expansion of the function about α.

Theorem 4.18. Taylor Series

If $f : A \subseteq \mathbb{C} \to \mathbb{C}$ is analytic inside a circle, centre α, of radius r, then at each point z inside this circle,

$$f(z) = \sum_{n=0}^{\infty} a_n(z-\alpha)^n \quad \text{where} \quad a_n = \frac{f^{(n)}(\alpha)}{n!} \quad \text{for all } n \in \mathbb{N}$$

That is, the given power series converges to $f(z)$ for all z satisfying $|z - \alpha| < r$. □

Proof

Suppose first of all that f is analytic inside the circle $\mathscr{C}$ given by $|z| = r$. Then by Cauchy's integral formula, if w is any point inside $\mathscr{C}$ so that $|w| < |z|$,

$$f(w) = \frac{1}{2\pi i}\int_{\mathscr{C}} \frac{f(z)}{z-w}\,dz = \frac{1}{2\pi i}\int_{\mathscr{C}} \frac{f(z)}{z}\frac{1}{1-w/z}\,dz$$

$$\Rightarrow \quad f(w) = \frac{1}{2\pi i}\int_{\mathscr{C}} \frac{f(z)}{z}\sum_{n=0}^{\infty}\left(\frac{w}{z}\right)^n dz = \sum_{n=0}^{\infty}\frac{w^n}{2\pi i}\int_{\mathscr{C}} \frac{f(z)}{z^{n+1}}\,dz$$

using the geometric series result (see Exercise 4.2.1). The fact that the series can be integrated termwise follows from Theorem 4.12 since the series is uniformly convergent within the given circle by Weierstrass's M test. (Let $|w/z| = \rho < 1$. Then $|w/z|^n = \rho^n$ and $\sum_{n=0}^{\infty} \rho^n$ converges.) It then follows by (4.3) that

$$f(w) = \sum_{n=0}^{\infty} f^{(n)}(0)w^n/n!$$

Now suppose that f is analytic inside the circle $|z - \alpha| = r$. Letting $z - \alpha = w$ in the above result shows that

$$f(z) = f(w+\alpha) = \sum_{n=0}^{\infty} \frac{f^{(n)}(\alpha)(z-\alpha)^n}{n!} \qquad (|z-\alpha| < r)$$

■

The power series expansion of f in 4.18 is the **Taylor series** expansion of f about α and is formally the same as in the real case. When $\alpha = 0$ the Taylor series is sometimes called the **Maclaurin series** expansion of f. Any Taylor series can be obtained from a Maclaurin series by a change of origin.

Important Notes

(i) It follows by Theorem 4.18 and the preceding analysis that any power series is the Taylor series expansion of its pointwise sum function.

(ii) Note that the Taylor series expansion of f about α converges to $f(z)$ within the circle whose radius is the distance from α to the nearest point, β, where f fails to be analytic. It can be shown that this is actually the largest circle centred at α such that the Taylor series coverges to $f(z)$ for all z inside it. It may happen that the Taylor series converges for $|z - \alpha| = \rho$, where $\rho > r$, but in this case the series will not converge to $f(z)$ (see Exercise 4.3.4).

We shall later prove Theorem 4.18 as a corollary of a more general result, Laurent's theorem (Theorem 4.21), the proof of which does not depend on Cauchy's integral results or on results concerning uniform convergence.

Example 4.9

Since it is formally the same as in the real case, the Taylor series expansion of a function $f: \mathbb{C} \to \mathbb{C}$ takes exactly the same form as the Taylor series expansion of the corresponding function $f: \mathbb{R} \to \mathbb{R}$.

(i) Let $f(z) = e^z$ for all $z \in \mathbb{C}$ so that f is entire. $f^{(n)}(z) = e^z$ for any $n \in \mathbb{Z}_{\geqslant 0}$, so that $f^{(n)}(0) = 1$ for any such n. Hence the Maclaurin series expansion of f is

$$e^z = 1 + z + \frac{z^2}{2!} + \ldots = \sum_{n=0}^{\infty} \frac{z^n}{n!}$$

and the series converges absolutely for all z. Then letting $z = w - i$ say, gives

$$e^{w-i} = \sum_{n=0}^{\infty} \frac{(w-i)^n}{n!} \quad \Rightarrow \quad e^z = e^i \sum_{n=0}^{\infty} \frac{(z-i)^n}{n!}$$

and this is the Taylor series expansion of f about i, which converges for all z. This expansion may also be found directly by using Theorem 4.18.

Similarly, the standard Maclaurin series expansions for the other elementary functions of a real variable hold for the analogous functions of a complex variable in the corresponding circle of convergence. In particular,

$$\sinh z = \sum_{n=0}^{\infty} \frac{z^{2n+1}}{(2n+1)!} \qquad \cosh z = \sum_{n=0}^{\infty} \frac{z^{2n}}{(2n)!} \tag{4.4}$$

and both series converge for all $z \in \mathbb{C}$. The Maclaurin series expansions of $\sin z$ and $\cos z$ follow easily from (4.4) using $\sin z = -i \sinh(iz)$ and $\cos z = \cosh(iz)$. Also,

$$\operatorname{Log}(1+z) = \sum_{n=0}^{\infty} (-1)^n \frac{z^{n+1}}{n+1}$$

and the series converges for $|z| < 1$, since $\operatorname{Log}(1+z)$ has a singular point at $z = -1$.

(ii) The derivatives of the function $f: \mathbb{C}\backslash\{0\} \to \mathbb{C}$ defined by $f(z) = 1/z$ are $f^{(n)}(z) = (-1)^n n! z^{-n-1}$ for all $n \in \mathbb{Z}_{\geqslant 0}$, so that $f^{(n)}(1) = (-1)^n n!$ for each $n \in \mathbb{N}$ and $f(1) = 1$. Hence the Taylor series expansion of f about 1, say, is

$$\frac{1}{z} = \sum_{n=0}^{\infty} (-1)^n (z-1)^n$$

Since f has a singular point at 0, the series converges for all z satisfying $|z-1| < 1$. Note that it diverges for $|z-1| \geqslant 1$.

(iii) Let $f(z) = (1+z)^\alpha$ where $\alpha \in \mathbb{R}$. Then $f'(z) = \alpha(1+z)^{\alpha-1}$, $f''(z) = \alpha(\alpha-1)(1+z)^{\alpha-2}, \ldots, f^{(n)}(\alpha) = \alpha(\alpha-1)\ldots(\alpha-n+1)(1+z)^{\alpha-n}$ for all $n \in \mathbb{N}$. Hence $f^{(n)}(0) = \alpha(\alpha-1)\ldots(\alpha-n+1)$ for all $n \in \mathbb{N}$. Then by 4.18, we obtain the usual **binomial series**

$$(1+z)^\alpha = 1 + \alpha z + \frac{\alpha(\alpha-1)}{2!} z^2 + \frac{\alpha(\alpha-1)(\alpha-2)}{3!} z^3 + \ldots$$

which, if $\alpha \notin \mathbb{N}$, converges for $|z| < 1$ since the given function has a branch point or pole at $z = -1$.

Theorem 4.18 can be used to give a simple proof of the fact that two absolutely convergent power series can be multiplied termwise to give another absolutely convergent power series, whose sum is the product of the previous two.

Theorem 4.19. Multiplication of Power Series

Suppose that $\sum_{n=0}^{\infty} a_n z^n = f(z)$, with radius of convergence R_1, and $\sum_{n=0}^{\infty} b_n z^n = g(z)$, with radius of convergence R_2. Then $f(z)g(z) = \sum_{n=0}^{\infty} c_n z^n$ where $c_n = \sum_{r=0}^{n} a_r b_{n-r}$ and the series has radius of convergence of at least $R = \min(R_1, R_2)$. □

Proof

Both f and g are analytic for $|z| < R$ and $a_n = f^{(n)}(0)/n!$ and $b_n = g^{(n)}(0)/n!$ by 4.17. The product fg is also analytic for $|z| < R$ and is represented in this region by a Taylor series $f(z)g(z) = \sum_{n=0}^{\infty} c_n z^n$, where

$$c_n = \frac{(fg)^{(n)}(0)}{n!} = \sum_{r=0}^{n} \frac{1}{n!} \frac{n!}{r!(n-r)!} f^{(r)}(0) g^{(n-r)}(0) = \sum_{r=0}^{n} a_r b_{n-r}$$

by Leibniz's formula for the nth derivative of a product. ■

Important Note

Since any power series is the Taylor series expansion of its pointwise sum function, the coefficients in a Taylor series expansion can be calculated by a number of different means, not just by 4.18. Note also that Taylor series can be differentiated and integrated termwise within their circle of convergence by 4.16 and 4.17.

Example 4.10

(i) The **error function**, erf : $\mathbb{C} \to \mathbb{C}$ is defined by

$$\operatorname{erf}(z) = \frac{2}{\sqrt{\pi}} \int_0^z e^{-w^2} \, dw$$

(Note that the integral does not depend on the choice of contour joining 0 to z, since the integrand is entire.) It follows by the Maclaurin series expansion of e^z that

$$e^{-w^2} = \sum_{n=0}^{\infty} (-1)^n \frac{w^{2n}}{n!}$$

and the series converges for all $w \in \mathbb{C}$. Then integrating termwise by 4.16 gives

$$\operatorname{erf}(z) = \frac{2}{\sqrt{\pi}} \sum_{n=0}^{\infty} \frac{(-1)^n}{n!} \int_0^z w^{2n} \, dw = \frac{2}{\sqrt{\pi}} \sum_{n=0}^{\infty} (-1)^n \frac{z^{2n+1}}{(2n+1)n!}$$

as the Maclaurin series expansion of $\operatorname{erf}(z)$, which converges for all $z \in \mathbb{C}$.

(ii) It follows from the Maclaurin expansions of e^z and $\cos z$ and Theorem 4.19 that

$$\begin{aligned} e^z \sec z &= \left(1 + z + \frac{z^2}{2!} + \frac{z^3}{3!} + \dots\right)\left(1 - \frac{z^2}{2!} + \frac{z^4}{4!} - \dots\right)^{-1} \\ &= \left(1 + z + \frac{z^2}{2!} + \frac{z^3}{3!} + \dots\right)\left(1 + \left(\frac{z^2}{2!} - \frac{z^4}{4!} + \dots\right)\right. \\ &\qquad \left. + \left(\frac{z^2}{2!} - \frac{z^4}{4!} + \dots\right)^2 \dots\right) \end{aligned}$$

using the binomial series result of Example 4.9. Hence

$$e^z \sec z = \left(1 + z + \frac{z^2}{2!} + \frac{z^3}{3!} + \frac{z^4}{4!} + \dots\right)\left(1 + \frac{z^2}{2} + \frac{5z^4}{24} + \dots\right)$$

$$\Rightarrow \quad e^z \sec z = 1 + z + z^2 + \frac{2z^3}{3} + \frac{z^4}{2} + \dots$$

and this is the Maclaurin series expansion of $e^z \sec z$, which converges for $|z| < \pi/2$ by 4.18.

Theorem 4.18 can be used to give another characterisation of zeros of a function.

Lemma 4.20. Zeros of a Function

Let f be analytic inside the circle $|z-\alpha|=r$, so that $f(z)=\sum_{n=0}^{\infty} a_n(z-\alpha)^n$ for $|z-\alpha|<r$. Then f has a zero of order m at α if and only if $a_n=0$ for $n<m$ and $a_m \neq 0$. □

Proof

Suppose that $a_n=0$ for $n<m$ with $a_m \neq 0$. Then

$$f(z)=(z-\alpha)^m \sum_{n=0}^{\infty} a_{n+m}(z-\alpha)^n$$

so that $f(z)=(z-\alpha)^m g(z)$ where g is analytic at α and $g(\alpha)=a_m \neq 0$, as required.

Now suppose that f has a zero of order m at α so that $f(z)=(z-\alpha)^m g(z)$ where $g(\alpha)\neq 0$ and g is analytic at α. Then by 4.18 and hypothesis, g has a Taylor series expansion about α of the form $g(z)=\sum_{n=m}^{\infty} a_n(z-\alpha)^{n-m}$, where $a_m \neq 0$, so that the result follows. ■

Note

Theorems 4.18 and 4.19 show that if f is analytic at α, then α is a zero of order m of f if and only if

$$f(\alpha)=f'(\alpha)=\ldots=f^{(m-1)}(\alpha)=0 \quad \text{and} \quad f^{(m)}(\alpha)\neq 0$$

Theorem 4.18 can be used to give an alternative proof of Liouville's theorem, as the following example demonstrates.

Example 4.11

Suppose that f is analytic inside the circle $|z-\alpha|=r$ and that $|f(z)|\leqslant M(\rho)$ on the circle $\mathscr{C}$, given by $|z-\alpha|=\rho<r$, where $M(\rho)$ is a positive constant which depends only on the radius of the circle. It follows by 4.18 that $f(z)=\sum_{n=0}^{\infty} a_n(z-\alpha)^n$ for $|z-\alpha|<r$. Then by (4.3),

$$|a_n|=\left|\frac{1}{2\pi i}\int_{\mathscr{C}} \frac{f(z)}{(z-\alpha)^{n+1}}\,dz\right| \leqslant \frac{1}{2\pi}\cdot\frac{M(\rho)}{\rho^{n+1}}\cdot 2\pi\rho=\frac{M(\rho)}{\rho^n}$$

by the ML lemma, 3.2. Hence if f is entire and $|f(z)|<M$ for all z, then $|a_n|\leqslant M/\rho^n$ for arbitrary ρ. Hence $a_n=0$ for $n>0$ and $|a_0|\leqslant M$, so that $f(z)=a_0$, a constant. This gives Liouville's theorem.

Exercise

4.3.1 Use Theorem 4.18 to find (i) the Maclaurin series expansion of $\sinh z$, and (ii) the Taylor series expansion of $1/z^2$ about $z=2$. State the circle of convergence in each case.

Exercise **4.3.2** Find the Maclaurin series expansion of $f(z) = 1/(1+z)$ using 4.18 and state the circle of convergence. Use this series to find

(i) the Taylor series expansion of $f(z)$ about 3

(ii) the Maclaurin series expansion of $1/(1+z)^2$

(iii) the Maclaurin series expansion of $\text{Log}\,(1+z)$

State the circle of convergence of the series in each case.

Exercise **4.3.3** Use the Maclaurin series expansion of $\cos z$ and the binomial series to find the first four non-zero terms in the Maclaurin series expansion of $\sec z$.

Exercise **4.3.4** Find the Taylor series expansion of $z^{1/2}$ about $-1-i$, and show that the series converges to $z^{1/2}$ for $|z+1+i| < 1$. Use the ratio test to show that the series converges for $|z+1+i| < \sqrt{2}$.

Exercise **4.3.5** Use the Maclaurin series expansion of $\sin z$ to find

(i) the Maclaurin series expansion of the **Fresnel sine integral** defined by

$$S(z) = \sqrt{\frac{2}{\pi}} \int_0^z \sin(w^2)dw$$

(ii) the Maclaurin series expansion of $f(z) = \sin z/(1-z)$

Hence find the nth derivative of $\sin z/(1-z)$ at $z = 0$.

Exercise **4.3.6** Let f be an entire function which satisfies $|f(z)| \leqslant k|z|^m$ for all $z \in \mathbb{C}$, where k and m are positive constants. Use the integral formula for the coefficients in the Maclaurin series expansion of f and the ML lemma to prove that

(i) if $m \in \mathbb{N}$ then $f(z) = a_m z^m$ for some a_m

(ii) if $m \notin \mathbb{N}$ then $f(z) = 0$ for all z

(Compare this result with the result of Exercise 3.3.5.)

Laurent Series

Very often it is necessary to expand a function about an isolated singular point, rather than a non-singular point. Such expansions, which are not Taylor series, are particularly useful in integration.

Example 4.12

It follows by Theorem 4.18 that

$$\frac{1}{1-z} = \sum_{n=0}^{\infty} z^n \qquad (|z| < 1)$$

$$\Rightarrow \quad \frac{1}{z(1-z)} = \frac{1}{z} + \sum_{n=0}^{\infty} z^n \qquad (0 < |z| < 1)$$

Although the second series converges for all z in the annular region $0 < |z| < 1$, it is an expansion of $f(z) = 1/z(1-z)$ about the isolated singular point 0, so is not a Taylor series.

Similarly, from Example 4.9(i), it follows that

$$e^{1/z} = \sum_{n=0}^{\infty} \frac{1}{n!z^n} \qquad \text{(for all } z \neq 0)$$

which again is the expansion of a function about an isolated singular point.

In general, if $\alpha \in \mathbb{C}$ is any isolated singular point of a function f, then f has a series expansion of the form

$$f(z) = \sum_{n=0}^{\infty} a_n(z-\alpha)^n + \sum_{n=1}^{\infty} \frac{b_n}{(z-\alpha)^n} \tag{4.5}$$

about α, where a_n and b_n are constants for all n, which converges for all z in some annular region $0 < |z - \alpha| < r_1$. The series in (4.5) is the **Laurent series** expansion of f about the isolated singular point α. Note that the constants a_n and b_n are not expressible in terms of values of derivatives of f at α since α is a singular point of f. However, they are expressible in terms of integrals, as in the case of the coefficients in a Taylor series. This is the content of the following important result.

Theorem 4.21. Laurent's Theorem

Let α be a singular point of $f\colon A \subseteq \mathbb{C} \to \mathbb{C}$. Let $\mathscr{C}_1$ and $\mathscr{C}_2$ be two concentric circles, centred at α, with radii r_1 and r_2 respectively, where $r_2 < r_1$. If f is analytic on $\mathscr{C}_1$ and $\mathscr{C}_2$ and throughout the annular region between the circles, then at each point in that region,

$$f(z) = \sum_{n=0}^{\infty} a_n(z-\alpha)^n + \sum_{n=1}^{\infty} \frac{b_n}{(z-\alpha)^n} \tag{4.5}$$

$$\text{where} \quad a_n = \frac{1}{2\pi i}\int_{\mathscr{C}_1} \frac{f(z)}{(z-\alpha)^{1+n}}\,dz \qquad (\text{for all } n \in \mathbb{Z}_{\geqslant 0})$$

$$\text{and} \quad b_n = \frac{1}{2\pi i}\int_{\mathscr{C}_2} \frac{f(z)}{(z-\alpha)^{1-n}}\,dz \qquad (\text{for all } n \in \mathbb{N}) \qquad \square$$

Important Note

It is convenient for the proof of 4.21 to have two concentric circles defining the domain of convergence. However, it follows by 4.21 and Lemma 3.10 that if f is analytic at every point inside and on the circle $\mathscr{C}_1$, except at α itself, so that α is an isolated singular point of f, then for $0 < |z - \alpha| < r_1$,

$$f(z) = \sum_{n=-\infty}^{\infty} c_n(z-\alpha)^n \quad \text{where} \quad c_n = \frac{1}{2\pi i}\int_{\mathscr{C}_1} \frac{f(z)}{(z-\alpha)^{n+1}}\,dz \tag{4.6}$$

for all $n \in \mathbb{Z}$. More generally, if f is analytic in the annular region $r_2 < |z - \alpha| < r_1$, then $\mathscr{C}_1$ in (4.6) is replaced by any simple closed contour lying inside the annular region.

Proof of 4.21

It is convenient to split the proof up into a number of parts.

Step 1

Let z be any point in the annular region and let $\mathscr{C}$ be a circle centred at z, which lies entirely within the annular region. Construct contours $\Gamma_k, k = 1, 2, 3$ as shown in Fig. 4.1. Note that by hypothesis and Cauchy's theorem, $\int_{\Gamma_k}(f(w)/(w-z))\,dw = 0$, for each k. Hence it follows that, since the net contribution to the integrals along the line segments is zero by Lemma 3.1,

$$\int_{\mathscr{C}_1} \frac{f(w)}{w-z}\,dw - \int_{\mathscr{C}_2} \frac{f(w)}{w-z}\,dw - \int_{\mathscr{C}} \frac{f(w)}{w-z}\,dw = \sum_{n=1}^{3} \int_{\Gamma_k} \frac{f(w)}{w-z}\,dw = 0 \tag{4.7}$$

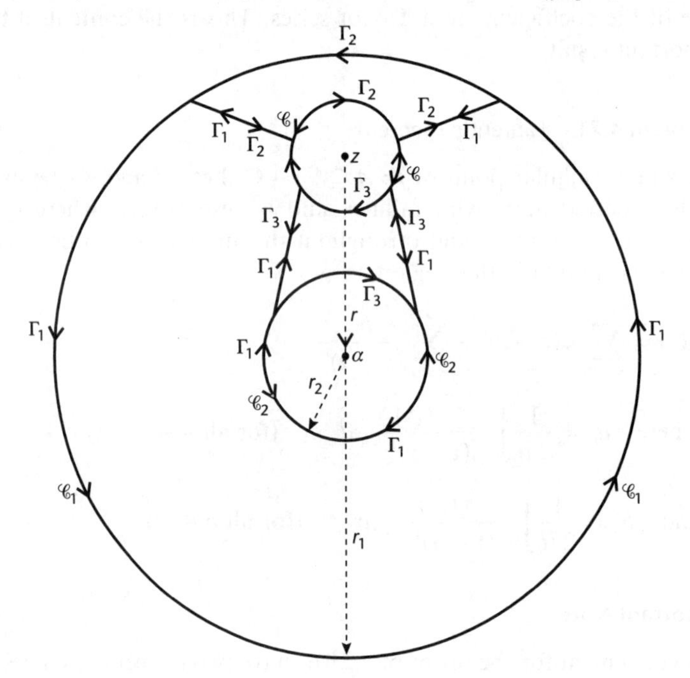

Figure 4.1

Also, by Cauchy's integral formula, $\int_{\mathscr{C}}(f(w)/(w-z))dw = 2\pi i f(z)$, so that by (4.7),

$$f(z) = \frac{1}{2\pi i}\int_{\mathscr{C}_1}\frac{f(w)}{w-z}dw - \frac{1}{2\pi i}\int_{\mathscr{C}_2}\frac{f(w)}{w-z}dw \tag{4.8}$$

In order to proceed further, it is necessary to expand the integrands in (4.8). It is possible to proceed using the technique used in the proof of 4.18 and uniform convergence results. Instead, we give an alternative approach which does not rely on uniform convergence results.

Step 2

Note that for any $\beta \in \mathbb{C}$, $\beta \neq 1$,

$$\frac{1}{1-\beta} = 1 + \beta + \beta^2 + \ldots + \beta^{k-1} + \frac{\beta^k}{1-\beta} \tag{4.9}$$

Letting $\beta = \dfrac{z-\alpha}{w-\alpha}$, $w \neq z$, in (4.9) gives

$$\frac{f(w)}{w-z} = \frac{f(w)}{w-\alpha}\frac{1}{1-(z-\alpha)/(w-\alpha)}$$

$$= \frac{f(w)}{w-\alpha} + \frac{f(w)}{(w-\alpha)^2}(z-\alpha) + \ldots + \frac{f(w)}{(w-\alpha)^k}(z-\alpha)^{k-1} + \frac{(z-\alpha)^k f(w)}{(w-\alpha)^k(w-z)}$$

$$\Rightarrow \quad \frac{1}{2\pi i}\int_{\mathscr{C}_1}\frac{f(w)}{w-z}dw = a_0 + a_1(z-\alpha) + \ldots + a_{k-1}(z-\alpha)^{k-1} + R_k(z) \tag{4.10}$$

where each a_n is given by the formula in the statement of the theorem and

$$R_k(z) = \frac{(z-\alpha)^k}{2\pi i}\int_{\mathscr{C}_1}\frac{f(w)}{(w-z)(w-\alpha)^k}dw \tag{4.11}$$

Similarly, letting $\beta = \dfrac{w-\alpha}{z-\alpha}$ in (4.9) gives

$$\frac{-f(w)}{w-z} = \frac{f(w)}{z-\alpha}\frac{1}{1-(w-\alpha)/(z-\alpha)}$$

$$= \frac{f(w)}{z-\alpha} + \frac{f(w)}{(z-\alpha)^2}(w-\alpha) + \ldots + \frac{f(w)}{(z-\alpha)^k}(w-\alpha)^{k-1} - \frac{f(w)(w-\alpha)^k}{(w-z)(z-\alpha)^k}$$

$$\Rightarrow \quad \frac{-1}{2\pi i}\int_{\mathscr{C}_2}\frac{f(w)}{w-z}dw = \frac{b_1}{z-\alpha} + \frac{b_2}{(z-\alpha)^2} + \ldots + \frac{b_k}{(z-\alpha)^k} + Q_k(z) \tag{4.12}$$

where the b_n are given by the formula in the theorem and

$$Q_k(z) = \frac{1}{2\pi i(z-\alpha)^k} \int_{\mathscr{C}_2} \frac{(w-\alpha)^k f(w)}{z-w} \, dw \tag{4.13}$$

Substituting (4.10) and (4.12) into (4.8) gives the required result as long as $\lim_{k\to\infty} R_k(z) = \lim_{k\to\infty} Q_k(z) = 0$.

Step 3

Let $r = |z-\alpha|$ as shown in Fig. 4.1. Then by the triangle inequality it follows that for any point w on $\mathscr{C}_1$,

$$|w-z| \geqslant |w-\alpha| - |z-\alpha| = r_1 - r$$

as can be seen in Fig. 4.1. Using the ML lemma, (4.11) then gives

$$|R_k(z)| = \frac{|z-\alpha|^k}{2\pi} \left| \int_{\mathscr{C}_1} \frac{f(w)}{(w-z)(w-\alpha)^k} \, dw \right| \leqslant \frac{r^k M}{2\pi(r_1 - r){r_1}^k}(2\pi r_1)$$

where $|f(w)| \leqslant M$ on $\mathscr{C}_1$. (Notice that f is bounded on $\mathscr{C}_1$ by 4.5.) Hence

$$|R_k(z)| \leqslant \left(\frac{r}{r_1}\right)^k \frac{Mr_1}{r_1 - r} \quad \text{where} \quad \frac{r}{r_1} < 1$$

(see Fig. 4.1). Since $\lim_{k\to\infty} (r/r_1)^k = 0$, it follows by the sandwich theorem for real sequences that $\lim_{k\to\infty} R_k(z) = 0$.

Similarly, it follows by (4.13) and the ML lemma that

$$|Q_k(z)| = \frac{1}{2\pi r^k} \left| \int_{\mathscr{C}_2} \frac{(w-\alpha)^k f(w)}{z-w} \, dw \right| \leqslant \frac{r_2^k M^*}{2\pi r^k (r - r_2)}(2\pi r_2)$$

where $|f(w)| \leqslant M^*$ on $\mathscr{C}_2$. Hence

$$|Q_k(z)| \leqslant \left(\frac{r_2}{r}\right)^k \frac{M^* r_2}{r - r_2} \quad \text{where} \quad \frac{r_2}{r} < 1$$

so that $\lim_{k\to\infty} Q_k(z) = 0$, as required. ■

Important Note

It may not be possible to expand a given function in any Laurent series about a non-isolated singular point. For example, Log z has no Laurent series expansion about the branch point at 0, since there is no annular region of the type described in Theorem 4.21, throughout which Log is analytic.

Taylor's theorem now follows easily from Laurent's theorem.

Proof of 4.18

Notice that Theorem 4.21 still holds if α is a non-singular point of f. If f is analytic at all points inside and on $\mathscr{C}_2$, including α, then

$$b_n = \frac{1}{2\pi i}\int_{\mathscr{C}_2} \frac{f(z)}{(z-\alpha)^{1-n}}\,dz = 0 \qquad \text{(for all } n \in \mathbb{N}\text{)}$$

by Cauchy's theorem. Also,

$$a_n = \frac{1}{2\pi i}\int_{\mathscr{C}_1} \frac{f(z)}{(z-\alpha)^{n+1}}\,dz = \frac{f^{(n)}(\alpha)}{n!}$$

by Cauchy's integral theorem for derivatives, as required. ∎

Important Note

Any Laurent series, given by (4.5), can be thought of as the sum of two power series, one in $z-\alpha$ and one in $(z-\alpha)^{-1}$, which converge absolutely with the domain of convergence of the Laurent series. It then follows from Theorem 4.15 that any Laurent series converges uniformly to its pointwise sum function within its domain of convergence. Hence by previous comments and Theorem 4.12:

(i) Any Laurent series can be multiplied by a constant, integrated or differentiated termwise within the domain of convergence.

(ii) Two Laurent series can be added or multiplied within the intersection of their domains of convergence.

In any case, the result is another convergent Laurent series.

In practice, it is not usually necessary to use Theorem 4.21 to calculate the Laurent series expansion of a given function about an isolated singular point. The following result states that the Laurent series expansion of a given function about a given point is unique, so the coefficients are not usually obtained by using their integral formulae.

Theorem 4.22. Uniqueness of Laurent Series

If the series

$$\sum_{n=-\infty}^{\infty} c_n(z-\alpha)^n = \sum_{n=0}^{\infty} a_n(z-\alpha)^n + \sum_{n=1}^{\infty} \frac{b_n}{(z-\alpha)^n}$$

(where a_k, b_k are constants for all k) converges to $f(z)$ at all points in some annular region about α, then it is the Laurent series expansion of $f(z)$ in that region. □

Proof

Let $\mathscr{C}$ be any circle lying in the given region, centred at α. Then by hypothesis and the above comments,

$$\frac{1}{2\pi i}\int_{\mathscr{C}} \frac{f(z)}{(z-\alpha)^{m+1}}\,dz = \sum_{n=-\infty}^{\infty} \frac{c_n}{2\pi i}\int_{\mathscr{C}} \frac{dz}{(z-\alpha)^{m-n+1}} \qquad \text{(for any } m \in \mathbb{Z})$$

Also, by definition

$$\int_{\mathscr{C}} \frac{dz}{(z-\alpha)^{m-n+1}} = \begin{cases} 2\pi i \text{ if } n = m \\ 0 \text{ if } n \neq m \end{cases}$$

Hence

$$c_m = \frac{1}{2\pi i}\int_{\mathscr{C}} \frac{f(z)}{(z-\alpha)^{m+1}}\,dz \quad \text{(for all } m \in \mathbb{Z})$$

as required. ∎

Example 4.13

(i) From the Maclaurin series for e^z it follows that

$$e^{1/z^2} = \sum_{n=0}^{\infty} \frac{1}{z^{2n} n!} \qquad (|z| > 0)$$

Also, for fixed $m \in \mathbb{N}$,

$$\frac{e^z}{z^m} = \frac{1}{z^m} + \ldots + \frac{1}{(m-1)!z} + \sum_{n=m}^{\infty} \frac{z^{n-m}}{n!} \qquad (|z| > 0)$$

These are the Laurent series expansions of the given functions about 0 by 4.22.

(ii) From the Maclaurin series for $\sin z$ it follows that

$$\sin z \sin(1/z) = \left(z - \frac{z^3}{3!} + \frac{z^5}{5!} - \frac{z^7}{7!} + \ldots\right)\left(\frac{1}{z} - \frac{1}{3!z^3} + \frac{1}{5!z^5} - \frac{1}{7!z^7} + \ldots\right)$$

for $|z| > 0$.

Hence

$$\begin{aligned}\sin z \sin(1/z) = {} & \left(1 + \frac{1}{(3!)^2} + \frac{1}{(5!)^2} + \ldots\right) \\ & - \left(\frac{1}{3!} + \frac{1}{3!5!} + \frac{1}{5!7!} + \ldots\right) z^2 \\ & + \left(\frac{1}{5!} + \frac{1}{3!7!} + \frac{1}{5!9!} + \ldots\right) z^4 - \ldots\end{aligned}$$

$$-\left(\frac{1}{3!}+\frac{1}{3!5!}+\frac{1}{5!7!}+\ldots\right)\frac{1}{z^2}$$
$$+\left(\frac{1}{5!}+\frac{1}{3!7!}+\frac{1}{5!9!}+\ldots\right)\frac{1}{z^4}-\ldots \qquad (|z|>0)$$

Once again, this is a Laurent series expansion of the given function about 0.

(iii) Let $f(z) = 1/(1+z^2)$. Then f has singular points at $\pm i$. Let $w = z - i$, so that

$$f(z) = \frac{1}{w(w+2i)} = \frac{1}{2iw}(1+w/2i)^{-1}$$

It then follows from the binomial series in Example 4.9(iii) and 4.22 that the Laurent series expansion of $f(z)$ about i is

$$f(z) = \frac{1}{2iw}\sum_{n=0}^{\infty}(-1)^n\left(\frac{w}{2i}\right)^n = \frac{1}{2i(z-i)}\sum_{n=0}^{\infty}(-1)^n\left(\frac{z-i}{2i}\right)^n$$

which converges for $|w/2i| < 1$, i.e. for $|z - i| < 2$.

(iv) It follows from Example 4.12 that if $\mathscr{C}$ is any simple closed contour lying within the annular region $0 < |z| < 1$,

$$\int_{\mathscr{C}}\frac{dz}{z(1-z)} = \int_{\mathscr{C}}\frac{dz}{z} + \sum_{n=0}^{\infty}\int_{\mathscr{C}} z^n\,dz = 2\pi i$$

by Cauchy's theorem and Cauchy's integral formula.

Partial fractions are often useful when finding Laurent series expansions of algebraic fractions, as the following example demonstrates. This example also demonstrates the fact that a given function can clearly have different Laurent series expansions in different regions of the complex plane.

Example 4.14

The function

$$f(z) = \frac{1}{z(z+1)(z+2)}$$

has singular points, which are simple poles, at 0, -1 and -2. By Laurent's theorem, f has Laurent series expansions about 0 valid in the regions $0 < |z| < 1$, $1 < |z| < 2$ and $|z| > 2$. Note that

$$f(z) = \frac{1}{2z} + \frac{1}{2(z+2)} - \frac{1}{z+1}$$

Using the binomial series, it follows that

$$\frac{1}{z+2} = \frac{1}{2(1+z/2)} = \frac{1}{2}\sum_{n=0}^{\infty}(-1)^n 2^{-n} z^n \qquad (|z/2| < 1)$$

$$\frac{1}{z+2} = \frac{1}{z(1+2/z)} = \frac{1}{z}\sum_{n=0}^{\infty}(-1)^n 2^n z^{-n} \qquad (|2/z| < 1)$$

$$\frac{1}{1+z} = \sum_{n=0}^{\infty}(-1)^n z^n \qquad (|z| < 1)$$

$$\frac{1}{1+z} = \frac{1}{z(1+1/z)} = \frac{1}{z}\sum_{n=0}^{\infty}(-1)^n z^{-n} \qquad (|1/z| < 1)$$

Hence it follows that

$$f(z) = \frac{1}{2z} + \sum_{n=0}^{\infty}(-1)^n(2^{-n-2} - 1)z^n \qquad (0 < |z| < 1)$$

$$f(z) = \frac{1}{2z} + \sum_{n=0}^{\infty}(-1)^n 2^{-n-2} z^n - \sum_{n=0}^{\infty}(-1)^n z^{-n-1} \qquad (1 < |z| < 2)$$

$$f(z) = \frac{1}{2z} + \sum_{n=0}^{\infty}(-1)^n(2^n - 1)z^{-n-1} \qquad (|z| > 2)$$

It follows by 4.22 that these are the Laurent series expansions of f about 0 in the given regions.

Laurent's theorem can be used to evaluate certain real integrals, by using the integral formulae for the coefficients in a known Laurent series, as the following example demonstrates.

Example 4.15

From the Maclaurin series expansion of e^z it follows that the Laurent series expansion of $e^{1/z}$ about 0 is

$$e^{1/z} = \sum_{n=0}^{\infty}\frac{1}{z^n n!} \qquad (|z| > 0)$$

(using Theorem 4.22). Comparing this series with (4.5) gives $a_0 = 1$, $a_n = 0$ for all $n > 1$ and $b_n = 1/n!$ for $n \in \mathbb{N}$. It follows by Theorem 4.21 that

$$b_n = \frac{1}{2\pi i}\int_{\mathscr{C}} \frac{e^{1/z}}{z^{1-n}}\,dz$$

where $\mathscr{C}$ is the circle $|z| = 1$, without loss of generality. Hence

$$\int_{\mathscr{C}} \frac{e^{1/z}}{z^{1-n}}\,dz = \frac{2\pi i}{n!} \qquad (n \in \mathbb{N})$$

On $|z| = 1$, $z = e^{i\theta}$, $-\pi \leqslant \theta \leqslant \pi$ say, so that by using the definition of a complex line integral,

$$\int_{-\pi}^{\pi} \frac{\exp(e^{-i\theta})}{e^{(1-n)\,i\theta}} ie^{i\theta}\,d\theta = \int_{-\pi}^{\pi} ie^{(\cos\theta - i\sin\theta + in\theta)}\,d\theta = \frac{2\pi i}{n!}$$

$$\Rightarrow \quad \int_{-\pi}^{\pi} e^{\cos\theta}(\cos(n\theta - \sin\theta) + i\sin(n\theta - \sin\theta))d\theta = \frac{2\pi}{n!}$$

Comparing real parts gives

$$\int_{-\pi}^{\pi} e^{\cos\theta}\cos(n\theta - \sin\theta)d\theta = \frac{2\pi}{n!} \qquad \text{(for all } n \in \mathbb{N})$$

Singular Points

Laurent series expansions can be used to give an easy classification of isolated singular points of functions. They can also be used to prove important theoretical results concerning certain types of isolated singular points. Recall that it follows from Theorem 2.8 that poles are isolated singular points. Clearly, removable singular points are also isolated.

Theorem 4.23. Classification of Isolated Singular Points

Let α be an isolated singular point of $f : A \subseteq \mathbb{C} \to \mathbb{C}$, so that f has a Laurent series expansion about α of the form (4.5), valid for $0 < |z - \alpha| < r$, for some r.

(i) The point α is a removable singular point of f if and only if $b_n = 0$ for all n in this Laurent series.

(ii) The point α is a pole of order m of $f (m \in \mathbb{N})$ if and only if $b_m \neq 0$ and $b_{m+1} = b_{m+2} = \ldots = 0$ in this Laurent series; that is, this Laurent series takes the form

$$f(z) = \sum_{n=0}^{\infty} a_n(z - \alpha)^n + \sum_{n=1}^{m} b_n(z - \alpha)^{-n}$$

(iii) The point α is an isolated essential singular point of f if and only if $b_n \neq 0$, except possibly for a finite number of values of n, in this Laurent series. □

Proof

(i) Let f have a Laurent series expansion about α of the form $f(z) = \sum_{n=0}^{\infty} a_n(z - \alpha)^n$, for $0 < |z - \alpha| < r$. Then $\lim_{z\to\alpha} f(z) = a_0$ since the series is uniformly convergent in the given region and

$\lim_{z\to\alpha}\sum_{n=0}^{\infty}a_n(z-\alpha)^n = a_0$. Hence α is a removable singular point of f. Now suppose α is a removable singular point of f, so that $\lim_{z\to\alpha}f(z)$ exists. Let $\lim_{z\to\alpha}f(z) = a_0$. Redefining $f(\alpha) = a_0$ makes f analytic at α (see Chapter 2) and so f has a Taylor series expansion about α, which converges for $|z-\alpha| < r$ say, as required.

(ii) Let f have a Laurent series expansion about α of the form

$$f(z) = \frac{b_m}{(z-\alpha)^m} + \frac{b_{m-1}}{(z-\alpha)^{m-1}} + \ldots + \frac{b_1}{(z-\alpha)} + \sum_{n=0}^{\infty} a_n(z-\alpha)^n$$

where $b_m \neq 0$, for $0 < |z-\alpha| < r$. Then $f(z) = (z-\alpha)^{-m}\sum_{n=0}^{\infty}d_n(z-\alpha)^n$, where $d_{m-n} = b_n$, $n = 1, \ldots, m$ and $d_{m+n} = a_n$, $n \in \mathbb{Z}_{\geqslant 0}$. Let $g(z) = \sum_{n=0}^{\infty}d_n(z-\alpha)^n$ for $0 < |z-\alpha| < r$. Then g is either analytic at α or has a removable singular point at α by (i). Then $\lim_{z\to\alpha}g(z) = d_0 = b_m \neq 0$ so that $\lim_{z\to\alpha}(z-\alpha)^m f(z) = b_m \neq 0$. Hence α is a pole of order m as required. Conversely, suppose α is a pole of order m of f, so that $(z-\alpha)^m f(z) = g(z)$ where g is analytic at α or has a removable singular point at α (see Chapter 2). Hence g either has a Taylor series expansion about α by 4.18, or a Laurent series expansion about α of the form (4.5) with $b_n = 0$ for all n by (i). Hence $g(z) = \sum_{n=0}^{\infty}d_n(z-\alpha)^n$ for $0 < |z-\alpha| < r$ say and f then has the required Laurent series expansion about α.

(iii) Let f have a Laurent series expansion of the form (4.5) for $0 < |z-\alpha| < r$, where $b_n \neq 0$, except possibly for a finite number of values of n. Then α is an isolated singular point of f which is not a removable singular point or a pole of f by (i) and (ii). Hence α is an isolated essential singular point of f. Conversely, if α is an isolated essential singular point of f, f has a Laurent series expansion about α of the form (4.5), valid in some region $0 < |z-\alpha| < r$. Since α is not a pole or removable singular point, this Laurent series takes the required form by (i) and (ii). ■

Definitions

Let α be an isolated singular point of $f: A \subseteq \mathbb{C} \to \mathbb{C}$ so that f has a Laurent series expansion about α of the form (4.5), valid for $0 < |z-\alpha| < r$ for some r.

(i) The series $\sum_{n=1}^{\infty}b_n/(z-\alpha)^n$ is the **principal part** of this Laurent series.

(ii) The coefficient of $1/(z-\alpha)$ in this series, i.e. b_1, is the **residue** of f at α. We shall denote it by $\text{Res}_\alpha f(z)$.

Note that if f is analytic at α or has a removable singular point at α, so that the principal part of the Laurent series expansion of f about α is 0, then $\text{Res}_\alpha f(z) = 0$. Residues play a very important role when it comes to integrating functions with singular points, as will be shown in the next chapter.

Example 4.16

(i) From Theorem 4.23 and Example 4.13(i), it follows that 0 is an isolated essential singularity of e^{1/z^2} and 0 is a pole of order m of e^z/z^m.

(ii) Note that 0 is an isolated singular point of $(\sin z)/z$. It also follows from the Maclaurin series expansion of $\sin z$ that

$$\frac{\sin z}{z} = \frac{1}{z}\left(z - \frac{z^3}{3!} + \frac{z^5}{5!} - \dots\right) = 1 - \frac{z^2}{3!} + \frac{z^4}{5!} - \dots \qquad (|z| > 0)$$

Hence 0 is a removable singular point of $(\sin z)/z$ by 4.23.

Example 4.17

Consider the function f given by $f(z) = 2/z(z-3)^2$. Note that f has isolated singular points at 0 and 3 (since f is not continuous at these points it is not differentiable there). By definition, 0 is a simple pole of f and 3 is a pole of order 2. It follows by the binomial series that

$$\frac{1}{(z-3)^2} = \frac{1}{(-3)^2(1-z/3)^2} = \frac{1}{9}\left(1 - \frac{z}{3}\right)^{-2}$$

$$\Rightarrow \quad \frac{1}{(z-3)^2} = \frac{1}{9}\left(1 + \frac{2z}{3} + \frac{(-2)(-3)}{2!}\left(\frac{z}{3}\right)^2 + \dots\right) = \frac{1}{9} + \frac{2z}{27} + \frac{z^2}{27} + \dots$$

for $|z/3| < 1$. Hence, by 4.22, the Laurent series expansion of f about 0, valid for $0 < |z| < 3$, is

$$f(z) = \frac{2}{9z} + \frac{4}{27} + \frac{2z}{27} + \dots$$

This verifies 4.23 in the case of the simple pole of f at 0. Note also that $\text{Res}_0 f(z) = 2/9$. To obtain the Laurent series expansion of f about 3, the simplest way to proceed is to let $w = z - 3$. Then

$$f(z) = \frac{2}{(w+3)w^2} = \frac{2}{3w^2}\left(1 + \frac{w}{3}\right)^{-1} = \frac{2}{3w^2}\left(1 - \frac{w}{3} + \frac{w^2}{9} - \frac{w^3}{27} + \dots\right)$$

for $0 < |w| < 3$ by the binomial series again. Hence, by 4.22, the Laurent series expansion of f about 3, valid for $0 < |z-3| < 3$ is

$$f(z) = \frac{2}{3(z-3)^2} - \frac{2}{9(z-3)} + \frac{2}{27} - \frac{2(z-3)}{81} + \dots$$

This verifies 4.23 in the case of the pole of order 2 of f at 3. Also, $\text{Res}_3 f(z) = -2/9$.

We now present some results concerning isolated singular points, which depend either explicitly or implicitly on Laurent series. We begin with a preliminary result.

Theorem 4.24. Poles of Meromorphic Functions

A meromorphic function has only a finite number of poles in any bounded subset A of $\mathbb{C}$. □

Proof

Suppose that a meromorphic function f has an infinite number of poles in A. Then the set of these poles has a limit point α by Theorem 4.4. The function f is not analytic at α, since otherwise f would be analytic in some neighbourhood of α which excludes all the poles, and this contradicts the fact that α is a limit point of the poles. Hence α is a non-isolated singular point of f and so cannot be a pole. This contradicts the fact that f is meromorphic. ■

Example 4.18

The function f defined by $f(z) = 1/\sin(1/z)$ has an infinite number of simple poles at the points $1/n\pi$, $n \in \mathbb{Z}$, $n \neq 0$. The set $\{1/n\pi : n \in \mathbb{Z}, n \neq 0\}$ is infinite and bounded and has limit point 0. This point is a non-isolated essential singular point of f, so f is not meromorphic.

Theorem 4.25. Weierstrass–Casorati Theorem

If $f: A \subseteq \mathbb{C} \to \mathbb{C}$ has an isolated essential singularity at α, then $f(z)$ approaches any given value arbitrarily closely in any neighbourhood of α. That is, given any $c \in \mathbb{C}$ and any $\varepsilon > 0$ and $\delta > 0$, there exists $z \in \mathbb{C}$ such that

$$|z - \alpha| < \delta \quad \text{and} \quad |f(z) - c| < \varepsilon$$ □

Proof

Suppose that the theorem is false. Then there exist $\delta,\ \varepsilon > 0$ such that $|f(z) - c| \geqslant \varepsilon$ for all z satisfying $|z - \alpha| < \delta$. Let $g(z) = 1/(f(z) - c)$. Then by hypothesis there is some deleted open neighbourhood $\mathcal{N}$ say, of α throughout which g is analytic and $|g(z)| \leqslant 1/\varepsilon$; that is, g is analytic and bounded in $\mathcal{N}$. Then g has a Laurent series expansion about α, of the form (4.5), valid in $\mathcal{N}$, where

$$b_n = \frac{1}{2\pi i}\int_{\mathscr{C}} \frac{g(z)}{(z - \alpha)^{1-n}}\,dz$$

with $\mathscr{C}$ any circle of radius r, centred at α and contained in $\mathcal{N}$. Hence by the ML lemma, $|b_n| \leqslant (2\pi r/2\pi\varepsilon)r^{n-1} = r^n/\varepsilon$ and so $b_n \to 0$ as $r \to 0$, for fixed n.

Then $b_n = 0$ for all $n \in \mathbb{N}$, so g has a removable singular point at α by 4.23. If $\lim_{z\to\alpha} g(z) = a_0 \neq 0$, then since $f(z) = c + 1/g(z)$, f is analytic at α, which is a contradiction. Then $a_0 = 0$. Also, $g(z)$ cannot be identically zero in $\mathcal{N}$ since this contradicts the fact that f is analytic in a deleted neighbourhood of α. Hence g has an expansion of the form $g(z) = \sum_{n=k}^{\infty} a_n(z-\alpha)^n$, $a_k \neq 0$, for some k, in $\mathcal{N}$. Then $(z-\alpha)^{-k}g(z)$ is analytic and non-zero at α, so $(z-\alpha)^k/g(z)$ has a Taylor series expansion about α. Hence, by Theorem 4.23, f has a pole of order k at α, which is a contradiction. ■

A much stronger result than Theorem 4.25, although much harder to prove, is Picard's theorem.

Theorem 4.26. Picard's Theorem

Let f have an isolated essential singular point at α. Then $f(z)$ assumes every finite value, with one possible exception, an infinite number of times, in any neighbourhood of α. □

Example 4.19

The function f defined by $f(z) = e^{1/z^2}$ has an isolated essential singular point at 0. Now $e^z = 1$ when $z = 2n\pi i$ for $n \in \mathbb{Z}$, so that $e^{1/z^2} = 1$ when $z^2 = -i/2n\pi$. As an infinite number of these points lie in any neighbourhood of the origin, $f(z)$ assumes the value 1 an infinite number of times in any neighbourhood of 0. Note, however, that $f(z)$ never assumes the value 0.

The behaviour of a function f at the point at infinity in the extended complex plane $\tilde{\mathbb{C}}$, can be investigated by examining the behaviour of $f(1/z)$ in a neighbourhood of 0.

Definitions

Function $f(z)$ is respectively **analytic**, has a **removable singular point**, has a **pole of order m** or has an **essential singular point at ∞ in $\tilde{\mathbb{C}}$** if and only if $f(1/z)$ is analytic, has a removable singular point, has a pole of order m or has an essential singular point at 0.

Example 4.20

(i) It is clear that any algebraic fraction is either analytic at ∞ or has a pole at ∞. For example, $f(z) = 1/z(1+z^2)$ has simple poles at 0 and $\pm i$. Also

$$f(1/z) = \frac{z}{1+1/z^2} = \frac{z^3}{1+z^2}$$

so $f(1/z)$ is analytic at 0. Hence f is analytic at ∞. Note that f has a zero of order 3 at ∞.

Now consider $g(z) = z^3/(z+1)$, which has a simple pole at -1. Then $g(1/z) = 1/z^2(1+z)$ has a pole of order 2 at 0. Hence g has a pole of order 2 at ∞.

(ii) Consider $f(z) = \sin z$, which is entire. Then $f(1/z) = \sin(1/z)$ has an isolated essential singularity at 0. Hence f has an isolated essential singularity at ∞.

It follows easily by Liouville's theorem that if a function is entire in the extended plane $\tilde{\mathbb{C}}$, then it must be constant.

Theorem 4.27. Entire Functions in the Extended Plane

Let f be entire in the extended plane $\tilde{\mathbb{C}}$. Then f is a constant. □

Proof

Since $f(z)$ is analytic at ∞, $f(1/z)$ is analytic at 0 and so there exists $r \in \mathbb{R}^+$ such that $f(1/z)$ is analytic, hence continuous, on the compact set $|z| \leqslant 1/r$. Then by 4.5, $f(1/z)$ is bounded for $|z| \leqslant 1/r$; that is, $f(z)$ is bounded for $|z| \geqslant r$. Since f is also continuous on the compact set $|z| \leqslant r$, it is also bounded on this set, by 4.5. Then f is bounded on $\mathbb{C}$ and the result follows by Liouville's theorem, 3.16. ■

Clearly, any algebraic fraction is meromorphic in $\tilde{\mathbb{C}}$. The converse of this result, which is an extension of 4.27, is also true.

Theorem 4.28. Meromorphic Functions in the Extended Plane

Let a function f be meromorphic in the extended complex plane. Then f is an algebraic fraction. □

Proof

Note that f is either analytic at ∞ or has a pole at ∞. Since poles are isolated, it is possible to construct a circle $\mathscr{C}$, centred at 0, such that ∞ is the only possible pole of f outside $\mathscr{C}$. There can only be a finite number of poles, $\alpha_1, \alpha_2, \ldots \alpha_n$ of f inside $\mathscr{C}$, by 4.24. If α_k is a pole of order m_k, then the principal part of the Laurent series expansion of f about α_k is of the form

$$g_k(z) = \frac{b_{1k}}{z-\alpha_k} + \frac{b_{2k}}{(z-\alpha_k)^2} + \ldots + \frac{b_{m_k k}}{(z-\alpha_k)^{m_k}}$$

where $b_{m_k k} \neq 0$. Since f is analytic or has a pole of order m say at ∞, $f(1/z)$ is analytic or has a pole of order m at 0, so that f has a Laurent series expansion of the form

$$f(z) = \sum_{n=0}^{\infty} \frac{a_n}{z^n} + b_1 z + b_2 z^2 + \ldots + b_m z^m \qquad (|z| > R \text{ say})$$

where each $b_k = 0$ or $b_m \neq 0$. In either case, the principal part of the Laurent series expansion of f about ∞ is of the form

$$h(z) = b_1 z + b_2 z^2 + \ldots + b_m z^m$$

for some non-zero b_m, or $b_k = 0$ for all k. Now consider the function ϕ defined by

$$\phi(z) = f(z) - \sum_{k=1}^{n} g_k(z) - h(z)$$

The function ϕ can clearly be made analytic at each α_k and is analytic at ∞ since $1/(z - \alpha_k)^N$ is analytic at ∞ for any non-zero α_k and $N \in \mathbb{N}$. Hence ϕ is entire in $\tilde{\mathbb{C}}$ and so is a constant by 4.27. Then $f(z)$ differs from $\sum_{k=1}^{n} g_k(z) + h(z)$ only by a constant and so is an algebraic fraction. ■

Exercise 4.4.1 Assuming the Maclaurin series expansions of $\sin z$ and $\cos z$:

(i) Find the first four terms in the Laurent series expansion of $\csc z$ in a region $0 < |z| < r$, stating the maximum value of r.

(ii) Find the Laurent series expansion of $(\cos z^2)/z^4$ about 0, valid for $z \neq 0$. Use this result to find the Laurent series expansion of $(\sin z^2)/z^3$ about 0, valid for $z \neq 0$.

Exercise 4.4.2 Assuming the Maclaurin series expansion for $\cos z$, find the Laurent series expansion of $z^2 \cos(z-1)^{-1}$ about 1, in the region $|z-1| > 0$.

Exercise 4.4.3 Assuming the Maclaurin series for e^z and the binomial series, find the Laurent series expansion of $e^z/(z^2-1)$

(i) about 0 in the region $|z| > 1$

(ii) about 1 in the region $0 < |z-1| < 2$

(iii) about 1 in the region $|z-1| > 2$

Exercise 4.4.4 Find the Laurent series expansions of $f(z) = 1/(z-2)(z-3)$ in the regions

(i) $2 < |z| < 3$

(ii) $|z| > 3$

(iii) $0 < |z-2| < 1$

(iv) $|z-2| > 1$

Exercise **4.4.5** Let f be given by $f(z) = 1/z^3 \cosh z$. Use the Maclaurin series expansion of $\cosh z$ and the binomial series to show that the Laurent series expansion of f about 0 is

$$f(z) = \frac{1}{z^3} - \frac{1}{2z} + \frac{5z}{24} + \ldots \qquad (0 < |z| < \pi/2)$$

Integrate termwise and use Cauchy's theorem and the fundamental theorem of calculus to show that

$$\int_{|z|=1} f(z)dz = -\pi i = 2\pi i \text{Res}_0 f(z)$$

Exercise **4.4.6** Use Laurent's theorem to show that $\exp(z + 1/z) = \sum_{n=-\infty}^{\infty} c_n z^n$, for $z \neq 0$, where $c_n = (1/\pi) \int_0^\pi \exp(2\cos\theta) \cos n\theta \, d\theta$ for all $n \in \mathbb{Z}$. Hence show that

$$\int_0^\pi \exp(2\cos\theta) d\theta = \pi \sum_{n=0}^{\infty} \frac{1}{(n!)^2}$$

Exercise **4.4.7** The **Bessel function** $J_n(z)$, $n \in \mathbb{Z}$, can be defined as the coefficient of w^n in the Laurent series of $e^{z(w-1/w)/2}$, valid in the region $|w| > 0$. Show that $J_n(z) = (1/\pi) \int_0^\pi \cos(n\theta - z\sin\theta) d\theta$.

Exercise **4.4.8** Locate the isolated singular points of the following functions. For each singular point of each function, find a Laurent series expansion of the function which converges in some deleted neighbourhood of the singular point. Hence classify the singular points and calculate the residue of each function at each singular point.

(i) $f(z) = ze^{1/z}$

(ii) $f(z) = \dfrac{e^z - 1}{z}$

(iii) $f(z) = \dfrac{1}{z^2(1 + z^2)}$

(iv) $f(z) = \dfrac{1}{\sinh^3 z}$

(v) $f(z) = \sin(z + 1/z)$

(vi) $f(z) = \dfrac{e^z}{(4 + z^2)^2}$

(vii) $f(z) = e^{\cosh(1/z)}$

(viii) $f(z) = \dfrac{1}{1 + e^z}$

Exercise **4.4.9** Let g be a function which is analytic at α. Expand g as a Taylor series to show that α is a removable singular point of the function f given by $f(z) = g(z)/(z - \alpha)$ if $g(\alpha) = 0$. Show also that if $g(\alpha) \neq 0$, the point α is a simple pole of f with residue $g(\alpha)$.

Exercise **4.4.10** For each of the functions listed in Exercise 4.4.8, decide whether or not the point at infinity in the extended complex plane is a singular point. Classify the point at infinity in the cases where it is a singular point. (You need not find any Laurent series expansions.)

Exercise **4.4.11** Let f be a non-constant function, entire on $\mathbb{C}$, with $f(z) \neq 0$ for any $z \in \mathbb{C}$. Prove that f has an isolated essential singularity at ∞ in $\tilde{\mathbb{C}}$.

Exercise **4.4.12** Prove that a function entire on $\mathbb{C}$ having a non-essential singular point at ∞ in $\tilde{\mathbb{C}}$ must be a polynomial.

Exercise **4.4.13** Use Picard's theorem to prove that a function f entire on $\mathbb{C}$ with $f(z) \neq 0$ and $f(z) \neq 1$ for any $z \in \mathbb{C}$ must be a constant.

5 The Residue Theorem and its Applications

This chapter is concerned with Cauchy's residue theorem, which is a generalisation of Cauchy's integral formula, given in Chapter 3. The residue theorem is a very powerful result which can be used to evaluate definite integrals and hence a large class of real definite integrals. A large number of real definite integrals which can be evaluated by this technique are difficult to evaluate by other means. The residue theorem can also be used to sum certain convergent series of real numbers, which can be difficult to sum by other means.

Historical Note

Cauchy's original application of the residue theorem was to the evaluation of certain real definite integrals.

Cauchy's Residue Theorem and Calculation of Residues

The residue theorem implies that an integral evaluated around a closed contour only depends on the behaviour of the integrand at any singular points inside the contour. In fact, the only quantities which affect the integral are the residues of the integrand at its singular points.

Recall that if α is an isolated singular point of $f: A \subseteq \mathbb{C} \to \mathbb{C}$, then f has a Laurent series expansion about α of the form

$$f(z) = \sum_{n=0}^{\infty} a_n (z-\alpha)^n + \sum_{n=1}^{\infty} \frac{b_n}{(z-\alpha)^n} \tag{5.1}$$

where a_n and b_n are constants for all n, which converges for all z in some region $0 < |z - \alpha| < r$. The coefficient of $1/(z-\alpha)$ in (5.1), i.e. b_1, is the **residue** of f at α and will be denoted by $\text{Res}_\alpha f(z)$.

Important Note

It is seldom necessary to calculate more than a small number of terms in this Laurent series expansion of f order to calculate $\text{Res}_\alpha f(z)$.

Theorem 5.1. The Residue Theorem

Let $\mathscr{C}$ be a simple closed contour within and on which $f: A \subseteq \mathbb{C} \to \mathbb{C}$ is analytic except at a finitc number of singular points, $z_1, z_2, \ldots, z_m$ inside $\mathscr{C}$. Then

$$\int_{\mathscr{C}} f(z)\,dz = 2\pi i \sum_{k=1}^{m} \mathrm{Res}_{z_k} f(z) \quad (= 2\pi i \times \text{sum of residues})$$ □

Proof

Step 1

Since there are a finite number of singular points, they are isolated, so it is possible to construct circles $\mathscr{C}_k$, $k = 1, 2, \dots m$, centred at z_k for each k, which all lie inside $\mathscr{C}$, with radii small enough so that no two of them intersect, as in Fig. 3.8. It then follows by the generalised deformation result, Lemma 3.20, that

$$\int_{\mathscr{C}} f(z)\,dz = \sum_{k=1}^{m} \int_{\mathscr{C}_k} f(z)\,dz \tag{5.2}$$

(Recall that this result depends on Cauchy's theorem.)

Step 2

The radius of each $\mathscr{C}_k$ can be chosen small enough so that each circle is within the domain of convergence of the Laurent series expansion of f about z_k, of the form (5.1). It follows by Laurent's theorem, 4.21, that if b_{nk} is the coefficient of $(z - z_k)^{-n}$ in each Laurent series, then

$$b_{nk} = \frac{1}{2\pi i} \int_{\mathscr{C}_k} \frac{f(z)}{(z - z_k)^{1-n}}\,dz \qquad (n \in \mathbb{N})$$

$$\Rightarrow \int_{\mathscr{C}_k} f(z)\,dz = 2\pi i b_{1k} = 2\pi i\,\mathrm{Res}_{z_k} f(z) \qquad \text{(for each } k\text{)} \tag{5.3}$$

Substituting (5.3) into (5.2) gives the required result. ■

Notes

(i) The result (5.3) can also be obtained by noting that each Laurent series can be integrated termwise around each of the circles, by the results of Chapter 4. All terms give zero, by Cauchy's theorem and the fundamental theorem of calculus, except for $2\pi i b_{1k}$, which follows from the definition. This is the term left after integration, hence the term 'residue' for b_{1k}.

(ii) Theorem 5.1 reduces to Cauchy's theorem as a special case. If f has no singular points inside and on $\mathscr{C}$, then the sum of the residues is 0, so that $\int_{\mathscr{C}} f(z)\,dz = 0$. Theorem 5.1 also includes Cauchy's integral formula as a special case. If f has one singular point, α say, inside $\mathscr{C}$, which is a simple pole, then $\int_{\mathscr{C}} f(z)\,dz = 2\pi i \mathrm{Res}_{\alpha} f(z)$ and $f(z) = g(z)/(z - \alpha)$, where g is analytic at α without loss of generality. Hence $\mathrm{Res}_{\alpha} f(z) = g(\alpha)$ and so $\int_{\mathscr{C}} (g(z)/(z - \alpha))\,dz = 2\pi i g(\alpha)$.

(iii) Remember that it is only necessary to calculate the residues at the singular points **enclosed** by the contour $\mathscr{C}$ in order to apply 5.1 to obtain $\int_{\mathscr{C}} f(z)\,dz$.

Example 5.1

(i) From the Maclaurin series expansion of e^z it follows that $z^2e^{1/z} = \sum_{n=0}^{\infty} z^{2-n}/n!$ for all $z \neq 0$, so that 0 is an essential singular point of $z^2e^{1/z}$ and $\text{Res}_0(z^2e^{1/z}) = 1/3!$. Clearly 0 is the only singular point of $z^2e^{1/z}$. Hence, if $\mathscr{C}$ is any simple closed contour enclosing the origin, then by 5.1, $\int_{\mathscr{C}} z^2e^{1/z}dz = \pi i/3$. Note however, that the real indefinite integral $\int x^2e^{1/x}dx$ cannot be evaluated in terms of elementary functions.

(ii) Similarly, it follows from the Maclaurin series expansion of sin z that

$$\sin(z^{-k}) = \sum_{n=0}^{\infty}(-1)^n z^{-(2n+1)k}/(2n+1)! \text{ for } k \in \mathbb{N}.$$

Hence 0 is an essential singular point of $\sin(z^{-k})$ and $\text{Res}_0(\sin(z^{-k})) = 1$ if $k = 1$ and 0 otherwise. Clearly, 0 is the only singular point of $\sin(z^{-k})$. Hence if $\mathscr{C}$ is any simple closed contour enclosing the origin, then

$$\int_{\mathscr{C}} \sin(z^{-k})\,dz = 0 \quad \text{if} \quad k \neq 1 \quad \text{and} \quad \int_{\mathscr{C}} \sin(1/z)\,dz = 2\pi i$$

Let $\mathscr{C}$ be the circle $|z| = 1$. Then $\mathscr{C}$ is parametrised by $z = e^{i\theta}$, $-\pi \leqslant \theta \leqslant \pi$ say. It follows by the above result and by definition that

$$\int_{-\pi}^{\pi} \sin(e^{-i\theta})ie^{i\theta}\,d\theta = 2\pi i \Rightarrow \int_{-\pi}^{\pi} \sin(\cos\theta - i\sin\theta)(\cos\theta + i\sin\theta)d\theta = 2\pi$$

$$\Rightarrow \int_{-\pi}^{\pi} (\sin(\cos\theta)\cosh(\sin\theta) - i\cos(\cos\theta)\sinh(\sin\theta))(\cos\theta + i\sin\theta)\,d\theta$$

$$= 2\pi$$

$$\Rightarrow \int_{-\pi}^{\pi} (\sin(\cos\theta)\cosh(\sin\theta)\cos\theta + \cos(\cos\theta)\sinh(\sin\theta)\sin\theta))\,d\theta$$

$$= 2\pi.$$

Since the calculation of residues is of fundamental importance, it is advantageous to look for methods of calculating them, other than the definition. The following result enables us, at least in theory, to calculate the residue at a pole without having to resort to the definition.

Lemma 5.2. The Residue at a Pole

Let α be a pole of order m of $f: A \subseteq \mathbb{C} \to \mathbb{C}$, so that $f(z) = g(z)/(z-\alpha)^m$, $m \in \mathbb{N}$, for some g, analytic at α with $g(\alpha) \neq 0$, without loss of generality. Then $\text{Res}_\alpha f(z) = g^{(m-1)}(\alpha)/(m-1)!$. □

Proof

Since g is analytic at α, it has, by 4.18, a Taylor series expansion in some neighbourhood of α of the form

$$g(z) = \sum_{n=0}^{\infty} \frac{g^{(n)}(\alpha)(z-\alpha)^n}{n!} \quad \Rightarrow \quad f(z) = \sum_{n=0}^{\infty} \frac{g^{(n)}(\alpha)(z-\alpha)^{n-m}}{n!} \tag{5.4}$$

Hence by definition,

$$\text{Res}_\alpha f(z) = \text{coefficient of } \frac{1}{z-\alpha} \text{ in } (5.4) = \frac{g^{(m-1)}(\alpha)}{(m-1)!}$$ ■

Notes

(i) If α is a simple pole of f then by 5.2,

$$\text{Res}_\alpha f(z) = g(\alpha) = \lim_{z \to \alpha} (z-\alpha) f(z)$$

This is the case in which 5.2 is most applicable.

(ii) If α is a pole of f of order greater than 2, then it is probably easier to use the definition to calculate $\text{Res}_\alpha f(z)$, rather than 5.2.

(iii) Lemma 5.2 also follows by Cauchy's integral formula for derivatives and the residue theorem (see Exercises 5.1).

Example 5.2

(i) Let $f(z) = \dfrac{1}{(z-1)^2(z^2+1)} = \dfrac{1}{(z-1)^2(z+i)(z-i)}$

Then f has isolated singular points at 1 and at $\pm i$ (since f is not continuous there). Note that by definition, $\pm i$ are simple poles and 1 is a pole of f of order 2. Then by 5.2,

$$\text{Res}_{\pm i} f(z) = \left. \frac{1}{(z-1)^2(z \pm i)} \right|_{z=\pm i} = \frac{1}{4}$$

$$\text{Res}_1 f(z) = \left. \frac{d}{dz}\left(\frac{1}{z^2+1}\right) \right|_{z=1} = \left. \frac{-2z}{(z^2+1)^2} \right|_{z=1} = -\frac{1}{2}$$

Alternatively, we can use the definition to find $\text{Res}_1 f(z)$. Let $w = z - 1$.

Then

$$f(z) = \frac{1}{w^2(w^2+2w+2)} = \frac{1}{2w^2}\left(1+w+\frac{w^2}{2}\right)^{-1}$$

$$\Rightarrow \quad f(z) = \frac{1}{2w^2}\left(1 - w - \frac{w^2}{2} + \left(w+\frac{w^2}{2}\right)^2 + \ldots\right)$$

$$= \frac{1}{2w^2}\left(1 - w + \frac{w^2}{2} + \ldots\right)$$

in some neighbourhood of $w = 0$ using the binomial series. Then by definition, $\mathrm{Res}_1 f(z) = -\frac{1}{2}$.

(ii) Consider $\int_{\mathscr{C}}((\mathrm{Log}\, z)/(z^2+1)^2)\, dz$, where $\mathscr{C}$ is the contour shown in Fig. 5.1. Let $f(z) = (\mathrm{Log}\, z)/(z^2+1)^2$. Then f has singular points at $\pm i$, 0 and by convention at all points on the negative real axis. Only i lies within $\mathscr{C}$ and is a pole of f of order 2. By 5.2,

$$\mathrm{Res}_i f(z) = \frac{d}{dz}\left(\frac{\mathrm{Log}\, z}{(z+i)^2}\right)\Bigg|_{z=i} = \left(\frac{1}{z(z+i)^2} - \frac{2\,\mathrm{Log}\, z}{(z+i)^3}\right)\Bigg|_{z=i} = \frac{\frac{1}{2}i\pi - 1}{4i}$$

Alternatively, $\mathrm{Res}_i f(z)$ can be found using the definition. Then by the residue theorem,

$$\int_{\mathscr{C}} \frac{\mathrm{Log}\, z}{(z^2+1)^2}\, dz = 2\pi i \cdot \frac{1}{4i}\left(\frac{i\pi}{2} - 1\right) = \frac{\pi}{2}\left(\frac{i\pi}{2} - 1\right)$$

It is often the case that a given function cannot easily be expressed in 'factorised' form, so 5.2 may not be immediately applicable. In such cases, where the pole is simple, the following result is sometimes useful.

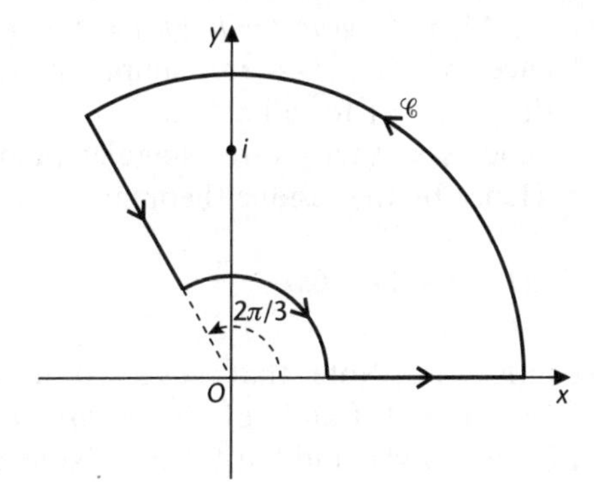

Figure 5.1

Lemma 5.3. The Residue at a Simple Pole

Suppose that $f(z)$ can be expressed in the form $f(z) = g(z)/h(z)$, where g and h are both analytic at α, with $g(\alpha) \neq 0$, $h(\alpha) = 0$ and $h'(\alpha) \neq 0$. Then α is a simple pole of f and $\operatorname{Res}_\alpha f(z) = g(\alpha)/h'(\alpha)$. □

Note

It follows by the hypothesis of 5.3 and Lemma 4.20 that h has a simple zero at α.

Proof

It follows by 4.18 and hypothesis that $f(z) = g(z)/(z - \alpha)H(z)$ where g and H are analytic and non-zero at α. Hence g/H is analytic and non-zero at α and so α is a simple pole of f. Then by 5.2, $\operatorname{Res}_\alpha f(z) = \lim_{z\to\alpha}(z - \alpha)g(z)/h(z)$ so that

$$\operatorname{Res}_\alpha f(z) = \lim_{z\to\alpha} g(z) \cdot \lim_{z\to\alpha} \frac{(z-\alpha)}{h(z) - h(\alpha)} = \frac{g(\alpha)}{h'(\alpha)}$$

by hypothesis and the definition of $h'(\alpha)$. ■

Example 5.3

(i) Let $f(z) = \coth z = \cosh z/\sinh z$. Then f has an infinite number of isolated singular points at $n\pi i$, $n \in \mathbb{Z}$, and these are the only singular points of f since $\sinh z = 0$ iff $\sin iz = 0$ iff $z = n\pi i$. Clearly, 5.2 is not directly appropriate here, unless L'Hôpital's rule is used. Let $g(z) = \cosh z$ and $h(z) = \sinh z$. Then g and h are entire and certainly analytic at $n\pi i$. Also, $g(n\pi i) = h'(n\pi i) = \cos n\pi \neq 0$ and $h(n\pi i) = \sin n\pi = 0$. Hence by 5.3, $n\pi i$ are simple poles of f and $\operatorname{Res}_{n\pi i} f(z) = \cosh z/\cosh z|_{z=n\pi i} = 1$ for all $n \in \mathbb{Z}$.

The circle $|z| = 3\pi/2$ encloses exactly three singular points of $\coth z$, those at $-i\pi$, 0 and $i\pi$. Hence by the residue theorem

$$\int_{|z|=3\pi/2} \coth z\, dz = 2\pi i(1 + 1 + 1) = 6\pi i$$

(ii) Consider $f(z) = z/\cos z(e^z - 1)$. Note that $\cos z = 0$ if and only if $z = (2n+1)\pi/2$, $n \in \mathbb{Z}$, and $e^z = 1$ if and only if $z = 2n\pi i$, $n \in \mathbb{Z}$. Then f has singular points at 0, $2n\pi i$, $n \neq 0$ and $(2n+1)\pi/2$. Note also that

$$f(z) = \frac{1}{(1 - z^2/2! + \ldots)(1 + z/2! + z^2/3! + \ldots)} = 1 - \frac{z}{2} + \frac{7z^2}{12} + \ldots$$

in some deleted neighbourhood of 0, so that 0 is a removable singular point of f and $\text{Res}_0 f(z) = 0$.

Let $g(z) = z/\cos z$ and $h(z) = e^z - 1$. Then g and h are analytic at $2n\pi i$, $g(2n\pi i) = 2n\pi i/\cosh 2n\pi \neq 0$ if $n \neq 0$, $h(2n\pi i) = 0$ and $h'(2n\pi i) = e^{2n\pi i} = 1 \neq 0$. Then by 5.3, $2n\pi i$, $n \neq 0$, are simple poles of f and $\text{Res}_{2n\pi i} f(z) = 2n\pi i/\cosh 2n\pi$.

Now let $g(z) = z/(e^z - 1)$ and $h(z) = \cos z$. Then the conditions of 5.3 are now satisfied for $\alpha = (2n+1)\pi/2$, so that $(2n+1)\pi/2$ are simple poles of f with

$$\text{Res}_{(2n+1)\pi/2} f(z) = \frac{(-1)^{n+1}(2n+1)\pi/2}{(e^{(2n+1)\pi/2} - 1)}$$

Notes

(i) It is usually clear which of 5.2 or 5.3 to use when calculating the residue at a simple pole. However, even in the case of simple poles, it is sometimes easier to resort to the definition to calculate the residue. Consider $f(z) = (z^2 \sinh z)/(e^z - 1)^4$. In this case, f has a simple pole at 0 by 5.4 below, but neither 5.2 or 5.3 is easy to apply.

(ii) There is no simple generalisation of 5.3 to the case of a pole of any order. The special case of a pole of order 2 is dealt with in Exercises 5.1. However, it is useful to bear in mind the following result, used in conjunction with 4.20.

Lemma 5.4. The Order of a Pole

Suppose that $f(z) = g(z)/h(z)$, where g and h are both analytic at α. Suppose also that α is a zero of order k of g and α is a zero of order m of h. Then α is a pole of f of order $m - k$ if $k < m$, α is a removable singular point of f if $k = m$ and α is a zero of order $k - m$ if $m < k$. □

Proof

By hypothesis $f(z) = (z - \alpha)^{k-m} G(z)/H(z)$ where G and H are analytic at α, so that G/H is also analytic at α, and $G(\alpha) \neq 0$ and $H(\alpha) \neq 0$. Then the results follow from the definitions. ■

For example, it follows by 4.20 that 0 is a zero of order 3 of $g(z) = z^2 \sinh z$ and a zero of order 4 of $h(z) = (e^z - 1)^4$. Then by 5.4, 0 is simple pole of $f(z) = (z^2 \sinh z)/(e^z - 1)^4$, as stated above.

Exercise **5.1.1** Use the residue theorem to evaluate the following integrals:

(i) $\displaystyle\int_{|z|=1} z\cos(1/z)\,dz$

(ii) $\displaystyle\int_{|z|=2} \frac{e^z}{z-1}\,dz$

Exercise **5.1.2** Evaluate $\int_{\mathscr{C}} z\sinh(1/z^2)\,dz$, where $\mathscr{C}$ is the circle $|z| = 1$, using the residue theorem. Let $z = e^{i\theta}$, $-\pi \leqslant \theta \leqslant \pi$ in this result to obtain the result of a particular real definite integral.

Exercise **5.1.3** Prove Lemma 5.2 using Cauchy's integral formula for derivatives and the residue theorem.

Exercise **5.1.4** Calculate the residue at each singular point of the following functions, using Lemma 5.2:

(i) $f(z) = \dfrac{1}{z^2 + a^2}$

(ii) $f(z) = \dfrac{z^2}{(z^2 + 4)^2}$

(iii) $f(z) = \dfrac{e^z}{z^2(z^2 - 1)}$

(iv) $f(z) = \dfrac{z^2 + 5z + 1}{(z + 1)^2(z^2 + 1)}$

Exercise **5.1.5** Calculate the residue at each singular point of the following functions, using Lemma 5.3 or the definition:

(i) $f(z) = \sec z$

(ii) $f(z) = z\coth z$

(iii) $f(z) = \dfrac{1}{z^n - 1}$, $n \in \mathbb{N}$

(iv) $f(z) = \dfrac{e^z}{\sin^2 z}$

(v) $f(z) = \dfrac{z^2 \sinh z}{(e^z - 1)^4}$

Exercise **5.1.6** Prove that if $f(z) = g(z)/h(z)$ where g and h are analytic at α, $g(\alpha) \neq 0$, $h(\alpha) = h'(\alpha) = 0$ and $h''(\alpha) \neq 0$, then α is a pole of f of order 2 and

$$\operatorname{Res}_\alpha f(z) = \frac{2g'(\alpha)}{h''(\alpha)} - \frac{2g(\alpha)h'''(\alpha)}{3(h''(\alpha))^2}$$

Exercise **5.1.7** Evaluate the following integrals using the residue theorem:

(i) $\displaystyle\int_{|z|=2} \frac{e^z}{z^3+z}\,dz$

(ii) $\displaystyle\int_{|z|=1} \frac{z+a}{z^n(z+b)}\,dz,\ n\in\mathbb{N}$

(iii) $\displaystyle\int_{\mathscr{C}} \frac{\sin z}{z^2\sinh z}\,dz$ where $\mathscr{C}$ is the quadrilateral with vertices ± 1, $3i\pi/2$ and $-i\pi/2$

Exercise **5.1.8** Evaluate $\displaystyle\int_{\mathscr{C}} \frac{\operatorname{Log}(z+i)}{z^2+1}\,dz$

where $\mathscr{C}$ is the semicircle in the upper half-plane, of radius $R>1$ and centre 0. Assuming that $\lim_{R\to\infty}\int_\Gamma f(z)\,dz = 0$, where Γ is the semicircular arc of $\mathscr{C}$, use this result to find

$$\int_0^\infty \frac{\operatorname{Log}(x^2+1)}{x^2+1}\,dx$$

Evaluation of Real Definite Integrals Using the Residue Theorem

The residue theorem can be used to great effect in the evaluation of certain real definite integrals. A suitable complex definite integral taken round a closed contour must be found which can be related to the given integral. It is useful to bear in mind that nearly every closed contour of practical use in this context consists of line segments and arcs of circles. We begin by investigating integrals of four commonly occuring types, which are the simplest to deal with.

Integrals of Type I

Relatively simple to deal with are real definite integrals which are related to the type

$$\int_0^{2\pi} f(\sin\theta,\cos\theta)\,d\theta \qquad \text{(I)}$$

The idea here is essentially to use the definition of a complex definite integral in reverse. By definition,

$$\sin\theta = \frac{1}{2i}(e^{i\theta}-e^{-i\theta}) \quad\text{and}\quad \cos\theta = \frac{1}{2}(e^{i\theta}+e^{-i\theta})$$

Hence letting $z = e^{i\theta}$ where $0 \leqslant \theta \leqslant 2\pi$ and noting $z'(\theta) = iz$,

$$\int_0^{2\pi} f(\sin\theta, \cos\theta)\, d\theta = \int_{\mathscr{C}} f\left(\frac{1}{2i}\left(z - \frac{1}{z}\right), \frac{1}{2}\left(z + \frac{1}{z}\right)\right)\frac{1}{iz}\, dz$$

where $\mathscr{C}$ is the circle $|z| = 1$. The transformed integral can be evaluated by the residue theorem in the normal way.

Example 5.4

(i) Consider $I = \int_0^{2\pi}(1/(5 + 4\sin\theta))\, d\theta$, which is of type I. Letting $z = e^{i\theta}$, $0 \leqslant \theta \leqslant 2\pi$, gives

$$I = \int_{\mathscr{C}} \frac{dz}{iz(5 - 2i(z - 1/z))} = \int_{\mathscr{C}} \frac{dz}{2z^2 + 5iz - 2} = \int_{\mathscr{C}} \frac{dz}{(2z + i)(z + 2i)}$$

where $\mathscr{C}$ is the circle $|z| = 1$. The integrand has simple poles at $-i/2$ and $-2i$. Only $-i/2$ lies within $\mathscr{C}$. It follows by 5.2 that

$$\operatorname{Res}_{-i/2}\left(\frac{1}{(2z + i)(z + 2i)}\right) = \left.\frac{1}{2(z + 2i)}\right|_{z=-i/2} = \frac{1}{3i}$$

Hence by the residue theorem, $\int_0^{2\pi}(1/(5 + 4\sin\theta))\, d\theta = 2\pi i/3i = 2\pi/3$.

(ii) Consider $\int_0^{2\pi}((\cos 2\theta)/(3\cos\theta + 5)^2)\, d\theta$, which is implicitly of type I. It is possible to substitute $z = e^{i\theta}$ directly but the evaluation is made a lot simpler by noting that

$$\int_0^{2\pi} \frac{\cos 2\theta}{(3\cos\theta + 5)^2}\, d\theta = \operatorname{Re}\left(\int_0^{2\pi} \frac{e^{2i\theta}}{(3\cos\theta + 5)^2}\, d\theta\right) = \operatorname{Re} I \quad \text{say}$$

Then letting $z = e^{i\theta}$ in I in the usual way gives

$$I = \int_{\mathscr{C}} \frac{z^2 dz}{iz(3(z + 1/z)/2 + 5)^2} = \frac{4}{i}\int_{\mathscr{C}} \frac{z^3 dz}{(3z + 1)^2(z + 3)^2}$$

where $\mathscr{C}$ is the circle $|z| = 1$. The integrand $f(z) = z^3/(3z + 1)^2(z + 3)^2$ has poles of order 2 at $-1/3$ and -3. Only $-1/3$ lies within $\mathscr{C}$. To find $\operatorname{Res}_{-1/3} f(z)$, we resort to the definition. Let $w = z + 1/3$.

$$f(z) = \frac{(w - 1/3)^3}{9w^2(w + 8/3)^2} = \frac{(-1/3)^3(1 - 3w)^3(1 + 3/8w)^{-2}}{9w^2(8/3)^2}$$

$$\Rightarrow \quad f(z) = \frac{-1}{27 \cdot 64w^2}(1 - 9w + \dots)\left(1 - \frac{3w}{4} + \dots\right)$$

in some deleted neighbourhood of $w = 0$. Hence

$$\operatorname{Res}_{-1/3} f(z) = \frac{-1}{27.64}\left(-9 - \frac{3}{4}\right) = \frac{13}{36.64}$$

Then by the residue theorem

$$I = 8\pi \text{Res}_{-1/3} f(z) = \frac{13\pi}{288} \Rightarrow \int_0^{2\pi} \frac{\cos 2\theta}{(3\cos\theta + 5)^2}\, d\theta = \text{Re}\, I = \frac{13\pi}{288}$$

Integrals of Type II

The next case we consider is a wide class of real improper integrals, which can all be evaluated using the same basic technique. We consider integrals related to the form

$$\int_{-\infty}^{\infty} f(x)\, dx \quad f \text{ an algebraic fraction with no real singular points} \qquad \text{(II)}$$

Important Note

It will be supposed in (II) that the degree of the denominator of f is at least 2 greater than the degree of the numerator of f. Then $\lim_{x\to\pm\infty} x^2 f(x)$ exists, so that by standard results from real analysis, the given real improper integral exists and is equal to its principal value, i.e.

$$\int_{-\infty}^{\infty} f(x)\, dx = \lim_{R,S\to\infty} \int_{-S}^{R} f(x)\, dx = \lim_{R\to\infty} \int_{-R}^{R} f(x)\, dx \qquad (5.5)$$

The idea is to choose a closed contour in the complex plane which encloses all relevant singular points of f and which includes an interval of the real axis. Since, in general, circles are the easiest contours to deal with, the method proceeds as follows.

Step 1

Consider $\int_{\mathscr{C}} f(z)\, dz$ where $\mathscr{C}$ is the semicircle in the upper half-plane of radius R, as shown in Fig. 5.2. The radius R is chosen large enough so that $\mathscr{C}$ encloses any singularities of f in the upper half-plane. This is possible since f has only a finite number of singular points. Let the semicircular arc of $\mathscr{C}$ be denoted by Γ, as shown.

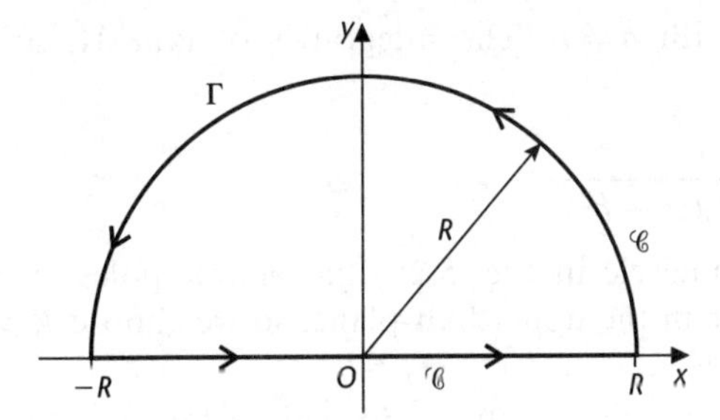

Figure 5.2

Since there are no singular points of f on $\mathscr{C}$, evaluate $\int_{\mathscr{C}} f(z)\,dz$ using the residue theorem. There is nothing special about the position of $\mathscr{C}$. We could equally well take a semicircle in the lower half-plane and ultimately get the same result.

Step 2

The next step is to make sure that $\lim_{R\to\infty}\int_\Gamma f(z)\,dz = 0$. Then $\lim_{R\to\infty}\int_{\mathscr{C}} f(z)\,dz = \lim_{R\to\infty}\int_{-R}^{R} f(x)\,dx$ and the desired integral is obtained, because of (5.5). The following result is used at this stage.

Lemma 5.5. Convergence of $\int_C f(z)\,dz$ as $R \to \infty$

Let C be an arc of a circle with radius R and centre 0. Let f be continuous on C, with $\lim_{z\to\infty} zf(z) = 0$. Then $\lim_{R\to\infty}\int_C f(z)\,dz = 0$. □

Proof

Note that $|z| = R$ on C. By hypothesis, given any real $\varepsilon > 0$, there exists a real $\delta > 0$ such that

$$|1/z| < \delta \Rightarrow |zf(z)| < \varepsilon \quad \text{so that} \quad 1/R < \delta \Rightarrow |f(z)| < \varepsilon/R \quad \text{on } C.$$

Hence by Lemma 3.2, given any $\varepsilon > 0$, there exists $\delta > 0$ such that

$$1/R < \delta \quad \Rightarrow \quad \left|\int_C f(z)\,dz\right| < \alpha\pi R(\varepsilon/R) = \alpha\pi\varepsilon$$

where $0 < \alpha \leqslant 2$. Hence, by definition, $\lim_{R\to\infty}\int_C f(z)\,dz = 0$. ■

It follows by 5.5 that for integrals of type II, $\lim_{R\to\infty}\int_\Gamma f(z)\,dz = 0$, as required.

Example 5.5

(i) Suppose we wish to evaluate

$$\int_0^\infty \frac{dx}{(x^2+a^2)(x^2+b^2)}$$

where $a, b > 0$ with $a \neq b$. The integral is of type II, so we consider $\int_{\mathscr{C}} f(z)\,dz$ where

$$f(z) = \frac{1}{(z^2+a^2)(z^2+b^2)}$$

and $\mathscr{C}$ is the semicircle in Fig. 5.2; f has simple poles at $\pm ai$ and $\pm bi$. Only ai and bi lie in the upper half-plane, so we choose $R > \max(a, b)$.

$$\text{Res}_{ai} f(z) = \frac{1}{2ai(b^2-a^2)} \quad \text{by 5.2 so that} \quad \text{Res}_{bi} f(z) = \frac{1}{2bi(a^2-b^2)}$$

Then by the residue theorem,

$$\int_{\mathscr{C}} f(z)\,dz = 2\pi i(\text{Res}_{ai} f(z) + \text{Res}_{bi} f(z)) = \frac{\pi}{ab(a+b)}$$

Note that

$$\lim_{z\to\infty} zf(z) = \lim_{z\to\infty} \frac{1}{(z+a^2/z)(z^2+b^2)} = 0$$

so that by 5.5, $\lim_{R\to\infty} \int_\Gamma f(z)\,dz = 0$. Hence, letting $R \to \infty$ gives

$$\int_{-\infty}^{\infty} f(x)\,dx = \frac{\pi}{ab(a+b)} \quad \Rightarrow \quad \int_0^\infty \frac{dx}{(x^2+a^2)(x^2+b^2)} = \frac{\pi}{2ab(a+b)}$$

since the integrand is even.

(ii) Consider $\int_0^\infty (1/(1+x^4))\,dx$, which is once again of type II. Let $f(z) = 1/(1+z^4)$ and consider $\int_{\mathscr{C}} f(z)\,dz$ where $\mathscr{C}$ is the semicircle in Fig. 5.2. The integrand f has isolated singular points when $z^4 = -1 = e^{-i\pi+2ki\pi}$, $k \in \mathbb{Z}$. Hence f has singular points $z_k = e^{-i\pi/4+ik\pi/2}$, $k = 0, 1, 2, 3$. It is easily checked that only z_1 and z_2 lie in the upper half plane, so let $R > 1$. It follows by 5.3 that all singular points are simple poles and

$$\text{Res}_{z_1} f(z) = \frac{1}{4z_1^3} = \frac{1}{4}e^{-3i\pi/4} \qquad \text{Res}_{z_2} f(z) = \frac{1}{4z_2^3} = \frac{1}{4}e^{-i\pi/4}$$

Hence by 5.1,

$$\int_{\mathscr{C}} f(z)\,dz = \frac{\pi i}{2}(\cos(\pi/4) - i\sin(\pi/4) - \sin(\pi/4) - i\cos(\pi/4)) = \frac{\sqrt{2}\pi}{2}$$

Clearly $\lim_{z\to\infty} zf(z) = 0$ so that by 5.5, $\lim_{R\to\infty} \int_\Gamma f(z)\,dz = 0$. Then letting $R \to \infty$ gives

$$\int_{-\infty}^{\infty} \frac{dx}{1+x^4} = \frac{\sqrt{2}\pi}{2} \quad \Rightarrow \quad \int_0^\infty \frac{dx}{1+x^4} = \frac{\sqrt{2}\pi}{4}$$

since the integrand is even.

Integrals of Type III

We next consider improper real integrals related to the form

$$\int_{-\infty}^{\infty} f(x)g(x)\,dx \quad f \text{ an algebraic fraction with no real singular points and } g(x) = \sin mx \text{ or } \cos mx \ (m \in \mathbb{R}^+) \qquad \text{(III)}$$

Important Note

It is supposed in (III) that the degree of the denominator of f is at least one greater than the degree of the numerator of f. The given real improper integral is absolutely convergent if the degree of the denominator is at least two greater than the numerator, by previous comments. If the degree of the denominator of f is only one greater than the numerator, then f' is an algebraic fraction with the degree of the denominator two greater than the numerator. Hence integration by parts shows that the given integral is (conditionally) convergent. Then in either case,

$$\int_{-\infty}^{\infty} f(x)g(x)\,dx = \lim_{R\to\infty}\int_{-R}^{R} f(x)g(x)\,dx$$

The same basic method employed for integrals of type II can clearly be employed in this case too. The only problem is with the convergence to 0 of the integral along the semicircular arc Γ. Generally $\int_\Gamma f(z)g(z)\,dz$ diverges as $R\to\infty$, but $\lim_{R\to\infty}\int_\Gamma e^{miz}f(z)\,dz = 0$. Hence the standard method is as follows.

Step 1

Consider $\int_{\mathscr{C}} e^{miz}f(z)\,dz$ where $\mathscr{C}$ is the standard semicircle of radius R shown in Fig. 5.2, with R chosen large enough so that $\mathscr{C}$ encloses any singular points of f in the upper half-plane. Since there are no singular points on $\mathscr{C}$, evaluate $\int_{\mathscr{C}} e^{miz}f(z)\,dz$ using the residue theorem.

Step 2

Show that $\lim_{R\to\infty}\int_\Gamma e^{miz}f(z)\,dz = 0$, so that letting $R\to\infty$ gives $\lim_{R\to\infty}\int_{\mathscr{C}} e^{miz}f(z)\,dz = \int_{-\infty}^{\infty} e^{mix}f(x)\,dx$. Then the required integral follows by comparing real or imaginary parts. In this case, the following result is used.

Theorem 5.6. Jordan's Lemma

Let Γ be the semicircular arc parametrised by $z = Re^{i\theta}$, $0\leqslant\theta\leqslant\pi$. Let f be continuous on Γ, with $\lim_{z\to\infty} f(z) = 0$. Then $\lim_{R\to\infty}\int_\Gamma e^{miz}f(z)\,dz = 0$ for $m\in\mathbb{R}^+$. □

Proof

Part 1

Using the given parametrisation of Γ gives

$$\left|\int_\Gamma e^{miz}\,dz\right| = \left|\int_0^\pi e^{miR(\cos\theta+i\sin\theta)}iRe^{i\theta}\,d\theta\right|$$

$$\leqslant \int_0^{\pi} |e^{miR\cos\theta}||e^{-mR\sin\theta}||iRe^{i\theta}|\,d\theta$$

$$\Rightarrow \quad \left|\int_\Gamma e^{miz}\,dz\right| \leqslant R\int_0^{\pi} e^{-mR\sin\theta}\,d\theta = 2R\int_0^{\pi/2} e^{-mR\sin\theta}\,d\theta \tag{5.6}$$

using (3.7) in the proof of 3.2 and noting that $e^{-mR\sin\theta}$ is symmetric about $\theta = \pi/2$ in the interval $[0, \pi]$.

Part 2

Note that $2\theta/\pi \leqslant \sin\theta$ for $\theta \in [0, \pi/2]$ since $2\theta/\pi = \sin\theta$ when $\theta = 0$ or $\pi/2$ in this interval and if $f(\theta) = 2\theta/\pi$ and $g(\theta) = \sin\theta$ then $f'(0) < g'(0)$. Hence inequality (5.6) becomes

$$\left|\int_\Gamma e^{miz}\,dz\right| \leqslant 2R\int_0^{\pi/2} e^{-2mR\theta/\pi}\,d\theta = \frac{\pi}{m}(1 - e^{-mR}) < \frac{\pi}{m} \tag{5.7}$$

Part 3

By hypothesis, $\lim_{z\to\infty} f(z) = 0$ so that given any real $\varepsilon > 0$, there exists a real $\delta > 0$ such that

$$|1/z| < \delta \quad \Rightarrow \quad |f(z)| < \varepsilon \quad \text{so that} \quad 1/R < \delta \quad \Rightarrow \quad |f(z)| < \varepsilon \quad \text{on } \Gamma$$

Then by (5.7), given any $\varepsilon > 0$, there exists $\delta > 0$ such that

$$1/R < \delta \quad \Rightarrow \quad \left|\int_\Gamma e^{miz} f(z)\,dz\right| < \varepsilon\left|\int_\Gamma e^{miz}\,dz\right| < \frac{\pi\varepsilon}{m}$$

Hence $\lim_{R\to\infty}\int_\Gamma e^{miz} f(z)\,dz = 0$ as required. ■

Example 5.6

(i) To evaluate

$$\int_{-\infty}^{\infty} \frac{x\sin mx}{x^2 + a^2}\,dx \qquad (m, a > 0)$$

which is of type III, we consider $\int_{\mathscr{C}} e^{miz} f(z)\,dz$ where $f(z) = z/(z^2 + a^2)$ and where $\mathscr{C}$ is the semicircle shown in Fig. 5.2; f has simple poles at $\pm ai$ and only ai lies in the upper half-plane, so let $R > a$. By 5.2, $\text{Res}_{ai}(e^{miz} f(z)) = e^{-ma}/2$, and so by 5.1, $\int_{\mathscr{C}} e^{miz} f(z)\,dz = \pi i e^{-ma}$. Also, $\lim_{z\to\infty} f(z) = \lim_{z\to\infty} 1/(z + a^2/z) = 0$, so that by Jordan's lemma, $\lim_{R\to\infty}\int_\Gamma e^{miz} f(z)\,dz = 0$. Hence

$$\lim_{R\to\infty}\int_{\mathscr{C}} e^{miz} f(z) = \pi i e^{-ma} = \int_{-\infty}^{\infty} e^{mix} f(x)\,dx$$

Equating imaginary parts gives

$$\int_{-\infty}^{\infty} \frac{x \sin mx}{x^2 + a^2}\, dx = \pi e^{-ma} \quad \Rightarrow \quad \int_{0}^{\infty} \frac{x \sin mx}{x^2 + a^2}\, dx = \frac{\pi}{2e^{ma}}$$

since the integrand is even.

(ii) To evaluate

$$\int_{0}^{\infty} \frac{\cos mx}{(x^2 + 1)^2}\, dx$$

we consider $\int_{\mathscr{C}} e^{miz} f(z)\, dz$ where $f(z) = 1/(z^2+1)^2$ and $\mathscr{C}$ is the semicircle shown in Fig. 5.2, where $R > 1$ since f has poles of order 2 at $\pm i$. Then only i lies inside $\mathscr{C}$. Using Lemma 5.2, $\mathrm{Res}_i(e^{miz} f(z)) = g'(i)$ where $g(z) = e^{miz}/(z+i)^2$, so that $\mathrm{Res}_i(e^{miz} f(z)) = (m+1)e^{-m}/4i$ and then by 5.1, $\int_{\mathscr{C}} e^{miz} f(z)\, dz = (m+1)e^{-m}\pi/2$. Letting $R \to \infty$, it follows by Jordan's lemma that

$$\int_{-\infty}^{\infty} e^{mix} f(x)\, dx = \frac{(m+1)e^{-m}\pi}{2} \quad \Rightarrow \quad \int_{0}^{\infty} \frac{\cos mx}{(x^2+1)^2}\, dx = \frac{(m+1)e^{-m}\pi}{4}$$

by comparing real parts and noting that the integrand is even.

Integrals of Type IV

It is also possible to evaluate convergent integrals of type III, except that f has real singular points, by amending the standard semicircle suitably. A closed contour is created which does not contain any real singular points, by constructing a small semicircle centred at each real singular point.

Important Note

Recall that it may happen that a given real improper integral, $\int_{-\infty}^{\infty} f(x)\, dx$ does not exist, whereas its Cauchy principal value $\lim_{R\to\infty} \int_{-R}^{R} f(x)\, dx$ does exist. Similarly, if f has a singular point at $x = c$, where $a < c < b$, it may happen that the improper integral

$$\int_a^b f(x)\, dx = \lim_{\varepsilon \to 0} \int_a^{c-\varepsilon} f(x)\, dx + \lim_{\eta \to 0} \int_{c+\eta}^{b} f(x)\, dx$$

does not exist, whereas its Cauchy principal value

$$\lim_{\varepsilon \to 0} \left(\int_a^{c-\varepsilon} f(x)\, dx + \int_{c+\varepsilon}^{b} f(x)\, dx \right)$$

does exist. In any case, we shall always take $\int_{-\infty}^{\infty} f(x)\, dx$ to mean the Cauchy principal value of the given improper integral. The notation reduces to its usual meaning when the integral converges in the usual sense.

Consider real convergent improper integrals, or more generally, integrals whose Cauchy principal value exists, which are related to the form

$\int_{-\infty}^{\infty} f(x)g(x)\,dx$	f an algebraic fraction with singular points on the real axis and $g(x) = \sin mx$ or $\cos mx$ ($m \in \mathbb{R}^+$) (IV)

Note

A necessary condition for convergence is that the degree of the denominator of f is at least one greater than the numerator of f.

The standard method of evaluating integrals of type IV is as follows.

Step 1

Suppose that f has real singular points at $a_1, \ldots, a_k$ where $a_1 < a_2 < \ldots < a_k$. Consider $\int_{\mathscr{C}} e^{miz} f(z)\,dz$ where $m \in R^+$ and $\mathscr{C}$ is the contour shown in Fig. 5.3, consisting of the large semicircular arc Γ of radius R, line segments of the real axis, and small semicircular arcs $\gamma_1, \ldots, \gamma_k$ of radii $r_1, \ldots, r_k$. The radius R is chosen large enough and the radii $r_1, \ldots, r_k$ are chosen small enough so that $\mathscr{C}$ encloses any singularities of f in the upper half-plane. Since there are no singular points of f on $\mathscr{C}$, evaluate $\int_{\mathscr{C}} e^{miz} f(z)\,dz$ using the residue theorem.

Step 2

Show that $\lim_{R\to\infty} \int_{\Gamma} e^{miz} f(z)\,dz = 0$ using Jordan's lemma as in type III.

Step 3

Find $\lim_{r_k \to 0} \int_{\gamma_k} e^{miz} f(z)\,dz$ for each k using Lemma 5.7 or Corollary 5.8.

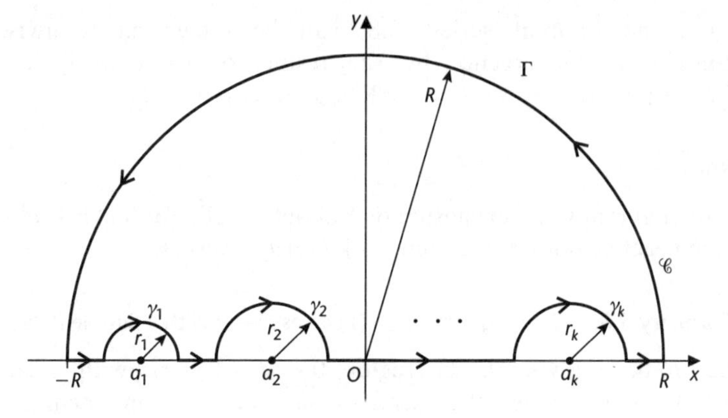

Figure 5.3

Step 4

Finally, letting $R \to \infty$ and $r_k \to 0$ for each k in $\mathscr{C}$ gives, provided the integrals converge,

$$\int_{\mathscr{C}} e^{miz} f(z)\,dz = \lim_{R\to\infty} \int_{-R}^{R} e^{mix} f(x)\,dx - i\pi \sum_{n=1}^{k} \operatorname{Res}_{a_n}(e^{miz} f(z))$$

The desired real integral is then obtained by equating real and imaginary parts.

Lemma 5.7. Convergence of $\int_\gamma F(z)\,dz$ as $r \to 0$

Let F be analytic in the region $0 < |z-a| < r_1$, with a singular point at $a \in \mathbb{R}$. Let γ be a semicircular arc of radius $r < r_1$, centred at a, parametrised by $z - a = re^{i\theta}$, where θ *decreases* from π to 0. If the Laurent series expansion of F about a, valid for $0 < |z-a| < r_1$, contains no even negative powers in its principal part, then $\lim_{r\to 0} \int_\gamma F(z)\,dz = -i\pi \operatorname{Res}_a F(z)$. □

Proof

By hypothesis, F has a Laurent series expansion about a of the form

$$F(z) = \sum_{n=0}^{\infty} \frac{b_{2m+1}}{(z-a)^{2m+1}} + \sum_{n=0}^{\infty} a_n (z-a)^n \quad \text{for} \quad 0 < |z-a| < r_1 \tag{5.8}$$

By definition,

$$\int_\gamma (z-a)^n dz = -ir^{n+1} \int_0^\pi e^{(n+1)i\theta} d\theta = \frac{r^{n+1}}{n+1}(1 + \cos n\pi)$$

where $n \in \mathbb{Z}$ with $n \neq -1$. Hence for $n \geqslant 0$ and $n < -1$ with n odd, $\lim_{r\to 0} \int_\gamma (z-a)^n dz = 0$. Also by definition,

$$\lim_{r\to 0} \int_\gamma (z-a)^{-1} dz = \lim_{r\to 0} \int_0^\pi -i d\theta = -i\pi$$

Since the Laurent series (5.8) can be integrated termwise within its domain of convergence by the results of Chapter 4, it follows that $\lim_{r\to 0} \int_\gamma F(z)\,dz = -i\pi b_1 = -i\pi \operatorname{Res}_a F(z)$ as required. ■

Note

If the Laurent series expansion of F about a, valid for $0 < |z-a| < r_1$, contains even negative powers, then $\lim_{r\to 0} \int_\gamma F(z)\,dz$ diverges.

Corollary 5.8. Convergence of $\int_\gamma F(z)\,dz$ as $r \to 0$ with a Simple Pole

Let F be analytic in the region $0 < |z-a| < r_1$, with a simple pole at $a \in \mathbb{R}$. Let γ be the semicircular arc as in Lemma 5.7. Then $\lim_{r\to 0} \int_\gamma F(z)\,dz = -i\pi \operatorname{Res}_a F(z)$. □

Example 5.7

(i) One of the easiest type IV integrals to deal with is the famous improper integral $\int_0^\infty ((\sin x)/x)\,dx$. Consider $\int_{\mathscr{C}} (e^{iz}/z)\,dz$ where $\mathscr{C}$ is the contour as shown in Fig. 5.4, since the integrand has only one singular point, a simple pole at 0. It follows by the residue theorem that $\int_{\mathscr{C}} (e^{iz}/z)\,dz = 0$. Also, by Jordan's lemma, $\lim_{R\to\infty} \int_\Gamma (e^{iz}/z)\,dz = 0$. $\mathrm{Res}_0(e^{iz}/z) = 1$ by 5.2, so by 5.8, $\lim_{r\to 0} \int_\gamma (e^{iz}/z)\,dz = -i\pi$. Then letting $R \to \infty$ and $r \to 0$ in $\mathscr{C}$ gives

$$\int_{\mathscr{C}} \frac{e^{iz}}{z}\,dz = 0 = \lim_{R\to\infty, r\to 0}\left(\int_{-R}^{-r} \frac{e^{ix}}{x}\,dx + \int_r^R \frac{e^{ix}}{x}\,dx\right) - i\pi$$

$$\Rightarrow \int_{-\infty}^{\infty} \frac{e^{ix}}{x}\,dx = i\pi \Rightarrow \int_{-\infty}^{\infty} \frac{\sin x}{x}\,dx = \pi \Rightarrow \int_0^\infty \frac{\sin x}{x}\,dx = \frac{\pi}{2}$$

by equating imaginary parts and noting that the resulting integrand is even. Compare this with Example 3.8.

(ii) Another convergent real improper integral of type IV is

$$\int_{-\infty}^{\infty} \frac{\cos \pi x}{1 - 4x^2}\,dx$$

(since the singular points at $x = \pm 1/2$ are removable). Following the standard technique, consider $\int_{\mathscr{C}} F(z)\,dz$ where $F(z) = e^{i\pi z}/(1 - 4z^2)$. Since F has only two singular points, which are simple poles at $\pm 1/2$, $\mathscr{C}$ is chosen to be the contour shown in Fig 5.5, where $r < 1/2$ and $R > 1$. It follows by the residue theorem that $\int_{\mathscr{C}} F(z)\,dz = 0$. And by Jordan's lemma, $\lim_{R\to\infty} \int_\Gamma F(z)\,dz = 0$. Also by Lemma 5.2, $\mathrm{Res}_{\pm 1/2} F(z) = 1/4i$. Hence letting $R \to \infty$ and $r \to 0$ in $\mathscr{C}$ and using 5.8 gives

$$\int_{\mathscr{C}} F(z)\,dz = 0 = \int_{-\infty}^{\infty} F(x)\,dx - i\pi\left(\frac{2}{4i}\right) \Rightarrow \int_{-\infty}^{\infty} \frac{\cos \pi x}{1 - 4x^2}\,dx = \frac{\pi}{2}$$

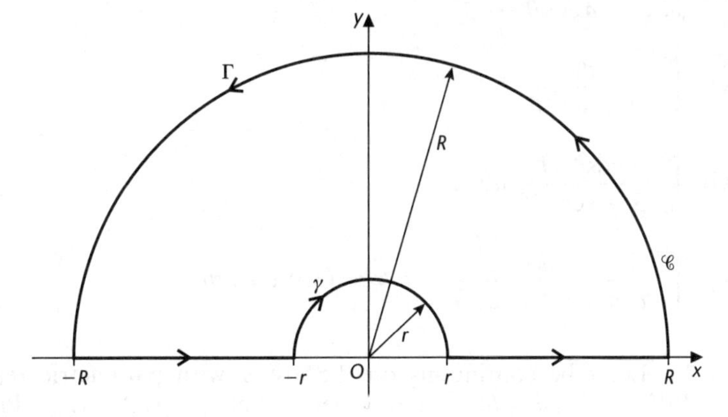

Figure 5.4

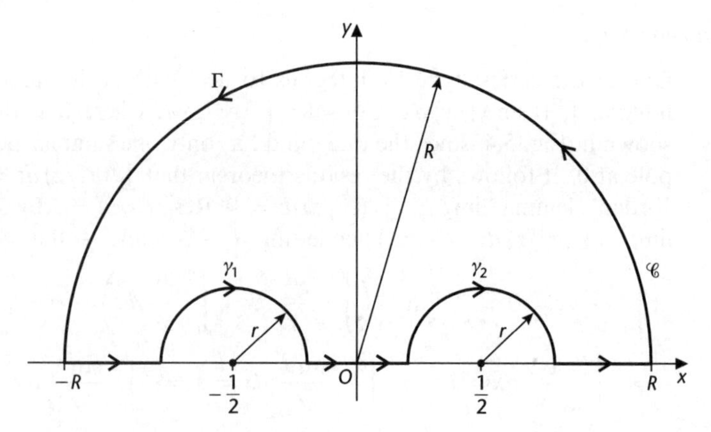

Figure 5.5

Exercise **5.2.1** Evaluate the following real integrals using the residue theorem:

(i) $\displaystyle\int_0^{\pi} \frac{d\theta}{a + b\cos\theta}, \ a > b > 0$

(ii) $\displaystyle\int_0^{2\pi} e^{2\cos\theta}\, d\theta$

(iii) $\displaystyle\int_0^{2\pi} \frac{d\theta}{(3\cos\theta + 5)^2}$

(iv) $\displaystyle\int_0^{\pi} \frac{\cos 2\theta}{1 - 2a\cos\theta + a^2}\, d\theta, \ |a| < 1$

(v) $\displaystyle\int_0^{\pi/2} \frac{d\theta}{1 + \sin^2\theta}$

(vi) $\displaystyle\int_0^{2\pi} \frac{\cos^2 3\theta}{5 - 4\cos 2\theta}\, d\theta$

(vii) $\displaystyle\int_0^{2\pi} \frac{d\theta}{a^2\sin^2\theta + b^2\cos^2\theta}, \ a, b > 0 \text{ with } a \neq b$

Exercise **5.2.2** Let f be continuous on the arc $\mathscr{C}$ with parametric representation $z = Re^{i\theta}$, $\alpha \leqslant \theta \leqslant \beta$, and let $\lim_{z\to\infty} zf(z) = \ell$. Prove that $\lim_{R\to\infty} \int_{\mathscr{C}} f(z)\, dz = (\beta - \alpha)\ell i$.

Exercise **5.2.3** Evaluate the following real convergent improper integrals where a, $b > 0$, $a \neq b$ and $n \in \mathbb{N}$, using the residue theorem:

(i) $\displaystyle\int_0^\infty \frac{x^2}{(x^2+a^2)(x^2+b^2)}\,dx$

(ii) $\displaystyle\int_0^\infty \frac{x^2}{(x^2+a^2)^2}\,dx$

(iii) $\displaystyle\int_0^\infty \frac{x^2}{x^6+a^6}\,dx$

(iv) $\displaystyle\int_0^\infty \frac{x^4}{(x^2+a^2)^3}\,dx$

(v) $\displaystyle\int_0^\infty \frac{dx}{(x^2+a^2)(x^2+b^2)^2}$

(vi) $\displaystyle\int_0^\infty \frac{x^2}{x^4+x^2+1}\,dx$

(vii) $\displaystyle\int_{-\infty}^\infty \frac{dx}{(x^2+x+1)^2}$

(viii) $\displaystyle\int_0^\infty \frac{dx}{(x^2+1)^n}$

Exercise **5.2.4** Evaluate the following real improper integrals, where a, $m > 0$ and $n \in \mathbb{N}$, using the residue theorem:

(i) $\displaystyle\int_0^\infty \frac{\cos^2 x}{x^2+a^2}\,dx$

(ii) $\displaystyle\int_0^\infty \frac{x^3 \sin mx}{(x^2+1)^2}\,dx$

(iii) $\displaystyle\int_{-\infty}^\infty \frac{x \sin x}{x^2+2x+2}\,dx$

(iv) $\displaystyle\int_0^\infty \frac{\cos mx}{(x^2+1)^3}\,dx$

(v) $\displaystyle\int_0^\infty \frac{x \sin x}{x^4+a^4}\,dx$

Exercise **5.2.5** Let F be analytic in the region $0 < |z-a| < r_1$, with a simple pole at a. Let $\mathscr{C}$ be the arc of a circle parametrised by $z - a = re^{i\theta}$ where $\alpha\pi \leqslant \theta \leqslant \beta\pi$ and $r < r_1$. Prove that $\lim_{r\to 0} \int_{\mathscr{C}} F(z)\,dz = \pi(\beta - \alpha)i\,\mathrm{Res}_a f(z)$.

Exercise **5.2.6** Evaluate the following real improper integrals using the residue theorem.

(i) $\displaystyle\int_0^\infty \frac{\sin x}{x(x^2+1)}\,dx$

(ii) $\displaystyle\int_0^\infty \frac{\cos(3\pi x/2)}{1-9x^2}\,dx$

(iii) $\displaystyle\int_0^\infty \frac{\sin x}{x(1+x^2)^2}\,dx$

(iv) $\displaystyle\int_0^\infty \frac{\sin^3 x}{x^3}\,dx$

(v) $\displaystyle\int_0^\infty \frac{\sin \pi x}{x(1-x^2)}\,dx$

Exercise **5.2.7** Use the residue theorem to evaluate $\int_{-\infty}^{\infty}(1/(1-x^3))\,dx$. (In this case, the Cauchy principal value exists, although the improper integral does not converge.)

Evaluation of Other Real Definite Integrals

The techniques used to evaluate integrals of types II to IV can be adapted to evaluate various other real convergent improper integrals. In every case the basic idea is to find a suitable contour, consisting of arcs of circles and line segments, that includes an interval of $\mathbb{R}$. Singularities on the real axis are avoided as in type IV. The residue theorem is used to evaluate the associated contour integral and then lemmas similar to 5.5, 5.6, 5.7 and 5.8 are used to reduce the contour integral to the desired real integral.

Example 5.8

Consider $\int_0^\infty(1/(1+x^3))\,dx$. This improper integral cannot be directly related to an integral of type II since the integrand is not even. Hence the techniques given so far fail. Instead, we adapt the method given for integrals of type II and use a sector of the semicircle instead. Hence consider $\int_{\mathscr{C}}(1/(1+z^3))\,dz$ where $\mathscr{C}$ is the contour as shown in Fig. 5.6; angle θ is fixed. (Note that the only real singularity of the integrand is -1.) Let $f(z)=1/(1+z^3)$. The idea is to choose θ suitably so that $\int_L f(z)\,dz$ can be related to the given integral. The integrand f has isolated singular points given by $z^3=-1=e^{-i\pi+2ki\pi}$, $k=0,1,2$. Hence the singular points of f are $z_k=e^{-i\pi/3+2ki\pi/3}$, $k=0,1,2$. Only $z_1=e^{i\pi/3}$ lies in the

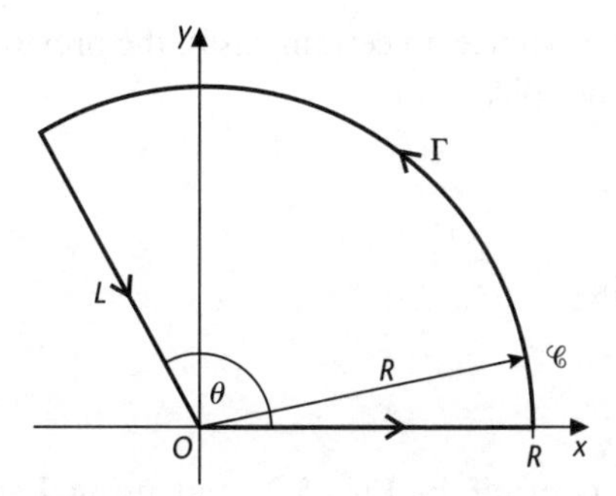

Figure 5.6

upper half-plane, so take $\pi/3 < \theta < \pi$ and $R > 1$ for $\mathscr{C}$. Note that by 5.3, $e^{i\pi/3}$ is a simple pole of f and

$$\operatorname{Res}_{e^{i\pi/3}} f(z) = \left.\frac{1}{3z^2}\right|_{z=e^{i\pi/3}} = \frac{e^{-2i\pi/3}}{3}$$

It follows by 5.1 that $\int_{\mathscr{C}} f(z)\,dz = \frac{2}{3}\pi i e^{-2i\pi/3}$. It also follows by 5.5 that $\lim_{R\to\infty} \int_\Gamma f(z)\,dz = 0$. Hence letting $R \to \infty$ gives

$$\int_{\mathscr{C}} \frac{dz}{1+z^3} = \frac{2\pi i e^{-2i\pi/3}}{3} = \lim_{R\to\infty} \int_0^R \frac{dx}{1+x^3} + \lim_{R\to\infty} \int_R^0 \frac{e^{i\theta}\,d\rho}{1+\rho^3 e^{3i\theta}}$$

since along the line segment L, $z = \rho e^{i\theta}$, with θ fixed and where ρ decreases from R to 0. Hence in order to obtain the desired integral, we choose θ such that $e^{3i\theta} = 1$, i.e., $\theta = 2\pi/3$. Then by the above it follows that

$$\int_0^\infty \frac{dx}{1+x^3} = \frac{2\pi i e^{-2i\pi/3}}{3(1-e^{2i\pi/3})} = \frac{2\pi i(-1-i\sqrt{3})}{3(3-i\sqrt{3})} = \frac{2\sqrt{3}\pi}{9}$$

This technique can be used to evaluate any integral of the form

$$\int_0^\infty \frac{dx}{(x^n+a^n)^m} \qquad (m, n \in \mathbb{N})$$

Integrals Involving Branch Points

Very often we wish to evaluate a real convergent improper integral using the residue theorem where the associated complex integral involves a particular branch of a multifunction. The techniques discussed so far can usually be adapted in such cases as long as any chosen contour does not enclose a branch point and so cross a branch cut. Hence, in certain cases, the previous techniques can be applied directly. For example,

$$\int_0^\infty \frac{\operatorname{Log}(x^2+1)}{x^2+1}\,dx$$

can be evaluated by considering

$$\int_{\mathscr{C}} \frac{\operatorname{Log}(z+i)}{z^2+1}\,dz$$

where $\mathscr{C}$ is the semicircular contour in Fig. 5.2, and using Lemma 5.5 (see Exercise 5.1.8. More generally, if a contour of a previous type crosses a branch cut, it can easily be amended so it does not. Specifically, convergent integrals related to the form $\int_0^\infty f(x)g(x)\,dx$ where f is an algebraic fraction and $g(x) = \operatorname{Log} x$ or x^λ where λ is real but not an integer, can usually be evaluated by using one of the four contours shown in Fig. 5.7. (The choice of contour depends on the behaviour of $f(x)$ and is sometimes a matter of personal preference.) The idea is to let $\theta \to \pi$ or $\theta \to 0, 2\pi$, etc. on L, $L\pi$, or L_2, as well as $r \to 0$ and $R \to \infty$ to obtain the desired real integral.

The following result is useful when dealing with this type of integral.

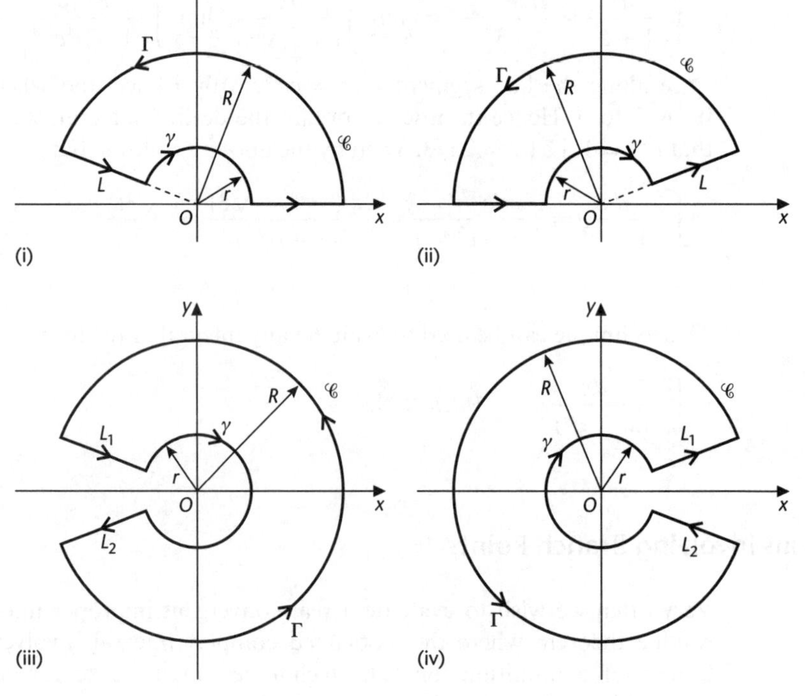

Figure 5.7

Lemma 5.9. Convergence of $\int_\gamma f(z)\,dz$ as $r \to 0$.

Let γ be an arc of a circle of radius r, centred at 0. Let f be continuous on γ with $\lim_{z\to 0} zf(z) = 0$. Then $\lim_{r\to 0}\int_\gamma f(z)\,dz = 0$. □

(Compare with Corollary 5.8. If 0 is a simple pole of f then $\lim_{z\to 0} zf(z) \neq 0$.)

Proof

Let the length of γ be $\alpha\pi r$ for some α, $0 < \alpha \leqslant 2$. By hypothesis, given any real $\varepsilon > 0$, there is a real $\delta > 0$ such that

$$0 < |z| < \delta \quad \Rightarrow \quad |zf(z)| < \varepsilon$$

On γ, $|z| = r$, so that as long as $r < \delta$ then $|f(z)| < \varepsilon/r$. It then follows by 3.2 that $r < \delta \Rightarrow |\int_\gamma f(z)\,dz| < (\alpha\pi r)\varepsilon/r = \alpha\pi\varepsilon$. Hence $\lim_{r\to 0}\int_\gamma f(z)\,dz = 0$, as required. ■

Example 5.9

Consider the convergent integral

$$\int_0^\infty \frac{\operatorname{Log} x}{x^2+a^2}\,dx \qquad (a \neq 0)$$

In order to evaluate this integral we consider $\int_{\mathscr{C}} f(z)\,dz$ where $f(z) = (\operatorname{Log} z)/(z^2+a^2)$ and $\mathscr{C}$ is the contour in Fig. 5.7(i), where $R > a$. This contour is chosen since 0 is a branch point of the multifunction log and, by our convention, the non-positive real axis is the branch cut for log, so that Log is singular along this line segment. Note, however, that if $z = \rho e^{i\pi}$ then $z^2 = \rho^2$.

The integrand f has simple poles at $\pm ai$ and only ai lies within $\mathscr{C}$. By 5.2, $\operatorname{Res}_{ai} f(z) = (\operatorname{Log} ai)/2ai = (\operatorname{Log} a + i\pi/2)/2ai$. Then by the residue theorem, $\int_{\mathscr{C}} f(z)\,dz = (\pi/a)(\operatorname{Log} a + \frac{1}{2}i\pi)$. It follows by L'Hôpital's rule that

$$\lim_{z\to\infty} zf(z) = \lim_{z\to 0}\frac{f(1/z)}{z} = \lim_{z\to 0}\frac{-\operatorname{Log} z}{a^2z+1/z} = \lim_{z\to 0}\frac{-z}{a^2z^2-1} = 0$$

Hence by 5.5, $\lim_{R\to\infty}\int_\Gamma f(z)\,dz = 0$. Also by L'Hôpital's rule,

$$\lim_{z\to 0} zf(z) = \lim_{z\to 0}\frac{\operatorname{Log} z}{z+a^2/z} = \lim_{z\to 0}\frac{1/z}{1-a^2/z^2} = \lim_{z\to 0}\frac{z}{z^2-a^2} = 0$$

Hence by 5.9, $\lim_{r\to 0}\int_\gamma f(z)\,dz = 0$. Note that along the line segment L, $z = \rho e^{i\theta}$ with θ fixed and ρ decreasing from R. Hence letting $R \to \infty$, $r \to 0$ and $\theta \to \pi$ on $\mathscr{C}$ gives

$$\int_{\mathscr{C}} f(z)\,dz = \int_0^\infty \frac{\operatorname{Log} x}{x^2+a^2}\,dx + \int_\infty^0 \frac{(\operatorname{Log}\rho + i\pi)e^{i\pi}}{\rho^2 e^{2i\pi}+a^2}\,d\rho = \frac{\pi}{a}\left(\operatorname{Log} a + \frac{i\pi}{2}\right)$$

$$\Rightarrow 2\int_0^\infty \frac{\operatorname{Log} x}{x^2+a^2}\,dx + i\pi\int_0^\infty \frac{dx}{x^2+a^2} = \frac{\pi}{a}\left(\operatorname{Log} a + \frac{i\pi}{2}\right) \Rightarrow \int_0^\infty \frac{\operatorname{Log} x}{x^2+a^2}\,dx = \frac{\pi \operatorname{Log} a}{2a}$$

comparing real parts. Note that comparing imaginary parts gives an elementary result!

Example 5.10

To evaluate $\int_0^\infty (x^{\lambda-1}/(1+x))\,dx$, where λ is real but not an integer, which converges for $0<\lambda<1$, we consider $\int_{\mathscr{C}} f(z)\,dz$ where $f(z)=z^{\lambda-1}/(1+z)$ and $\mathscr{C}$ is the contour in Fig. 5.7(iv), where $R>1$. This contour is chosen since f has a branch point at 0 and if $z=\rho e^{i\theta}$, then $z=\rho$ if $\theta=2n\pi$ but $z=-\rho$ for $\theta=(2n+1)\pi$, $n\in\mathbb{Z}$. Note also that f has a simple pole at -1 and this lies on the contour in Fig. 5.7(ii). Hence, in this case, $z^{\lambda-1}$ is chosen as a branch of a multifunction whose branch cut is the non-negative real axis.

There is only one singular point of f inside $\mathscr{C}$ and that is the simple pole at -1. By 5.2, $\operatorname{Res}_{-1} f(z)=e^{(\lambda-1)\log(-1)}=-e^{i\pi\lambda}$ and so by the residue theorem, $\int_{\mathscr{C}} f(z)\,dz=-2\pi i e^{i\pi\lambda}$. Note that $\lim_{z\to\infty} zf(z)=\lim_{z\to\infty} z^\lambda/(1+z)=0$ if and only if $\lambda<1$. Then by 5.5, $\lim_{R\to\infty}\int_\Gamma f(z)\,dz=0$ if and only if $\lambda<1$. Also $\lim_{z\to 0} zf(z)=0$ if and only if $\lambda>0$. Then by 5.9, $\lim_{r\to 0}\int_\gamma f(z)\,dz=0$ if and only if $\lambda>0$. Note that $z=\rho e^{i\theta}$ on the line segments L_1 and L_2, where θ is fixed and ρ increases to R on L_1 and decreases from R on L_2. Then letting $R\to\infty$, $r\to 0$, $\theta\to 0$ on L_1, and $\theta\to 2\pi$ on L_2,

$$\int_0^\infty \frac{x^{\lambda-1}}{1+x}\,dx+\int_\infty^0 \frac{\rho^{\lambda-1}e^{2i\pi(\lambda-1)}e^{2i\pi}}{1+\rho e^{2i\pi}}\,d\rho=-2\pi i e^{\lambda i\pi}$$

$$\Rightarrow \int_0^\infty \frac{x^{\lambda-1}}{1+x}\,dx=\frac{-2\pi i e^{\lambda i\pi}}{1-e^{2i\lambda\pi}}=\frac{2\pi i}{e^{i\lambda\pi}-e^{-i\lambda\pi}}=\frac{\pi}{\sin\lambda\pi}\qquad(0<\lambda<1)$$

Integrals with an Infinite Number of Singular Points

Techniques described so far will fail on real improper integrals whose integrands have an infinite number of non-real singular points, since any semicircle will never enclose all the singular points, no matter how large the radius. Instead a rectangle is chosen. Possibly indented by small semicircles to avoid singular points, the rectangle should include an interval of the real axis and enclose a finite number of singular points of the integrand.

Example 5.11

Consider the integral $\int_{-\infty}^\infty (e^{\lambda x}/(1+e^x))\,dx$, where λ is to be determined for convergence. To evaluate this integral, we consider $\int_{\mathscr{C}} f(z)\,dz$ where $f(z)=e^{\lambda z}/(1+e^z)$ and $\mathscr{C}$ is the rectangle as shown in Fig. 5.8. This rectangle is chosen since $e^{z+2\pi i}=e^z$ and it is chosen to be non-symmetric to avoid the need to take the principal value of the given real integral. The only singular point of f

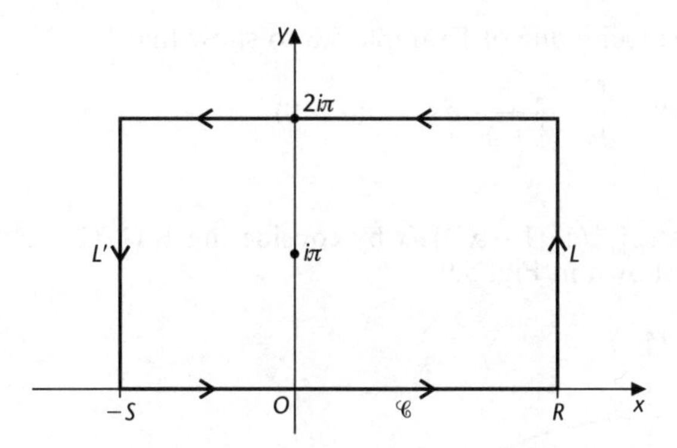

Figure 5.8

inside $\mathscr{C}$ is πi and this is a simple pole with $\text{Res}_{i\pi} f(z) = -e^{\lambda i\pi}$, by Lemma 5.3. Hence by the residue theorem, $\int_{\mathscr{C}} f(z)\,dz = -2\pi i e^{\lambda i\pi}$. Now on the line segment L, $z = R + iy$, $0 \leqslant y \leqslant 2\pi$. Hence,

$$\int_L f(z)\,dz = \int_0^{2\pi} \frac{ie^{\lambda(R+iy)}}{1+e^{R+iy}}\,dy$$

It follows by the triangle inequality that

$$|1 + e^R e^{iy}| \geqslant |1 - |e^R e^{iy}|| = e^R - 1$$

Hence by 3.2, $|\int_L f(z)\,dz| \leqslant 2\pi e^{\lambda R}/(e^R - 1) \to 0$ as $R \to \infty$ if and only if $\lambda < 1$. Similarly, $|\int_{L'} f(z)\,dz| \leqslant 2\pi e^{-\lambda S}/(1 - e^{-S}) \to 0$ as $S \to \infty$ if and only if $\lambda > 0$. Then letting $R \to \infty$ and $S \to \infty$ in $\mathscr{C}$ gives

$$\int_{\mathscr{C}} f(z)\,dz = \int_{-\infty}^{\infty} \frac{e^{\lambda x}}{1+e^x}\,dx + \int_{\infty}^{-\infty} \frac{e^{\lambda(x+2i\pi)}}{1+e^{x+2i\pi}}\,dx = -2\pi i e^{\lambda i\pi}$$

$$\Rightarrow \quad \int_{-\infty}^{\infty} \frac{e^{\lambda x}}{1+e^x}\,dx = \frac{2\pi i e^{\lambda i\pi}}{e^{2\lambda i\pi} - 1} = \frac{\pi}{\sin \lambda\pi} \qquad (0 < \lambda < 1)$$

Note that letting $y = e^x$ in this result gives $\int_0^\infty (y^{\lambda-1}/(1+y))\,dy = \pi/\sin\lambda\pi$ for $0 < \lambda < 1$, which is the result of Example 5.10.

Exercise

5.3.1 Use the technique employed in Example 5.8 to evaluate the following real convergent improper integrals ($a \neq 0$):

(i) $\displaystyle\int_0^\infty \frac{dx}{x^4 + a^4}$ (ii) $\displaystyle\int_0^\infty \frac{dx}{1+x^n}$, $n \subset \mathbb{N}$, $n > 1$ (iii) $\displaystyle\int_0^\infty \frac{dx}{(1+x^3)^2}$

Exercise **5.3.2** Use the technique of Example 5.8 to show that

$$\frac{1}{a}\int_0^\infty \frac{\sin x}{x+a}\,dx = \int_0^\infty \frac{e^{-x}}{x^2+a^2}\,dx \qquad (a>0)$$

Exercise **5.3.3** Evaluate $\int_0^\infty (1/(1-x^3))\,dx$ by considering $\int_{\mathscr{C}} (1/(1-z^3))\,dz$ where $\mathscr{C}$ is the contour shown in Fig. 5.9.

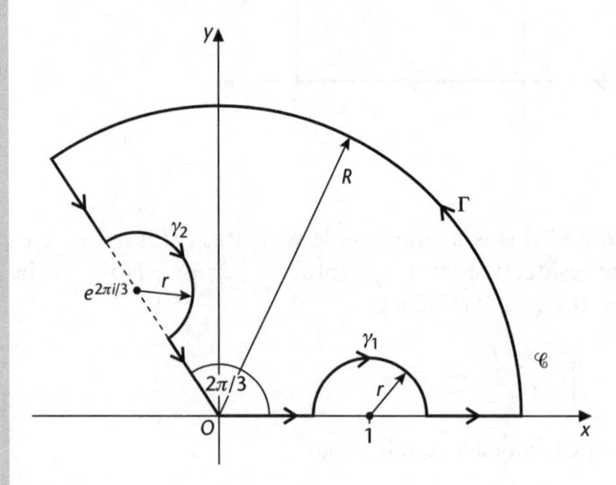

Figure 5.9

Exercise **5.3.4** Evaluate the following real convergent improper integrals, where λ is real but not an integer. State the range of values of λ necessary and sufficient for convergence where appropriate.

(i) $\displaystyle\int_0^\infty \frac{x^\lambda}{x^2+1}\,dx$

(ii) $\displaystyle\int_0^\infty \frac{\operatorname{Log} x}{(1+x^2)^2}\,dx$

(iii) $\displaystyle\int_0^\infty \frac{x^\lambda}{(x^2+a^2)^2}\,dx$

(iv) $\displaystyle\int_0^\infty \frac{(\operatorname{Log} x)^2}{1+x^2}\,dx$

(v) $\displaystyle\int_0^\infty \frac{x^\lambda \operatorname{Log} x}{1+x}\,dx$

(vi) $\displaystyle\int_0^\infty \frac{dx}{\sqrt{x}(x^3+1)}$

Exercise **5.3.5** Evaluate $\int_{\mathscr{C}} \left(1/z^{\lambda}(1-z)\right) dz$, $0<\lambda<1$, where $\mathscr{C}$ is the contour as shown in Fig. 5.10. Hence evaluate $\int_0^{\infty} \left(1/x^{\lambda}(1+x)\right) dx$ and $\int_0^{\infty} \left(1/x^{\lambda}(1-x)\right) dx$.

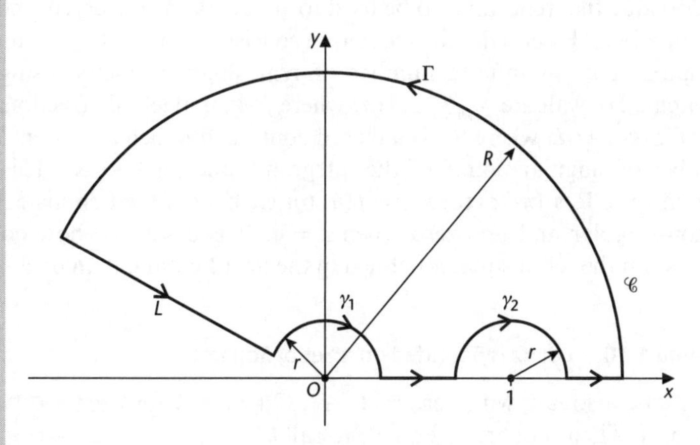

Figure 5.10

Exercise **5.3.6** Use the technique of Example 5.9 to show that

$$\int_0^{\infty} x^{\lambda-1} \sin x \, dx = \Gamma(\lambda) \sin(\pi\lambda/2) \qquad (0<\lambda<1)$$

where the **gamma function** is defined by $\Gamma(\lambda) = \int_0^{\infty} e^{-x} x^{\lambda-1} dx$.

Exercise **5.3.7** Evaluate the following convergent integrals.

(i) $\displaystyle\int_{-\infty}^{\infty} \frac{e^{\lambda x}}{\cosh x} dx$ and hence $\displaystyle\int_0^{\infty} \frac{\cosh \lambda x}{\cosh x} dx$, $-1<\lambda<1$

(ii) $\displaystyle\int_0^{\infty} \frac{\sin \lambda x}{\sinh x} dx$, $\lambda>0$

(iii) $\displaystyle\int_0^{\infty} \frac{x}{\sinh x} dx$

Exercise ***5.3.8** (Cauchy). Evaluate $\int_{\mathscr{C}} (z/(\lambda - e^{-iz}))\, dz$, $0<\lambda<1$, where $\mathscr{C}$ is the rectangle with vertices $\pm\pi$ and $\pm\pi + iR$, $R>1$. Hence show that

$$\int_0^{\pi} \frac{x \sin x}{\lambda^2 - 2\lambda \cos x + 1} dx = \frac{\pi}{\lambda} \operatorname{Log}(1+\lambda)$$

Exercise ***5.3.9** Adapt the technique used in Example 5.11 to evaluate

$$\int_0^{\infty} \frac{\sin x}{e^x + 1} dx$$

Summation of Series Using the Residue Theorem

The residue theorem can also be used to sum certain convergent infinite series of real numbers. Essentially the technique consists of evaluating an integral, whose integrand has an infinite number of real singular points, using the residue theorem. To evaluate $\sum_{n=-\infty}^{\infty} f(n)$ where f is an algebraic fraction, we consider $\int_{\mathscr{C}_k} \pi f(z) \cot \pi z\, dz$ where $\mathscr{C}_k$ is a closed contour for each $k \in \mathbb{N}$ enclosing a finite number of singular points of the integrand, and let $k \to \infty$. This integrand is chosen since $\text{Res}_n(\pi f(z) \cot \pi z) = f(n)$ for each $n \in \mathbb{Z}$ by Lemma 5.3, as long as f is non-singular and non-zero when $z = n$. It is easier to show convergence as $k \to \infty$ on the set of squares defined in the next Lemma, than on a set of circles.

Lemma 5.10. $\cot \pi z$ is Bounded on a Set of Squares

Let $\mathscr{C}_k$ be a square with vertices $(k + 1/2)(\pm 1 \pm i)$ for each $k \in \mathbb{N}$ as shown in Fig. 5.11. Then $|\cot \pi z| < 2$ on $\mathscr{C}_k$ for all k. □

Proof

Let $\alpha = k + 1/2$. On the line segments L_1 and L_3, $z = \pm\alpha + iy$ respectively, where $|y| \leqslant \alpha$. Note that $\cos \alpha\pi = 0$ and $\sin \alpha\pi = (-1)^k$. Hence on L_1 and L_3

$$|\cot \pi z| = \left|\frac{\cos \alpha\pi \cosh \pi y \mp i \sin \alpha\pi \sinh \pi y}{\pm \sin \alpha\pi \cosh \pi y + i \cos \alpha\pi \sinh \pi y}\right| = |\tanh \pi y| < 1$$

for all y, independent of k. On the line segments L_2 and L_4 $z = x \pm i\alpha$ respectively, where $|x| \leqslant \alpha$. Hence on these lines

$$|\cot \pi z| = \left|\frac{e^{i\pi z} + e^{-i\pi z}}{e^{i\pi z} - e^{-i\pi z}}\right| = \left|\frac{e^{i\pi x} e^{\mp\alpha\pi} + e^{-i\pi x} e^{\pm\alpha\pi}}{e^{i\pi x} e^{\mp\alpha\pi} - e^{-i\pi x} e^{\pm\alpha\pi}}\right|$$

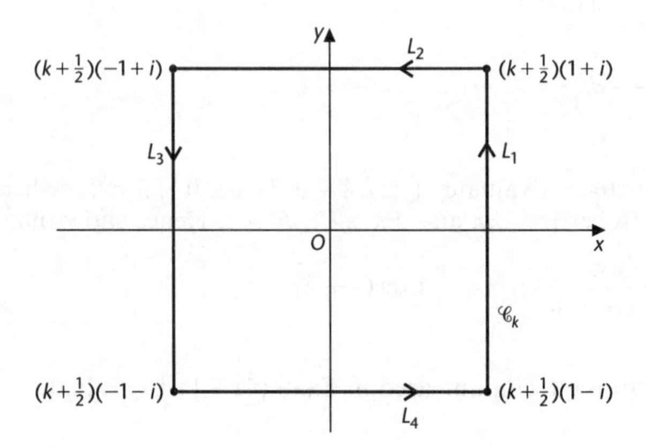

Figure 5.11

Also, it follows by the triangle inequality that

$$|e^{i\pi x}e^{\mp\alpha\pi} + e^{-i\pi x}e^{\pm\alpha\pi}| \leqslant |e^{\mp\alpha\pi}| + |e^{\pm\alpha\pi}| = e^{\alpha\pi} + e^{-\alpha\pi}$$

and

$$|e^{i\pi x}e^{\mp\alpha\pi} - e^{-i\pi x}e^{\pm\alpha\pi}| \geqslant |e^{\mp\alpha\pi} - e^{\pm\alpha\pi}| = e^{\alpha\pi} - e^{-\alpha\pi}$$

It then follows that

$$|\coth \pi z| \leqslant \frac{e^{\alpha\pi} + e^{-\alpha\pi}}{e^{\alpha\pi} - e^{-\alpha\pi}} = \coth \alpha\pi \leqslant \coth \frac{3}{2}\pi < 2$$

on these lines, independent of k, since $\coth x$ is monotonic decreasing for $x >$ ■

Theorem 5.11. Convergence of $\int_{\mathscr{C}_k} f(z) \cot \pi z \, dz$ as $k \to \infty$

Let $\mathscr{C}_k$ be a square with vertices $(k + 1/2)(\pm 1 \pm i)$ for each $k \in \mathbb{N}$. Let f be meromorphic, with f continuous on $\mathscr{C}_k$ for $k > k_1$ say, and $\lim_{z\to\infty} zf(z) = 0$. Then $\lim_{k\to\infty} \int_{\mathscr{C}_k} f(z) \cot \pi z \, dz = 0$. □

Proof

On each square $\mathscr{C}_k$, $|z| \geqslant k + 1/2 > k$ and by hypothesis, given any $\varepsilon > 0$, there exists $\delta > 0$ such that

$$\frac{1}{|z|} < \delta \quad \Rightarrow \quad |zf(z)| < \varepsilon$$

Then choosing $k > 1/\delta$ shows that given any $\varepsilon > 0$ there exists k such that $|f(z)| < \varepsilon/(k + 1/2)$ on $\mathscr{C}_k$. The length of $\mathscr{C}_k$ is $8(k + 1/2)$, so that by 3.2 and 5.10,

$$\frac{1}{k} < \delta \quad \Rightarrow \quad \left| \int_{\mathscr{C}_k} f(z) \cot \pi z \, dz \right| \leqslant \frac{8(k + 1/2)2\varepsilon}{k + 1/2} = 16\varepsilon$$

Hence $\lim_{k\to\infty} \int_{\mathscr{C}_k} f(z) \cot \pi z \, dz = 0$, as required. ■

It is due to this convergence result that the residue theorem can be used to evaluate $\sum_{n=-\infty}^{\infty} f(n)$ for certain functions f. In particular, we have the following result.

Theorem 5.12. The Sum of an Infinite Series

Let f be an algebraic fraction such that $\lim_{z\to\infty} zf(z) = 0$. Let f be non-zero at $n \in \mathbb{Z}$ and have non-integer singular points $z_1, \ldots, z_m$. Then

$$\sum_{n=-\infty}^{\infty} f(n) = -\sum_{n=1}^{m} \mathrm{Res}_{z_n}(\pi f(z) \cot \pi z)$$ □

Proof

Consider $\int_{\mathscr{C}_k} \pi f(z) \cot \pi z \, dz$ where $\mathscr{C}_k$ is a member of the set of squares in 5.11, with k chosen large enough so that all the singular points of f lie within $\mathscr{C}_k$. By hypothesis and Lemma 5.3, $F(z) = \pi f(z) \cot \pi z$ has simple poles at $n \in \mathbb{Z}$ and

$$\text{Res}_n F(z) = \left. \frac{\pi f(z) \cos \pi z}{\pi \cos \pi z} \right|_{z=n} = f(n)$$

Also by hypothesis, F has non-integer singular points $z_1, z_2, \ldots, z_m$. Hence by 5.1,

$$\int_{\mathscr{C}_k} F(z)\, dz = 2\pi i \left(\sum_{n=-k}^{k} f(n) + \sum_{n=1}^{m} \text{Res}_{z_n} F(z) \right)$$

The result then follows by letting $k \to \infty$ and using 5.11. ■

Notes

(i) If $\lim_{z\to\infty} zf(z) \neq 0$ then the infinite series will not converge as in the case of the harmonic series $\sum_{n=1}^{\infty} 1/n$.

(ii) The above result is easily adapted in the case where at least one of the singular points of f is an integer, as demonstrated in the second of the following examples.

Example 5.12

(i) Let $f(z) = 1/(2z+1)^2$. Then f has a pole of order two at $-1/2$ and $\lim_{z\to\infty} zf(z) = 0$. Also by Lemma 5.2,

$$\text{Res}_{-1/2}(\pi f(z) \cot \pi z) = \left. \frac{d}{dz} \left(\frac{\pi \cot \pi z}{4} \right) \right|_{z=-1/2} = -\frac{\pi^2 \csc^2(-\pi/2)}{4} = -\frac{\pi^2}{4}$$

Hence by Theorem 5.12,

$$\sum_{n=-\infty}^{\infty} f(n) = \frac{\pi^2}{4} \Rightarrow \sum_{n=1}^{\infty} \frac{1}{(2n-1)^2} + \sum_{n=0}^{\infty} \frac{1}{(2n+1)^2} = \frac{\pi^2}{4}$$

Letting $n = m + 1$ gives

$$\sum_{n=1}^{\infty} \frac{1}{(2n-1)^2} = \sum_{m=0}^{\infty} \frac{1}{(2m+1)^2} \quad \text{hence} \quad \sum_{n=0}^{\infty} \frac{1}{(2n+1)^2} = \frac{\pi^2}{8}$$

(ii) Let $f(z) = 1/z^2$ so that f has a pole of order 2 at 0 and $\lim_{z\to\infty} zf(z) = 0$. Theorem 5.12 is not directly applicable in this case but the technique is easily adapted. F defined by $F(z) = \pi f(z) \cot \pi z$ has a pole of order 3 at 0 and simple poles at $n \in \mathbb{Z}$, $n \neq 0$. As in the proof of 5.12, $\text{Res}_n F(z) = f(n)$ for $n \neq 0$. Also, in some neighbourhood of 0,

$$F(z) = \frac{\pi}{z^2}\left(\frac{1}{\pi z} - \frac{\pi z}{3} - \frac{\pi^3 z^3}{45} + \dots\right)$$

so that $\text{Res}_0 F(z) = -\pi^2/3$. Now let $\mathscr{C}_k$ be a square with vertices $(k+1/2)(\pm 1 \pm i)$. Then by 5.1,

$$\int_{\mathscr{C}_k} F(z)\,dz = 2\pi i\left(\sum_{\substack{n=-k \\ n\neq 0}}^{k} f(n) - \frac{\pi^2}{3}\right)$$

Letting $k \to \infty$ and using 5.11 gives

$$\sum_{n=-\infty}^{-1} \frac{1}{n^2} + \sum_{n=1}^{\infty} \frac{1}{n^2} = \frac{\pi^2}{3} \quad \Rightarrow \quad \sum_{n=1}^{\infty} \frac{1}{n^2} = \frac{\pi^2}{6}$$

It is also possible to sum certain convergent alternating sign series using the residue theorem. To sum $\sum_{n=-\infty}^{\infty} (-1)^n f(n)$ where f is an algebraic fraction, we consider $\int_{\mathscr{C}_k} \pi f(z) \csc \pi z\,dz$, where $\mathscr{C}_k$ is a square as in 5.10, since if f is non-singular and non-zero at $n \in \mathbb{Z}$, $\text{Res}_n(\pi f(z) \csc \pi z) = (-1)^n f(n)$. The proofs of the following results are very similar to the proofs of 5.10, 5.11 and 5.12; they are left as an exercise.

Theorem 5.13. Convergence of $\int_{\mathscr{C}_k} f(z) \csc \pi z\,dz$ as $k \to \infty$

Let $\mathscr{C}_k$ be a square with vertices $(k+1/2)(\pm 1 \pm i)$ for each $k \in \mathbb{N}$. Let f be meromorphic, with f continuous on $\mathscr{C}_k$ for $k > k_1$ say, and $\lim_{z\to\infty} zf(z) = 0$. Then $\lim_{k\to\infty} \int_{\mathscr{C}_k} f(z) \csc \pi z\,dz = 0$. □

Theorem 5.14. The Sum of an Alternating Sign Series

Let f be an algebraic fraction such that $\lim_{z\to\infty} zf(z) = 0$. Let f be non-zero at $n \in \mathbb{Z}$ with non-integer singular points $z_1, \dots, z_m$. Then

$$\sum_{n=-\infty}^{\infty} (-1)^n f(n) = -\sum_{n=1}^{m} \text{Res}_{z_n}(\pi f(z) \csc \pi z)$$ □

The technique used to prove 5.12 and 5.14 can be adapted to sum other series, as demonstrated in the following example.

Example 5.13

Let $f(z) = (\text{csch}\, \pi z)/z^3$ so that f is meromorphic, $\lim_{z\to\infty} zf(z) = 0$, f has a pole of order 4 at 0 and simple poles at in, $n \in \mathbb{Z}$, $n \neq 0$. Let $F(z) = \pi f(z) \csc \pi z$.

Then F has simple poles at n and in, $n \in \mathbb{Z}$, $n \neq 0$ and a pole of order 5 at 0. It follows by Lemma 5.3 that

$$\mathrm{Res}_n F(z) = (-1)^n f(n) \qquad \mathrm{Res}_{in} F(z) = \left.\frac{\pi}{\pi z^3 \sin \pi z \cosh \pi z}\right|_{z=in} = (-1)^n f(n)$$

Also, in some neighbourhood of 0,

$$F(z) = \frac{\pi}{z^3}\left(\frac{1}{\pi z} + \frac{\pi z}{6} + \frac{7\pi^3 z^3}{360} + \ldots\right)\left(\frac{1}{\pi z} - \frac{\pi z}{6} + \frac{7\pi^3 z^3}{360} - \ldots\right)$$

$$\text{Hence} \quad \mathrm{Res}_0 F(z) = \frac{7\pi^3}{360} - \frac{\pi^3}{36} + \frac{7\pi^3}{360} = \frac{\pi^3}{90}$$

It then follows by the residue theorem that

$$\int_{\mathscr{C}_k} F(z)\, dz = 2\pi i \left(\frac{\pi^3}{90} + 2 \sum_{\substack{n=-k \\ n \neq 0}}^{k} (-1)^n f(n)\right)$$

where $\mathscr{C}_k$ is a square with vertices $(k + 1/2)(\pm 1 \pm i)$. Hence by 5.13 it follows that

$$4\sum_{n=1}^{\infty} (-1)^n f(n) = \frac{-\pi^3}{90} \quad \Rightarrow \quad \sum_{n=1}^{\infty} (-1)^{n+1} \frac{\operatorname{csch} n\pi}{n^3} = \frac{\pi^3}{360}$$

Partial Fraction Expansions

Besides summing certain series of real numbers, the residue theorem can also be used to expand certain functions as a series of partial fractions involving their poles. Once again, this idea is originally due to Cauchy. The following result is the simplest of its type but can be generalised to include non-simple poles.

Theorem 5.15. Partial Fraction Expansion of a Meromorphic Function with Simple Poles

Let f be a meromorphic function with only simple poles $z_1, z_2, z_3, \ldots$ with $|z_1| \leqslant |z_2| \leqslant |z_3| \leqslant \ldots$. Let $\mathrm{Res}_{z_n} f(z) = b_n$ for each n. Let $(\mathscr{C}_k)$ be a sequence of nested squares with vertices $a_k(\pm 1 \pm i)$ such that $a_k \to \infty$ as $k \to \infty$, which do not pass through any pole of f. Suppose that f is analytic at 0 and that there exists $M \in \mathbb{R}^+$, independent of k, such that $|f(z)| \leqslant M$ on $\mathscr{C}_k$ for each k. Then except at the poles of f,

$$f(z) = f(0) + \sum_n b_n \left(\frac{1}{z - z_n} + \frac{1}{z_n}\right) \qquad \square$$

Proof

Step 1

Suppose that α is not a pole of f and define g by $g(z) = f(z)/(z - \alpha)$. Then g has simple poles at z_n for each n, and at α, with $\text{Res}_{z_n} g(z) = b_n/(z_n - \alpha)$ and $\text{Res}_{\alpha} g(z) = f(\alpha)$ by 5.2. Then

$$\frac{1}{2\pi i}\int_{\mathscr{C}_k} g(z)\,dz = f(\alpha) + \sum_n \frac{b_n}{z_n - \alpha} \tag{5.9}$$

by 5.1, where $\sum_n$ is taken over all poles of f inside $\mathscr{C}_k$ and k is chosen large enough so that α lies inside $\mathscr{C}_k$. Letting $\alpha = 0$, which lies inside all the $\mathscr{C}_k$, in (5.9) gives

$$\frac{1}{2\pi i}\int_{\mathscr{C}_k} \frac{f(z)}{z}\,dz = f(0) + \sum_n \frac{b_n}{z_n} \tag{5.10}$$

Subtracting (5.10) from (5.9) gives

$$\frac{\alpha}{2\pi i}\int_{\mathscr{C}_k} \frac{f(z)}{z(z-\alpha)}\,dz = f(\alpha) - f(0) + \sum_n b_n\left(\frac{1}{z_n - \alpha} - \frac{1}{z_n}\right) \tag{5.11}$$

Step 2

Note that the length of each square $\mathscr{C}_k$ is $8a_k$ and that $|z| \geqslant a_k$ on each $\mathscr{C}_k$, so that by the triangle inequality, $|z - \alpha| \geqslant a_k - |\alpha|$ on each $\mathscr{C}_k$. Then by hypothesis and Lemma 3.2,

$$\left|\int_{\mathscr{C}_k} \frac{f(z)}{z(z-\alpha)}\,dz\right| \leqslant \frac{8Ma_k}{a_k(a_k - |\alpha|)} = \frac{8M}{a_k - |\alpha|}$$

Then by hypothesis, $\int_{\mathscr{C}_k}(f(z)/z(z-\alpha))\,dz$ and the result follows from (5.11). ■

Example 5.14

Let $f(z) = \tan z$. Then f has simple poles at $(n + 1/2)\pi$, $n \in \mathbb{Z}$ and f is analytic at 0. Note that by Lemma 5.3, $\text{Res}_{(n+1/2)\pi} f(z) = -1$ for all n. Also, using the technique of the proof of 5.10, it is easily seen that $|f(z)| < 2$ on any square $\mathscr{C}_k$ with vertices $k(\pm 1 \pm i)$, $k \in \mathbb{N}$ (see Exercise 5.4.5). Clearly, f is analytic on each $\mathscr{C}_k$. Then by 5.15,

$$\tan z = -\sum_{n=-\infty}^{\infty}\left(\frac{1}{z - (n+1/2)\pi} + \frac{1}{(n+1/2)\pi}\right)$$

$$\Rightarrow \quad \tan z = -\sum_{n=0}^{\infty}\left(\frac{1}{z - (n+1/2)\pi} + \frac{1}{(n+1/2)\pi}\right)$$

$$-\sum_{n=1}^{\infty}\left(\frac{1}{z + (n-1/2)\pi} + \frac{1}{(-n+1/2)\pi}\right)$$

$$\Rightarrow \quad \tan z = -\sum_{n=0}^{\infty}\left(\frac{1}{z-(n+1/2)\pi}+\frac{1}{(n+1/2)\pi}\right)$$
$$-\sum_{m=0}^{\infty}\left(\frac{1}{z+(m+1/2)\pi}-\frac{1}{(m+1/2)\pi}\right)$$

letting $n = m + 1$ in the second series. Then simplifying gives

$$\tan z = \sum_{n=0}^{\infty}\frac{2z}{(n+1/2)^2\pi^2 - z^2}$$

except at the simple poles of tan. Letting $z \to 0$ in this result,

$$\lim_{z\to 0}\frac{\tan z}{z} = 1 = \sum_{n=0}^{\infty}\frac{2}{(n+1/2)^2\pi^2} \quad \Rightarrow \quad \sum_{n=0}^{\infty}\frac{1}{(2n+1)^2} = \frac{\pi^2}{8}$$

which is the result of Example 5.12(i). Letting $z = \pi/4$ gives

$$\sum_{n=0}^{\infty}\frac{1}{4(2n+1)^2 - 1} = \frac{\pi}{8}$$

Note

Suppose that the conditions of 5.15 are satisfied, except that 0 is also a simple pole of f. In this case, by definition, $\lim_{z\to 0} zf(z) = k$ say, so define g by $g(z) = f(z) - k/z$ with $g(0) = \lim_{z\to 0}(f(z) - k/z)$. Then g is analytic at 0 and is bounded on the same sequence of nested squares as f, so that 5.15 can be applied to g (see Exercises 5.4).

Exercise 5.4.1 Find the sum for each of the following convergent series:

(i) $\displaystyle\sum_{n=-\infty}^{\infty}\frac{1}{(n-a)^2},\ a \notin \mathbb{Z}$

(ii) $\displaystyle\sum_{n=0}^{\infty}\frac{1}{n^2+a^2},\ a > 0$

(iii) $\displaystyle\sum_{n=1}^{\infty}\frac{1}{n^4+a^4},\ a > 0$

(iv) $\displaystyle\sum_{n=1}^{\infty}\frac{1}{n^4}$

Exercise 5.4.2 Prove that $|\csc \pi z| \leqslant 1$ on the square $\mathscr{C}_k$ with vertices $(k+1/2)(\pm 1 \pm i)$ for each k. Use this result to prove Theorem 5.13 and hence prove Theorem 5.14.

Exercise **5.4.3** Find the sum of each of the following convergent series:

(i) $\displaystyle\sum_{n=0}^{\infty} \frac{(-1)^n}{n^2+a^2},\ a>0$

(ii) $\displaystyle\sum_{n=-\infty}^{\infty} \frac{(-1)^n}{(2n+1)^3}$

(iii) $\displaystyle\sum_{n=1}^{\infty} \frac{(-1)^n}{(n^2+1)^2}$

Exercise **5.4.4** Use the technique of Example 5.13, with

$$f(z) = \frac{\operatorname{sech}(z+1/2)\pi}{2z+1}$$

to sum the infinite series

$$\sum_{n=0}^{\infty} \frac{(-1)^n}{(2n+1)\cosh(n+1/2)\pi}$$

Exercise **5.4.5** Prove that $|\tan \pi z| < 2$ on the square $\mathscr{C}_k$ with vertices $k(\pm 1 \pm i)$ for each $k \in \mathbb{N}$. Now let f be an algebraic fraction with $\lim_{z\to\infty} zf(z) = 0$. Prove that $\lim_{k\to\infty} \int_{\mathscr{C}_k} f(z)\tan \pi z\, dz = 0$. Hence use the residue theorem to prove that if $f(n+1/2) \neq 0$ and f has singular points $z_1, z_2, \ldots, z_m \neq n+1/2$, for any $n \in \mathbb{Z}$ then

$$\sum_{n=-\infty}^{\infty} f(n+1/2) = \sum_{n=1}^{m} \operatorname{Res}_{z_n}(\pi f(z)\tan \pi z)$$

Exercise **5.4.6** Use Theorem 5.15 to prove the following results:

(i) $\displaystyle \cot z = \frac{1}{z} + \sum_{n=1}^{\infty} \frac{2z}{z^2 - n^2\pi^2}$, except at the poles of $\cot z$

(ii) $\displaystyle \pi \operatorname{csch} \pi z = \frac{1}{z} + \sum_{n=1}^{\infty} \frac{(-1)^n 2z}{z^2+n^2}$, except at the poles of $\operatorname{csch} \pi z$

(Compare with Exercise 5.4.3(i).

Exercise **5.4.7** Let $f(z) = (\sin az)/(\sin \pi z)$ where $0 < a < \pi$ and let $\mathscr{C}_k$ be a square with vertices $(k+1/2)(\pm 1 \pm i)$ for each $k \in \mathbb{N}$. Show that $|f(z)| \leqslant M$ on each $\mathscr{C}_k$ for some $M \in \mathbb{R}^+$, independent of k. Hence apply Theorem 5.15 to prove that, except at the poles of f,

$$\frac{\sin az}{\sin \pi z} = \frac{a}{\pi} + \frac{2z^2}{\pi} \sum_{n=1}^{\infty} \frac{(-1)^n \sin an}{n(z^2-n^2)}$$

Exercise **5.4.8** Let $f(z) = g(z) + \sum_{n=1}^{m} a_n/z^n$ where g is analytic in the region $|z| < R$ for some R, so that f has a pole of order m at 0. Prove that if $\mathscr{C}$ is any simple closed contour inside $|z| < R$ and α is a point inside $\mathscr{C}$, then

$$\frac{1}{2\pi i}\int_{\mathscr{C}} \frac{f(z)}{z-\alpha}\,dz = f(\alpha) - \sum_{n=1}^{m} \frac{a_n}{\alpha^n}$$

6 Conformal Transformations

Recall that functions which map $A \subseteq \mathbb{C}$ to $\mathbb{C}$ map points in the plane to points in the plane and so map curves to curves and regions to regions. Some simple examples illustrating this idea were given in Chapter 1 and another example follows. It is usually clear, from the function under consideration, which is the best choice of coordinate system in each plane.

Example 6.1

Let $w = f(z) = z + 1/z$, $z \neq 0$. Let $w = u + iv$ and $z = re^{i\theta}$. Then

$$w = u + iv = re^{i\theta} + \frac{1}{r}e^{-i\theta} = \left(r + \frac{1}{r}\right)\cos\theta + i\left(r - \frac{1}{r}\right)\sin\theta$$

$$\Rightarrow \quad u(r, \theta) = \left(r + \frac{1}{r}\right)\cos\theta, \; v(r, \theta) = \left(r - \frac{1}{r}\right)\sin\theta \tag{6.1}$$

When $r = a \neq 1$ (a constant), for example,

$$\frac{u^2}{(a + 1/a)^2} + \frac{v^2}{(a - 1/a)^2} = \cos^2\theta + \sin^2\theta = 1 \tag{6.2}$$

and when $r = 1$, $u = 2\cos\theta$, $v = 0$. Hence f maps the circle $|z| = a \neq 1$ to the ellipse with equation (6.2) and maps the unit circle to the segment of the u-axis given by $-2 \leqslant u \leqslant 2$.

It also follows from (6.1) that in the case $a < 1$, if $r < a$ then $|u| > (a + 1/a)|\cos\theta|$ and $|v| > |a - 1/a||\sin\theta|$. (This is so since $1/(ar) > 1$ and $a - r > 0 \;\Rightarrow\; \pm(r - a) + (a - r)/(ar) > 0 \;\Rightarrow\; r + 1/r > a + 1/a$ and $1/r - r > 1/a - a$.) Hence f maps the inside of the circle $|z| = a < 1$ to the outside of the ellipse, as shown in Fig. 6.1. The inside of the unit circle is mapped to the whole w-plane with the segment of the u-axis, given by $-2 \leqslant u \leqslant 2$ excluded. For $a > 1$, f is not an injection inside the circle $|z| = a$ and f maps this region to the whole w-plane.

In this chapter, we are concerned with these geometrical aspects of a function mapping $A \subseteq \mathbb{C}$ to $\mathbb{C}$ and think of such a function mapping some figure or region in one plane to a figure or region in another plane. This idea has many practical applications. Very often a physical problem defined in some complicated region or with a complicated boundary, as for example in fluid mechanics, can be solved by using a complex function to map the region or

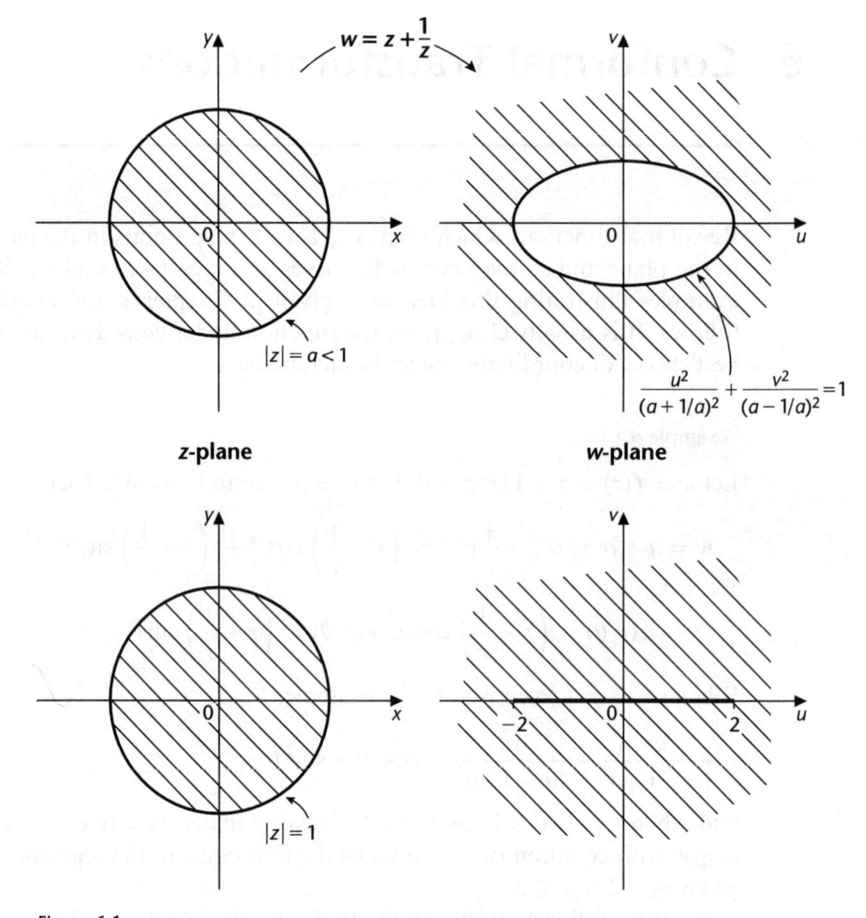

Figure 6.1

boundary to a simpler region or boundary on which the problem may be solved. The inverse function is then used to provide the solution of the original problem. A number of practical applications are studied in Chapter 9.

Conformal Transformations

Recall that if $f: A \subseteq \mathbb{C} \to \mathbb{C}$ and $w = f(z)$ with $w = u + iv$ and $z = x + iy$, then f is equivalent to the mapping from points in the z-plane to points in the w-plane given by $u = u(x, y)$ and $v = v(x, y)$. If f is analytic in a region $\mathscr{R}$, then u and v are continuous functions of x and y with continuous partial derivatives of all orders in $\mathscr{R}$. In this case, any curve in $\mathscr{R}$ which has a continuously turning tangent is mapped to a curve in the w-plane with the same property. Hence analyticity is a desirable property, but this alone does not guarantee the existence of even a local inverse. For example, f given by $f(z) = z^2$ is entire but is clearly not an injection in the region $\{z : |z| < a\}$ for any $a \in \mathbb{R}^+$. The following result, the proof of which gives another application of Rouché's theorem, gives a sufficient condition for the existence of a local inverse.

Theorem 6.1. An Inverse Function Theorem

Let f be analytic at α with $f'(\alpha) \neq 0$. Then an inverse function f^{-1} exists in some open neighbourhood of $f(\alpha)$. □

Proof

We need to show that there is an open neighbourhood $\mathcal{N}$ of $f(\alpha)$ such that, for any given $w \in \mathcal{N}$, we can specify a *unique* solution to the equation $w = f(z)$. Let $\beta = f(\alpha)$ and let g be defined by $g(z) = f(z) - \beta$, so that $g(\alpha) = 0$ and $g'(\alpha) \neq 0$. Then g has a simple zero at α, and since zeros are isolated, there exists $\rho \in \mathbb{R}^+$ such that g is analytic for $|z - \alpha| \leqslant \rho$ and $g(z) \neq 0$ for $0 < |z - \alpha| \leqslant \rho$. Let $\mathscr{C}$ be the circle $|z - \alpha| = \rho$, $m = \min_{\mathscr{C}} |g(z)|$ and γ any point such that $|\gamma| < m$. Then since $g(z) \neq 0$ on $\mathscr{C}$ and $|-\gamma| < m \leqslant |g(z)|$ on $\mathscr{C}$, it follows by Rouché's theorem that g and $g - \gamma$ have the same number of zeros inside $\mathscr{C}$. But g has only one zero, a simple zero at α, inside $\mathscr{C}$. Hence $g(z) = \gamma$ has exactly one solution inside $\mathscr{C}$ for $|\gamma| < m$. Let $w = \beta + \gamma$. Then $f(z) - \beta = w - \beta$, i.e. $w = f(z)$, has exactly one solution inside $\mathscr{C}$ for $|w - \beta| < m$, as required. ■

Clearly, the existence of a local inverse of a function is a desirable property and we make the following definition.

Definition

A function $f: A \subseteq \mathbb{C} \to \mathbb{C}$ is a **conformal transformation** or **conformal mapping** on a region $\mathscr{R} \subseteq A$ if f is analytic on $\mathscr{R}$ with $f'(z) \neq 0$ for any $z \in \mathscr{R}$.

Note

It follows by the chain rule that the composition of two conformal transformations is conformal. Hence, in practice, very often a desired conformal transformation is constructed from a composition of simpler ones.

Notice that in general, even if f is conformal on a region $\mathscr{R}$, f need not be an injection on $\mathscr{R}$. For instance, f given by $f(z) = z^2$ maps $\{z : 0 < |z| < 1\}$ conformally to $\{w : 0 < |w| < 1\}$ but f is clearly not an injection on the given region. However, the following result is the partial converse of Theorem 6.1.

Theorem 6.2. An Analytic Injection is Conformal

If f is analytic and an injection on a region $\mathscr{R}$, then $f'(z) \neq 0$ for any $z \in \mathscr{R}$, so that f is conformal on $\mathscr{R}$. □

Proof

Suppose that $f'(\alpha) = 0$ for some $\alpha \in \mathscr{R}$. Then g defined by $g(z) = f(z) - f(\alpha)$ has a zero of order $n \geqslant 2$ at α. Also by hypothesis, there is a circle $\mathscr{C}$, lying within $\mathscr{R}$, given by $|z - \alpha| = \rho$ say, on which g does not vanish and inside of which $g'(z) = 0$ only at α. If $0 < |\gamma| < m = \min_{\mathscr{C}} |g(z)|$, then by Rouché's theorem, $g - \gamma$ has n zeros inside $\mathscr{C}$. These zeros are simple since $g'(z) = 0$ only at α inside $\mathscr{C}$. Hence $f(z) = f(\alpha) + \gamma$ at two or more points inside $\mathscr{C}$. This contradicts the fact that f is an injection on $\mathscr{R}$. ■

Conformal transformations have an important geometrical property. They preserve the angle of intersection between two curves and the sense of orientation, as stated in the next theorem.

Convention

Extending the convention introduced in Chapter 1, we shall adopt the convention that if $w = f(z)$ then the image of any point P, curve $\mathscr{C}$ or region $\mathscr{R}$ in the z-plane, under f in the w-plane, is denoted by P', $\mathscr{C}'$ or $\mathscr{R}'$ respectively.

Theorem 6.3. Conformal Mappings Preserve Angles Between Curves

Let f be conformal on a region $\mathscr{R}$. Let $\mathscr{C}_1$ and $\mathscr{C}_2$ be any two curves in $\mathscr{R}$ which intersect at $\alpha \in \mathscr{R}$. Let ψ be the acute angle between the tangents to $\mathscr{C}_1$ and $\mathscr{C}_2$ at α. Then ψ is also the angle between the tangents to $\mathscr{C}_1'$ and $\mathscr{C}_2'$, the images of $\mathscr{C}_1$ and $\mathscr{C}_2$ respectively under f, at $\beta = f(\alpha)$, with the same orientation, as shown in Fig. 6.2. □

Proof

Let ψ_1 and ψ_2 be the angles the tangents to $\mathscr{C}_1$ and $\mathscr{C}_2$ at α respectively make with the x-axis (Fig. 6.2). Let z_1 and z_2 be points on $\mathscr{C}_1$ and $\mathscr{C}_2$ respectively, both distance r from α. Then $z_1 - \alpha = re^{i\theta_1}$ and $z_2 - \alpha = re^{i\theta_2}$ say. As $r \to 0$, $\theta_1 \to \psi_1$ and $\theta_2 \to \psi_2$. Let $w_k = f(z_k)$, $k = 1, 2$ and let $w_1 - \beta = R_1 e^{i\phi_1}$ and $w_2 - \beta = R_2 e^{i\phi_2}$ say. Since $f'(\alpha)$ exists and is non-zero, let $f'(\alpha) = \rho e^{i\lambda}$ say, where $\rho \neq 0$. Then by definition,

$$f'(\alpha) = \lim_{z_1 \to \alpha} \frac{w_1 - \beta}{z_1 - \alpha} = \lim_{z_1 \to \alpha} \frac{R_1}{r} e^{i(\phi_1 - \theta_1)} = \rho e^{i\lambda}$$

$$\Rightarrow \quad \lim_{z_1 \to \alpha} (\phi_1 - \theta_1) = \lambda \quad \Rightarrow \quad \lim_{w_1 \to \beta} \phi_1 = \psi_1 + \lambda$$

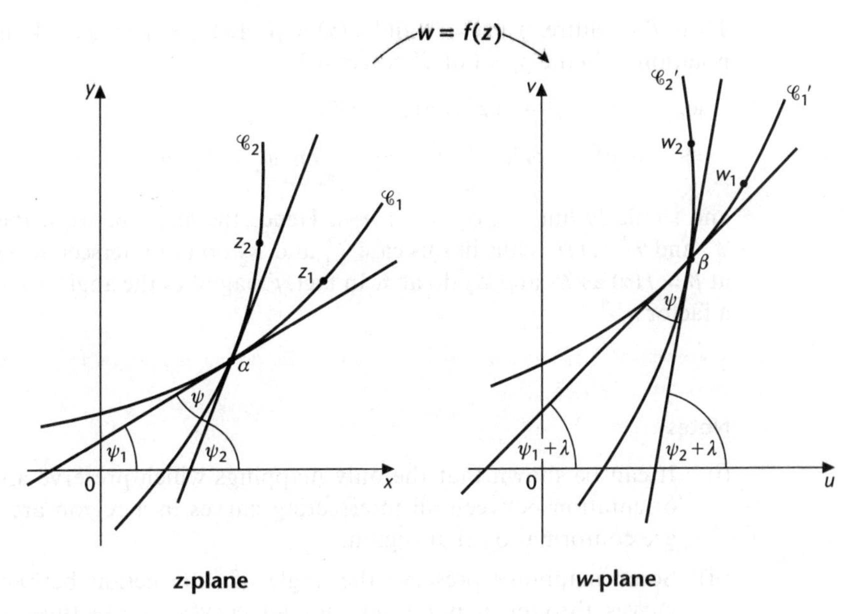

Figure 6.2

Hence the tangent to $\mathscr{C}_1'$ makes an angle $\psi_1 + \lambda$ with the real axis, and similarly the tangent to $\mathscr{C}_2'$ makes an angle $\psi_2 + \lambda$ with the real axis, as shown in Fig. 6.2. Hence the angle between the tangents to $\mathscr{C}_1'$ and $\mathscr{C}_2'$ is given by

$$(\psi_2 + \lambda) - (\psi_1 + \lambda) = \psi_2 - \psi_1 = \psi$$

Also by above, the angle between $\mathscr{C}_1'$ and $\mathscr{C}_2'$ has the same orientation as the angle between $\mathscr{C}_1$ and $\mathscr{C}_2$ as required. ■

Note

If a function f is such that the image of an angle between curves in the z-plane changes sign under f, then f is **isogonal**. In this case, the magnitude of the angle is preserved but the orientation is changed. For example, f given by $f(z) = \bar{z}$ is isogonal since it reflects curves in the real axis, but f is clearly not conformal since it is differentiable nowhere.

Even if f is analytic in a region, if $f'(\alpha) = 0$ at some point in this region, then f will not preserve angles between curves passing through α, as indicated in the following example.

Example 6.2

Consider the function $f : \mathbb{C} \to \mathbb{C}$ given by

$$w = f(z) = \gamma(z - \alpha)^n + \beta \qquad (\gamma \neq 0, n \in \mathbb{N}, n > 1)$$

Then f is entire, $f'(\alpha)=0$ and $f(\alpha)=\beta$. Let $\gamma=\rho e^{i\lambda}$ say. Using the same notation as in the proof of Theorem 6.3,

$$w_1-\beta=f(z_1)-f(\alpha)=\gamma(z_1-\alpha)^n$$
$$\Rightarrow \quad R_1e^{i\phi_1}=\rho r^n e^{i(\lambda+n\theta_1)} \quad \Rightarrow \quad \lim_{w_1\to\beta}\phi_1=\lambda+n\psi_1$$

and similarly $\lim_{w_2\to\beta}\phi_2=\lambda+n\psi_2$. Hence the angle between the tangents to $\mathscr{C}_1'$ and $\mathscr{C}_2'$ is $n\psi$. Thus in this case $\mathscr{C}_1'$ and $\mathscr{C}_2'$ do not intersect at the same angle at $\beta=f(\alpha)$ as $\mathscr{C}_1$ and $\mathscr{C}_2$ do at α. In fact, f magnifies the angle of intersection by a factor n.

Notes

(i) It can be shown that the only mappings which preserve angles and the orientation between all intersecting curves in a region are those which are conformal on that region.

(ii) Some mappings preserve the angle of intersection between particular curves through a point but not all curves intersecting at that point. Such mappings are clearly not conformal.

Definition

Let $f: A\subseteq\mathbb{C}\to\mathbb{C}$. A point α at which $f'(\alpha)=0$ is a **critical point** of f.

We have indicated that the angle of intersection of two curves at a critical point of f is not the same as the angle of intersection of their images under f. Also, f may not have a local inverse at such a point.

Example 6.3

Consider the mapping $f:\mathbb{C}\to\mathbb{C}$ given by $w=f(z)=z^3$. Then f is conformal on any region not containing 0 and 0 is a critical point of f. Note that f maps the circle $\mathscr{C}$, with equation $|z|=a$, to the circle $\mathscr{C}'$, with equation $|w|=a^3$. Also, f maps the line segment L, with equation $y=x$, $x\geqslant 0$, $y\geqslant 0$, i.e. $z=re^{i\pi/4}$, to the line segment L', given by $w=r^3e^{3i\pi/4}$. In addition, f maps the point $P(a/\sqrt{2}, a/\sqrt{2})$ to the point $P'(-a^3/\sqrt{2}, a^3/\sqrt{2})$. This is indicated in Fig. 6.3.

The gradient of L at P is 1 and since $x^2+y^2=a^2 \Rightarrow dy/dx=-x/y$, the gradient of $\mathscr{C}$ at P is -1. Hence $\mathscr{C}$ and L are orthogonal at P. Similarly, the gradient of L' at P' is -1 and the gradient of $\mathscr{C}'$ at P' is 1. Hence $\mathscr{C}'$ and L' are orthogonal at P'. In other words, the angle of intersection of $\mathscr{C}$ and L at P is preserved under f.

By Example 6.2, the angle of intersection of any two curves passing through 0 is magnified by a factor of 3 by f. For instance, f maps the line segment L_1,

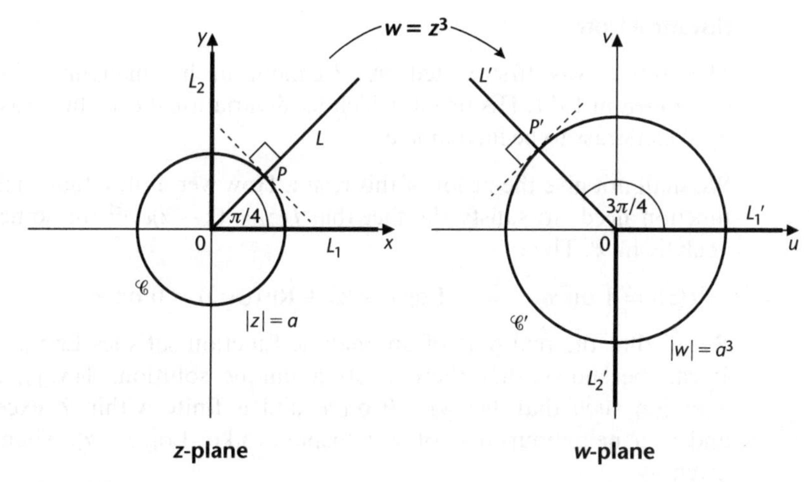

Figure 6.3

given by $z = x$, $x \geqslant 0$, to the line segment L_1', given by $w = u$, $u \geqslant 0$ and the line segment L_2, given by $z = iy$, $y \geqslant 0$, to the line segment L_2', given by $w = iv$, $v \leqslant 0$, as shown in Fig. 6.3.

Important Note

Suppose that an analytic injection f maps a simple closed contour $\mathscr{C}$ to a simple closed contour $\mathscr{C}'$. Then f must map the inside of $\mathscr{C}$, say, to the inside or outside of $\mathscr{C}'$. Hence to show that f maps the inside of $\mathscr{C}$ to the inside of $\mathscr{C}'$ say, in a particular case, it is sufficient to find the image of a particular point inside $\mathscr{C}$, under f. Clearly this is not the case if f is not an injection; recall Example 6.1.

The Existence of Conformal Transformations

An obvious question to ask is, Given a region $\mathscr{R}$ in the z-plane and a region $\mathscr{R}'$ in the w-plane, does there exist a conformal transformation which maps $\mathscr{R}$ to $\mathscr{R}'$? It suffices to consider whether it is possible to map any given region $\mathscr{R}$ conformally onto $\{w : |w| < 1\}$, since if f maps $\mathscr{R}$ to $\{\zeta : |\zeta| < 1\}$ and g maps $\{\zeta : |\zeta| < 1\}$ to $\mathscr{R}'$, then $g \circ f$ maps $\mathscr{R}$ to $\mathscr{R}'$. The main existence theorem of this type is the following.

Theorem 6.4. The Riemann Mapping Theorem

If $\mathscr{R}$ is the region bounded by a simple closed contour $\mathscr{C}$, then there exists a unique analytic injection f such that f maps $\mathscr{R}$ to $\{w : |w| < 1\}$, $f(\alpha) = 0$ for some chosen $\alpha \in \mathscr{R}$ and $f'(\alpha) \in \mathbb{R}^+$. □

Historical Note

This result was first stated by Riemann in his inaugural dissertation at Göttingen in 1851. His proof, which used variational calculus, was later shown by Weierstrass to be incomplete.

We shall not give the proof of this result. However, notice that such an analytic function needs to satisfy the fact that $f(z) = (z - \alpha)e^{\phi(z)}$ for some function ϕ, analytic in $\mathscr{R}$. Then

$$|f(z)| = 1 \text{ on } \mathscr{C} \quad \Rightarrow \quad \operatorname{Log}|z - \alpha| + \operatorname{Re}(\phi(z)) = 0 \text{ on } \mathscr{C}$$

Recall that the real part of an analytic function satisfies Laplace's equation. It can be shown that there exists a unique solution, $\Phi(x, y)$, of Laplace's equation such that $\Phi(x, y) = 0$ on $\mathscr{C}$ and is finite within $\mathscr{C}$ except at $z = \alpha$, and in a neighbourhood of α it 'behaves like' $\operatorname{Log}|z - \alpha|$. Then $\operatorname{Re}(\phi(z))$ is given by

$$\Phi(x, y) = \operatorname{Log}|z - \alpha| + \operatorname{Re}(\phi(z))$$

and $\operatorname{Im}(\phi(z))$ is then obtained from the Cauchy–Riemann equations. $\operatorname{Im}(\phi(z))$ contains an arbitrary additive constant, k say, and k can be chosen so that $\operatorname{Im}(\phi(\alpha)) = 0$, hence $e^{\phi(\alpha)} = f'(\alpha) \in \mathbb{R}^+$.

Exercise 6.1.1

(a) Show that f given by $f(z) = \sin z$ maps the line $x = \pi/4$ in the z-plane to one half of a hyperbola in the w-plane.

(b) Find the image in the w-plane of the lines $x = k$, where k is a constant, under the mapping given by $f(z) = 1 + e^z$.

Exercise 6.1.2 Show that the families of curves $y = cx^2$ and $x^2 + 2y^2 = k$, where c and k are constants, are orthogonal at every point of intersection. Define the mapping f by $f(z) = (x^2 + 2y^2) + iy/x^2$ for all z such that $\operatorname{Re} z \neq 0$, so that the images of these families under f in the w-plane are lines which also cut orthogonally. Show that f is only conformal, however, along one line in the z-plane.

Exercise 6.1.3 Find the critical points of the entire function f given by $f(z) = (z - 1)^2(z^2 + 1)^2$.

Exercise 6.1.4 Let $f(z) = z + 1/z$, $z \neq 0$. Find the image of the semicircular arc $\mathscr{C}$ given by $|z| = 1$, $0 \leqslant \operatorname{Arg} z \leqslant \pi$, the line segment L_1 given by $z = x$, $x \geqslant 1$, and the line segment L_2 given by $z = re^{i\pi/4}$, $r > 0$. Show that the angle between the tangents to $\mathscr{C}$ and L_1 at their point of intersection is not preserved by f but the angle between the tangents to $\mathscr{C}$ and L_2 at their point of intersection is preserved by f. Explain this result.

Exercise **6.1.5** Let f be conformal in a region and $u+iv=f(x+iy)$. Show that the Jacobian

$$J=\frac{\partial(u,v)}{\partial(x,y)}$$

is non-zero at all points in this region.

Exercise **6.1.6**

$$\text{Let } f(z)=\frac{z-i}{z+i} \qquad (z\neq -i)$$

Show that f is conformal on the region given by $|z+i|<2$ and $\operatorname{Re} z$, $\operatorname{Im} z>0$. Find the image of this region under f.

Exercise **6.1.7** The circles $|z-1|=\sqrt{2}$ and $|z+1|=\sqrt{2}$ divide the z-plane into four regions, one of which contains the origin. Show that if $w=f(z)=2z/(1-z^2)$ then f is analytic and an injection on this region and that it maps this region to $\{w:|w|<1\}$.

Exercise **6.1.8** Let $w=F(z)=f(z)+ig(z)$ for some choice of functions f and g. Show that F maps the real axis to the curve defined parametrically by $u=f(t)$ and $v=g(t)$ in the w-plane. Hence find a mapping which maps the real axis to the circle $|w|=1$.

Exercise **6.1.9** Show that there is no conformal transformation which maps $\mathbb{C}$ to $\{w:|w|<1\}$.

Bilinear Transformations

Bilinear transformations are among the most useful conformal transformations and they have many practical applications. They map circles to circles or lines, and lines to lines or circles.

Important Note

It is convenient in this area to work in the extended plane $\tilde{\mathbb{C}}$. Then any line is a circle passing through the point at infinity. In this section, the term 'circle' will be used in this wider sense.

Bilinear transformations are the composition of the following elementary conformal transformations.

I. **Translation**. The function $f:\mathbb{C}\to\mathbb{C}$ defined by

$$f(z)=z+\alpha$$

where α is a constant, is merely a translation of the origin and is conformal on $\mathbb{C}$.

II. **Magnification**. The function $f: \mathbb{C} \to \mathbb{C}$ defined by

$$f(z) = \rho z$$

where ρ is a non-zero real constant, is also conformal on $\mathbb{C}$. If ρ is positive then f multiplies the modulus of z by ρ and leaves its argument unchanged. If ρ is negative, f gives a reflection through the origin, followed by such a 'magnification'.

III. **Rotation**. The function $f: \mathbb{C} \to \mathbb{C}$ defined by

$$f(z) = e^{i\theta} z$$

where θ is a real constant, is conformal on $\mathbb{C}$. Let $w = f(z)$. Then $|w| = |z|$ and $\arg w = \arg z + \theta$. Hence f produces rotation through angle θ about the origin in the positive sense.

IV. **Inversion**. The function $f: \mathbb{C}\backslash\{0\} \to \mathbb{C}$ defined by

$$f(z) = \frac{1}{z}$$

is conformal on $\mathbb{C}\backslash\{0\}$. Letting $z = re^{i\theta}$ and $w = f(z)$ gives $|w| = 1/r$ and $\arg w = -\theta$, so f produces a geometric inversion.

Note that mappings I to III preserve the shape of any contour in the z-plane. It is only mapping IV that can affect shape.

Definition

The function $f: \mathbb{C} \to \mathbb{C}$ defined by $f(z) = \alpha z + \beta$, where α and β are complex constants ($\alpha \neq 0$), is a **linear transformation**.

Note

It is conventional in this area to call such a function a linear transformation, but in a strictly algebraic sense f is an **affine** transformation and only linear if $\beta = 0$.

It follows that any linear transformation produces a possible magnification, rotation and reflection, followed by a translation, so it does not alter the shape of any contour. A bilinear transformation is the quotient of two such functions.

Definition

The function $f: \mathbb{C}\backslash\{-\delta/\gamma\} \to \mathbb{C}$ defined by

$$f(z) = \frac{\alpha z + \beta}{\gamma z + \delta} \tag{6.3}$$

where α, β, γ and δ are complex constants with $\alpha\delta \neq \beta\gamma$ is a **bilinear transformation**.

Historical Note

The bilinear transformation was first studied by Möbius (1790–1868) and is sometimes called the Möbius transformation. It is also sometimes called the linear fractional transformation.

Notes

(i) If $\alpha\delta = \beta\gamma$ in (6.3) then $f(z) = \alpha/\gamma = \beta/\delta$ and so f is merely a constant. If $\gamma = 0$ then f is a linear transformation. It is assumed that $\gamma \neq 0$ in points (ii) to (vi).

(ii) The bilinear transformation is conformal on $\mathbb{C}\backslash\{-\delta/\gamma\}$ since

$$f'(z) = \frac{\alpha\delta - \beta\gamma}{(\gamma z + \delta)^2}$$

Note that $-\delta/\gamma$ is a simple pole of f.

(iii) It is convenient to extend the domain and range of f to the extended plane $\tilde{\mathbb{C}}$, so that $f: \tilde{\mathbb{C}} \to \tilde{\mathbb{C}}$ with $f(-\delta/\gamma) = \infty$ and since $\lim_{z\to\infty} f(z) = \alpha/\gamma$ in $\mathbb{C}$, $f(\infty) = \alpha/\gamma$.

(iv) If $w = \dfrac{\alpha z + \beta}{\gamma z + \delta}$ then $z = \dfrac{\beta - \delta w}{\gamma w - \alpha}$

so that if $f: \tilde{\mathbb{C}} \to \tilde{\mathbb{C}}$ then f is an injection, with $f^{-1}: \tilde{\mathbb{C}} \to \tilde{\mathbb{C}}$ defined by

$$f^{-1}(z) = \frac{\beta - \delta z}{\gamma z - \alpha}$$

It can be shown that the bilinear transformation is the most general type of elementary function which is an injection and conformal on the extended complex plane.

(v) Let $w_1 = \gamma z + \delta, w_2 = \dfrac{1}{w_1}$ and $w = \dfrac{\alpha}{\gamma} + \dfrac{(\beta\gamma - \alpha\delta)w_2}{\gamma}$

then $w = \dfrac{\alpha z + \beta}{\gamma z + \delta}$

hence any bilinear transformation is the composition of a possible magnification, rotation, reflection and translation, followed by an inversion and then another possible magnification, rotation, reflection and translation. Since the geometrical inversion of a circle in $\tilde{\mathbb{C}}$ is a circle, it follows that any bilinear transformation maps circles to circles. We give a proof of this result below.

(vi) It is easy to show that the composition of any two bilinear transformations is bilinear.

We now show formally that bilinear transformations map circles to circles. The equation of any circle in the (x, y)-plane can be written in the form

$$a(x^2 + y^2) + 2px + 2qy + c = 0 \tag{6.4}$$

where a, p, q and c are real constants with $a \neq 0$. If $a = 0$ then (6.4) is the equation of any line. If $a \neq 0$ then (6.4) can be written

$$(x + p/a)^2 + (y + q/a)^2 = r^2 \quad \text{where} \quad r^2 = p^2/a^2 + q^2/a^2 - c/a \tag{6.5}$$

so that (6.4) is the equation of a circle with centre $(-p/a, -q/a)$ and radius r. Letting $z = x + iy$, (6.4) becomes

$$az\overline{z} + Bz + \overline{Bz} + c = 0 \tag{6.6}$$

where $B = p - iq$. Equation (6.6) represents any circle in $\tilde{\mathbb{C}}$. Since $r^2 > 0$, $B\overline{B} > ac$.

Theorem 6.5. Bilinear Transformations Map Circles to Circles

Any bilinear transformation maps a circle in $\tilde{\mathbb{C}}$ to a circle in $\tilde{\mathbb{C}}$. □

Proof

Under any linear transformation, $w = (z - \beta)/\alpha$, equation (6.6) becomes

$$a\alpha\overline{\alpha}w\overline{w} + (a\alpha\overline{\beta} + \alpha B)w + (a\overline{\alpha}\beta + \overline{\alpha B})\overline{w} + (a\beta\overline{\beta} + (\beta B + \overline{\beta B}) + c) = 0$$

an equation of the same form as (6.6). Under the inversion $w = 1/z$ (6.6) becomes

$$a + B\overline{w} + \overline{B}w + cw\overline{w} = 0$$

which is again an equation of the same form as (6.6). Since any bilinear transformation is the composition of linear transformations and inversions, the result follows. ■

Important Note

It follows by (6.5) that the centre and radius of a circle in $\mathbb{C}$ are uniquely determined by three points on its circumference. Hence any three points in $\tilde{\mathbb{C}}$ determine a unique circle. The circle will be a line if one of these points is the point at infinity. Thus to construct a particular bilinear transformation f, which maps a given circle in $\tilde{\mathbb{C}}$ to a given circle in $\tilde{\mathbb{C}}$, it suffices to choose f so that it maps three chosen points on one circle to three chosen points on the other. If a transformation which maps a particular circle in $\mathbb{C}$ to a particular circle in $\mathbb{C}$ is required, then a linear transformation suffices.

Example 6.4

Suppose we wish to find a bilinear transformation which maps the circle $|z-i|=1$ to the circle $|w|=2$. Since $|w/2|=1$, the linear transformation $w=f(z)=2z-2i$, which magnifies the first circle, and translates its centre, is a suitable choice. (Note that there is no unique choice of bilinear transformation satisfying the given criteria.) Since $f(i)=0$, f maps the inside of the first circle to the inside of the second.

Suppose now we wish to find a bilinear transformation g which maps the inside of the first circle to the outside of the second circle. Let $g(z)=(\alpha z+\beta)/(\gamma z+\delta)$. We choose $g(i)=\infty$, so that $g(z)=(\alpha z+\beta)/(z-i)$ without loss of generality. Three points on the first circle are 0, $1+i$ and $2i$, and $g(0)=i\beta$, $g(1+i)=\alpha(1+i)+\beta$ and $g(2i)=2\alpha-i\beta$. All three of these points must lie on $|w|=2$, so the simplest choice is $\alpha=0$ and $\beta=2$. Then $g(z)=2/(z-i)$.

Suppose now we wish to find a bilinear transformation h which maps the circle $|z-i|=1$ to the real line. Since 0, $1+i$ and $2i$ lie on the given circle and the given line passes through 0, 1 and ∞, we simply choose h so that $h(0)=0$, $h(1+i)=1$ and $h(2i)=\infty$ say. Then $h(z)=z/(iz+2)$. Note that $h(i)=i$, so h maps the region given by $|z-i|<1$ to the upper half-plane. This can be shown formally by letting $z=x+iy$. Then

$$h(z)=\frac{2x-i(x^2+(y-1)^2-1)}{x^2+(y-2)^2}$$

and $|z-i|<1 \quad \Rightarrow \quad x^2+(y-1)^2<1.$

Cross Ratios

Given three distinct points $z_1, z_2, z_3 \in \tilde{\mathbb{C}}$, it is easily checked that the bilinear transformation g which maps these points to 0, 1 and ∞ respectively is given by

$$g(z)=\frac{(z_2-z_3)(z-z_1)}{(z_2-z_1)(z-z_3)} \tag{6.7}$$

The cross ratio of these three points and another distinct point z_4 is then defined as the point $g(z_4)$

Definition

The **cross ratio** of four distinct points $z_k \in \tilde{\mathbb{C}}$, $k = 1, 2, 3, 4$, is denoted and defined by

$$(z_1, z_2, z_3, z_4) = \frac{(z_1 - z_4)(z_3 - z_2)}{(z_1 - z_2)(z_3 - z_4)}$$

Notice that, for example, in $\tilde{\mathbb{C}}$, $(z_1, z_2, z_3, \infty) = (z_3 - z_2)/(z_1 - z_2)$. The next result follows easily by the above construction.

Lemma 6.6. Invariance of a Cross Ratio

The cross ratio of four points in $\tilde{\mathbb{C}}$ is invariant under any bilinear transformation. □

Proof

Let z_k, $k = 1, 2, 3, 4$ be any four distinct points in $\tilde{\mathbb{C}}$ and let f be any bilinear transformation with $w_k = f(z_k)$. Let $g(z) = (z_1, z_2, z_3, z)$ as in (6.7). Then $g \circ f^{-1}$ maps w_1, w_2 and w_3 to 0, 1 and ∞ respectively. Hence

$$(w_1, w_2, w_3, w_4) = (g \circ f^{-1})(w_4) = g(z_4) = (z_1, z_2, z_3, z_4)$$ ■

Suppose that z_1, z_2 and z_3 are fixed points in $\tilde{\mathbb{C}}$. Then the invariance of the cross ratio of z_k, $k = 1, 2, 3$ and any point z under any bilinear transformation f is expressed as

$$(z_1, z_2, z_3, z) = (f(z_1), f(z_2), f(z_3), f(z)) \tag{6.8}$$

If the image points $w_k = f(z_k)$, $k = 1, 2, 3$ are given, f is uniquely determined and (6.8) is easily solved for $w = f(z)$. Since any circle in $\tilde{\mathbb{C}}$ is determined by three points, cross ratios can be used to determine a bilinear transformation mapping a given circle to a given circle but this may not be the easiest method in every case.

Example 6.5

Suppose that we require the bilinear transformation which maps the points 1, i and -1 to the points 0, 2 and ∞. By (6.8) it follows that the required transformation f is given by

$$\frac{w - 0}{2 - 0} = \frac{(1 + i)(z - 1)}{(i - 1)(z + 1)} = \frac{i(1 - z)}{(1 + z)} \quad \Rightarrow \quad w = f(z) = \frac{2i(1 - z)}{(1 + z)}$$

Note that f maps the circle passing through 1, i and -1, i.e. $|z| = 1$, to the 'circle' passing through 0, 2 and ∞, i.e. the real line. Since $f(0) = 2i$, f maps the inside of $|z| = 1$ to the upper half-plane. Finally, $f(-i) = -2$. These facts are summarised in Fig. 6.4.

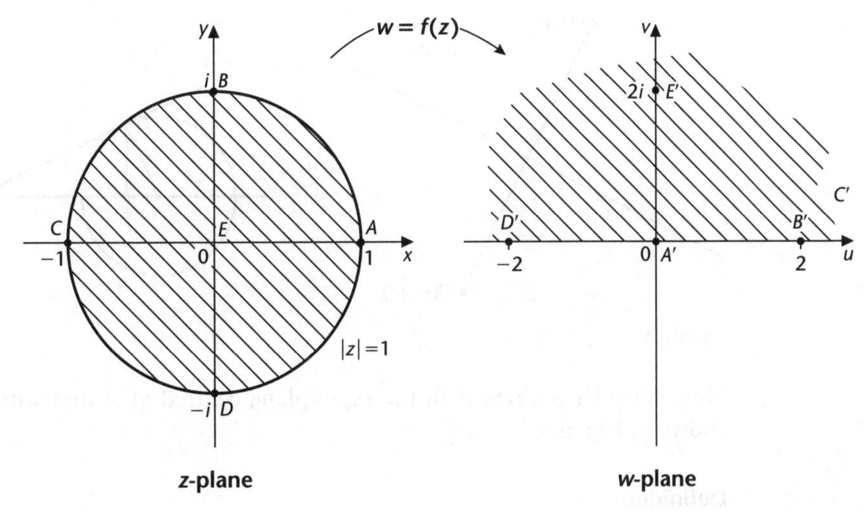

Figure 6.4

Inverse Points

Since bilinear transformations map circles in $\tilde{\mathbb{C}}$ to circles in $\tilde{\mathbb{C}}$, there is a close connection between such transformations and both linear and circular inversion.

Definition

If L is a line in the (x, y)-plane, then two points P and Q are **inverse points with respect to L** if P is the reflection of Q in L. This is a generalisation of the idea that (x, y) and $(x, -y)$ are inverse points with respect to the x-axis.

Let the line L have equation $2px + 2qy + c = 0$ as in (6.4). If $P(x, y)$ and $Q(x^*, y^*)$ are inverse points with respect to L, the line segment PQ is perpendicular to L, and if PQ intersects L at A, then $PA = QA$, as shown in Fig. 6.5. Since the gradient of L is $-p/q$, the gradient of PQ is q/p. Also, A is the midpoint of PQ and lies on L. It then follows by elementary geometry that

$$\frac{y - y^*}{x - x^*} = \frac{q}{p} \quad \text{and} \quad p(x + x^*) + q(y + y^*) + c = 0$$

Letting $z = x + iy$ and $z^* = x^* + iy^*$ and using the notation of (6.6), it follows that

> z and z^* are inverse points with respect to the line $Bz + \overline{Bz} + c = 0$ if and only if $Bz^* + \overline{Bz} + c = 0$

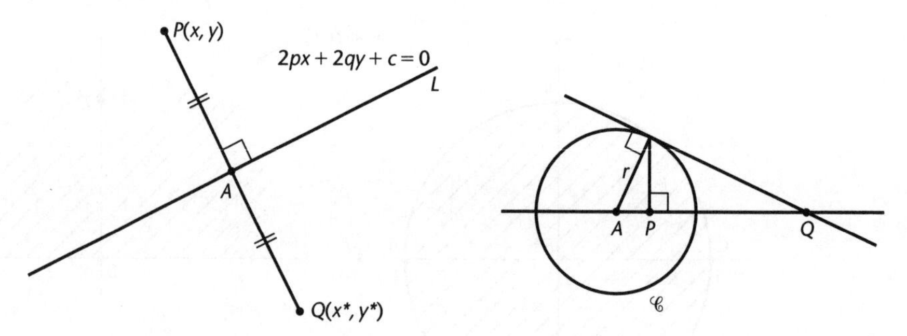

Figure 6.5

Now consider a circle $\mathscr{C}$ in the (x, y)-plane centred at A and with radius r, as shown in Fig. 6.5.

Definition

Two points P and Q, collinear with A, satisfying $(AP) \cdot (AQ) = r^2$ are **inverse points with respect to the circle $\mathscr{C}$.**

Note

As in the case of inverse points with respect to a line, it can be shown that any circle passing through P and Q is orthogonal to $\mathscr{C}$.

Let the centre of circle $\mathscr{C}$ be $\alpha \in \mathbb{C}$, P be $z \in \mathbb{C}$ and Q be $z^* \in \mathbb{C}$. Then by definition

$$\begin{aligned} &|z - \alpha|\,|z^* - \alpha| = r^2 \quad \text{and} \quad \operatorname{Arg}(z - \alpha) = \operatorname{Arg}(z^* - \alpha) \\ &\Leftrightarrow \quad (z^* - \alpha)(\overline{z - \alpha}) = r^2 \end{aligned} \tag{6.9}$$

Suppose that $\mathscr{C}$ is the circle with equation (6.6). It follows by (6.4) and (6.5) that $\mathscr{C}$ has centre $\alpha = -\overline{B}/a$ and radius $r = (B\overline{B} - ac)^{1/2}/a$. Substituting these values into (6.9) gives

$$\left(z^* + \frac{\overline{B}}{a}\right)\left(\overline{z} + \frac{B}{a}\right) = \frac{B\overline{B} - ac}{a^2} \quad \Leftrightarrow \quad az^*\overline{z} + Bz^* + \overline{Bz} + c = 0$$

Combining this result with the result for the straight line, we obtain the following result.

Theorem 6.7. Construction of Inverse Points

Two points z and z^* are inverse points with respect to the circle $az\overline{z} + Bz + \overline{Bz} + c = 0$ in $\tilde{\mathbb{C}}$ iff $az^*\overline{z} + Bz^* + \overline{Bz} + c = 0$ □

The fact that this geometrical inversion is preserved by bilinear transformations, a result sometimes called the **symmetry principle**, follows easily from Theorem 6.5.

Theorem 6.8. Transformation of Inverse Points

Any bilinear transformation maps two points which are inverse with respect to a circle in $\tilde{\mathbb{C}}$ into two points which are inverse with respect to the transformed circle in $\tilde{\mathbb{C}}$. □

Proof

If z and z^* are inverse points with respect to the circle with equation (6.6) then, by Theorem 6.7, z and z^* satisfy

$$az^*\bar{z} + Bz^* + \overline{Bz} + c = 0 \tag{6.10}$$

It follows by the proof of 6.5 that if f is any bilinear transformation and $w = f(z)$, then the image of (6.10) under f is the same as the image of (6.6) under f except that w is replaced by w^* and $\bar{w}$ is left unaltered. Hence, by 6.7, w and w^* are inverse points with respect to the transformed circle. ■

It follows by 6.8 that if z and z^* are inverse points with respect to a circle $\mathscr{C}$ in $\tilde{\mathbb{C}}$, there is a bilinear transformation f which maps $\mathscr{C}$ to the real axis such that $w^* = \bar{w}$ where $w = f(z)$. Theorem 6.8 can be used to construct *all* bilinear transformations which map a given circle in $\tilde{\mathbb{C}}$ to a given circle in $\tilde{\mathbb{C}}$.

Example 6.6 Bilinear Transformations Which Map $|z| \leqslant \rho$ to $|w| \leqslant 1$

Suppose we wish to find all bilinear transformations which map $|z| \leqslant \rho$ to $|w| \leqslant 1$. (Note that the simplest choice is $w = z/\rho$.) Let f be any such transformation. Then there exists $\mu \in \mathbb{C}$ with $|\mu| < \rho$ such that $f(\mu) = 0$. By 6.7, $z = \mu$ and $z^* = \rho^2/\bar{\mu}$ are inverse points with respect to $|z| = \rho$, and 0 and ∞ are inverse points with respect to $|w| = 1$. Hence by 6.8, $f(\rho^2/\bar{\mu}) = \infty$. Let $w = f(z) = (\alpha z + \beta)/(\gamma z + \delta)$. Then $f(\mu) = 0 \quad \Rightarrow \quad \beta = -\alpha\mu$ and $f(\rho^2/\bar{\mu}) = \infty$ $\Rightarrow \quad \delta = -\gamma\rho^2/\bar{\mu}$. Hence $f(z) = \alpha\bar{\mu}(z - \mu)/\gamma(\bar{\mu}z - \rho^2)$. Also, $f(\rho)$ say is a point on $|w| = 1$ so that $|f(\rho)| = |\alpha\bar{\mu}/\gamma\rho| = 1 \quad \Rightarrow \quad \alpha\bar{\mu} = \gamma\rho e^{i\theta}$ for some $\theta \in \mathbb{R}$. Then

$$w = f(z) = \frac{\rho e^{i\theta}(z - \mu)}{(\bar{\mu}z - \rho^2)} \qquad (|\mu| < \rho)$$

Taking $|\mu| > \rho$, f maps $|z| \geqslant \rho$ to $|w| \leqslant 1$ so that f maps $|z| \leqslant \rho$ to $|w| \geqslant 1$.

Exercise

6.2.1 Prove that the composition of any two bilinear transformations is also bilinear. Hence prove that the set of all bilinear transformations with the operation of composition forms a group.

Exercise **6.2.2** Show that the transformation $w = f(z) = (z - a)/(z - b)$, $a, b \in \mathbb{R}$, maps the z-plane with the sets of points $\{x \in \mathbb{R} : x \geqslant a > 0\}$ and $\{x \in \mathbb{R} : x \leqslant b < 0\}$ excluded, to the w-plane with the set of points $\{u \in \mathbb{R} : u \geqslant 0\}$ excluded.

Exercise **6.2.3** Show that there are at most two fixed points of the bilinear transformation $f(z) = (\alpha z + \beta)/(\gamma z + \delta)$ unless f is the identity transformation. If f is a bilinear transformation with one fixed point $a \in \mathbb{C}$ and $w = f(z)$, show that f is defined by

$$\frac{1}{w - a} = \frac{1}{z - a} + k$$

where k is a constant.

Exercise **6.2.4** Show that $f(z) = (1 + iz)/(i + z)$ maps the semicircle consisting of the arc $|z| = 1, 0 \leqslant \operatorname{Arg} z \leqslant \pi$ and the line segment $z = x, -1 \leqslant x \leqslant 1$, to its reflection in the x-axis.

Exercise **6.2.5** Find a bilinear transformation which maps the circle $|z - 1| = 1$ to (i) the circle $|w + i| = 2$, (ii) the line given parametrically by $w = (1 + i)t$, $t \in \mathbb{R}$.

Exercise **6.2.6** Find the bilinear transformation f such that $f(1) = i$, $f(i) = 0$ and $f(-1) = -i$. Show that f maps $|z| \leqslant 1$ to $\operatorname{Re} w \geqslant 0$.

Exercise **6.2.7** Find a bilinear transformation mapping the region bounded by the circles $|z - 1| = 1$ and $|z - i| = 1$ to the first quadrant of the w-plane.

Exercise **6.2.8** Find the bilinear transformation f such that $f(-1) = 0$, $f(i) = 1$ and $f(1) = \infty$. Find the image of the region given by $|z - 1| < 2$ and $|z + 1| < 2$ under f.

Exercise **6.2.9** Use the invariance of cross ratios to find the bilinear transformation which maps

(i) $0, -1 - i$ and -2 to $0, 1$ and ∞ respectively

(ii) $i, 0$ and $-i$ to $1, i$ and -1 respectively

(iii) $|z| = 1$ to $|w - 1| = 1$, and 0 and 1 to $1/2$ and 0 respectively

Exercise **6.2.10** Use cross ratios to find the bilinear transformation which maps ∞, 0 and 1 to i, 1 and $e^{i\pi/4}$ respectively. Show that this transformation maps $\operatorname{Im} z \geqslant 0$ to $|w| \leqslant 1$.

Exercise **6.2.11** Prove that the cross ratio of any four different points on a circle is real. Is the converse true?

Exercise **6.2.12** Let $ABCD$ be a cyclic quadrilateral. By formulating the result in terms of the cross ratio of A, B, C and D and then applying a suitable bilinear transformation, prove that

$$AB \cdot CD + BC \cdot AD = AC \cdot BD$$

Exercise **6.2.13** Let f be the transformation which maps $z \in \tilde{\mathbb{C}}$ to its inverse point with respect to a given circle. Show that f is the composition of a reflection in the real axis and a bilinear transformation, so that f is isogonal but not conformal.

Exercise **6.2.14** Find the most general bilinear transformation which maps Im $z \geqslant 0$ onto $|w| \leqslant 1$.

Exercise ***6.2.15** Use Schwarz's lemma Exercises 3.3.6 to prove that if f is analytic and an injection on $\{z : |z| \leqslant 1\}$ and maps $|z| \leqslant 1$ to $|w| \leqslant 1$, with $f(\mu) = 0$ for some $\mu \in \mathbb{C}$ with $|\mu| < 1$, then f is a bilinear transformation.

Special Elementary Transformations

We now investigate the effect of certain elementary functions on curves and regions in the z-plane.

The Transformation $f(z) = z^n$, $n \in \mathbb{Z}$

Consider the function $f : \mathbb{C} \to \mathbb{C}$ defined by $f(z) = z^n$, $n \in \mathbb{Z}$, $n \neq 0$. Note that f is not differentiable at 0 if $n < 0$ and f has a critical point at 0 if $n > 1$. Hence f is conformal on $\mathbb{C}\backslash\{0\}$. Let $w = f(z)$, $w = Re^{i\phi}$ and $z = re^{i\theta}$. Then

$$w = z^n = r^n e^{ni\theta} \quad \Rightarrow \quad R = r^n \quad \text{and} \quad \phi = n\theta \tag{6.11}$$

Hence, for example, f maps the circle $|z| = a$ to the circle $|w| = a^n$ and, in particular, maps the unit circle to the unit circle. It is easily seen that if $n > 0$, then f maps $|z| \leqslant a$ to $|w| \leqslant a^n$, and if $n < 0$, f maps $|z| \leqslant a$ to $|w| \geqslant a^n$. Also by (6.11), f maps any radial half-line, given by $\theta = c$, to a radial half-line given by $\phi = nc$.

Note that, in the case $n = 2$, $w = u + iv = f(z) = z^2$ gives $u(x, y) = x^2 - y^2$ and $v(x, y) = 2xy$ so that f maps orthogonal families of rectangular hyperbolae in the z-plane, given by $x^2 - y^2 = c$ and $2xy = c$, c a constant, to the orthogonal families of lines, $u = c$ and $v = c$ in the w-plane.

Example 6.7

Consider the third quadrant of the z-plane together with the boundary lines L_1, given by $z = x \leqslant 0$, and L_2, given by $z = iy$, $y \leqslant 0$, as shown in Fig. 6.6. Let f be defined by $w = f(z) = z^2$. Then $f(0) = 0, f(-1) = 1, f(-i) = -1$, and f maps L_1 to $w = u \geqslant 0$ and L_2 to $w = u \leqslant 0$, as shown. If $z = re^{i\theta}$ with $r > 0$ and $-\pi < \theta < -\pi/2$, then $w = Re^{i\phi}$ with $R > 0$ and $-2\pi < \phi < -\pi$, i.e. $0 < \phi < \pi$. Hence f maps the third quadrant to the upper half-plane, as shown.

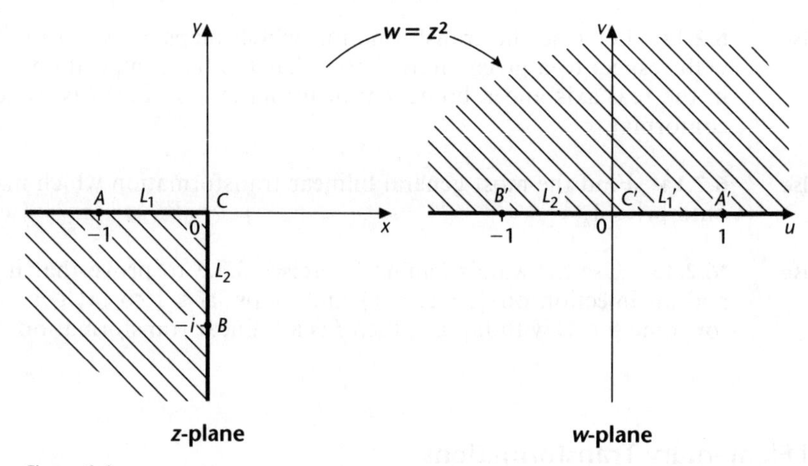

Figure 6.6

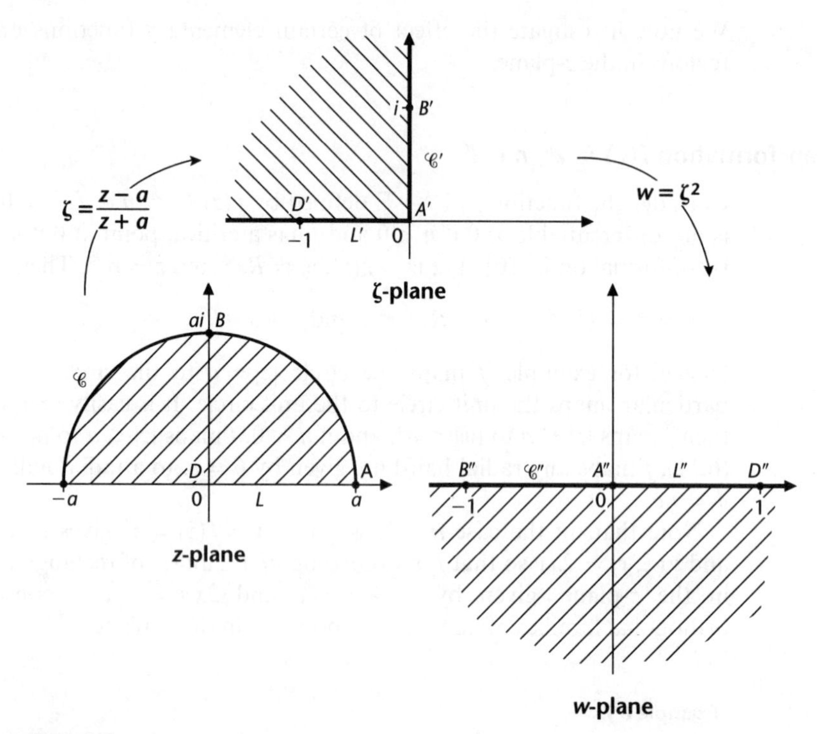

Figure 6.7

Example 6.8 Mapping Semicircles to Circles

Although bilinear transformations do not map semicircles to circles directly, a transformation which maps a given semicircle to a circle is easily found. Figure 6.7 shows a particular semicircle in the z-plane. We begin by using a

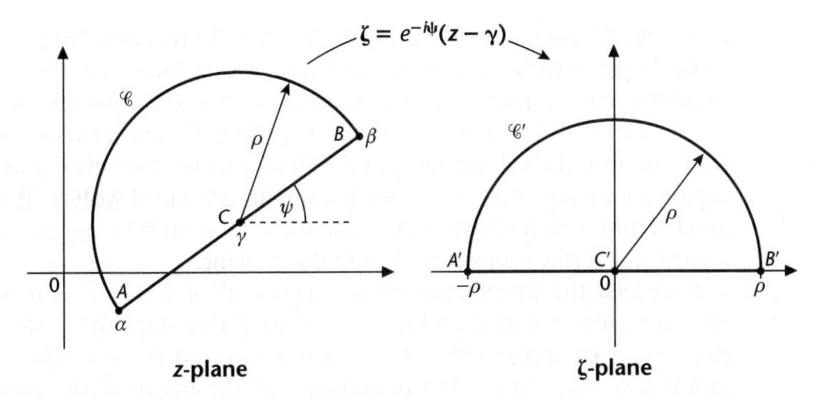

Figure 6.8

bilinear transformation to map this semicircle to a quarter-plane. It is easily checked that the transformation $\zeta = f(z) = (z - a)/(z + a)$ maps the line segment L, given by $z = x$, $-a \leqslant x \leqslant a$, to the line segment L', given by $\text{Re}\,\zeta \leqslant 0$, $\text{Im}\,\zeta = 0$, with $f(a) = 0, f(0) = -1$ and $f(-a) = \infty$. Also, f maps the semicircular arc $\mathscr{C}$ given by $|z| = a$, $\text{Im}\,z \geqslant 0$, to the line segment $\text{Im}\,\zeta \geqslant 0$, $\text{Re}\,\zeta = 0$ with $f(ia) = i$, since $z = ae^{i\theta} \quad \Rightarrow \quad \zeta = i \tan \theta/2$, where $0 \leqslant \theta \leqslant \pi$. Since $f(ia/2) = (-3 + 4i)/5$, f maps the inside of the semicircle to the second quadrant, as shown.

We now use an elementary transformation to map the quarter-plane to a half-plane. As in Example 6.7, it is easily checked that $w = g(\zeta) = \zeta^2$ maps the second quadrant to the lower half-plane, as shown. Hence $w = (z - a)^2/(z + a)^2$ maps the inside of the given semicircle to the lower half-plane. Similarly, $w = (z + a)^2/(z - a)^2$ maps the inside of the semicircle to the upper half-plane.

It is then easy to construct a bilinear transformation which maps the lower half-plane to the inside of the circle, centred at the origin and of radius b say. For example, the bilinear transformation $w = h(z) = b(1 - iz)/(1 + iz)$ maps $\text{Im}\,z \leqslant 0$ to $|w| \leqslant b$ with $h(0) = b$, $h(1) = -ib$, $h(-i) = 0$ and $h(\infty) = -b$.

More generally, consider the semicircle $\mathscr{C}$ given in Fig. 6.8. The translation $w = z - \gamma$ translates γ to 0, then the rotation $\zeta = e^{-i\psi}w$ rotates the semicircle through angle ψ clockwise. Hence the linear transformation $\zeta = e^{-i\psi}(z - \gamma)$ maps $\mathscr{C}$ to the semicircle $\mathscr{C}'$ as shown. Note that $\alpha = \gamma - \rho e^{i\psi}$ and $\beta = \gamma + \rho e^{i\psi}$. Hence, by previous results, the transformation which maps the semicircle $\mathscr{C}$ in Fig. 6.8 to the upper half-plane, say, is

$$w = \left(\frac{\zeta + \rho}{\zeta - \rho}\right)^2 = \left(\frac{e^{-i\psi}(z - \gamma) + \rho}{e^{-i\psi}(z - \gamma) - \rho}\right)^2 = \left(\frac{z - \alpha}{z - \beta}\right)^2$$

Clearly, f given by $f(z) = z^n$, $n \in \mathbb{Z}$ is not an injection on $\mathbb{C}$ for $n \neq \pm 1$. However, f can be made an injection either by restricting its domain or by constucting a Riemann surface from the w-plane. For example, if the domain of f is restricted to $\mathscr{D}_1 = \{z = re^{i\theta} : r \neq 0, -\pi/n < \theta < \pi/n\}$, then f is an injection and f and f^{-1} are conformal on their domains. In this case f maps the region $\mathscr{D}_1$ to the w-plane with the non-positive real axis excluded, i.e. the w-plane with a **cut** along the non-positive real axis, as in Fig. 6.9. If the domain of f is restricted to $\mathscr{D}_2 = \{z = re^{i\theta} : r \neq 0,\ 0 < \theta < 2\pi/n\}$, then f is an injection, f and f^{-1} are conformal on their domains, and f maps $\mathscr{D}_2$ to the w-plane cut along the non-negative real axis. And so on. Such domains are called **wedges**. If the z-plane is divided into n such wedges then each one corresponds to one of the n sheets for a Riemann surface constructed from the w-plane.

Note that the transformation $w = f(z) = z^n$, $n \in \mathbb{N}$, $n \neq 1$, maps the wedge $\{z = re^{i\theta} : 0 \leqslant \theta \leqslant \pi/n\}$ to $\operatorname{Im} w \geqslant 0$. Since f also maps the circle $|z| = a$ to the circle $|w| = a^n$ and $f(re^{i\theta}) = Re^{i\phi}$ where $r < a$ and $0 < \theta < \pi/n \quad \Rightarrow \quad R < a^n$ and $0 < \phi < \pi$, f maps the boundary and the inside of the sector shown in Fig. 6.10 to the boundary and the inside of the semicircle. Then by the results of Example 6.8, it is easy to construct a transformation mapping a given sector to a circle.

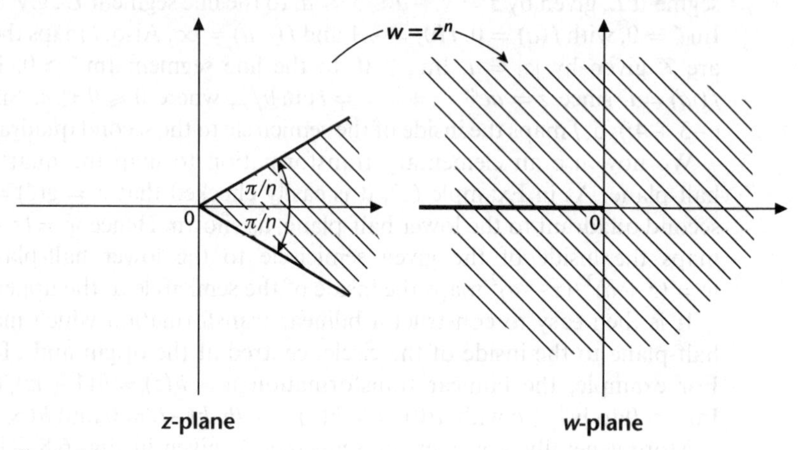

Figure 6.9

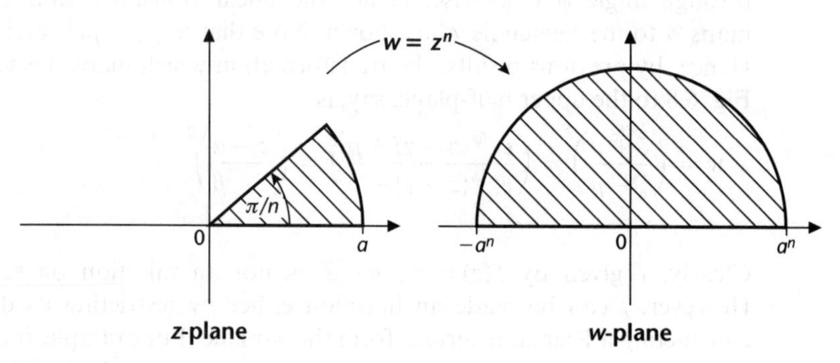

Figure 6.10

The Transformation $f(z) = z^{1/2}$

Note that, in general, the function f defined by $f(z) = z^{1/n}$, $n \in \mathbb{N}$ is the inverse of the function g defined by $g(z) = z^n$, provided that the domain of g is suitably restricted. Hence f, as a transformation, has the reverse effect of applying g. We consider the particular case of the function f defined by $f(z) = z^{1/2}$, where, according to our convention, the principal branch of the multifunction is taken, i.e. $z = re^{i\theta} \Rightarrow z^{1/2} = \sqrt{r}e^{i\theta/2}$, $-\pi < \theta \leqslant \pi$. The domain of f will be taken to be $\mathscr{D} = \{z = re^{i\theta} : r \neq 0, \ -\pi < \theta < \pi\}$. Then $f : \mathscr{D} \to \mathscr{D}_1$ where $\mathscr{D}_1 = \{w = Re^{i\phi} : R \neq 0, -\pi/2 < \phi < \pi/2\}$ and f is an injection and conformal on $\mathscr{D}$.

Note

With the above convention, the branch cut for f has been excluded from the domain of f. Care has to be taken when finding the image of a branch cut. The image depends very much on the convention used.

Let $w = u + iv$ and $z = x + iy$ as usual. Then $w = f(z) = z^{1/2}$ gives

$$z = w^2 \quad \Rightarrow \quad x = u^2 - v^2 \quad \text{and} \quad y = 2uv \quad \Rightarrow \quad 4u^2(u^2 - x) = y^2$$

Hence, with our convention, f maps the parabola $y^2 = 4a^2(a^2 - x)$ in the z-plane, with focus at 0 and vertex $x = a^2$, to the line $u = a > 0$ in the w-plane.

Example 6.9

Let $\zeta = f(z) = z^{1/2}$, $z \in \mathscr{D}$. Then f maps the parabola $y^2 = 4(1 - x)$ to $\operatorname{Re}\zeta = 1$ and maps the region given by $y^2 > 4(1 - x)$ to $\operatorname{Re}\zeta > 1$, as shown in Fig. 6.11. Suppose we wish to find the function which maps the shaded region in the z-plane in Fig. 6.11 to $|w| < 1$. Recall that some bilinear transformation, $w = g(\zeta)$ will map $\operatorname{Re}\zeta \geqslant 1$ to $|w| \leqslant 1$. Let g be given by $g(\zeta) = (\alpha\zeta + \beta)/(\gamma\zeta + \delta)$. Suppose that $g(2) = 0$, the centre of the circle $|w| = 1$. The points 2 and 0 are inverse points with respect to the line $\operatorname{Re}\zeta = 1$, so g maps them to the inverse points 0 and ∞ with respect to the circle $|w| = 1$. Hence $g(2) = 0$ and $g(0) = \infty \Rightarrow g(\zeta) = \alpha(\zeta - 2)/\zeta$. Since $g(1)$ must be a point on $|w| = 1$, $|\alpha| = 1$, so let $\alpha = -1$ without loss of generality, so that $g(1) = 1$. Then $g(\zeta) = 2/\zeta - 1$. Hence altogether $w = h(z) = (g \circ f)(z) = 2/z^{1/2} - 1$ maps the shaded region and boundary in the z-plane to $|w| \leqslant 1$, as shown in Fig. 6.11.

The Transformation $f(z) = az + b/z$, $a, b \in \mathbb{R}$

The function $f : \tilde{\mathbb{C}} \to \tilde{\mathbb{C}}$ defined by $f(z) = az + b/z$, where a and b are non-zero real constants, maps concentric circles centred at the origin to confocal ellipses, and it maps radial half-lines to confocal hyperbolae. A particular case of f is also useful in aerodynamics. Note that f is conformal in any region not containing $\pm\sqrt{b/a}$.

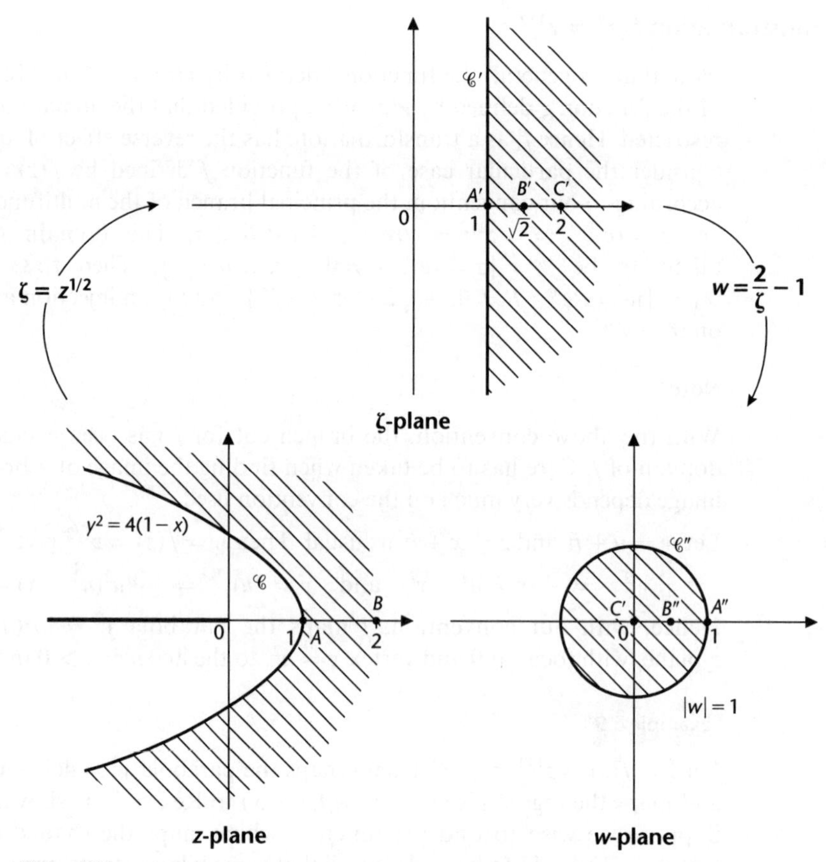

Figure 6.11

Let $w = u + iv$ and $z = re^{i\theta}$. Then $w = f(z)$ gives

$$u(r, \theta) = ar\cos\theta + b\cos\theta/r \quad \text{and} \quad v(r, \theta) = ar\sin\theta - b\sin\theta/r$$

$$\Rightarrow \quad \frac{u^2}{p^2} + \frac{v^2}{q^2} = 1 \tag{6.12}$$

where $p = ar + b/r$ and $q = ar - b/r$. If r is a constant, (6.12) is the equation of an ellipse, centre 0, with foci $\pm\sqrt{p^2 - q^2}$. Hence f maps concentric circles, with centre 0, to confocal ellipses. In particular, if $a = b$, f maps $|z| = 1$ to $w = u$, $-2a \leqslant u \leqslant 2a$. This is a generalisation of Example 6.1. Notice also that

$$\frac{u^2}{\cos^2\theta} - \frac{v^2}{\sin^2\theta} = p^2 - q^2 = 4ab \tag{6.13}$$

If θ is a constant, (6.13) is the equation of a hyperbola, so that f maps half-lines to confocal hyperbolae.

Clearly f is not an injection on $\tilde{\mathbb{C}}$. For example, if $a = b$, the circles $|z| = k$

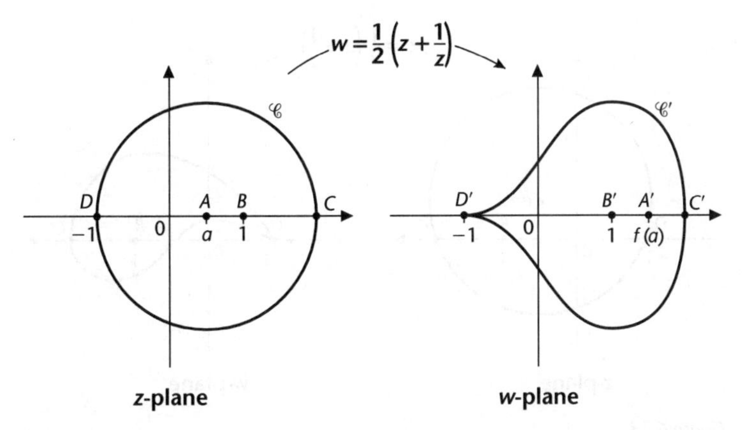

Figure 6.12

and $|z| = 1/k$ are mapped to the same ellipse by f. In this case, it is easily seen that f is an injection on either of the regions given by $|z| < 1$ and $|z| > 1$, and f maps both of these regions to the w-plane with cut $-2a \leqslant u \leqslant 2a$ on the real axis.

We consider now in particular, the transformation

$$w = f(z) = \frac{1}{2}\left(z + \frac{1}{z}\right) \tag{6.14}$$

Notice that it has critical points at ± 1. Consider a circle $\mathscr{C}$ in the z-plane, with centre A on the real axis, say at $a \in \mathbb{R}$, passing through -1 and enclosing 1 as shown in Fig. 6.12. With f defined as in (6.14), $f(-1) = -1$ and $f(1) = 1$. By Taylor's theorem, $f(z) + 1 = -\frac{1}{2}\sum_{n=2}^{\infty}(z+1)^n$, for $|z+1| < 1$. Hence it follows, by using the method of Example 6.2, that the angle between the tangents to $\mathscr{C}$ in the second and third quadrants and the real axis at -1, i.e. $\pi/2$, is doubled under f. That is, the angle between the tangents to $\mathscr{C}'$ in the second and third quadrants and the real axis at -1 is π, so that -1 is a cusp point on $\mathscr{C}'$, as shown in Fig. 6.12. Note that $f(\bar{z}) = \overline{f(z)}$ so that f preserves the symmetry of $\mathscr{C}$ about the real axis. The general shape of $\mathscr{C}'$ is shown in the figure.

In general, if $\mathscr{C}$ is a circle passing though -1 and enclosing 1, with centre $\alpha \in \mathbb{C}$ in the upper half-plane, then the image of $\mathscr{C}$ under f will be the same general shape as in the previous case but will no longer be symmetric about the real axis, as shown in Fig. 6.13. In either case, f maps the outside of the circle $\mathscr{C}$ to the outside of $\mathscr{C}'$.

Historical Note

The curve $\mathscr{C}'$ in these cases gives a crude model of an aircraft wing. Transformations such as (6.14) were first used by Joukowski in aerodynamics. Curves such as $\mathscr{C}'$ are known as **Joukowski aerofoils** and (6.14) is an example of a **Joukowski transformation**.

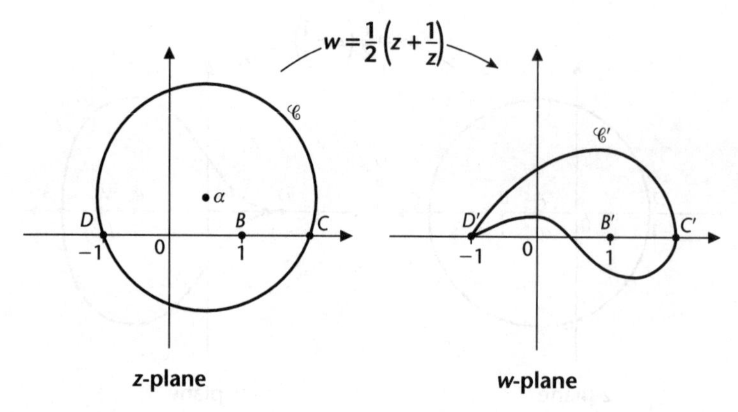

Figure 6.13

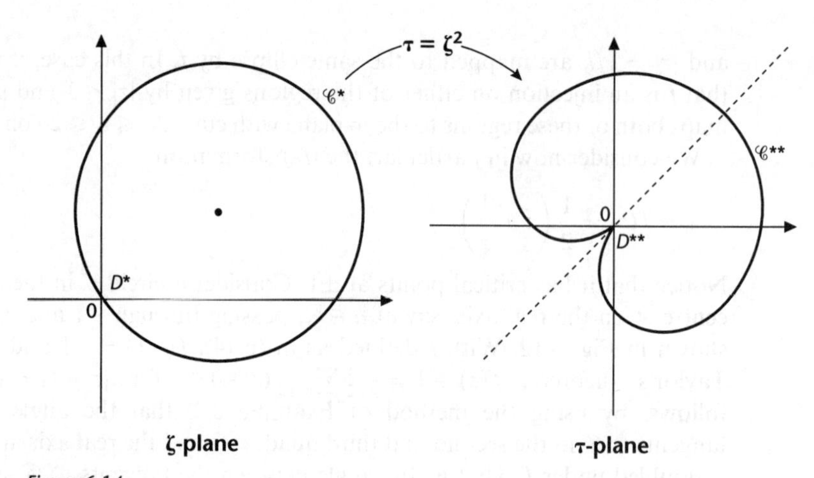

Figure 6.14

Note that if $\zeta = g(z) = (z+1)/(z-1)$, $\tau = h(\zeta) = \zeta^2$ and $w = g(\tau)$, then $w = (g \circ h \circ g)(z) = f(z)$. Hence the transformation f given by (6.14) is the composition of the bilinear transformation g, h and then g again. Let $\mathscr{C}$ be any circle passing through -1 and enclosing 1 as in Fig. 6.13. The transformation g maps $\mathscr{C}$ to a circle $\mathscr{C}^*$, passing through the origin, but not through ± 1, in the ζ-plane (Fig. 6.14). It can be shown that h maps the circle $\mathscr{C}^*$ to the cardioid $\mathscr{C}^{**}$ in the τ-plane, which does not pass through 1 (Fig. 6.14). Finally, g maps the cardioid $\mathscr{C}^{**}$ to the aerofoil $\mathscr{C}'$ as shown in Fig. 6.13 (see Exercises 6.3).

Exponential and Logarithmic Transformations

Consider the function $f: \mathbb{C} \to \mathbb{C}$ defined by $f(z) = e^z$. Note that f is conformal on $\mathbb{C}$ but not an injection on $\mathbb{C}$. Recall that the range of f is $C\backslash\{0\}$. If the domain of f is restricted to $\mathscr{D} = \{z = x + iy \in \mathbb{C} : -\pi < y \leqslant \pi\}$ say, then f is an injection. The domain of f can be restricted to any such strip of width 2π in order to make f an injection.

Let $z = x + iy$ and $w = Re^{i\phi}$. Then if $w = f(z) = e^z$,

$$R(x, y) = e^x \quad \text{and} \quad \phi(x, y) = y \tag{6.15}$$

(without loss of generality). It follows by (6.15) that f maps any horizontal line given by $y = c$, where c is a constant, to the radial half-line $\phi = c$ and any vertical line given by $x = c$ to the circle $|w| = e^c$. Hence if $c_2 > c_1$ and $c_2 - c_1 < 2\pi$, f maps the infinite strip bounded by the lines $y = c_1$ and $y = c_2$ to the wedge of angle $\psi = c_2 - c_1$ (Fig. 6.15). Also f maps the infinite strip bounded by the lines $x = c_1$ and $x = c_2$ with $c_2 > c_1$ to the annulus bounded by the circles $|w| = e^{c_1}$ and $|w| = e^{c_2}$ as shown. In particular, $\operatorname{Re} z > c_1$ is mapped to the outside of $|w| = e^{c_1}$ and $\operatorname{Re} z < c_2$ is mapped to the inside of the centreless circle $|w| = e^{c_2}$.

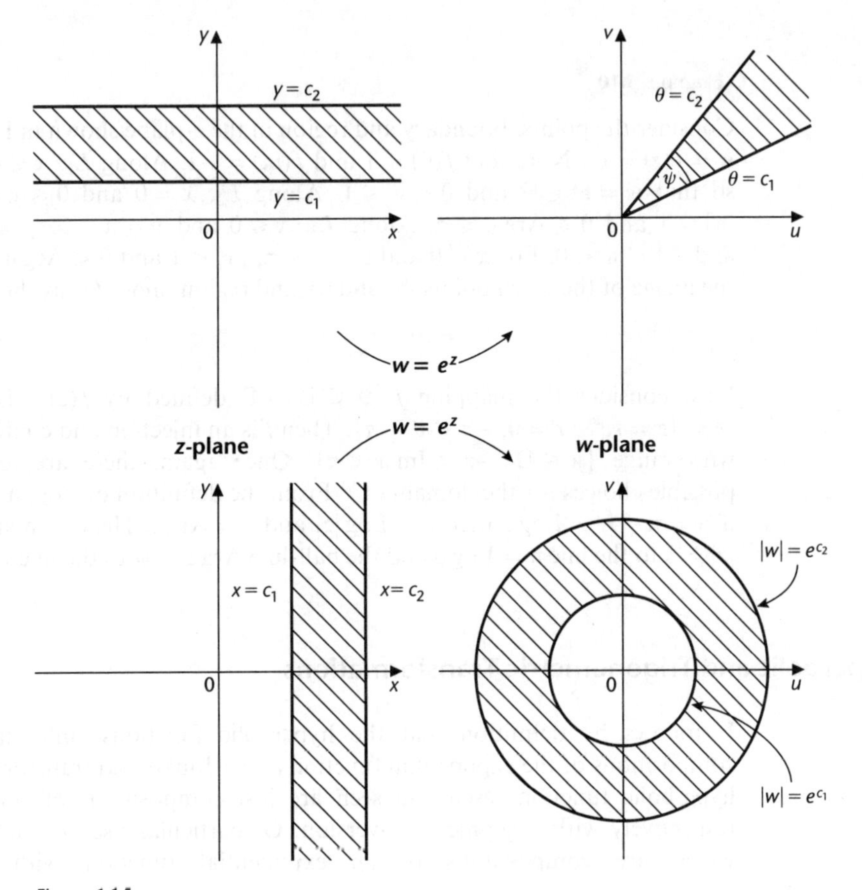

Figure 6.15

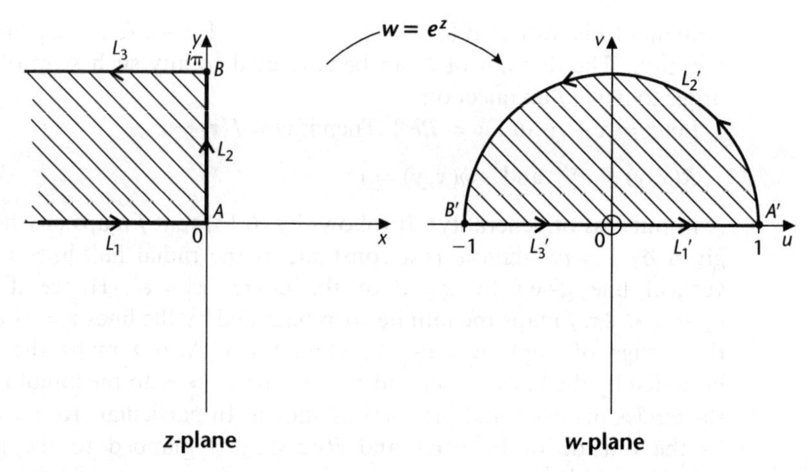

Figure 6.16

Example 6.10

Consider the points, boundary and region in the z-plane shown in Fig. 6.16. Let $w = f(z) = e^z$. Note that $f(0) = 1$ and $f(i\pi) = -1$. Along L_1, $x \leqslant 0$ and $y = 0$ so that $w = e^x \in \mathbb{R}$ and $0 < w \leqslant 1$. Along L_2, $x = 0$ and $0 \leqslant y \leqslant \pi$ so that $|w| = 1$ and $0 \leqslant \operatorname{Arg} w \leqslant \pi$. Along L_3, $x \leqslant 0$ and $y = \pi \quad \Rightarrow \quad w = -e^x \in \mathbb{R}$ and $-1 \leqslant w < 0$. For $x < 0$ and $0 < y < \pi$, $|w| < 1$ and $0 < \operatorname{Arg} w < \pi$. Hence the image of the given points, boundary and region under f is as shown.

Now consider the mapping $f : \mathscr{D} \subseteq \mathbb{C} \to \mathbb{C}$ defined by $f(z) = \operatorname{Log} z$, where $\mathscr{D} = \{z = re^{i\theta} : r \neq 0, -\pi < \theta < \pi\}$. Then f is an injection and conformal on $\mathscr{D}$, with range $\{w \in \mathbb{C} : -\pi < \operatorname{Im} w < \pi\}$. Once again, there are lots of other possible choices for the domain of f. From the definition of Log, it follows that if $w = u + iv = \operatorname{Log} z$ then $u = \operatorname{Log}|z|$ and $v = \operatorname{Arg} z$. Hence f maps the circle $|z| = a$ to the line $u = \operatorname{Log} a$ and the half-line $\operatorname{Arg} z = \psi$ to the line $v = \psi$.

Hyperbolic and Trigonometric Transformations

It follows by definition that the hyperbolic functions sinh and cosh are compositions of the exponential function and a Joukowski transformation. The hyperbolic functions csch and sech are just compositions of sinh and cosh respectively with a geometric inversion. Of particular use are tanh and coth, which are compositions of an exponential function with a bilinear transformation.

Example 6.11

Consider the points, boundary lines and infinite strip in Fig. 6.17 and the transformation $w = f(z) = -\tanh(z/2)$. Note that

$$w = \frac{e^{-z/2} - e^{z/2}}{e^{-z/2} + e^{z/2}} = \frac{1 - e^z}{1 + e^z}$$

so that $w = h(\zeta) = (1 - \zeta)/(1 + \zeta)$ where $\zeta = g(z) = e^z$. By previous work, g maps the infinite strip to the half-plane as shown. It is easy to check that the bilinear transformation h maps $\operatorname{Re}\zeta \geqslant 0$ to $|w| \leqslant 1$. Hence $f = h \circ g$ maps the given points, boundary and region to the points, boundary and region in the w-plane as shown in Fig. 6.17.

The trigonometric functions by definition are compositions of a rotation and a hyperbolic function. They can be investigated directly in the usual way. To be

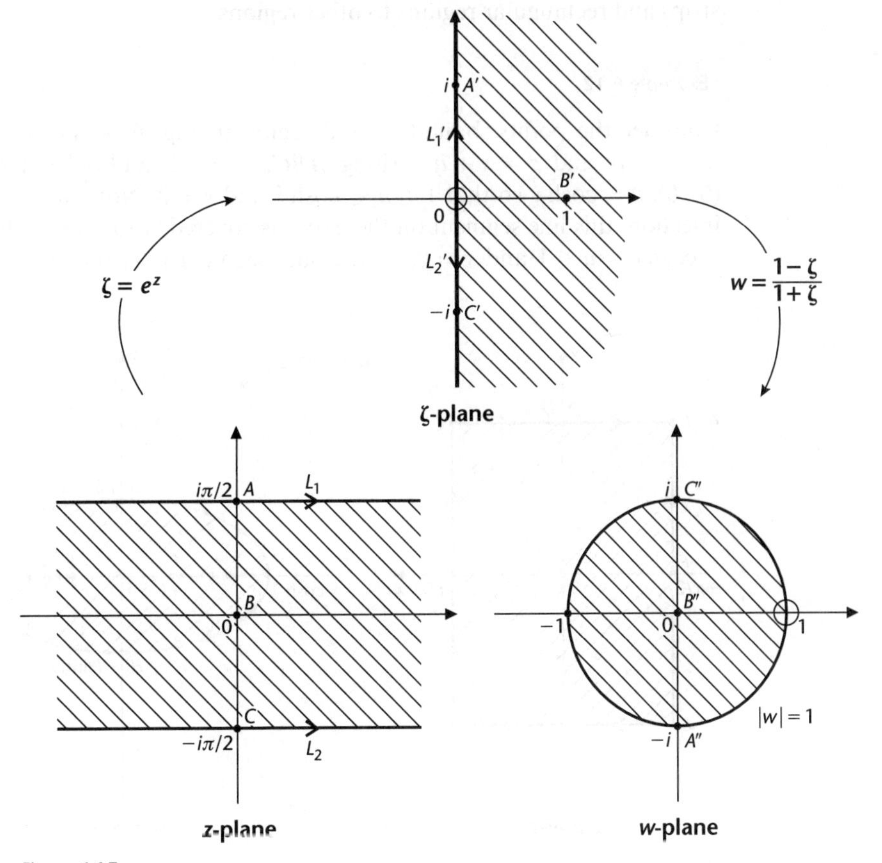

Figure 6.17

specific, consider the transformation $w = f(z) = \sin z$. Letting $z = x + iy$ and $w = u + iv$,

$$u(x, y) = \sin x \cosh y \quad \text{and} \quad v(x, y) = \cos x \sinh y \tag{6.16}$$

It follows by (6.16) that f maps the line $y = k$ to the ellipse

$$\frac{u^2}{\cosh^2 k} + \frac{v^2}{\sinh^2 k} = 1$$

with centre 0 and foci at ± 1, independent of k. Hence f maps parallel horizontal lines to confocal ellipses. Also by (6.16), f maps the line $x = k$ to the hyperbola

$$\frac{u^2}{\sin^2 k} - \frac{v^2}{\cos^2 k} = 1$$

with centre 0 and foci at ± 1. Hence f maps parallel vertical lines to confocal hyperbolae.

In general, the functions sin and cos are particularly useful for mapping strips and rectangular regions to other regions.

Example 6.12

Consider the points, boundary and region in Fig. 6.18. Let $w = f(z) = \sin z$, $w = u + iv$ and $z = x + iy$. Along ABC, $x = \pi/2$ and $-\lambda \leqslant y \leqslant \lambda$. Then by (6.16), $u = \cosh y$ so that $1 \leqslant u \leqslant \cosh \lambda$ and $v = 0$. Note that since f is not an injection, this line segment on the u-axis is covered twice. Similarly, along GFE, $-\cosh \lambda \leqslant u \leqslant 1$ and $v = 0$, so this line segment is also covered twice. Along

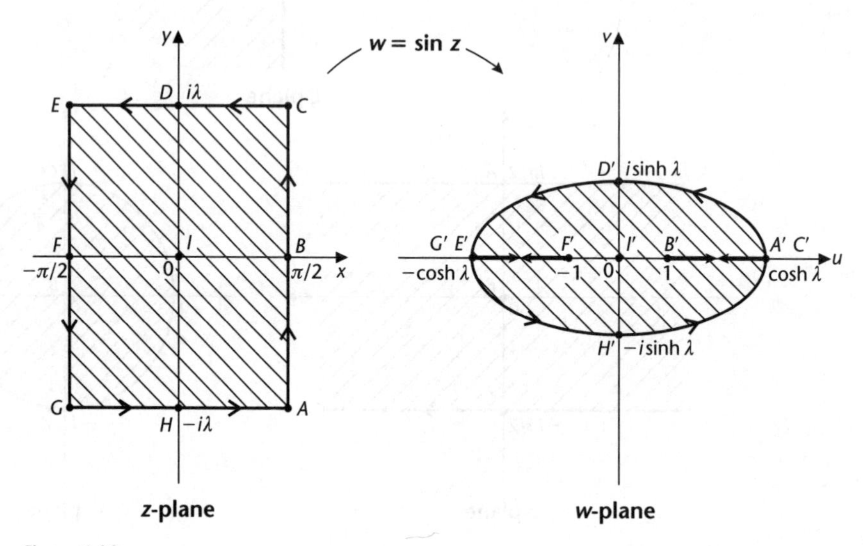

Figure 6.18

EDC, $-\pi/2 \leqslant x \leqslant \pi/2$ and $y = \lambda$ so that by (6.16) $-\cosh\lambda \leqslant u \leqslant \cosh\lambda$ and $0 \leqslant v \leqslant \sinh\lambda$. Also

$$\frac{u^2}{\cosh^2\lambda} + \frac{v^2}{\sinh^2\lambda} = \sin^2 x + \cos^2 x = 1 \tag{6.17}$$

Similarly, along GHA, $-\cosh\lambda \leqslant u \leqslant \cosh\lambda$ and $-\sinh\lambda \leqslant v \leqslant 0$ with u and v satisfying (6.17). Hence the given rectangle is mapped to the ellipse with equation (6.17) in the w-plane, together with the two line segments indicated in Fig. 6.18. Note that any point (x, y) inside the given rectangle satisfies $-\pi/2 < x < \pi/2$, $-\lambda < y < \lambda$, so that $-\cosh\lambda < u < \cosh\lambda$ and $-\sinh\lambda < v < \sinh\lambda$. Hence f maps the inside of the rectangle to the inside of the given contour in the w-plane.

Exercise 6.3.1 Find the image of the wedge $0 \leqslant \operatorname{Arg} z \leqslant \pi/3$ under $f(z) = iz^3$.

Exercise 6.3.2 Find a transformation which maps the semicircle given by $|z| = a$, $-\pi/2 \leqslant \operatorname{Arg} z \leqslant \pi/2$, and its inside, to $|w| \leqslant 1$.

Exercise 6.3.3 Show that $w = (z+a)^2/(z-a)^2$ maps the semicircle given by $|z| = a$, $0 \leqslant \operatorname{Arg} z \leqslant \pi$ and its inside, to $\operatorname{Im} w \geqslant 0$. Hence show that

$$w = \frac{(z^4+1)^2 - i(z^4-1)^2}{(z^4+1)^2 + i(z^4-1)^2}$$

maps the inside and boundary of the sector given by $|z| \leqslant 1$, $0 \leqslant \operatorname{Arg} z \leqslant \pi/4$ to $|w| \leqslant 1$.

Exercise 6.3.4 Let $w = f(z) = (1 - 1/z^2)^{1/2}$ where g, given by $g(z) = z^{1/2}$, is defined by $g(re^{i\theta}) = \sqrt{r}e^{i\theta/2}$, with domain $\{z = re^{i\theta} : 0 < \theta < 2\pi\}$. Show that f maps the half-plane $\operatorname{Re} z > 0$ with cut $\{z = x \in \mathbb{R} : x \geqslant 1\}$, to $\operatorname{Im} w > 0$.

Exercise 6.3.5 Show that, with the convention used in the text, $w = i(z^{1/2} - a)$ maps the given points, curve and shaded region in the z-plane shown in Fig. 6.19 to the given points, curve and shaded region in the w-plane.

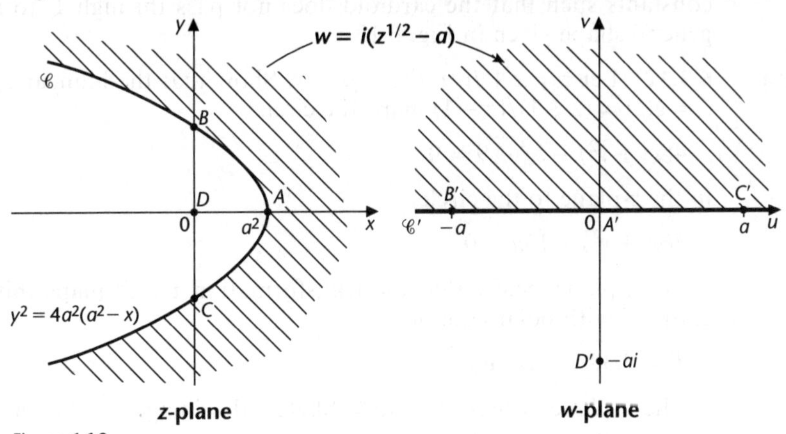

Figure 6.19

Exercise 6.3.6 Show that the transformation $f(z) = az + b/z$ where $a, b \in \mathbb{R}$ maps the family of radial half-lines given by $\operatorname{Arg} z = k$, where k is any constant, to a family of confocal hyperbolae.

Exercise 6.3.7 Find a transformation which maps the circle $|z| = 4$ to the ellipse $|w - 2| + |w + 2| = 6$.

Exercise 6.3.8 Find the image of the z-plane, cut along the real axis for $-1 \leqslant x \leqslant 1$, under $f(z) = z - (z^2 - 1)^{1/2}$, where the branch cut for f is the given cut in the z-plane.

Exercise 6.3.9 Find the image of the points, boundary and region in the z-plane shown in Fig. 6.20 under $w = f(z) = \frac{1}{2}(z + 1/z)$.

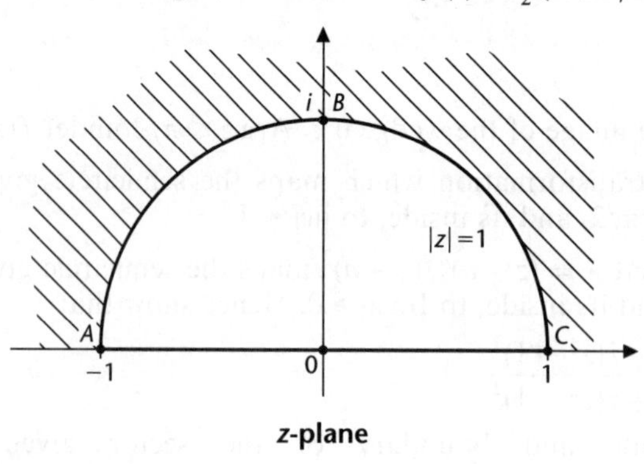

Figure 6.20

Exercise 6.3.10 Let $\mathscr{C}$ be a circle in the z-plane passing through 0, with centre $\alpha \in \mathbb{C}$. Show that $w = f(z) = z^2$ maps $\mathscr{C}$ to the cardioid with polar equation $R = 2|\alpha|^2(1 + \cos(\phi - 2\omega))$ where $w = Re^{i\phi}$ and $\omega = \operatorname{Arg} \alpha$.

Exercise 6.3.11 Show that $w = f(z) = (z + 1)/(z - 1)$ maps the cardioid with polar equation $r = 2a^2(1 + \cos(\theta - \theta_1))$, where $z = re^{i\theta}$ and a and θ_1 are real constants such that the cardioid does not pass through 1, to an aerofoil of general shape given in Fig. 6.13.

Exercise 6.3.12 Let $z = x + iy$ and $\zeta = u + iv$. Show that the bilinear transformation $\zeta = g(z) = (z + 1)/(z - 1)$ maps the circle

$$4(x^2 + y^2) - 5x - 9 = 0$$

in the z-plane to the circle

$$5(u^2 + v^2) - 13u = 0$$

in the ζ-plane. Show that the transformation $\tau = \zeta^2$ maps this circle to the cardioid with polar equation

$$R = 3.38(1 + \cos\phi)$$

in the τ-plane, where $\tau = Re^{i\phi}$. Sketch the image of this curve under the transformation $w = g(\tau)$.

Exercise **6.3.13** Show that $w = e^{z^{1/2}}$ maps the region in the z-plane shown in Fig. 6.21 to the inside of the annulus bounded by the circles $|w| = 1$ and $|w| = e$.

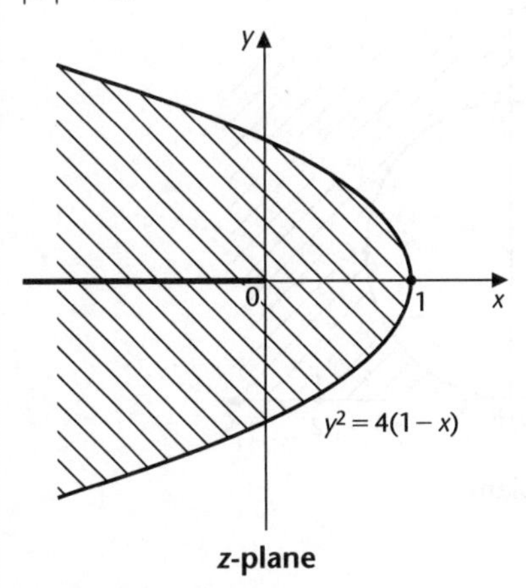

Figure 6.21

Exercise **6.3.14** Show that, under the transformation

$$w = g(z) = \operatorname{Log}\left(\frac{z-1}{z+1}\right)$$

the region given by $x^2 + (y-1)^2 < 2$, $y > 0$ is mapped to an infinite strip in the w-plane of width $3\pi/4$.

Exercise **6.3.15** Find the image of the region given by $|z| < 1$ under the transformation

$$w = \left(\frac{1-z}{1+z}\right)^{2ai/\pi} \qquad (a \in \mathbb{R}^+)$$

Exercise **6.3.16** Find the image of the points, boundary and region shown in Fig. 6.22 under $\zeta = g(z) = e^z$. Hence find the image of these points, boundary and region under $w = f(z) = \coth(z/2)$.

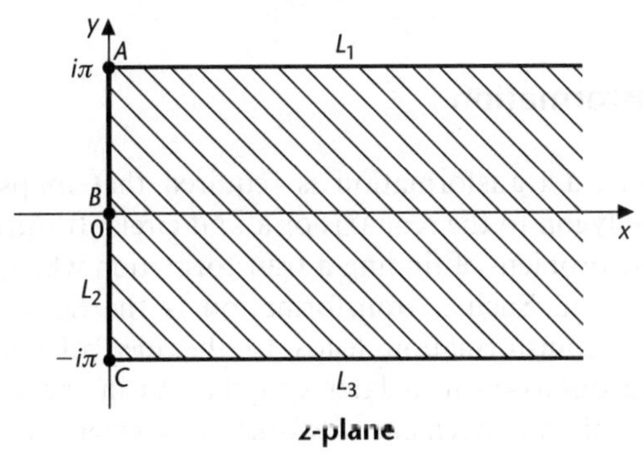

Figure 6.22

Exercise **6.3.17** By using a sequence of simpler transformations, find the image of the points, boundary and region shown in Fig. 6.23 under $w = \coth(\pi/z)$.

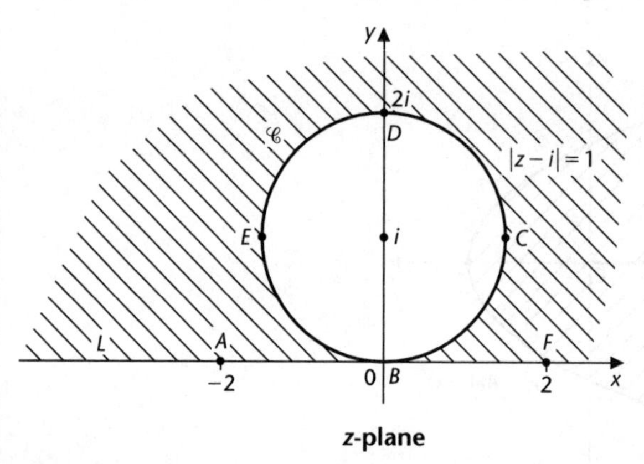

Figure 6.23

Exercise **6.3.18** Find a transformation which maps the semi-infinite strip given by $-\pi/2 < x < \pi/2$, $y > 0$ to the first quadrant. (*Hint*: Consider $f(z) = \sin z$.)

Exercise **6.3.19** By considering a sequence of simpler transformations, find the image of the region given in Exercise 6.3.13 under $w = f(z) = \tan^2(\pi z^{1/2}/4)$. (*Hint*: $\tan^2(z/2) = (1 - \cos z)/(1 + \cos z)$.)

Exercise **6.3.20** Find the image of thc boundary and region in Exercise 6.3.16 under $w = \operatorname{Log}(\coth z/2)$; use your answer to Exercise 6.3.16 to help you with this question.

Exercise **6.3.21** Let $w = f(z) = c\cos(k \operatorname{Log} z)$, where $c^2 = a^2 - b^2$, a, b and k are real and $\cosh(k\pi/2) = a/c$. Show that the region in the z-plane given by $x > 0$ and $1 < |z| < e^{\pi/k}$, is mapped by f into the inside of the ellipse $(u/a)^2 + (v/b)^2 = 1$ in the w-plane cut along the lines joining c to a and $-c$ to $-a$.

The Schwarz–Christoffel Transformation

In many applications, a transformation is required that maps a given (not necessarily closed) polygon to the real axis or a unit circle. It turns out that it is better to consider the problem of finding a transformation which maps the real axis to a given polygon. Such a transformation is the Schwarz–Christoffel transformation. This transformation leads to the idea of Jacobian elliptic functions, which are discussed in a later chapter. As in the case of bilinear transformations, it is often convenient to work in the extended complex plane in this area. In this section, any closed polygon will be assumed to be simple.

Consider first of all, the transformation

$$w = f(z) = \frac{K\pi}{\psi}(z-a)^{\psi/\pi} + C \tag{6.18}$$

where K and C are complex constants and ψ and a are real constants $(K, \psi \neq 0)$. Note that unless $\psi = n\pi$, $n \in \mathbb{Z}$, f has a branch point at a. By choosing a suitable branch cut, f is conformal on the set $\{z \in \mathbb{C} : \operatorname{Im} z \geqslant 0,\ z \neq a\}$, with $f'(z) = K(z-a)^{\psi/\pi - 1}$. Taking the simplest case of (6.18), where $K = \psi/\pi$ and $C = 0$ and letting $z - a = re^{i\theta}$, gives $w = f(z) = r^{\psi/\pi}e^{i\psi\theta/\pi}$. Hence f maps the real axis to two lines passing through the origin, inclined at angle ψ as shown in Fig. 6.24. Also, if z is any point in the upper half-plane, then $0 < \operatorname{Arg}(z-a) < \pi \quad \Rightarrow \quad 0 < \operatorname{Arg} w < \psi$. Hence f maps the upper half-plane to the wedge as shown.

In general, the transformation (6.18) performs the same task as in this simple case, with a possible rotation and translation. This simple idea is the basis for the following result.

Theorem 6.9. The Schwarz–Christoffel (S–C) Transformation

Let $\mathscr{P}$ be an n-sided polygon in the w-plane with vertices $w_1, w_2, \ldots, w_n$ and interior angles $\psi_1, \psi_2, \ldots, \psi_n$ respectively. A transformation f which maps the real axis in the z-plane to $\mathscr{P}$ such that $f(x_r) = w_r$, $x_r \in \mathbb{R}$, $r = 1, \ldots, n$ as shown in Fig. 6.25, is the **Schwarz–Christoffel (S–C) transformation** given by

$$w = f(z) = K\int_{z_0}^{z} \prod_{r=1}^{n} (\zeta - x_r)^{\alpha_r}\, d\zeta + C \quad \text{where} \quad \alpha_r = \psi_r/\pi - 1 \tag{6.19}$$

for some choice of (complex) constants, K, C and z_0. □

Convention

Note that with f defined as in (6.19), f' in general is a multifunction with branch points x_r, $r = 1, \ldots, n$. The simplest way to define f' as a function, so that f' is

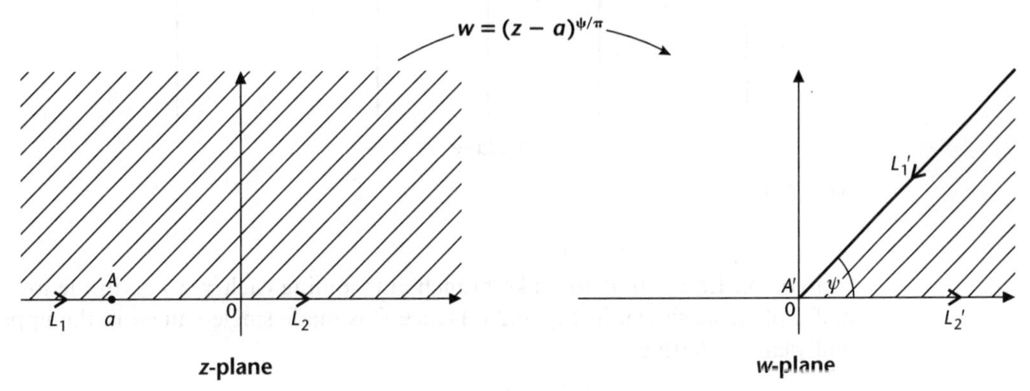

Figure 6.24

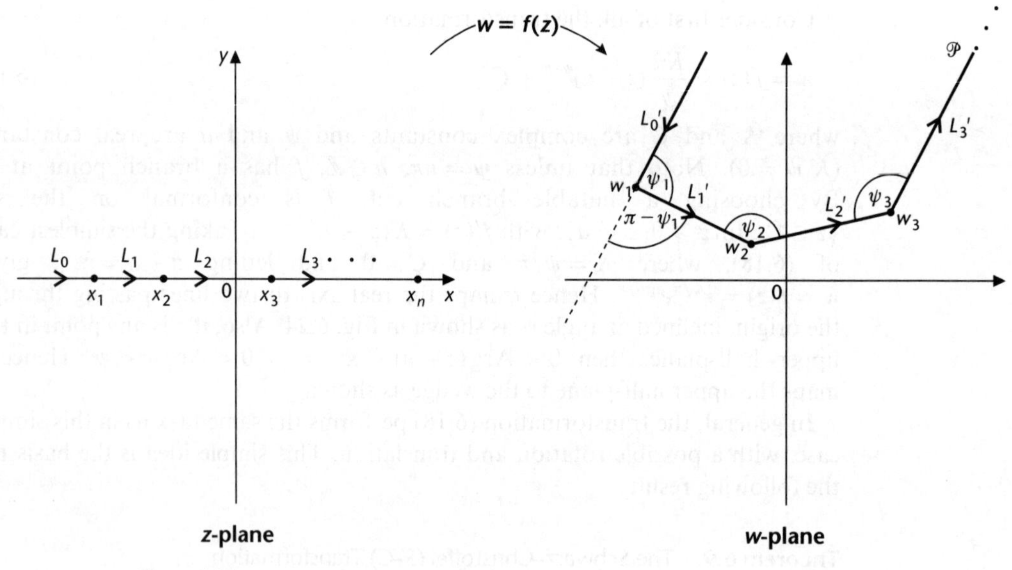

Figure 6.25

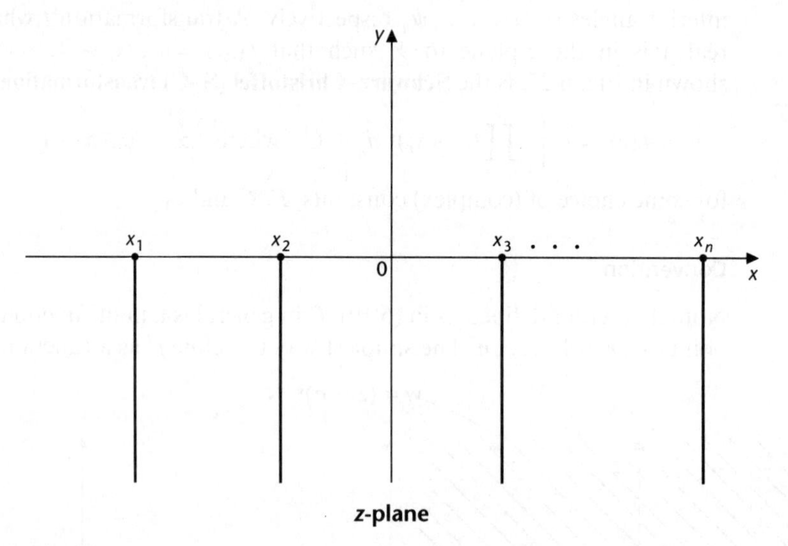

Figure 6.26

analytic on $\operatorname{Im} z > 0$, is to make branch cuts joining each x_r to ∞ in the lower half z-plane as shown in Fig. 6.26. Hence f' is made single-valued in the upper half-plane by letting

$$(z - x_r)^{\alpha_r} = |z - x_r|^{\alpha_r} e^{i\alpha_r \theta_r} \quad \text{where} \quad -\pi/2 < \theta_r < 3\pi/2 \quad \text{for each } r$$

Then by the fundamental theorem of calculus, f is an analytic function on any strip $x_r < \operatorname{Re} z < x_{r+1}$. This is the convention we shall adopt and is the one adopted by most authors.

Proof of 6.9

Step 1

It follows from the proof of Theorem 6.3 that if a function g is conformal on the whole real line, then $\operatorname{Arg}(g'(x))$ is the angle that the tangent to the image of the real line passing through $g(x)$ makes with the real axis. Hence if $\operatorname{Arg}(g'(x))$ is constant over a line segment L of the real axis, g maps L to a line segment inclined at angle $\operatorname{Arg}(g'(x))$ to the real axis.

Step 2

Consider the image of the real line under f defined by (6.19). It follows that

$$f'(z) = K\prod_{r=1}^{n}(z - x_r)^{\alpha_r} \quad \Rightarrow \quad \arg(f'(z)) = \arg K + \sum_{r=1}^{n} \alpha_r \arg(z - x_r) \quad (6.20)$$

Along the line segment L_0 in Fig. 6.25, $z = x \in \mathbb{R}$ with $x < x_1$. By (6.20), $\operatorname{Arg}(f'(x))$ is constant along L_0, so f maps this line segment to the line segment L_0' as shown. Along L_1, $z = x$ with $x_1 < x < x_2$ so again $\operatorname{Arg}(f'(x))$ is constant and f maps L_1 to L_1'. As x increases through x_1, $\operatorname{Arg}(x - x_1)$ decreases from π to 0 while $\operatorname{Arg}(x - x_r)$ for $r > 1$ remains unchanged. Hence by (6.20), $\arg(f'(x))$ decreases by $\alpha_1\pi = \psi_1 - \pi$ and so increases by $\pi - \psi_1$ as shown in Fig. 6.25. It is shown in step 3 that f is continuous for all $x \in \mathbb{R}$, and so if $f(x_1) = w_1$ then L_0' and L_1' intersect at w_1 and are inclined at interior angle ψ_1. In general, f maps each line segment L_r to a line segment L_r' and as x passes though x_r, $\arg(f'(x))$ increases by $\pi - \psi_r$.

Step 3

It remains to show that f is continuous everywhere. It follows from (6.20) that $f'(z) = (z - x_1)^{\alpha_1}F(z)$ where $F(z)$ is a product of factors independent of x_1. Since F is analytic at x_1, it follows by Taylor's theorem that

$$f'(z) = (z - x_1)^{\alpha_1}(F(x_1) + F'(x_1)(z - x_1) + \ldots)$$

in some open neighbourhood $\mathcal{N}_1$ of x_1 where, from (6.19), $-1 < \alpha_1 \leqslant 1$. Hence

$$f'(z) = (z - x_1)^{\alpha_1}F(x_1) + (z - x_1)^{1+\alpha_1}G(z)$$

say with G analytic in $\mathcal{N}_1$. Since $\alpha_1 > -1$, $(z - x_1)^{1+\alpha_1}G(z)$ is continuous in $\mathcal{N}_1$. Hence if z_1 is any point in $\mathcal{N}_1$, $\int_{z_1}^{z}(\zeta - x_1)^{1+\alpha_1}G(\zeta)d\zeta$ is continuous in $\mathcal{N}_1$, and

$$\int_{z_1}^{z}(\zeta - x_1)^{\alpha_1}d\zeta = \frac{1}{1+\alpha_1}((z - x_1)^{1+\alpha_1} - (z_1 - x_1)^{1+\alpha_1})$$

which is continuous in $\mathcal{N}_1$. Hence $\int_{z_1}^{z} f'(\zeta)d\zeta$ is continuous in $\mathcal{N}_1$. Similarly, for each r, there exists a neighbourhood $\mathcal{N}_r$ of x_r such that f is continuous on $\mathcal{N}_r$. The result then follows. ■

Historical Note

The S–C transformation was discovered independently by Schwarz (1869) and Christoffel (1867, 1871). It turns out that in most applications of the S–C transformation, the integral cannot be evaluated in terms of elementary functions.

Notes

(i) The transformation f given by (6.19) is conformal except at the n points $x_1, \ldots, x_n$.

(ii) The constant z_0 is usually taken as 0 for convenience. (If $z_0 \neq 0$, we can just relabel C.) Note that $0 < \psi_r \leqslant 2\pi$ gives $-1 < \alpha_r \leqslant 1$ for all r in (6.19).

(iii) If in (6.19), $K = 1$, $C = 0$ and $z_0 = 0$, a polygon $\mathscr{P}'$ similar to the desired polygon $\mathscr{P}$ is obtained. The constant K is adjusted to give the correct magnification and rotation of $\mathscr{P}'$, and C is chosen to give the correct linear translation.

(iv) In the case of $n \leqslant 3$, the interior angles of a given polygon are enough to determine any similar figure (think of the case of a triangle). This includes open polygons as long as they are closed in $\tilde{\mathbb{C}}$. For $n \geqslant 4$ the interior angles are not enough to determine any similar figure. For example, if all the interior angles of a four-sided closed polygon are $\pi/2$, then the polygon is a rectangle, but not necessarily a square. Hence in (6.19), any *three* of x_r, $r = 1, \ldots, n$ can be assigned arbitrary values, but the others must be chosen so that f maps the real line to a polygon similar to the given one. Another way to think of this is to note that the form of any S–C transformation is unaffected by a bilinear transformation (see the proof of Theorem 6.13), and a chosen bilinear transformation of the real line onto itself allows us to map three arbitrary values of the x_r onto three prescribed values. For $n \geqslant 5$, in general, unless the given polygon is highly symmetrical, the choice of the remaining $n - 3$ values of x_r to obtain the desired polygon is very difficult.

(v) It follows by the proof of 6.9 that if the desired polygon is not closed and has one vertex, w_n say, at ∞ in $\tilde{\mathbb{C}}$ then the factor in $(\zeta - x_n)^{\alpha_n}$ is omitted in (6.19). This corresponds to the convention that $\psi_n = \pi$, so that $\alpha_n = 0$.

(vi) Clearly, the sum of the exterior angles of any **closed** polygon is 2π. Hence in (6.19),

$$\sum_{r=1}^{n}(\pi - \psi_r) = 2\pi \quad \Rightarrow \quad \sum_{r=1}^{n}(\psi_r/\pi - 1) = \sum_{r=1}^{n}\alpha_r = -2 \tag{6.21}$$

Note then that the sum of the interior angles of a closed polygon is $\sum_{r=1}^{n}\psi_r = (n-2)\pi$.

(vii) It can be proved that *any* function which maps the real line to a polygon must be of the form (6.19).

Since, by note (iv) above, three of the x_r in (6.19) can be assigned arbitrary values, treating f as defined on $\tilde{\mathbb{C}}$, it is often convenient to choose $x_n = \infty$.

Lemma 6.10. The S–C Transformation When $x_n = \infty$

If $x_n = \infty$ in (6.18) then the factor in $(\zeta - x_n)^{\alpha_n}$ is omitted without loss of generality. □

Proof

Letting $K = K'(-x_n)^{-\alpha_n}$ in (6.19) gives

$$w = f(z) = K' \int_{z_0}^{z} \prod_{r=1}^{n-1} (\zeta - x_r)^{\alpha_r} \left(1 - \frac{\zeta}{x_n}\right)^{\alpha_n} d\zeta + C$$

and $\lim_{x_n \to \infty}(1 - \zeta/x_n) = 1$. ■

The following result shows that, if $\mathscr{P}$ is closed, the S–C transformation maps the upper half-plane to the inside of $\mathscr{P}$.

Theorem 6.11. The S–C Transformation Maps $\operatorname{Im} z > 0$ to the Inside of a Closed Polygon

The S–C transformation f, given by (6.19), maps $\operatorname{Im} z > 0$ to the inside of $\mathscr{P}$ if $\mathscr{P}$ is closed. With domain $\operatorname{Im} z \geqslant 0$, f is an injection. □

Part Proof

Let $z \in \mathbb{C}$ such that $\operatorname{Im} z > 0$. We prove the result by showing that the number of zeros of $f(z) - \gamma$ is 0 if γ is outside $\mathscr{P}$ and is 1 if γ is inside $\mathscr{P}$. Consider the contour $\mathscr{C}$ consisting of a semicircle of radius R, indented by small semicircles, each of radius ε and centred at each x_r as shown in Fig. 6.27.

Consider the function g given by $g(z) = f(z) - \gamma$, where γ does not lie on the

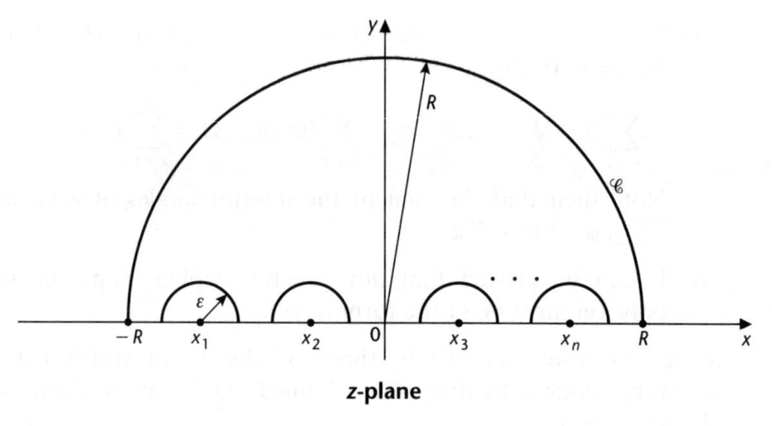

Figure 6.27

perimeter of $\mathscr{P}$. Then by previous analysis, g is analytic inside and on $\mathscr{C}$. By 6.9, f maps $\mathscr{C}$ to $\mathscr{P}^*$ say, where $\mathscr{P}^*$ is $\mathscr{P}$ indented at each of the vertices. Then by Theorem 3.21, if Z is the number of zeros of g inside $\mathscr{C}$ and $w = f(z)$,

$$Z = \frac{1}{2\pi i}\int_{\mathscr{C}} \frac{g'(z)}{g(z)}\,dz = \frac{1}{2\pi i}\int_{\mathscr{P}^*} \frac{dw}{w-\gamma}$$

Hence by Cauchy's integral formula $Z = 1$ if γ is inside $\mathscr{P}^*$, and by Cauchy's theorem $Z = 0$ if γ is outside $\mathscr{P}^*$. The result then follows by letting $R \to \infty$ and $\varepsilon \to 0$ on $\mathscr{C}$. ■

Important Note

In the case of an open polygon, which is closed in $\tilde{\mathbb{C}}$, the above proof is easily adapted to show that f maps $\operatorname{Im} z > 0$ to the 'inside' of the polygon, determined by which of the angles between sucessive line segments of the polygon is chosen as the interior angle.

Example 6.13

Suppose that we wish to map $\operatorname{Im} z \geqslant 0$ to the boundary and region shown in Fig. 6.28, using a transformation f such that $f(0) = 0$ and $f(1) = ia$ where $a \in \mathbb{R}$. By Theorem 6.9, the required transformation is of the form

$$w = f(z) = K\int_0^z (\zeta - 0)^{\pi/2\pi - 1}(\zeta - 1)^{3\pi/2\pi - 1}\,d\zeta + C$$

$$\Rightarrow \quad w = f(z) = K\int_0^z \zeta^{-1/2}(1-\zeta)^{1/2}\,d\zeta + C$$

without loss of generality. Now letting $\zeta = \sin^2\tau$ gives (as in real variable calculus)

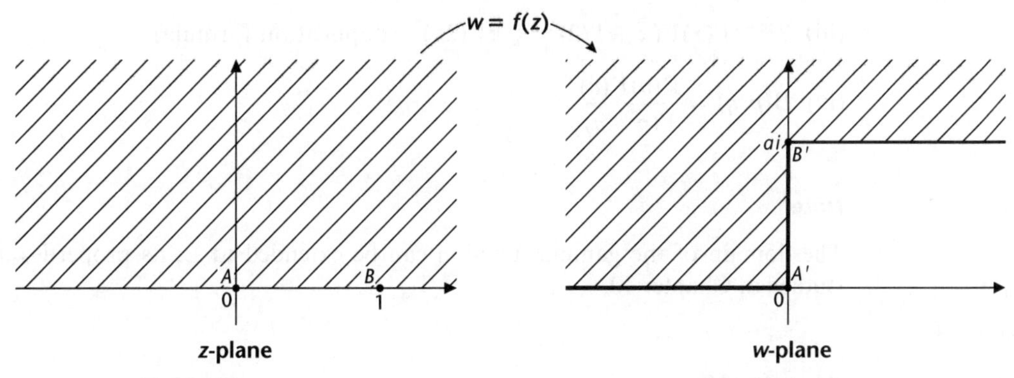

Figure 6.28

$$\int \zeta^{-1/2}(1-\zeta)^{1/2}d\zeta = 2\int \cos^2 \tau d\tau = \int (1+\cos 2\tau)d\tau = \tau + \sin\tau\cos\tau$$
$$= \sin^{-1}\zeta^{1/2} + \zeta^{1/2}(1-\zeta)^{1/2}$$

$$\Rightarrow \quad w = f(z) = K(\sin^{-1} z^{1/2} + z^{1/2}(1-z)^{1/2}) + C$$

Since we require $f(0) = 0$, $C = 0$ and then $f(1) = ai$ gives $K = 2ai/\pi$. Hence

$$w = f(z) = \frac{2ai}{\pi}\left(\sin^{-1} z^{1/2} + z^{1/2}(1-z)^{1/2}\right)$$

When constructing explicit S–C transformations, the following results, concerning real beta and gamma functions, listed here for convenience, are often useful in calculations.

Definitions

The **gamma function** is denoted and defined by

$$\Gamma(x) = \int_0^\infty t^{x-1}e^{-t}dt \qquad (\text{for all } x \in \mathbb{R}^+) \tag{6.22}$$

The **beta function** is denoted and defined by

$$\mathrm{B}(p,q) = \int_0^1 t^{p-1}(1-t)^{q-1}dt \qquad (\text{for all } p, q \in \mathbb{R}^+) \tag{6.23}$$

Theorem 6.12. Properties of Gamma and Beta Functions

(i) $\Gamma(x+1) = x\Gamma(x) \quad \Rightarrow \quad \Gamma(n+1) = n!\ \forall n \in \mathbb{N}$ and $\Gamma(1) = 1$

(ii) $\Gamma(x)\Gamma(1-x) = \dfrac{\pi}{\sin \pi x} \quad \Rightarrow \quad \Gamma(1/2) = \sqrt{\pi}$

(iii) $2^{2x-1}\Gamma(x)\Gamma(x+1/2) = \sqrt{\pi}\Gamma(2x)$ (duplication formula)

(iv) $B(p,q) = \dfrac{\Gamma(p)\Gamma(q)}{\Gamma(p+q)}$ □

Note

The domain of the gamma function can be extended to $\mathbb{C}$; its properties are studied in Chapter 10.

Example 6.14

Suppose we require a transformation which maps $\operatorname{Im} z > 0$ onto the inside of a triangle with interior angles $\alpha\pi$, $\beta\pi$ and $\gamma\pi$ at vertices P, Q and R respectively. Following note (iv) after 6.9, the simplest choice in the S–C transformation is to let $x_1 = 0$, $x_2 = 1$ and $x_3 = \infty$. Then by 6.9, 6.10 and 6.11 the required transformation is of the form

$$w = f(z) = K\int_0^z \zeta^{\alpha-1}(\zeta-1)^{\beta-1}\,d\zeta + C$$

Suppose we wish to map $\operatorname{Im} z > 0$ onto the inside of a triangle with vertex P at the origin and side PQ along the real axis. Any other triangle is obtained from this one by using a simple translation and rotation. In this case, we simply choose $C = 0$ so that $f(0) = 0$ and $K = (-1)^{\beta-1}$ so that $f(1)$ is a point on the real axis, as shown in Fig. 6.29. Then $PQ = f(1) = \int_0^1 x^{\alpha-1}(1-x)^{\beta-1}dx = B(\alpha,\beta)$. It then follows by Theorem 6.12 that

$$PQ = \frac{\Gamma(\alpha)\Gamma(\beta)}{\Gamma(\alpha+\beta)} = \frac{\Gamma(\alpha)\Gamma(\beta)}{\Gamma(1-\gamma)} = \frac{\sin\pi\gamma}{\pi}\Gamma(\alpha)\Gamma(\beta)\Gamma(\gamma)$$

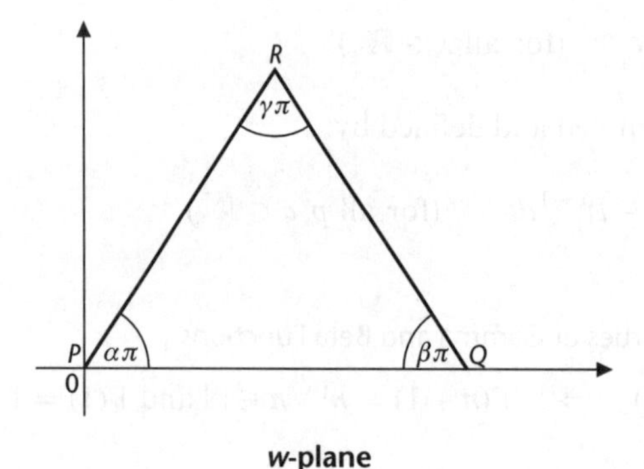

Figure 6.29

Hence, by the sine rule,

$$\frac{QR}{\sin \pi\alpha} = \frac{PQ}{\sin \pi\gamma} \quad \Rightarrow \quad QR = \frac{\sin \pi\alpha}{\pi}\Gamma(\alpha)\Gamma(\beta)\Gamma(\gamma)$$

$$\frac{PR}{\sin \pi\beta} = \frac{PQ}{\sin \pi\gamma} \quad \Rightarrow \quad PR = \frac{\sin \pi\beta}{\pi}\Gamma(\alpha)\Gamma(\beta)\Gamma(\gamma)$$

Then R is the point $e^{i\alpha\pi}\dfrac{\sin \pi\beta}{\pi}\Gamma(\alpha)\Gamma(\beta)\Gamma(\gamma)$.

Example 6.15

Show that the transformation

$$w = f(z) = \int_0^z \frac{d\zeta}{\zeta^{1/2}(1-\zeta^2)^{1/2}}$$

maps $\operatorname{Im} z > 0$ onto the inside of a square with vertices at 0, a, ia and $(1+i)a$, where $a = \Gamma^2(1/4)/2\sqrt{2\pi}$.

Solution

The given transformation is obtained from (6.19) by choosing $x_1 = -1$, $x_2 = 0$, $x_3 = 1$ and $x_4 = \infty$, $\alpha_r = -1/2$ for each $r = 1, 2, 3$, $K = -i$ and $C = 0$. Hence, by 6.9, 6.10 and 6.11, f maps $\operatorname{Im} z \geqslant 0$ to a rectangle and its inside, and since $f(0) = 0$, one vertex of this rectangle is at the origin. Let $f(1) = a \in \mathbb{R}$. Note that $f(-z) = if(z)$ (making a trivial substitution in the integral) so that $f(-1) = ia$ and the rectangle is a square. Hence $f(\infty) = (1+i)a$. Finally note that

$$a = \int_0^1 \frac{dx}{\sqrt{x(1-x^2)}} = \frac{1}{2}\int_0^1 t^{-3/4}(1-t)^{-1/2}dt = \frac{1}{2}\mathrm{B}\left(\frac{1}{4}, \frac{1}{2}\right)$$

letting $t = x^2$. Hence

$$a = \frac{\Gamma(1/4)\Gamma(1/2)}{2\Gamma(3/4)} \qquad \text{(by 6.12)}$$

and since $\Gamma(1/4)\Gamma(3/4) = \sqrt{2}\pi\Gamma(1/2)$ by the duplication formula, $a = \Gamma^2(1/4)/2\sqrt{2\pi}$.

The S–C transformation can also be used to map $|z| \leqslant 1$ to a polygon and its inside.

Theorem 6.13. The S–C Transformation Maps the Unit Circle to a Polygon

The S–C transformation

$$w = f(z) = K\int_0^z \prod_{r=1}^{n} (\zeta - z_r)^{\psi_r/\pi - 1} d\zeta + C \tag{6.24}$$

maps the unit circle $|z| = 1$ to an n-sided polygon $\mathscr{P}$ with interior angles ψ_r and vertices w_r, $r = 1, \ldots n$, such that $f(z_r) = w_r$ where z_r are any n distinct points on the circle. □

Note

The S–C transformation (6.24) takes the same form as (6.19) except that in this case, $z_r \notin \mathbb{R}$ in general and one of these points cannot be chosen as ∞.

Proof

It is easily checked that the bilinear transformation $\tau = g(z) = i(1 - z)/(1 + z)$ maps $|z| = 1$ to $\operatorname{Im}\tau = 0$, and maps $|z| < 1$ to $\operatorname{Im}\tau > 0$. Then if $t_r = g(z_r) \in \mathbb{R}$ for each r,

$$g'(z) = \frac{-2i}{(1+z)^2} \quad \text{and} \quad \tau - t_r = \frac{-2i(z - z_r)}{(1+z)(1+z_r)}$$

It then follows by Theorem 6.9 that the transformation

$$w = f(z) = K'\int_0^z \prod_{r=1}^{n} \frac{(-2i(\zeta - z_r))^{\alpha_r}}{(1+\zeta)^{\alpha_r}(1+z_r)^{\alpha_r}} \cdot \frac{-2i}{(1+\zeta)^2} d\zeta + C \tag{6.25}$$

maps $|z| = 1$ to the desired polygon. It follows by (6.21) that

$$(1+\zeta)^{-2} \prod_{r=1}^{n} (1+\zeta)^{-\alpha_r} = (1+\zeta)^{-2-\sum_{r=1}^{n}\alpha_r} = 1$$

Hence, letting $K = -2iK' \prod_{r=1}^{n} (-2i)^{\alpha_r}(1+z_r)^{-\alpha_r}$ in (6.25) gives (6.24) as required. ■

Note

Since the bilinear transformation used in the above proof maps the inside of the unit circle to the upper half-plane, if the desired polygon $\mathscr{P}$ is closed, then the transformation (6.24) maps the inside of the unit circle to the inside of $\mathscr{P}$.

Exercise

6.4.1 In each case, use the S–C transformation to find the explicit transformation which maps the boundary and region in the w-plane, shown in Fig. 6.30, to the real axis and upper half z-plane, such that 0 is mapped to 0 and A is mapped to 1.

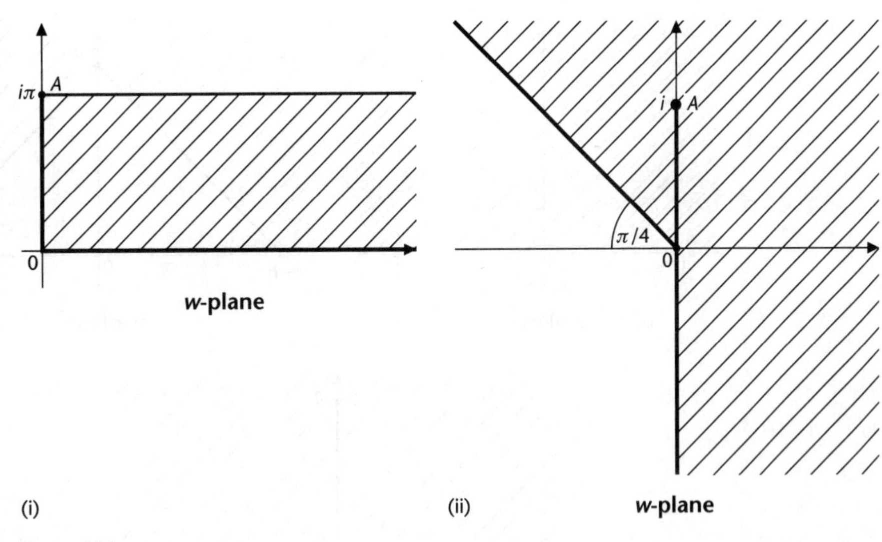

Figure 6.30

Exercise **6.4.2** Find the transformation, in integral form, which maps the given points and $\operatorname{Im} z \geqslant 0$ to the given points, boundary and region in the w-plane shown in Fig. 6.31.

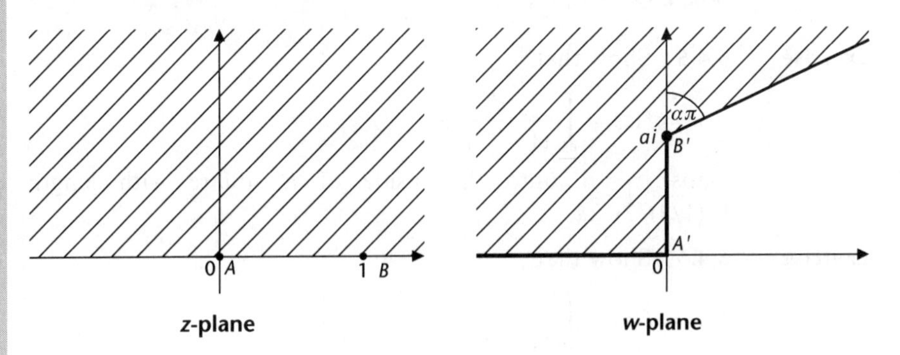

Figure 6.31

Exercise **6.4.3** Find the transformation which maps the given points and $\operatorname{Im} z \geqslant 0$ to the given points, boundary and region in the w-plane, shown in Fig. 6.32. By letting $b \to 0$ in this result, find the explicit transformation which maps the cut upper half w-plane shown in Fig. 6.32 to $\operatorname{Im} z \geqslant 0$.

Exercise **6.4.4** Find the transformation, in integral form, which maps -1, 1 and ∞ to 0, 2 and $1+i$ respectively, and $\operatorname{Im} z > 0$ to the inside of the isosceles triangle with vertices 0, 2 and $1+i$.

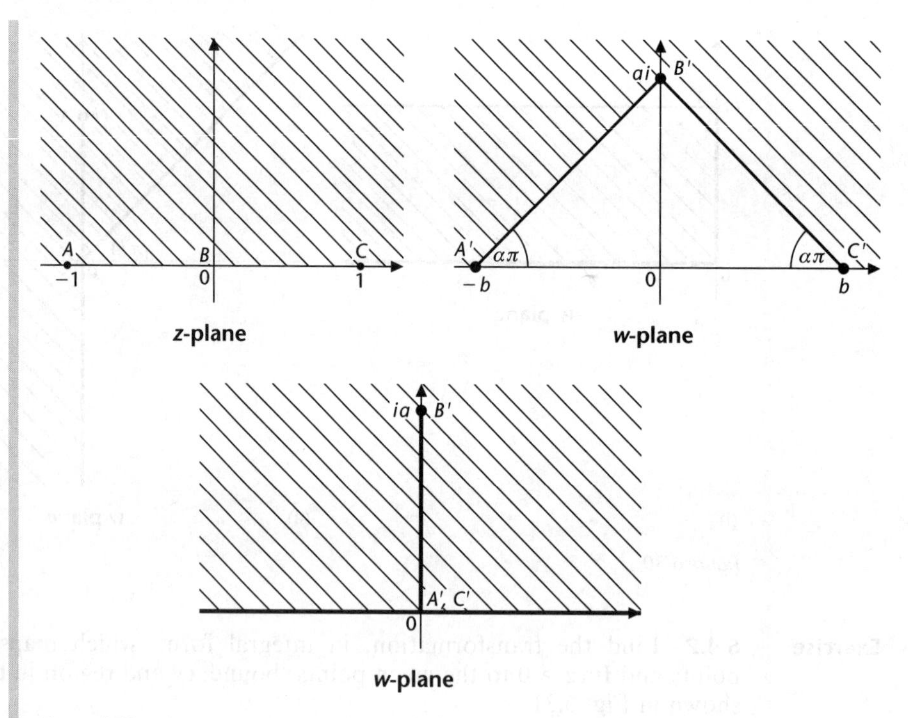

Figure 6.32

Exercise 6.4.5 Show that

$$w = f(z) = \int_0^z \frac{d\zeta}{(1-\zeta^4)^{1/2}}$$

maps $|z| < 1$ onto the inside of a square with length of diagonal $\Gamma^2(1/4)/2\sqrt{2\pi}$.

Exercise 6.4.6 Show that

$$w = f(z) = \int_0^z \frac{d\zeta}{(1-k^2\zeta^2)^{1/2}(1-\zeta^2)^{1/2}}$$

maps the given points and $\operatorname{Im} z \geqslant 0$ to the given points, rectangle and its inside shown in Fig. 6.33, where $0 < k < 1$, $K = f(1)$ and

$$K' = \int_1^{1/k} \frac{dx}{\sqrt{(x^2-1)(1-k^2x^2)}}$$

Exercise 6.4.7 Show that

$$w = f(z) = \int_0^z \frac{d\zeta}{(1-\zeta^6)^{1/3}}$$

maps $|z| < 1$ to the inside of a regular hexagon, centred at 0. Find the length of each side of the hexagon.

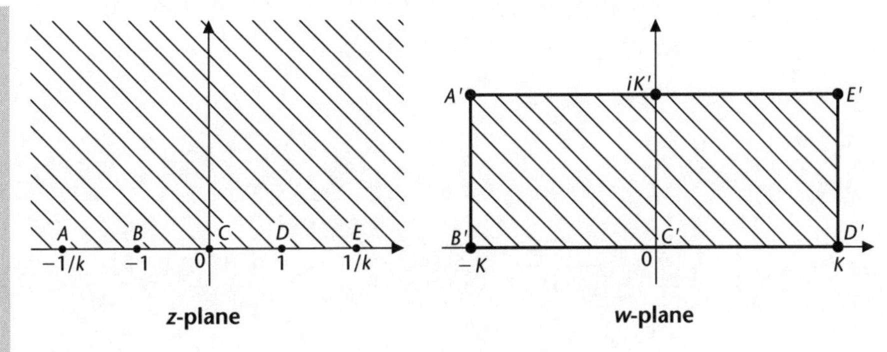

Figure 6.33

7 Linear Ordinary Differential Equations

This chapter shows how some of the results we have already obtained can be applied to the study of linear ordinary differential equations involving functions mapping $\mathbb{C}$ to $\mathbb{C}$. It is not intended as an introduction to differential equations and it is assumed that the reader is familiar with the elementary methods of solution for real variables. Nor is this chapter intended to be a comprehensive account of linear differential equations. We are particularly concerned with the solution of second-order linear equations in series, for which $\mathbb{C}$, rather than $\mathbb{R}$, is the natural setting. We also study solutions of certain linear second-order equations in terms of contour integrals in the complex plane.

For convenience, we shall adopt the following convention.

Convention

If $w = f(z)$ then $dw/dz = w'(z)$, $d^2w/dz^2 = w''(z)$ and in general, $d^kw/dz^k = w^{(k)}(z)$, $k \in \mathbb{N}$. We shall also denote $w^{(k)}(z)$ by $w^{(k)}$ when no ambiguity can occur.

Definition

A **linear ordinary differential equation (l.o.d.e.) of order *n*** in $w = f(z)$, where f is to be determined, is an equation of the form

$$\sum_{k=0}^{n} a_k(z)w^{(k)} = F(z) \tag{7.1}$$

where a_k, $k = 0, 1, \ldots, n$ and F are given functions and $w^{(0)} = w$.

In order that (7.1) has a solution, $w = f(z)$, in a given region $\mathscr{R} \subseteq \mathbb{C}$, f must be analytic in $\mathscr{R}$, so that by 3.14, all the derivatives of f exist and are analytic in $\mathscr{R}$. The following result is a special case of the standard existence and uniqueness result for nth order equations. This result may be proved, for example, by Picard iteration, using essentially the same steps as for real variables.

Theorem 7.1. Existence and Uniqueness of Solutions

If the a_k and F are analytic, with $a_n(z) \neq 0$, for $|z - \alpha| < \rho$ say, then there exists a unique solution, $w = f(z)$, to (7.1), analytic for $|z - \alpha| < \rho$, for which $w^{(k)}(\alpha)$, $k = 0, \ldots, n - 1$, have assigned values. □

Definition

Equation (7.1) with $F(z) \equiv 0$ is the **reduced** or **homogeneous** l.o.d.e. of order n.

Second-Order Linear Equations

We restrict our attention primarily to second-order linear equations, because this is the first non-trivial case, a lot of the methods employed can be extended to higher-order equations and a large number of special functions which are of practical use are solutions of second-order linear equations. Any l.o.d.e. of the second order can be written in the form

$$w'' + p(z)w' + q(z)w = F(z) \tag{7.2}$$

for given functions p, q and F, without loss of generality. We consider first of all the reduced equation

$$w'' + p(z)w' + q(z)w = 0 \tag{7.3}$$

Definition

Two functions, f_1 and f_2, of a complex variable are **linearly dependent** on a region $\mathscr{R} \subseteq \mathbb{C}$ if and only if $f_1(z) = \alpha f_2(z)$ for $z \in \mathscr{R}$, for some constant $\alpha \in \mathbb{C}$, or $f_k(z) \equiv 0$, $k = 1$ or 2, on $\mathscr{R}$.

The following standard result can be proved in exactly the same way as in the real variables. See Chapter 2 in M. Braun, *Differential Equations and Their Applications*, 3rd edn, Springer-Verlag, 1983.

Theorem 7.2. The General Solution of the Reduced Equation

Let p and q in (7.3) be analytic in the region $|z - \alpha| < \rho$. If $w = w_1(z)$ and $w = w_2(z)$ are two particular linearly independent solutions of (7.3), defined in this region, then any solution of (7.3), in this region, is of the form $w = \sum_{n=1}^{2} \alpha_n w_n(z)$ for some choice of constants α_1 and α_2. □

Equation (7.3) is easily solvable when both p and q are constants. In this case, if $\beta = \beta_1$ and $\beta = \beta_2$ are the roots of the quadratic equation $\beta^2 + p\beta + q = 0$, then by 7.2, the general solution of (7.3) is easily seen to be

$$w = \alpha_1 e^{\beta_1 z} + \alpha_2 e^{\beta_2 z} \; (\beta_1 \neq \beta_2) \quad \text{or} \quad w = (\alpha_1 + \alpha_2 z)e^{\beta_1 z} \; (\beta_1 = \beta_2)$$

In either case, the general solution of (7.3) is entire.

Certain l.o.d.e.'s of the second order can be reduced to a constant coefficient equation by a simple substitution. For example, the **Euler equation**

$$z^2 w''(z) + pzw'(z) + qw(z) = 0 \tag{7.4}$$

where p and q are constants, can be reduced to a linear equation by using the substitution $z = e^{\zeta}$. Equation (7.4) then reduces to

$$w''(\zeta) + (p - 1)w'(\zeta) + qw(\zeta) = 0$$

Consider now the general second-order l.o.d.e., (7.2). If the general solution of the reduced equation (7.3) is known, and a particular solution of (7.2) can be obtained, then as in the real case, the following result gives the general solution of (7.2).

Theorem 7.3. The General Solution of a l.o.d.e.

Let p and q be analytic in the region $|z - \alpha| < \rho$. Then any solution to (7.2), defined for $|z - \alpha| < \rho$, is of the form $w = g(z) + h(z)$ where $w = g(z)$ satisfies (7.3) and $w = h(z)$ is a particular solution to (7.2). □

Also, as for real variables, if the general solution to the reduced equation (7.3) is known, a particular solution of (7.2) can be found by using **variation of parameters**. See Chapter 2 in Braun's *Differential Equations and Their Applications*. Then by Theorem 7.3, the general solution of (7.2) is obtained.

The Solution of Linear Second-Order Equations in Series

We now consider the general reduced equation

$$w'' + p(z)w' + q(z)w = 0 \tag{7.3}$$

One way of obtaining formal solutions to (7.3) is to assume that $w = f(z)$ is expressible as a power series or a more general series which converges on some region, then to substitute this series into (7.3) and compare coefficients. Sometimes this series is related to the Taylor or Laurent series expansion of a known elementary function. In this case, the solution obtained for (7.3) is expressible in closed form. In many cases, the series obtained is not expressible in terms of elementary functions and provides a definition of a new function.

Suppose first of all that p and q are analytic for $|z - \alpha| < \rho$. Then by Theorem 7.1, (7.3) has a unique solution, $w = f(z)$, which is analytic for $|z - \alpha| < \rho$, for which $f(\alpha) = \beta$ and $f'(\alpha) = \beta'$ say. Hence, in this case, by Theorem 4.18, any solution of (7.3) must be expressible as a Taylor series expansion about α, which converges for $|z - \alpha| < \rho$.

Example 7.1

Suppose we wish to find the general solution to **Airy's equation**

$$w'' = zw \tag{7.5}$$

By Theorem 7.1, any solution of (7.5) is entire and so can be expanded in a Taylor series about the origin say, which is convergent everywhere. Hence we let $w = \sum_{n=0}^{\infty} a_n z^n$ where the a_n are constants to be determined. By Theorem 4.17, termwise differentiation is valid within the domain of convergence and so is valid everywhere in this case. Then substituting into (7.5) gives

$$\sum_{n=2}^{\infty} n(n-1)a_n z^{n-2} = \sum_{n=0}^{\infty} a_n z^{n+1}$$

for all z. Hence the coefficient of every power of z must vanish. The constant term gives $a_2 = 0$ and the coefficient of z^{n-2} gives

$$n(n-1)a_n = a_{n-3} \qquad \text{(for all } n \geqslant 3) \tag{7.6}$$

The recurrence relation (7.6) gives $a_3 = a_0/3 \cdot 2$, $a_4 = a_1/4 \cdot 3$, $a_5 = 0$, $a_6 = a_0/6 \cdot 5 \cdot 3 \cdot 2, \ldots$. In general,

$$a_n = \frac{k(n)}{n(n-1)(n-3)(n-4)(n-6)(n-7)\ldots b}$$

where $k(n)$ is a_2, a_1 or a_0, and $b = 2$, 3 or 4, depending on n. Since a_0 and a_1 are arbitrary constants, the general solution of (7.5) is, by Theorem 7.2,

$$w = a_0\left(1 + \frac{z^3}{3 \cdot 2} + \frac{z^6}{6 \cdot 5 \cdot 3 \cdot 2} + \ldots\right) + a_1\left(z + \frac{z^4}{4 \cdot 3} + \frac{z^7}{7 \cdot 6 \cdot 4 \cdot 3} + \ldots\right)$$

$$\text{i.e.} \quad w = a_0\left(1 + \sum_{n=1}^{\infty} \frac{z^{3n}}{3n(3n-1)(3n-3)(3n-4)\ldots 3 \cdot 2}\right)$$

$$+a_1\left(z + \sum_{n=1}^{\infty} \frac{z^{3n+1}}{(3n+1)3n(3n-2)(3n-3)\ldots 4 \cdot 3}\right)$$

Let the first power series have terms $A_n(z)$. Then

$$\left|\frac{A_{n+1}(z)}{A_n(z)}\right| = \frac{|z|^3}{(3n+3)(3n+2)} \quad \Rightarrow \quad \lim_{n\to\infty}\left|\frac{A_{n+1}(z)}{A_n(z)}\right| = 0 \qquad \text{(for all } z)$$

Hence the series converges for all z by the ratio test, as expected. Similarly, the second series converges for all z. Neither series is the expansion of an elementary function, so each defines a new special function.

Notes

(i) It is important to simplify the identity involving powers of z, which occurs upon substituting a power series for w into a given equation, as much as possible before trying to calculate the coefficients with this approach.

(ii) It may happen that a solution obtained by this method is only valid in some region of convergence of the power series but the general solution actually exists on a larger subset of $\mathbb{C}$. In particular, if a power series is the expansion of a known function, that function may be a solution of the given equation on a region larger than the region of convergence of the power series. In general, if the power series expansion about the origin is only valid in a certain neighbourhood of the origin, power series expansions about a different point may provide the solution of the equation valid on a different subset of $\mathbb{C}$.

Definition

A point $\alpha \in \mathbb{C}$ at which p and q are analytic is an **ordinary point** of (7.3). Otherwise α is a **singular point** of (7.3).

If α is an ordinary point of (7.3), then there exist two linearly independent Taylor series solutions of (7.3) about α.

Theorem 7.4. The Existence of Taylor Series Solutions

If α is an ordinary point of (7.3), and p and q are analytic for $|z-\alpha|<\rho$, then (7.3) has two linearly independent Taylor series solutions which converge for $|z-\alpha|<\rho$. □

Proof

Step 1

We take $\alpha = 0$ without loss of generality. By hypothesis, $p(z)=\sum_{n=0}^{\infty} p_n z^n$ and $q(z)=\sum_{n=0}^{\infty} q_n z^n$ for $|z|<\rho$, for some constants p_n and q_n. Letting $w=\sum_{n=0}^{\infty} a_n z^n$ in (7.3) gives

$$\sum_{n=2}^{\infty} n(n-1)a_n z^{n-2} + \left(\sum_{n=0}^{\infty} p_n z^n\right)\left(\sum_{n=1}^{\infty} n a_n z^{n-1}\right) + \left(\sum_{n=0}^{\infty} q_n z^n\right)\left(\sum_{n=0}^{\infty} a_n z^n\right) = 0$$

for $|z|<\rho'$ say, where $\rho' \leqslant \rho$, as long as the power series for w converges for $|z|<\rho'$, for some ρ'. By comparing coefficients of z^{n-2}, this leads to the general recurrence relation

$$\begin{aligned} n(n-1)a_n + p_0(n-1)a_{n-1} + (p_1(n-2)+q_0)a_{n-2} + (p_2(n-3)+q_1)a_{n-3} \\ + \ldots + (p_{n-2}+q_{n-3})a_1 + q_{n-2}a_0 = 0 \qquad (n \geqslant 2) \end{aligned} \tag{7.7}$$

where $a_{-1} = q_{-1} = 0$. This is a consistent recurrence relation for the coefficients a_n, $n \geqslant 2$, with a_1 and a_0 arbitrary, and so provides two linearly independent formal Taylor series solutions of (7.3) about 0.

Step 2

We show that these formal series solutions converge for $|z|<\rho$. Let $\mathscr{C}$ be the circle $|z| = r < \rho$ and let $|p(z)| \leqslant M_1$ and $|q(z)| \leqslant M_2$ on $\mathscr{C}$. By Theorems 4.18, 3.13 and 3.2,

$$|p_n| = |p^{(n)}(0)/n!| = \frac{1}{2\pi}\left|\int_{\mathscr{C}} \frac{p(z)}{z^{n+1}}\,dz\right| \leqslant M_1/r^n \quad \text{on } \mathscr{C}$$

and similarly for q_n. Then letting $M = \max(M_1, M_2 r)$ gives

$$|p_n| \leqslant M/r^n \quad \text{and} \quad |q_n| \leqslant M/r^{n+1} \quad \text{on } \mathscr{C} \quad (n \geqslant 0) \tag{7.8}$$

Let $b_0 = |a_0|$ and $b_1 = |a_1|$. The recurrence relation (7.7), the triangle inequality and (7.8) give

$$2a_2 = -p_0 a_1 - q_0 a_0 \quad \Rightarrow \quad 2|a_2| \leqslant b_1|p_0| + b_0|q_0| \leqslant b_1 M + b_0 M/r$$

$$\Rightarrow |a_2| \leqslant b_2 \quad \text{where} \quad 2b_2 = M(2b_1 + b_0/r) \tag{7.9}$$

Similarly,

$$2 \cdot 3|a_3| \leqslant 2|a_2||p_0| + b_1(|p_1| + |q_0|) + b_0|q_1|$$

$$\Rightarrow \quad |a_3| \leqslant b_3 \text{ where } 2 \cdot 3b_3 = 3b_2 M + 2b_1 M/r + b_0 M/r^2 = 3b_2 M + 2b_2/r,$$

using (7.9). Continuing this process, in general, for $n > 2$,

$$|a_n| \leqslant b_n \quad \text{where} \quad (n-1)nb_n = nb_{n-1}M + (n-2)(n-1)b_{n-1}r^{-1}$$

$$\Rightarrow \quad \frac{b_n}{b_{n-1}} = \frac{M}{n-1} + \frac{n-2}{nr} \to \frac{1}{r} \quad \text{as} \quad n \to \infty$$

and so by the ratio test, $\sum_{n=0}^{\infty} b_n z^n$ converges for $|z| < r$. Since $|a_n| \leqslant b_n$ for all n, it follows by the comparison test that the radius of convergence of $\sum_{n=0}^{\infty} a_n z^n$ cannot be less than r. Since r is any number less than ρ, $\sum_{n=0}^{\infty} a_n z^n$ converges for $|z| < \rho$. ■

Solutions in a Neighbourhood of a Regular Singular Point

From now on, we investigate series solutions of (7.3) in a neighbourhood of the origin without loss of generality. It is clear that not all second-order reduced l.o.d.e.'s possess a general solution expressible in terms of Taylor series expansions about the origin. For example, it is easily checked that the Euler equation

$$4z^2 w'' + 4zw' - w = 0$$

has general solution $w = \alpha_1 z^{1/2} + \alpha_2 z^{-1/2}$. Comparing this equation with (7.3), it is seen that p and q are not analytic at the origin. Hence it is not surprising that any particular solution of this equation is not analytic at the origin. The general solution not only has no Taylor series expansion about the origin, it has no Laurent series expansion either, since the origin is a branch point in this case. Hence we are led to consider series solutions of the form

$$w = z^{\lambda} \sum_{n=0}^{\infty} a_n z^n = \sum_{n=0}^{\infty} a_n z^{n+\lambda} \tag{7.10}$$

where $\lambda \in \mathbb{C}$ is an exponent to be determined.

Note

We can take $a_0 \neq 0$ in (7.10) without loss of generality since if $a_0 = 0$, we can just relabel λ. If $\lambda = 0$ or $\lambda \in \mathbb{N}$ in (7.10), then (7.10) is just the Maclaurin series expansion of w. If $\lambda \in \mathbb{Z}^-$, then (7.10) is the Laurent series expansion of w about the origin. If $\lambda \in \mathbb{Q}\backslash\mathbb{Z}$, then w has a branch point at the origin. This illustrates the great advantage of studying l.o.d.e.'s over $\mathbb{C}$ as opposed to $\mathbb{R}$. If $\lambda \in \mathbb{C}\backslash\mathbb{R}$, then (7.3) has a solution of the form $w = e^{\lambda \operatorname{Log} z} \sum_{n=0}^{\infty} a_n z^n$.

Historical Note

The process of finding series solutions of (7.3) of the form (7.10) was first introduced by Fuchs and later refined by Frobenius. This fuller treatment first appeared in 1873.

Definition

Series of the form (7.10) are often called **Frobenius series.**

Example 7.2

Consider the reduced l.o.d.e.

$$z(z-1)w'' + (z-3)w' - 4w = 0 \tag{7.11}$$

Comparing (7.11) with the general equation (7.3), it is seen that p and q are not analytic at the origin. Just suppose for the moment that (7.11) has at least one solution expressible as a series of the form (7.10). Letting $w = \sum_{n=0}^{\infty} a_n z^\alpha$, where $\alpha = n + \lambda$ for convenience and $a_0 \neq 0$, in (7.11), gives formally

$$z(z-1)\sum_{n=0}^{\infty} \alpha(\alpha-1)a_n z^{\alpha-2} + (z-3)\sum_{n=0}^{\infty} \alpha a_n z^{\alpha-1} - 4\sum_{n=0}^{\infty} a_n z^\alpha \equiv 0$$

$$\Rightarrow \quad \sum_{n=0}^{\infty}(\alpha+2)(\alpha-2)a_n z^\alpha \equiv \sum_{n=0}^{\infty} \alpha(\alpha+2)a_n z^{\alpha-1} \tag{7.12}$$

Equating coefficients of the *lowest power* of z, in this case $z^{\lambda-1}$, gives $\lambda(\lambda+2) = 0$. This is known as the **indicial equation**. Hence $\lambda = 0$ or $\lambda = -2$. In general, equating coefficients of $z^{\alpha-1}$, i.e. $z^{n+\lambda-1}$, in (7.12) gives the recurrence relation

$$(n+\lambda+1)(n+\lambda-3)a_{n-1} = (n+\lambda+2)(n+\lambda)a_n \qquad (n \geqslant 1) \tag{7.13}$$

For $\lambda = 0$, (7.13) reduces to

$$a_n - \frac{(n+1)(n-3)}{n(n+2)} a_{n-1}$$

and so gives $a_1 = -4a_0/3$, $a_2 = -3a_1/8 = a_0/2$, $a_3 = 0 \Rightarrow a_n = 0$, for $n > 3$.

This gives the polynomial solution

$$w = a_0(1 - 4z/3 + z^2/2) \qquad (\text{for all } z \in \mathbb{C})$$

For $\lambda = -2$, (7.13) gives

$$a_n = \frac{(n-1)(n-5)}{n(n-2)} a_{n-1} \qquad (n \geqslant 1)$$

Hence $a_1 = 0$, a_2 is arbitrary, $a_3 = -4a_2/3$, $a_4 = -3a_3/8 = a_2/2$, $a_5 = 0 \Rightarrow a_n = 0$ for $n > 5$. This gives a second solution,

$$w = a_0/z^2 + a_2(1 - 4z/3 + z^2/2)$$

which includes the first as a special case. This must be the general solution of the given equation by Theorem 7.2, valid for all $z \neq 0$.

An obvious question to ask is, What type of behaviour must the functions p and q have in a neighbourhood of 0 in order that (7.3) has two linearly independent Frobenius series solutions, which converge in some deleted neighbourhood of 0? A clue comes from the general reduced Euler equation, which has two such (trivial) linearly independent solutions. The following definitions are based on the behaviour of p and q in a neighbourhood of 0 in this case. It turns out that, in a large number of cases, this type of behaviour is sufficient to guarantee the existence of two convergent linearly independent Frobenius series solutions.

Definitions

The point 0 is a **regular singular point** of (7.3) if it is a singular point with $zp(z)$ and $z^2q(z)$ both analytic at 0, that is, 0 is at most a simple pole of p and a pole of order 2 of q. Otherwise 0 is an **irregular singular point**.

Theorem 7.5. The Existence of Frobenius Series Solutions in a Neighbourhood of a Regular Singularity

Let 0 be a regular singular point of (7.3), so that $zp(z) = \sum_{n=0}^{\infty} p_n z^n$ and $z^2q(z) = \sum_{n=0}^{\infty} q_n z^n$, for $|z| < \rho$ say, where the p_n and q_n are constants with p_0, q_0 and q_1 not all zero. Let λ_1 and λ_2 be the two roots of the **indicial equation**

$$F(\lambda) = \lambda(\lambda - 1) + p_0\lambda + q_0 = 0 \tag{7.14}$$

Then if $|\lambda_1 - \lambda_2|$ is not a positive integer or 0, equation (7.3) has two linearly independent series solutions of the form (7.10) which converge for $0 < |z| < \rho$. □

Proof

Step 1

Let $w = \sum_{n=0}^{\infty} a_n z^{n+\lambda}$ with $a_0 \neq 0$ in (7.3). Then

$$z^2 w'' + z^2 p(z) w' + z^2 q(z) w = 0$$

$$\Rightarrow \sum_{n=0}^{\infty} a_n (n+\lambda)(n+\lambda-1) z^{n+\lambda} + \left(\sum_{n=0}^{\infty} p_n z^n\right)\left(\sum_{n=0}^{\infty} a_n (n+\lambda) z^{n+\lambda}\right)$$

$$+ \left(\sum_{n=0}^{\infty} q_n z^n\right)\left(\sum_{n=0}^{\infty} a_n z^{n+\lambda}\right) = 0 \tag{7.15}$$

for $0 < |z| < \rho' \leqslant \rho$, provided that $\sum_{n=0}^{\infty} a_n z^n$ converges for $|z| < \rho'$ for some ρ'. Equating coefficients of the lowest power of z, i.e. z^λ, in (7.15) gives the indicial equation (7.14), and (7.15) can be written as

$$a_0 F(\lambda) z^\lambda + (a_1 F(1+\lambda) + (\lambda p_1 + q_1) a_0) z^{\lambda+1} + \dots$$

$$+ \left(a_n F(n+\lambda) + \sum_{k=0}^{n-1} ((k+\lambda) p_{n-k} + q_{n-k}) a_k\right) z^{n+\lambda} + \dots = 0$$

This gives (7.14) and the recurrence relation

$$F(n+\lambda) a_n = -\sum_{k=0}^{n-1} ((\lambda+k) p_{n-k} + q_{n-k}) a_k \qquad (n \geqslant 1) \tag{7.16}$$

Since, by hypothesis, there exist two distinct roots, λ_1 and λ_2, of (7.14) which do not differ by an integer, so that $F(n+\lambda) \neq 0$ for some $n \neq 0$, there exist two linearly independent formal series solutions of (7.3) of the form (7.10), obtained from (7.16) with $\lambda = \lambda_1$ and $\lambda = \lambda_2$.

Step 2

We show that if either of the formal series solutions does not terminate, then $\sum_{n=0}^{\infty} a_n z^n$ converges for $|z| < \rho$. The proof is very similar to the proof of Theorem 7.4. Consider the formal solution given by $w = z^{\lambda_1} \sum_{n=0}^{\infty} a_n z^n$. Since by (7.14), $\lambda_1 + \lambda_2 = 1 - p_0$ it follows by (7.16) and $F(\lambda_1) = 0$ that

$$n(n + \lambda_1 - \lambda_2) a_n = -\sum_{k=0}^{n-1} ((\lambda_1 + k) p_{n-k} + q_{n-k}) a_k \qquad (n \geqslant 1) \tag{7.17}$$

Let $c = |\lambda_1 - \lambda_2|$, $d = |\lambda_1|$, $b_n = |a_n|$ for $0 \leqslant n < c$ and m the least natural number greater than c. Then by (7.17) and the triangle inequality,

$$m(m-c)|a_m| \leqslant |m(m + \lambda_1 - \lambda_2) a_m| \leqslant \sum_{k=0}^{m-1} ((d+k)|p_{m-k}| + |q_{m-k}|) b_k \tag{7.18}$$

Now let $|zp(z)| \leqslant M_1$ and $|z^2 q(z)| \leqslant M_2$ on the circle $\mathscr{C}$ given by $|z| = r < \rho$ and let $M = \max(M_1, M_2)$. Then as in the proof of 7.4,

$$|p_n| \leqslant M/r^n \quad \text{and} \quad |q_n| \leqslant M/r^n \quad \text{on } \mathscr{C} \tag{7.19}$$

It follows by (7.18) and (7.19) that

$$|a_m| \leqslant b_m \quad \text{where} \quad m(m-c)b_m = M\sum_{k=0}^{m-1}(d+k+1)b_k/r^{m-k}$$

and then, in the same way,

$$|a_n| \leqslant b_n \quad \text{where} \quad n(n-c)b_n = M\sum_{k=0}^{n-1}(d+k+1)b_k/r^{n-k} \quad (n \geqslant m) \tag{7.20}$$

It follows by (7.20) that

$$n(n-c)b_n = M(d+n)b_{n-1}/r + M\sum_{k=0}^{n-2}(d+k+1)b_k/r^{n-k}$$

$$\Rightarrow \quad n(n-c)b_n = M(d+n)b_{n-1}/r + (n-1)(n-1-c)b_{n-1}/r$$

$$\Rightarrow \quad \frac{b_n}{b_{n-1}} = \frac{M(n+d)}{n(n-c)r} + \frac{(n-1)(n-1-c)}{n(n-c)r} \quad (n \geqslant m) \Rightarrow \lim_{n\to\infty}\frac{b_n}{b_{n-1}} = \frac{1}{r}$$

Hence the radius of convergence of $\sum_{n=0}^{\infty} b_n z^n$ is r and since $|a_n| \leqslant b_n$ for all n, and r is any number less than ρ, $\sum_{n=0}^{\infty} a_n z^n$ converges for $|z| < \rho$. Similarly for the second solution. Note that since the two values of λ do not differ by an integer, at least one of the solutions has a branch point at the origin, so the general solution is not analytic at 0. ■

Example 7.3

The equation

$$4zw'' + 2w' + w = 0 \tag{7.21}$$

has a regular singular point at 0, since in this case, $zp(z) = 1/2$ and $z^2q(z) = z/4$. It follows by Theorem 7.5 that (7.21) has two linearly independent Frobenius series solutions which converge for all z except possibly for $z = 0$. Letting $w = \sum_{n=0}^{\infty} a_n z^{\alpha}$, where $\alpha = n + \lambda$ and $a_0 \neq 0$ in (7.21) gives

$$\sum_{n=0}^{\infty} 2a_n\alpha(2\alpha-1)z^{\alpha-1} + \sum_{n=0}^{\infty} a_n z^{\alpha} \equiv 0$$

Equating coefficients of the lowest power of z, i.e. $z^{\lambda-1}$, gives the indicial equation

$$\lambda(2\lambda-1) = 0 \quad \Rightarrow \quad \lambda = 0 \quad \text{or} \quad \lambda = 1/2$$

Equating coefficients of $z^{n+\lambda-1}$ gives the recurrence relation

$$2(n+\lambda)(2n+2\lambda-1)a_n + a_{n-1} = 0 \qquad (n \geqslant 1) \tag{7.22}$$

For $\lambda = 0$, (7.22) gives

$$a_n = \frac{-a_{n-1}}{2n(2n-1)} \quad \Rightarrow \quad a_1 = \frac{-a_0}{2} \quad a_2 = \frac{-a_1}{4 \cdot 3} = \frac{a_0}{4!} \quad a_3 = \frac{-a_2}{6 \cdot 5} = \frac{-a_0}{6!} \quad \ldots$$

$$\Rightarrow a_n = \frac{(-1)^n a_0}{(2n)!} \qquad (n \geqslant 1)$$

For $\lambda = 1/2$, (7.22) gives

$$a_n = \frac{-a_{n-1}}{(2n+1)2n} \quad \Rightarrow \quad a_1 = \frac{-a_0}{3!} \quad a_2 = \frac{-a_1}{5 \cdot 4} = \frac{a_0}{5!} \quad \ldots$$

$$\Rightarrow a_n = \frac{(-1)^n a_0}{(2n+1)!} \qquad (n \geqslant 1)$$

Hence the general solution to (7.21) is

$$w = a \sum_{n=0}^{\infty} \frac{(-1)^n z^n}{(2n)!} + b z^{1/2} \sum_{n=0}^{\infty} \frac{(-1)^n z^n}{(2n+1)!}$$

and both series converge for all z by the ratio test, as expected. Recalling the standard Maclaurin series expansions of the elementary functions, it is seen that the general solution can be expressed in the closed form

$$w = a \cos(z^{1/2}) + b \sin(z^{1/2})$$

which is analytic and hence a valid solution of (7.21) for all $z \in \mathbb{C}$ except along a chosen branch cut of $G(z) = z^{1/2}$, which must include $z = 0$.

Notes

(i) When trying to express series solutions in closed form, in terms of elementary functions, it is essential to remember the standard Maclaurin expansions of the exponential, logarithmic, trigonometric and hyperbolic functions as well as the binomial series.

(ii) In general, the recurrence relation obtained from the given linear equation will involve a_k for more than two values of k. In these cases, the recurrence relation determines a_n as an explicit function of n only if it is of a special form. See Exercise 7.1.8.

Important Note

In certain cases, there may still be two linearly independent series solutions of (7.3) of the form (7.10) when the roots of the indicial equation (7.14) are distinct but differ by an integer. If $F(n + \lambda) = 0$ for some $n \neq 0$ in (7.16), it may happen that the right-hand side of (7.16) is also 0 for this n, as in Example 7.2. Then two distinct Frobenius series solutions are obtained.

It follows by the proof of Theorem 7.5 that if the indicial equation has a repeated root, then (7.3) has only one Frobenius series solution. We can obtain some idea of the form of the general solution in this case by considering the general Euler equation (7.4). If the indicial equation has a repeated root $\lambda = \lambda_1$ in this case, then the general solution is $w = (\alpha_1 + \alpha_2 \operatorname{Log} z)z^{\lambda_1}$.

If a series solution of a given equation can be expressed in closed form, reduction in order will give a second, linearly independent solution. As in the real case, if $w = w_1(z) \not\equiv 0$ is a known solution of (7.3), a second, linearly independent solution, $w = w_2(z)$, is given by

$$w = w_2(z) = w_1(z) \int \frac{e^{-\int p(z)dz}}{w_1^2(z)} dz \tag{7.23}$$

This is left as an exercise.

More generally, if the indicial equation has equal roots or roots differing by an integer, which gives an indeterminacy in the recurrence relations, equation (7.23) can be used to determine the form that the general solution of (7.3) takes. Let 0 be a regular singular point of (7.3), so that by Theorem 7.5, there is at least one series solution of the form $w = w_1(z) = \sum_{n=0}^{\infty} a_n z^{n+\lambda_1}$, which converges and is analytic in a region $\mathscr{R} \subseteq \mathbb{C}$, defined by $0 < |z| < \rho$ for some ρ, with the line segment $-\rho < \operatorname{Re} z < 0$ excluded, to avoid any problem with possible branch cuts. Let $a_0 = 1$ without loss of generality. Now

$$zp(z) = \sum_{n=0}^{\infty} p_n z^n \quad \Rightarrow \quad \int p(z)dz = p_0 \operatorname{Log} z + \sum_{n=1}^{\infty} p_n \frac{z^n}{n} + c \tag{7.24}$$

where c is any constant, in the region $\mathscr{R}$, by Theorem 4.16. Then, using the Maclaurin series expansion of e^{-z} and Theorem 4.22,

$$e^{-\int p(z)dz} = e^{-c} z^{-p_0} (1 - p_1 z + (p_1^2 - p_2)z^2/2 - \ldots)$$

in $\mathscr{R}$. It also follows from the binomial series and 4.22 that

$$\frac{1}{w_1^2(z)} = z^{-2\lambda_1} \left(\sum_{n=0}^{\infty} a_n z^n \right)^{-2} = z^{-2\lambda_1} (1 - 2a_1 z + (3a_1^2 - 2a_2)z^2 + \ldots)$$

Then using (7.24) with $c = 0$, without loss of generality, gives

$$\frac{e^{-\int p(z)dz}}{w_1^2} = z^{-(2\lambda_1 + p_0)} (1 + b_1 z + b_2 z^2 + \ldots) \tag{7.25}$$

where $b_1 = -(p_1 + 2a_1)$, $b_2 = 3a_1^2 - 2a_2 + 2a_1 p_1 + p_1^2 - p_2$, ... in the region $\mathscr{R}$, by Theorem 4.19. Suppose that the roots of the indicial equation (7.14) are equal or differ by an integer, so that $\lambda_2 = \lambda_1 - m$ for some m, where $m = 0$ or $m \in \mathbb{N}$, without loss of generality. Then $2\lambda_1 + p_0 = 1 + m$ and by (7.23) and (7.25), a second solution of (7.3), defined and analytic in $\mathscr{R}$, is given by

$$w = w_2(z) = w_1(z) \int \sum_{n=0}^{\infty} b_n z^{n-m-1} dz$$

where $b_0 = 1$. Hence, by 4.16 and 4.19,

$$w_2(z) = w_1(z)\left(z^{-m}\sum_{n\neq m} b_n^* z^n + b_m \operatorname{Log} z\right) = b_m w_1(z)\operatorname{Log} z + z^{\lambda_2}\sum_{n=0}^{\infty} c_n z^n$$

for some constants b_n^* and c_n, in the region $\mathscr{R}$.

In the case of equal roots, $m = 0$ so there is always a term in $\operatorname{Log} z$ present in the general solution of (7.3). Very often in practice, a solution to (7.3) which is analytic at the origin is sought, so if $\lambda_1 \in \mathbb{N}$, the first solution is appropriate and the second solution need not be found. If $m \neq 0$, it may happen that $b_m = 0$, in which case there are two linearly independent solutions of (7.3) of the form (7.10). We have now proved the following result.

Theorem 7.6. The General Solution of the Reduced Equation in a Neighbourhood of a Regular Singularity

If 0 is a regular singular point of (7.3), then the general solution of (7.3), defined and analytic in a region $\mathscr{R}$ defined by $0 < |z| < \rho$ for some ρ, with the line segment $-\rho < \operatorname{Re} z < 0$ excluded, is of the form $w = \alpha_1 w_1(z) + \alpha_2 w_2(z)$ where

$$w_1(z) = z^{\lambda_1}\sum_{n=0}^{\infty} a_n z^n \quad \text{and} \quad w_2(z) = z^{\lambda_2}\sum_{n=0}^{\infty} b_n z^n + K w_1(z)\operatorname{Log} z \tag{7.26}$$

for some constants K, a_n and b_n, where λ_1 and λ_2 are the roots of the indicial equation (7.14). □

The converse of Theorem 7.6 is also true.

Theorem 7.7. Regular Singularities of the Reduced Equation

Suppose that (7.3) has general solution $w = \alpha_1 w_1(z) + \alpha_2 w_2(z)$ where $w_1(z)$ and $w_2(z)$ are given by (7.26) in the region $\mathscr{R}$ and where $\lambda_1 = \lambda_2$ if $K \neq 0$ without loss of generality. Then 0 is a regular singular point of (7.3). □

Proof

Since $w = w_1(z)$ and $w = w_2(z)$ are solutions of (7.3), it follows by eliminating $q(z)$ that

$$(w_2 w_1'' - w_1 w_2'') + p(z)(w_2 w_1' - w_1 w_2') = 0$$

$$\Rightarrow \quad p(z) = -\frac{d}{dz}\left(\operatorname{Log}\left(w_1 w_2' - w_1' w_2\right)\right) = -\frac{d}{dz}\left(\operatorname{Log}\left(w_1^2 \frac{d}{dz}(w_2/w_1)\right)\right)$$

$$\Rightarrow \quad p(z) = \frac{-2w_1'}{w_1} - \frac{(w_2/w_1)''}{(w_2/w_1)'} \tag{7.27}$$

It follows from (7.26), 4.17 and 4.19 that, in the region $\mathscr{R}$,

$$\frac{w_2}{w_1} = K\,\mathrm{Log}\,z + z^{\lambda_2 - \lambda_1}\sum_{n=0}^{\infty} c_n z^n \quad \Rightarrow \quad \frac{d}{dz}\left(\frac{w_2}{w_1}\right) = \frac{K}{z} + z^{\lambda_2 - \lambda_1 - 1}\sum_{n=0}^{\infty} c'_n z^n$$

for some choice of constants c_n and c'_n. Then by 4.17 again,

$$\frac{d^2}{dz^2}\left(\frac{w_2}{w_1}\right) = \frac{-K}{z^2} + z^{\lambda_2 - \lambda_1 - 2}\sum_{n=0}^{\infty} c''_n z^n$$

in $\mathscr{R}$, for some constants c''_n. Hence from (7.26) and (7.27), using 4.17 and 4.19 again, bearing in mind that we have taken $\lambda_1 = \lambda_2$ if $K \neq 0$, it follows that

$$p(z) = \frac{-2}{z}\sum_{n=0}^{\infty} d_n z^n + \frac{1}{z}\sum_{n=0}^{\infty} e_n z^n$$

in $\mathscr{R}$, for some choice of constants d_n and e_n, so that $zp(z)$ is analytic at 0. Then, from (7.3) and 4.17 and 4.19 again,

$$q(z) = \frac{-w''_1}{w_1} - p(z)\frac{w'_1}{w_1} = \frac{1}{z^2}\sum_{n=0}^{\infty} f_n z^n - p(z)\sum_{n=0}^{\infty} d_n z^n$$

in $\mathscr{R}$, for some choice of constants f_n. It follows that $z^2 q(z)$ is analytic at 0. Hence 0 is a regular singular point of (7.3). ■

Note

If 0 is an irregular singular point of (7.3), then there may exist one series solution of the form (7.10), but in general, there are no such series solutions. In some cases, the reason why such a series solution does not exist is that the general solution has an essential singularity at 0. For example, the equation $z^2 w'' - w' = 0$, for which 0 is an irregular singular point, has general solution $w = \alpha_1 \int e^{-1/z} dz + \alpha_2$. In general, it can be shown that if 0 is an isolated singular point of p and q in (7.3), then two linearly independent solutions take the form

$$w_1(z) = z^{\lambda_1}\sum_{n=-\infty}^{\infty} a_n z^n \qquad w_2(z) = z^{\lambda_2}\sum_{n=-\infty}^{\infty} b_n z^n$$

$$\text{or} \quad w_2 = z^{\lambda_1}\sum_{n=-\infty}^{\infty} b_n z^n + K w_1\,\mathrm{Log}\,z$$

Exercise

7.1.1 Find the general solution of each of the following equations in terms of a linear combination of two power series expansions about the origin. Express each solution in closed form and state its domain of validity:

(i) $(1 - z)w'' + (1 + z)w' - 2w = 0$

(ii) $(1 - z^2)w'' - zw' + w = 0$

Exercise **7.1.2** Find the general solution of each of the following equations in terms of a linear combination of two power series expansions about the origin. Find the region of convergence of each power series.

(i) $w'' - 2zw' + 4w = 0$

(ii) $(1 + z^2)w'' + 3zw' + w = 0$

Exercise **7.1.3** Find the general solution of the Legendre equation

$$(1 - z^2)w'' - 2zw' + 6w = 0$$

in terms of a polynomial and a power series expansion about the origin. Find the region of convergence of the power series. Determine the particular solution statisfying $w(0) = 1$ and $w'(0) = 0$.

Exercise **7.1.4** Show that one solution of **Legendre's equation**

$$(1 - z^2)w'' - 2zw' + m(m + 1)w = 0 \qquad (m \in \mathbb{N})$$

is a polynomial of degree m. The polynomial solution with coefficient of z^m equal to $(2m)!/2^m(m!)^2$ defines the **Legendre polynomial** $P_m(z)$. Find expressions for $P_m(z)$, $m = 0, 1, \ldots 4$.

Exercise **7.1.5** Find the general solution of the equation

$$2zw'' + (2z + 1)w' + 2w = 0$$

in terms of two series and state its domain of validity.

Exercise **7.1.6** Find the general solution of each of the following equations in closed form by first finding two Frobenius series solutions. In each case, find the subset of $\mathbb{C}$ on which the solution is valid.

(i) $z^2w'' + 2z(1 - z)w' - 2w = 0$

(ii) $4z^2w'' + 4zw' + (4z^2 - 1)w = 0$

(iii) $z(1 - z)w'' + 3(1 - 2z)w' - 6w = 0$

(iv) $zw'' + 2w' - 4zw = 0$

(v) $z^2(1 - z^2)w'' - 2z(1 + z^2)w' + 2w = 0$

Exercise **7.1.7** Find two linearly independent series solutions of

$$3zw'' + (1 - z)w' - w = 0$$

and find the region on which the general solution is valid. Express the series solution satisfying $w(0) = 0$ in closed form.

Exercise **7.1.8** Solve the equation

$$zw'' + (2 + az)w' + (a + bz)w = 0$$

where a and b are constants, in terms of a power series.
(*Hint*: Put $a_n = c_n/n!$ in the general recurrence relation.)

Exercise **7.1.9** Let $z = 0$ be an irregular singular point of the equation

$$w'' + p(z)w' + q(z)w = 0$$

(i) If $z^2p(z)$ and $z^2q(z)$ are analytic at 0, show that there is one and only one series solution of the equation of the form (7.10).

(ii) If $zp(z)$ and $z^3q(z)$ are analytic at 0, show that there is no series solution of the equation of the form (7.10).

Exercise **7.1.10** Find one solution, in closed form, of the equation

$$z^3(1-z)w'' + z(1+2z-2z^2)w' + w = 0$$

by first finding a Frobenius series solution. Find a second, linearly independent solution using (7.23). State the region on which the general solution is valid.

Exercise **7.1.11** Find two Frobenius series solutions of the equation

$$z^2w'' + zw' + (1+\alpha z)w = 0 \qquad (\alpha \in \mathbb{C})$$

and state the region on which the general solution is valid.

Exercise **7.1.12** (Reduction in order). Prove that if $w = w_1(z) \not\equiv 0$ is a known solution of (7.3), then a second, linearly independent solution of (7.3) is given by (7.23).
(*Hint*: Consider $F(z) = w_1w_2' - w_1'w_2$ and find the first-order l.o.d.e satisfied by $F(z)$.)

The Method of Frobenius

Although (7.26) gives the form of the general solution of (7.3) in the case of equal roots of the indicial equation or roots differing by an integer, reduction in order does not, in general, provide a practical method of determining the second solution in these cases. The following method, due to Frobenius, is such a method.

Case 1: Equal Roots in the Indicial Equation

Let $z = 0$ be a regular singular point of (7.3) and suppose that, on substituting $w = \sum_{n=0}^{\infty} a_n z^{n+\lambda}$ in (7.3), all recurrence relations are satisfied except the indicial equation (7.14), so that all coefficients of powers of z vanish, except for the lowest power. Then if the indicial equation has a repeated root, $\lambda = \lambda_1$ say,

$$z^2w'' + z^2p(z)w' + z^2q(z)w = a_0(\lambda - \lambda_1)^2 \tag{7.28}$$

Also, $w(z,\lambda) = \sum_{n=0}^{\infty} a_n(\lambda)z^{n+\lambda}$ where λ has yet to be assigned a value. The coefficient a_0 is arbitrary and does not depend on λ. Then formally,

$$\frac{\partial w}{\partial \lambda} = w_\lambda(z,\lambda) = \left(\sum_{n=0}^{\infty} a_n(\lambda)z^{n+\lambda}\right) \operatorname{Log} z + \sum_{n=0}^{\infty} a_n'(\lambda)z^{n+\lambda}$$

and from (7.28),

$$z^2 w''_\lambda(z) + z^2 p(z) w'_\lambda(z) + z^2 q(z) w_\lambda(z)$$
$$= a_0 z^\lambda \operatorname{Log} z (\lambda - \lambda_1)^2 + 2a_0 z^\lambda (\lambda - \lambda_1) = 0 \quad \text{when} \quad \lambda = \lambda_1$$

Then $w = w_1(z) = w(z, \lambda_1)$ is a solution of (7.3) and

$$w = w_2(z) = w_\lambda(z, \lambda_1) = w_1(z) \operatorname{Log} z + z^{\lambda_1} \sum_{n=0}^{\infty} a'_n(\lambda_1) z^n$$

is a second solution of (7.3). The formal manipulations are justified by the fact this is the desired solution by Theorem 7.6. This leads to the following result.

Theorem 7.8. Equal Roots in the Indicial Equation

Let 0 be a regular singular point of (7.3) and let the indicial equation (7.14) have a repeated root $\lambda = \lambda_1$. Let $w(z, \lambda) = \sum_{n=0}^{\infty} a_n(\lambda) z^{n+\lambda}$ for some constants $a_n(\lambda)$. Then two linearly independent solutions of (7.3) are $w = w_1(z) = w(z, \lambda_1)$ and

$$w = w_2(z) = w_\lambda(z, \lambda_1) = w_1(z) \operatorname{Log} z + z^{\lambda_1} \sum_{n=0}^{\infty} a'_n(\lambda_1) z^n$$

These are analytic in the region $\mathscr{R}$, given in 7.6. □

Notation

In this section, we shall let $\Phi(n) = \sum_{k=1}^{n} (1/k)$, $\Phi(0) = 1$.

Example 7.4 Bessel's Equation of Order 0

Consider the equation

$$zw'' + w' + zw = 0 \tag{7.29}$$

By 7.5, equation (7.29) possesses at least one Frobenius series solution which converges everywhere, except possibly for $z = 0$. Letting $w = \sum_{n=0}^{\infty} a_n z^\alpha$, $\alpha = n + \lambda$, in the usual way, gives

$$\sum_{n=0}^{\infty} \alpha^2 a_n z^{\alpha - 1} + \sum_{n=0}^{\infty} a_n z^{\alpha + 1} \equiv 0$$

The coefficient of $z^{\lambda - 1}$ gives the indicial equation $\lambda^2 = 0$ so that $\lambda = 0$ is a repeated root. The coefficient of z^λ gives $a_1 = 0$ and the coefficient of $z^{n+\lambda-1}$ gives

$$(n + \lambda)^2 a_n = -a_{n-2} \qquad (\text{for all } n \geqslant 2) \tag{7.30}$$

Hence $a_{2n+1} = 0$, $n \geqslant 0$, and when $\lambda = 0$, $n^2 a_n = -a_{n-2}$ so that

$$a_{2n} - \frac{-a_{2n-2}}{(2n)^2} - \frac{a_{2n-4}}{(2n)^2 (2n-2)^2} = \ldots = \frac{(-1)^n a_0}{2^{2n} (n!)^2}$$

Taking $a_0 = 1$ without loss of generality gives the solution

$$w = w_1(z) = \sum_{n=0}^{\infty} \frac{(-1)^n z^{2n}}{2^{2n}(n!)^2}$$

This solution of (7.29) is denoted by $w = J_0(z)$ and is the **Bessel function of the first kind of order 0**. The series converges for all z.

It follows from (7.30) that, for any λ,

$$a_{2n}(\lambda) = \frac{-a_{2n-2}}{(2n+\lambda)^2} = \ldots = \frac{(-1)^n a_0}{(2n+\lambda)^2(2n-2+\lambda)^2 \ldots (2+\lambda)^2} \tag{7.31}$$

Important Note

In general, it is essential to take logarithms before attempting to differentiate these coefficients with respect to λ.

Noting that some values may be complex, (7.31) gives

$$\mathrm{Log}\,(a_{2n}(\lambda)) = \mathrm{Log}\,((-1)^n a_0) - 2\,\mathrm{Log}\,(2n+\lambda) - 2\,\mathrm{Log}\,(2n+\lambda-2) - \ldots - 2\,\mathrm{Log}\,(\lambda+2)$$

$$\Rightarrow \quad \frac{a'_{2n}(\lambda)}{a_{2n}(\lambda)} = -\frac{2}{2n+\lambda} - \frac{2}{2n+\lambda-2} - \ldots - \frac{2}{\lambda+2}$$

$$\Rightarrow \quad a'_{2n}(0) = -a_{2n}(0)\Phi(n) = \frac{(-1)^{n+1}\,\Phi(n) a_0}{2^{2n}(n!)^2}$$

Then from Theorem 7.8, a second solution of (7.29) is

$$w = w_2(z) = w_1(z)\,\mathrm{Log}\,z + \sum_{n=0}^{\infty} \frac{(-1)^{n+1}\,\Phi(n) z^{2n}}{2^{2n}(n!)^2}$$

and the series in this solution converges for all z.

Case 2: Roots of the Indicial Equation Differing by an Integer

Let λ_1 and $\lambda_2 = \lambda_1 - m$, for some $m \in \mathbb{Z}$, be the roots of (7.14) and suppose that $\lambda = \lambda_2$ gives an inconsistency in the recurrence relation (7.16). Suppose that, on substituting $w = \sum_{n=0}^{\infty} a_n z^{n+\lambda}$ into (7.3), the recurrence relation (7.16) is satisfied but the indicial equation (7.14) is not. Then

$$z^2 w'' + z^2 p(z) w' + z^2 q(z) w = a_0 z^{\lambda}(\lambda - \lambda_2)(\lambda - \lambda_2 - m)$$

and $w(z,\lambda) = \sum_{n=0}^{\infty} a_n(\lambda) z^{n+\lambda}$ where λ has yet to be assigned a value. Since a_0 is arbitrary, choosing $a_0 = \lambda - \lambda_2$ and formally differentiating with respect to λ gives

$$z^2 w''_{\lambda}(z) + z^2 p(z) w'_{\lambda}(z) + z^2 q(z) w_{\lambda}(z)$$
$$= z^{\lambda}\mathrm{Log}\,z(\lambda-\lambda_2)^2(\lambda-\lambda_2-m) + 2z^{\lambda}(\lambda-\lambda_2)(\lambda-\lambda_2-m) + z^{\lambda}(\lambda-\lambda_2)^2$$

Hence $w = w_{\lambda}(z,\lambda_2)$ is a solution of (7.3) and we have the following result.

Theorem 7.9. Roots of the Indicial Equation Differing by an Integer

Let 0 be a regular singular point of (7.3). Suppose that the roots of the indicial equation (7.14) are λ_1 and λ_2, where $\lambda_2 = \lambda_1 - m$ for some $m \in \mathbb{Z}$, $m \neq 0$, and that $\lambda = \lambda_2$ leads to an inconistency in the recurrence relation (7.16). Let $w(z, \lambda) = \sum_{n=0}^{\infty} a_n(\lambda) z^{n+\lambda}$, with $a_0 = \lambda - \lambda_2$. Then two linearly independent formal solutions of (7.3) are $w = w(z, \lambda_1)$ and $w = w_\lambda(z, \lambda_2)$. □

Example 7.5

Consider the equation

$$zw'' - zw' - w = 0 \tag{7.32}$$

Letting $w = \sum_{n=0}^{\infty} a_n z^\alpha$, where $\alpha = n + \lambda$ in (7.32) gives

$$\sum_{n=0}^{\infty} \alpha(\alpha - 1) a_n z^{\alpha - 1} \equiv \sum_{n=0}^{\infty} (\alpha + 1) a_n z^\alpha$$

The coefficient of $z^{\lambda - 1}$ gives the indicial equation $\lambda(\lambda - 1) = 0$ so that $\lambda = 0$ or 1. In general, the coefficient of $z^{n+\lambda-1}$ gives

$$(n + \lambda - 1) a_n = a_{n-1} \qquad \text{(for all } n \geqslant 1) \tag{7.33}$$

The recurrence relation (7.33) is consistent when $\lambda = 1$ but involves a division by 0 when $\lambda = 0$. Hence, using the notation of 7.9, let $\lambda_1 = 1$ and $\lambda_2 = 0$. With $\lambda = \lambda_1 = 1$, (7.33) reduces to $a_n = a_{n-1}/n \Rightarrow a_n = a_0/n!$. Taking $a_0 = \lambda_1 - \lambda_2 = 1$ provides the solution

$$w = w_1(z) = z\left(1 + \sum_{n=1}^{\infty} \frac{z^n}{n!}\right) = ze^z \qquad \text{(for all } z)$$

to (7.32). Following 7.9, to obtain a second, linearly independent solution to (7.32), we let $a_0(\lambda) = \lambda$ in (7.33) and find $w_\lambda(z, 0)$ where $w = \sum_{n=0}^{\infty} a_n(\lambda) z^{n+\lambda}$. Equation (7.33) then gives $a_0(0) = 0$, $a_1(0) = 1$, $a_n(0) = 1/(n-1)!$ for all $n \geqslant 2$, $a_0'(0) = 1$, $a_1'(0) = 0$ and in general,

$$a_n(\lambda) = \frac{1}{(n + \lambda - 1)(n + \lambda - 2)\ldots(1 + \lambda)} \qquad (n \geqslant 2)$$

$$\Rightarrow \operatorname{Log} a_n(\lambda) = -\operatorname{Log}(n + \lambda - 1) - \operatorname{Log}(n + \lambda - 2) - \ldots - \operatorname{Log}(1 + \lambda)$$

$$\Rightarrow \frac{a_n'(\lambda)}{a_n(\lambda)} = \frac{-1}{n + \lambda - 1} - \frac{1}{n + \lambda - 2} - \ldots - \frac{1}{1 + \lambda}$$

$$\Rightarrow a_n'(0) = \frac{-\Phi(n-1)}{(n-1)!} \qquad (n \geqslant 2)$$

Hence, by Theorem 7.9, a second solution to (7.32) is

$$w = w_2(z) = w_\lambda(z, 0) = \operatorname{Log} z \sum_{n=0}^{\infty} a_n(0) z^n + \sum_{n=0}^{\infty} a_n'(0) z^n$$

$$\Rightarrow \quad w = w_2(z) = w_1(z) \operatorname{Log} z + 1 - \sum_{n=2}^{\infty} \frac{\Phi(n-1) z^n}{(n-1)!}$$

It follows, by 7.6 or the ratio test, that this solution is defined for all z such that $z \neq 0$ and $z \notin \mathbb{R}^-$.

Exercise 7.2.1 Find two linearly independent solutions of the equation

$$zw'' + kw' - w = 0$$

using the method of Frobenius when (i) the constant k is not an integer, (ii) $k = 1$. State the region on which the general solution is defined. Find the general solution in closed form when $k = 1/2$.

Exercise 7.2.2 Find a Frobenius series solution of the equation

$$4z^2 w'' + (1 - 2z)w = 0$$

Find a second, linearly independent solution, using the method of Frobenius.

Exercise 7.2.3 Find a Frobenius series solution of the equation

$$zw'' + (1 - z)w' - w = 0$$

and express the solution in closed form. Find a second, linearly independent solution using the method of Frobenius.

Exercise 7.2.4 Find a Frobenius series solution of the equation

$$z(z - 1)w'' + 3zw' + w = 0$$

and express this solution in closed form. Find a second, linearly independent solution, using the method of Frobenius. Express this solution in closed form.

Exercise 7.2.5 Find a Frobenius series solution of the equation

$$zw'' - 3w' + zw = 0$$

Find a second, linearly independent solution.

Equations with Assigned Singularities

It is possible to study the structure of reduced linear second-order differential equations in the extended complex plane, having a number of assigned regular singular points. Consider once again the reduced l.o.d.e.

$$w''(z) + p(z)w'(z) + q(z)w(z) = 0 \tag{7.3}$$

Let $z = 1/\zeta$. Then (7.3) becomes

$$w''(\zeta) + \left(\frac{2}{\zeta} - \frac{p(1/\zeta)}{\zeta^2}\right)w'(\zeta) + \frac{q(1/\zeta)}{\zeta^4}w(\zeta) = 0 \tag{7.34}$$

and $\zeta = 0$ is an ordinary point of (7.34) if $2/\zeta - p(1/\zeta)/\zeta^2$ and $q(1/\zeta)/\zeta^4$ are analytic at $\zeta = 0$; $\zeta = 0$ is a regular singular point of (7.34) if $2 - p(1/\zeta)/\zeta$ and $q(1/\zeta)/\zeta^2$ are analytic at 0. Hence we have the following result.

Lemma 7.10. Regular Singularities at ∞

In the extended complex plane $\tilde{\mathbb{C}}$, equation (7.3) has an ordinary point at ∞ if $2z - z^2p(z)$ and $z^4q(z)$ are analytic at ∞. It has a regular singular point at ∞ if $zp(z)$ and $z^2q(z)$ are analytic at ∞. □

It is easily seen, using the definition and 7.10, that the equation with just one singular point in $\tilde{\mathbb{C}}$, a regular singular point at 0, is

$$zw'' + 2w' = 0$$

with general solution $w = a/z + b$.

We present two further examples.

Example 7.6 An Equation with Two Regular Singular Points

Suppose we wish to find the equation with two singular points in $\tilde{\mathbb{C}}$, which are regular singular points at 0 and ∞. Since 0 is a regular singular point,

$$p(z) = \frac{p_{-1}}{z} + p_0 + p_1 z + \dots \quad \text{and} \quad q(z) = \frac{q_{-2}}{z^2} + \frac{q_{-1}}{z} + q_0 + \dots$$

$$\Rightarrow \quad zp(z) = p_{-1} + z\sum_{n=0}^{\infty} p_n z^n \quad \text{and} \quad z^2q(z) = q_{-2} + q_{-1}z + z^2\sum_{n=0}^{\infty} q_n z^n$$

for some constants p_n and q_n, in some neighbourhood of 0. Then since ∞ is a regular singular point, it follows by 7.10 that $p_n = 0$, $n \geqslant 0$ and $q_n = 0$, $n \geqslant -1$. Hence the required equation is

$$z^2w''(z) + p_{-1}zw'(z) + q_{-2}w(z) = 0$$

an Euler type equation.

Example 7.7 An Equation with Three Regular Singular Points, at 0, 1 and ∞

Note that any equation with three regular singular points, at a, b and c, can be transformed to this case by use of a bilinear transformation. Since 0 and 1 are regular singular points,

$$p(z) = \frac{A}{z} + \frac{B}{z-1} + P(z) \qquad q(z) = \frac{C}{z^2} + \frac{D}{z} + \frac{E}{(z-1)^2} + \frac{F}{(z-1)} + Q(z)$$

where P and Q are analytic at 0 and 1 and $A, \ldots, F$ are constants. Hence

$$zp(z) = A + \frac{B}{1 - 1/z} + zP(z)$$

$$z^2 q(z) = C + Dz + \frac{E}{(1 - 1/z)^2} + \frac{Fz}{1 - 1/z} + z^2 Q(z)$$

Since ∞ is a regular singular point, $P(z) \equiv 0$, $Q(z) \equiv 0$ and $D + F = 0$, by 7.10. Hence the form of the required equation is

$$w''(z) + \left(\frac{A}{z} + \frac{B}{z-1}\right) w'(z) + \left(\frac{C}{z^2} - \frac{D}{z(z-1)} + \frac{E}{(z-1)^2}\right) w(z) = 0$$

We can reduce this form further by letting $w = z^\alpha (z-1)^\beta W$, for some α and β. Then

$$W''(z) + \left(\frac{A + 2\alpha}{z} + \frac{B + 2\beta}{z - 1}\right) W'(z) + \left(\frac{\alpha(\alpha - 1) + \alpha A + C}{z^2}\right.$$
$$\left. + \frac{2\alpha\beta + \alpha B + \beta A - D}{z(z-1)} + \frac{\beta(\beta - 1) + \beta B + E}{(z-1)^2}\right) W(z) = 0$$

Hence choosing α and β to satisfy

$$\alpha^2 + (A - 1)\alpha + C = 0 \quad \text{and} \quad \beta^2 + (B - 1)\beta + E = 0$$

and relabelling, the form of the desired equation becomes

$$z(1 - z)W''(z) + (c - Gz)W'(z) - HW(z) = 0 \tag{7.35}$$

where c, G and H are constants. Letting $\zeta = 1/z$ in (7.35) gives

$$\zeta^2(\zeta - 1)W''(\zeta) + ((2 - c)\zeta^2 + (G - 2)\zeta)W'(\zeta) - HW(\zeta) = 0$$

Seeking a series solution of this equation of the form $W(\zeta) = \sum_{n=0}^{\infty} a_n \zeta^{n+\lambda}$ and comparing coefficients of z^λ gives the indicial equation

$$\lambda^2 + (1 - G)\lambda + H = 0$$

Let the roots of this equation be a and b. Then $a + b = G - 1$ and $ab = H$. Hence (7.35) reduces to

$$z(1 - z)W''(z) + (c - (a + b + 1)z)W'(z) - abW(z) = 0 \tag{7.36}$$

where a, b and c are constants. This is the **hypergeometric equation**.

Special Functions

Many special functions are defined as series solutions of linear differential equations. A large number of these equations arise by solving some linear partial differential equation by separation of variables. We give a number of commonly occurring examples in this section.

For example, **Legendre's equation**

$$(1 - z^2)w'' - 2zw' + n(n+1)w = 0 \qquad (n \in \mathbb{N}) \tag{7.37}$$

occurs in the separation of variables of Laplace's equation in spherical polar coordinates. By 7.4, it has two linearly independent Taylor series solutions, which converge for at least $|z| < 1$. Recall that one solution is a polynomial of degree n (see Exercise 7.1.4). The other series solution converges for $|z| < 1$ (see Exercise 7.1.3). It is easily shown that the polynomial of degree n, denoted and defined by

$$P_n(z) = \frac{1}{2^n n!} D^n(z^2 - 1)^n \tag{7.38}$$

where D^n denotes the nth derivative with respect to z, satisfies (7.37) (see Exercises 7.3). This is **Rodrigues' definition** of the **Legendre polynomial of degree *n***. By equation (7.23), a second, linearly independent solution of (7.37) is the **Legendre function of the second kind**, denoted and defined by

$$Q_n(z) = P_n(z) \int_z^\infty \frac{d\zeta}{(\zeta^2 - 1)P_n^2(\zeta)} \tag{7.39}$$

where the path of integration does not cross the branch cut $\{x \in \mathbb{R} : -1 \leqslant x \leqslant 1\}$.

Another important equation is **Bessel's equation**

$$z^2 w'' + zw' + (z^2 - v^2)w = 0 \tag{7.40}$$

which arises, for instance, in the separation of variables of Laplace's equation in cylindrical polar coordinates. Using the standard method, it is easily seen that (7.40) has a series solution

$$J_v(z) = \sum_{n=0}^{\infty} \frac{(-1)^n (z/2)^{2n+v}}{n!\Gamma(n+v+1)} \tag{7.41}$$

This defines the **Bessel function of the first kind of order** v, where Γ is the gamma function defined in Chapter 6. By Theorem 7.5 or the ratio test, this power series expansion of $J_v(z)$ converges for all z. If $v \notin \mathbb{Z}$, a second, linearly independent solution of (7.40) is $J_{-v}(z)$. However, if $v = m \in \mathbb{Z}$, $1/\Gamma(-m+n+1) = 0$ for $n = 0, \ldots, m-1$, so that

$$J_{-m}(z) = \sum_{n=m}^{\infty} \frac{(-1)^n (z/2)^{2n-m}}{n!(n-m)!} = \sum_{k=0}^{\infty} \frac{(-1)^{k+m}(z/2)^{2k+m}}{(k+m)!k!} = (-1)^m J_m(z)$$

and the two solutions are linearly dependent. In this case, the roots of the indicial equation are equal if $m = 0$ and differ by an integer if $m \neq 0$. The method of Frobenius gives a second linearly independent solution as

$$Y_m(z) = \frac{2}{\pi}(\operatorname{Log}(z/2) + \gamma)J_m(z) - \frac{1}{\pi}\sum_{n=0}^{m-1} \frac{(m-n-1)!(z/2)^{2n-m}}{n!}$$
$$- \frac{1}{\pi}\sum_{n=0}^{\infty} \frac{(-1)^n[\Phi(n) + \Phi(n+m)](z/2)^{2n+m}}{n!(n+m)!}$$

where $\gamma = -\Gamma'(1)$ is **Euler's constant**. This is the **Bessel function of the second kind of order *m***. Note that 0 is a singular point of $Y_m(z)$.

Many special functions are related to solutions of the hypergeometric equation (7.36). One solution of (7.36) is the **hypergeometric function**, denoted and defined by

$$F(a, b; c; z) = \frac{\Gamma(c)}{\Gamma(a)\Gamma(b)} \sum_{n=0}^{\infty} \frac{\Gamma(a+n)\Gamma(b+n)z^n}{\Gamma(c+n)n!} \qquad (7.42)$$

This series converges for $|z| < 1$. If c is not an integer, then a second, linearly independent solution is

$$W = z^{1-c}F(a-c+1, b-c+1; 2-c; z) \qquad (0 < |z| < 1)$$

Exercise 7.3.1 Show that ∞ is a regular singular point of Legendre's equation.

7.3.2 Show that ∞ is a regular singular point of the equation

$$z^3w'' + z^2w' + w = 0$$

and find a series solution valid in a neighbourhood of ∞.

Exercise 7.3.3 Find the reduced l.o.d.e. of second order with one singular point, a regular singular point at (i) 0, (ii) ∞. Find the general solution of the equation in each case.

Exercise 7.3.4 Show that any reduced l.o.d.e. of second order having regular singular points at -1, 1 and ∞ can be reduced to the form

$$(1-z^2)w'' + (c-(a+b+1)z)w' - abw = 0$$

Show that Legendre's equation is a special case.

Exercise 7.3.5 Show that any reduced l.o.d.e. of second order having regular singular points at 0, 1, α and ∞ can be reduced to the form

$$w'' + \left(\frac{b}{z} + \frac{c}{z-1} + \frac{d}{z-\alpha}\right)w' + \frac{r(z-s)w}{z(z-1)(z-\alpha)} = 0$$

where b, c, d, r and s are constants.

Exercise 7.3.6 Show that the Legendre polynomial, $P_n(z)$, defined by (7.38), satisfies Legendre's equation.
(*Hint*: Find a second order differential equation satisfied by $W = (z^2-1)^n$ then find the nth derivative of this equation using Leibniz's theorem for the nth derivative of a product.)

Exercise **7.3.7** Use Cauchy's integral formula for derivatives and Rodrigues' definition of $P_n(z)$ to show that

$$P_n(z) = \frac{1}{2^{n+1}\pi i}\int_{\mathscr{C}} \frac{(\zeta^2-1)^n}{(\zeta-z)^{n+1}}\,d\zeta \qquad \text{(Schläfli's integral form)}$$

where $\mathscr{C}$ is any simple closed contour enclosing $\zeta = z$. Let $\mathscr{C}$ be the circle $|\zeta - z| = |z^2-1|^{1/2}$ and deduce that

$$P_n(z) = \frac{1}{\pi}\int_0^{\pi} (z + (z^2-1)^{1/2}\cos\theta)^n\,d\theta \qquad \text{(Laplace's integral)}$$

Exercise **7.3.8** Use Rodrigues' definition of $P_n(z)$ to show that

$$P_n(z) = \sum_{k=0}^{N} \frac{(-1)^k(2n-2k)!z^{n-2k}}{2^n k!(n-k)!(n-2k)!} \qquad \begin{cases} N = n/2 & n \text{ even} \\ N = (n-1)/2 & n \text{ odd} \end{cases}$$

This is a polynomial solution of Legendre's equation produced by solving the equation in series.

Exercise **7.3.9** Find $P_0(z)$ and $P_1(z)$ using Rodrigues' definition. Deduce that

$$Q_0(z) = \frac{1}{2}\operatorname{Log}\left(\frac{z+1}{z-1}\right) \quad \text{and} \quad Q_1(z) = zQ_0(z) - 1$$

where $Q_n(z)$ is defined by (7.39).

Exercise **7.3.10** Solve Bessel's equation in series to show that two linearly independent solutions are $w = J_\nu(z)$ and $w = J_{-\nu}(z)$ if $\nu \notin \mathbb{Z}$, where $J_\nu(z)$ is defined by the series (7.41).

Exercise **7.3.11** Find a Frobenius series solution of Bessel's equation of order 1, i.e.

$$z^2w'' + zw' + (z^2-1)w = 0$$

Find a second, linearly independent solution, using the method of Frobenius.

Exercise **7.3.12** Use the series definition of $J_\nu(z)$ and results of Chapter 4 to show that

$$\frac{d}{dz}(z^\nu J_\nu(z)) = z^\nu J_{\nu-1}(z)$$

Exercise **7.3.13** Show that the equation

$$z^2w'' + (1-2a)zw' + (b^2c^2z^{2c} + (a^2 - \nu^2c^2)w = 0$$

where a, b, c and ν are constants with $\nu \notin \mathbb{Z}$, has two linearly independent solutions, $w = z^aJ_\nu(bz^c)$ and $w = z^aJ_{-\nu}(bz^c)$. Hence find the general solution of the equation $w'' + zw = 0$.

Exercise **7.3.14** Use the series definition of $J_n(z)$, $n \in \mathbb{Z}$, to show that

$$e^{z(t-1/t)/2} = \sum_{n=-\infty}^{\infty} J_n(z)t^n \qquad (t \in \mathbb{C},\ t \neq 0)$$

Hence use Laurent's theorem to show that

$$J_n(z) = \frac{1}{2\pi i}\int_{|\zeta|=1} e^{z(\zeta-1/\zeta)/2}\zeta^{-n-1}\,d\zeta$$

Deduce that $J_n(z) = (1/\pi)\int_0^{\pi} \cos(n\theta - z\sin\theta)\,d\theta$ (Bessel's integral).

Exercise **7.3.15** Find a Frobenius series solution of the hypergeometric equation hence show that one solution is $w = F(a,\ b;\ c;\ z)$. Show that if c is non-integer, then a second, linearly independent solution is $W = z^{1-c}F(a-c+1;\ b-c+1;\ 2-c;\ z)$.

Exercise **7.3.16** Use the series definition of the hypergeometric function to show that

(i) $F(1,\ 1;\ 1;\ z) = (1-z)^{-1}$, $|z| < 1$

(ii) $F(-m,\ 1;\ 1;\ -z) = (1+z)^m$, $m \in \mathbb{N}$

(iii) $zF(1/2,\ 1;\ 3/2;\ -z^2) = \tan^{-1} z$, $|z| < 1$

Exercise **7.3.17**

(a) Show that $w = \mathrm{Log}\,(1-z)$ satisfies the equation

$$(1-z)w'' - w' = 0$$

Hence show that $zF(1,\ 1;\ 2;\ z) = \mathrm{Log}\,(1-z)^{-1}$, $|z| < 1$.

(b) Let $\zeta = (1+z)/2$ in Legendre's equation to show that

$$P_n(z) = (-1)^n F\left(n+1,\ -n;\ 1;\ \frac{1+z}{2}\right)$$

(*Hint*: Use the fact that $P_n(-1) = (-1)^n$.)

Exercise **7.3.18** Let $z = \zeta/b$ in the hypergeometric equation and let $b \to \infty$ to show that one solution of the equation

$$zw'' + (c-z)w' - aw = 0$$

where a and c are constants, is the **confluent hypergeometric function**

$$w = F(a;\ c;\ z) = \frac{\Gamma(c)}{\Gamma(a)}\sum_{n=0}^{\infty}\frac{\Gamma(a+n)z^n}{\Gamma(c+n)n!} \qquad (|z| < 1)$$

Contour Integral Solutions of Differential Equations

A Frobenius series is not the only form of a solution of a reduced l.o.d.e. which may be assumed *a priori* when solving such an equation. Other assumed forms for the solution may give distinct information about the character of the solutions of the l.o.d.e. Of particular relevance in complex analysis is the idea of finding solutions of reduced l.o.d.e.'s in terms of contour integrals. This approach provides alternative definitions of special functions defined by such equations and avoids the problems of convergence encountered with infinite series.

Once again, we consider the reduced second-order l.o.d.e. and in particular, first of all, the **Laplace linear equation**

$$(a_2 z + b_2)w'' + (a_1 z + b_1)w' + (a_0 z + b_0)w = 0 \tag{7.43}$$

where a_k and b_k, $k = 0$, 1 and 2 are constants. Equation (7.43) has a particularly simple solution in terms of a contour integral. It includes the confluent hypergeometric equation as a special case and Bessel's equation can also be reduced to this form. We assume a contour integral solution of the form $w(z) = \int_{\mathscr{C}} e^{z\zeta} P(\zeta) d\zeta$ where the function P and the contour $\mathscr{C}$, which is independent of z, are to be determined. Then formally,

$$w'(z) = \int_{\mathscr{C}} \zeta e^{z\zeta} P(\zeta) d\zeta \qquad w''(z) = \int_{\mathscr{C}} \zeta^2 e^{z\zeta} P(\zeta) d\zeta$$

and (7.43) gives

$$\int_{\mathscr{C}} e^{z\zeta} P(\zeta)(zQ(\zeta) + R(\zeta)) d\zeta = 0 \tag{7.44}$$

where $Q(\zeta) = a_2\zeta^2 + a_1\zeta + a_0$ and $R(\zeta) = b_2\zeta^2 + b_1\zeta + b_0$.

Note

It follows from the definition of a definite integral and results from real calculus that differentiation under the integral sign is valid as long as P is analytic on $\mathscr{C}$. This is a special case of Theorem 10.14.

It is easily seen that (7.44) reduces to

$$\int_{\mathscr{C}} \frac{d}{d\zeta}\left(e^{z\zeta}\phi(\zeta)\right) d\zeta = 0 \quad \Rightarrow \quad \left[e^{z\zeta}\phi(\zeta)\right]_{\mathscr{C}} = 0 \tag{7.45}$$

where $\phi(\zeta) = P(\zeta)Q(\zeta)$ and $\phi'(\zeta) = P(\zeta)R(\zeta)$, so that

$$\frac{\phi'(\zeta)}{\phi(\zeta)} = \frac{R(\zeta)}{Q(\zeta)} \quad \Rightarrow \quad \operatorname{Log}\phi(\zeta) = \int \frac{R(\zeta)}{Q(\zeta)} d\zeta \quad \text{and} \quad P(\zeta) = \frac{\phi(\zeta)}{Q(\zeta)} \tag{7.46}$$

Hence $P(\zeta)$ is given by (7.46) and $\mathscr{C}$ can be chosen to be any (not necessarily closed) contour in the ζ-plane for which (7.45) is true. We give a list of possible choices of contour, but this list is by no means exhaustive.

Choices of Contour

(i) There may exist a finite non-closed contour $\mathscr{C}$ such that ϕ is non-singular at all points along $\mathscr{C}$ and is zero at its endpoints.

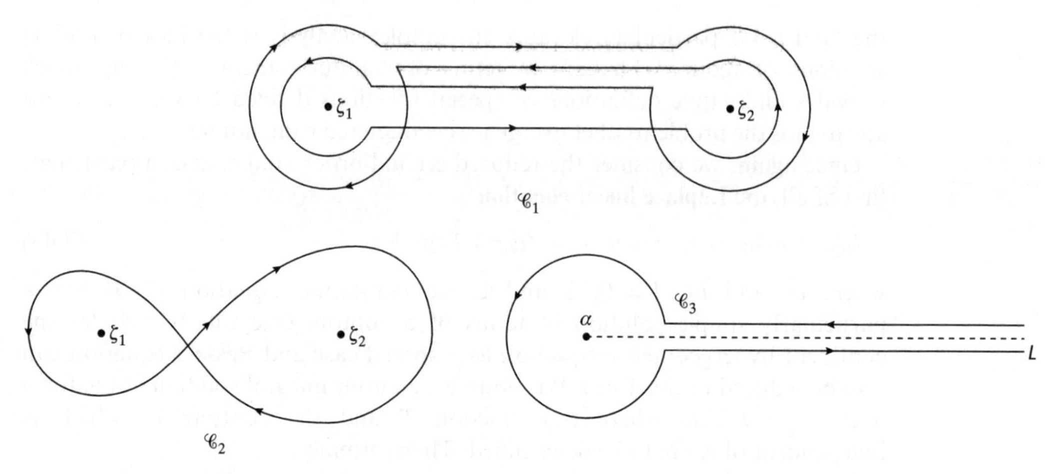

Figure 7.1

(ii) If there exists a real number a for which $\phi(a) = 0$ and ϕ is non-singular for all real numbers greater than a, then if $\operatorname{Re} z < 0$ with $\lim_{t\to\infty} e^{zt}\phi(t) = 0$, $t \in \mathbb{R}$, $\mathscr{C}$ can be taken to be the segment of the real axis, $x \geqslant a$. Etc.

(iii) Any closed contour $\mathscr{C}$ can be chosen provided that $e^{z\zeta_1}\phi(\zeta_1) = e^{z\zeta_2}\phi(\zeta_2)$ for any chosen initial point ζ_1 and endpoint ζ_2. Note, however, that $\mathscr{C}$ must enclose at least one singular point of ϕ, since otherwise $\int_{\mathscr{C}} e^{z\zeta}P(\zeta)\,d\zeta = 0$ by Cauchy's theorem and only the trivial solution, $w(z) \equiv 0$, of (7.43) is obtained. For example, if ϕ has two singular points, ζ_1 and ζ_2, two possible choices of contour, $\mathscr{C}_1$ and $\mathscr{C}_2$, are shown in Fig. 7.1.

(iv) If α is a singular point of ϕ and $\lim_{\zeta\to\infty} e^{z\zeta}\phi(\zeta) = 0$ along a line segment L joining α to the point at infinity then a possible choice of contour is $\mathscr{C}_3$, shown in Fig. 7.1.

Note

In general, two linearly independent contour integral solutions of (7.43) can only be obtained by placing some restriction on the values of z.

Example 7.8

Consider the equation

$$zw'' + (a + b + z)w' + aw = 0 \tag{7.47}$$

where a and b are real constants. Letting $w = \int_{\mathscr{C}} e^{z\zeta}P(\zeta)d\zeta$ gives

$$\int_{\mathscr{C}} e^{z\zeta}P(\zeta)[z\zeta(\zeta + 1) + ((a + b)\zeta + a)]d\zeta = \int_{\mathscr{C}} \frac{d}{d\zeta}(e^{z\zeta}\phi(\zeta))d\zeta = 0$$

where, by (7.46),

$$\frac{\phi'(\zeta)}{\phi(\zeta)} = \frac{(a+b)\zeta + a}{\zeta(\zeta+1)} = \frac{a}{\zeta} + \frac{b}{1+\zeta} \quad \Rightarrow \quad \operatorname{Log}\phi(\zeta) = a\operatorname{Log}\zeta + b\operatorname{Log}(1+\zeta)$$

$$\Rightarrow \quad \phi(\zeta) = \zeta^a(1+\zeta)^b \quad \text{and} \quad P(\zeta) = \zeta^{a-1}(1+\zeta)^{b-1}$$

without loss of generality. Hence a solution to (7.47) is

$$w = \int_{\mathscr{C}} e^{z\zeta}\zeta^{a-1}(1+\zeta)^{b-1}d\zeta \quad \text{where} \quad \left[e^{z\zeta}\zeta^a(1+\zeta)^b\right]_{\mathscr{C}} = 0$$

For example, if $a > 0$ and $b > 0$ then $\phi(0) = \phi(-1) = 0$ and ϕ is analytic on $(-1, 0)$. In this case, $\mathscr{C}$ can be chosen to be the closed interval $[-1, 0]$ of the real axis, so that with $t \in \mathbb{R}$, a solution to (7.47) is

$$w = \int_{-1}^{0} e^{zt}t^{a-1}(1+t)^{b-1}dt = (-1)^{a-1}\int_0^1 e^{-zt}t^{a-1}(1-t)^{b-1}dt$$

If $a > 0$ and $\operatorname{Re} z < 0$, then $\phi(0) = 0$ and ϕ is analytic on the positive real axis with $\lim_{t\to\infty} e^{zt}\phi(t) = 0$, $t \in \mathbb{R}^+$. In this case, a solution of (7.47) is $w = \int_0^\infty e^{zt}t^{a-1}(1+t)^{b-1}dt$.

Example 7.9

Consider **Laguerre's equation**

$$zw'' + (1-z)w' + nw = 0 \qquad (n \in \mathbb{N}) \tag{7.48}$$

Assuming a solution of the form $w(z) = \int_{\mathscr{C}} e^{z\zeta}P(\zeta)\,d\zeta$ gives

$$\int_{\mathscr{C}} e^{z\zeta}P(\zeta)[z\zeta(\zeta-1) + (\zeta+n)]d\zeta = \int_{\mathscr{C}} \frac{d}{d\zeta}\left(e^{z\zeta}\phi(\zeta)\right)d\zeta = 0$$

where, by (7.46),

$$\frac{\phi'(\zeta)}{\phi(\zeta)} = \frac{\zeta+n}{\zeta(\zeta-1)} = \frac{-n}{\zeta} + \frac{1+n}{\zeta-1}$$

$$\Rightarrow \quad \operatorname{Log}\phi(\zeta) = -n\operatorname{Log}\zeta + (1+n)\operatorname{Log}(\zeta-1)$$

$$\Rightarrow \quad \phi(\zeta) = \zeta^{-n}(\zeta-1)^{1+n} \quad \text{and} \quad P(\zeta) = \zeta^{-n-1}(\zeta-1)^n$$

without loss of generality. Then a solution to (7.48) is

$$w = K\int_{\mathscr{C}} e^{z\zeta}\zeta^{-n-1}(\zeta-1)^n d\zeta \tag{7.49}$$

where $\mathscr{C}$ is any contour such that $\left[e^{z\zeta}\zeta^{-n}(\zeta-1)^{1+n}\right]_{\mathscr{C}} = 0$ and where K is a constant. It is clear that a possible choice of $\mathscr{C}$ in this case is the circle $|\zeta| = 1$, which encloses 0.

The **Laguerre polynomial**, $L_n(z)$, is a solution of (7.48) of the form (7.49) where $\mathscr{C}$ is the circle $|\zeta| = 1$, with the property that $L_n(0) = n!$. Note that by the residue theorem

$$\int_{\mathscr{C}} \frac{(\zeta-1)^n}{\zeta^{n+1}}\, d\zeta = 2\pi i \operatorname*{Res}_{\zeta=0} \left(\zeta^{-n-1}(\zeta-1)^n\right) = 2\pi i$$

$$\Rightarrow \quad L_n(0) = n! = 2\pi i K \quad \Rightarrow \quad L_n(z) = \frac{n!}{2\pi i}\int_{\mathscr{C}} \frac{e^{z\zeta}(\zeta-1)^n}{\zeta^{n+1}}\, d\zeta \tag{7.50}$$

This result can be used to prove various results concerning Laguerre polynomials. For example, let $G(z,u) = \sum_{n=0}^{\infty} u^n L_n(z)/n!$, where $u \in \mathbb{C}$. Then by (7.50) and Theorem 4.16,

$$G(z,u) = \sum_{n=0}^{\infty} \frac{u^n}{2\pi i}\int_{\mathscr{C}} \frac{e^{z\zeta}(\zeta-1)^n}{\zeta^{n+1}}\, d\zeta = \frac{1}{2\pi i}\int_{\mathscr{C}} \frac{e^{z\zeta}}{\zeta}\sum_{n=0}^{\infty}\left(\frac{u(\zeta-1)}{\zeta}\right)^n d\zeta$$

within the domain of convergence. Then for $|u(\zeta-1)| < |\zeta|$, summing the geometric series gives

$$G(z,u) = \frac{1}{2\pi i}\int_{\mathscr{C}} \frac{e^{z\zeta}}{\zeta}\cdot\frac{1}{1-u(\zeta-1)/\zeta}\, d\zeta = \frac{1}{2\pi i(1-u)}\int_{\mathscr{C}} \frac{e^{z\zeta}\, d\zeta}{\zeta + u/(1-u)}$$

The integrand has a simple pole at $\zeta = -u/(1-u)$ so that

$$G(z,u) = \frac{1}{1-u}\operatorname*{Res}_{\zeta=-u/(1-u)}\left(\frac{e^{z\zeta}}{\zeta + u/(1-u)}\right) = \frac{e^{-uz/(1-u)}}{1-u}$$

$$\Rightarrow \quad \sum_{n=0}^{\infty} \frac{u^n L_n(z)}{n!} = \frac{e^{-uz/(1-u)}}{1-u}$$

This is the **generating function** for $L_n(z)$.

Now let $\tau = z(1-\zeta)$ in (7.50). Then

$$L_n(z) = \frac{n!}{2\pi i}\int_{\mathscr{C}} \frac{e^z e^{-\tau}(-1)^n \tau^n}{z^n(z-\tau)^{n+1}(-z)z^{-n-1}}\, d\tau = e^z \frac{n!}{2\pi i}\int_{\mathscr{C}} \frac{e^{-\tau}\tau^n}{(\tau-z)^{n+1}}\, d\tau$$

$$\Rightarrow \quad L_n(z) = e^z \frac{d^n}{dz^n}(e^{-z}z^n) \qquad (n \in \mathbb{N})$$

using Cauchy's integral formula for derivatives. This is **Rodrigues' formula** for $L_n(z)$, which shows clearly that $L_n(z)$ is a polynomial of degree n.

It can be shown that in the case of the equation

$$r(z)w'' + p(z)w' + q(z)w = 0 \tag{7.51}$$

where $r(z)$ is a polynomial of degree m and $p(z)$ and $q(z)$ are polynomials of degree not exceeding m, substituting $w = \int_{\mathscr{C}} e^{z\zeta}P(\zeta)d\zeta$ into (7.51) gives a linear differential equation of degree m for $P(\zeta)$. Hence the method is not of much practical use if $m > 1$. However, there are equations of special form for which contour integral solutions of a particular form are readily found.

We shall consider the reduced l.o.d.e.

$$r(z)w'' + s(z)w' + kw = 0 \tag{7.52}$$

where $r(z)$ is a quadratic in z, $s(z)$ is linear and k is a constant. This equation includes the hypergeometric equation and Legendre's equation as special cases. In this case, we assume a solution of the form $w(z) = \int_{\mathscr{C}} (\zeta - z)^{\alpha+1} P(\zeta)d\zeta$ for some choice of contour $\mathscr{C}$, constant α and function P. Then

$$w'(z) = \int_{\mathscr{C}} -(\alpha+1)(\zeta - z)^{\alpha} P(\zeta)d\zeta \quad \text{and} \quad w''(z) = \int_{\mathscr{C}} \alpha(\alpha+1)(\zeta - z)^{\alpha-1} P(\zeta)d\zeta$$

and (7.52) gives

$$\int_{\mathscr{C}} P(\zeta)\left[\alpha(\alpha+1)(\zeta - z)^{\alpha-1} r(z) - (\alpha+1)(\zeta - z)^{\alpha} s(z) + k(\zeta - z)^{\alpha+1}\right] d\zeta = 0$$

Since $r(z)$ is a polynomial of degree 2, it follows by Taylor's theorem that $r(\zeta) = r(z) + (\zeta - z)r'(z) + (\zeta - z)^2 r''(z)/2$. Then

$$\int_{\mathscr{C}} P(\zeta)\left[\alpha(\alpha+1)(\zeta - z)^{\alpha-1} r(\zeta) + (\alpha+1)(\zeta - z)^{\alpha} f(z) + c(\zeta - z)^{\alpha+1}\right] d\zeta = 0$$

where $-f(z) = \alpha r'(z) + s(z)$ and $c = k - \alpha(\alpha+1)r''(z)/2$. Since $f(z)$ is a polynomial of degree 1, $f(\zeta) = f(z) + (\zeta - z)f'(z)$. Then

$$\int_{\mathscr{C}} P(\zeta)\left[\alpha(\alpha+1)(\zeta - z)^{\alpha-1} r(\zeta) + (\alpha+1)(\zeta - z)^{\alpha} f(\zeta)\right] d\zeta = 0 \tag{7.53}$$

by choosing α such that $(\alpha+1)f'(z) = c$. Hence altogether (7.52) reduces to (7.53) where

$$-f(z) = \alpha r'(z) + s(z) \quad \text{and} \quad k = (\alpha+1)(\alpha r''(z)/2 + f'(z)) \tag{7.54}$$

These equations determine the constant α. Then (7.53) reduces to

$$\int_{\mathscr{C}} \frac{d}{d\zeta}((\zeta - z)^{\alpha}\phi(\alpha))d\zeta = \int_{\mathscr{C}} \alpha(\zeta - z)^{\alpha-1}\phi(\zeta) + (\zeta - z)^{\alpha}\phi'(\zeta)d\zeta = 0$$

where $\phi(\zeta) = (\alpha+1)r(\zeta)P(\zeta)$ and $\phi'(\zeta) = (\alpha+1)f(\zeta)P(\zeta)$. Hence

$$\frac{\phi'(\zeta)}{\phi(\zeta)} = \frac{f(\zeta)}{r(\zeta)} \quad \text{and} \quad P(\zeta) = \frac{\phi(\zeta)}{(\alpha+1)r(\zeta)} \qquad (\alpha \neq -1) \tag{7.55}$$

Then $w(z) = \int_{\mathscr{C}} (\zeta - z)^{\alpha+1} P(\zeta)d\zeta$ is a solution of the given equation provided that a contour $\mathscr{C}$ is chosen so that $[(\zeta - z)^{\alpha}\phi(\zeta)]_{\mathscr{C}} = 0$.

Example 7.10

Consider the hypergeometric equation

$$z(1-z)w'' + (c-(a+b+1)z)w' - abw = 0 \tag{7.36}$$

In this case, $r(z) = z(1-z)$, $s(z) = c-(a+b+1)z$ and $k = -ab$. Supposing that (7.36) has a solution of the form $w(z) = \int_{\mathscr{C}} (\zeta - z)^{\alpha+1} P(\zeta) d\zeta$, equations (7.54) give

$$f(z) = (2\alpha + a + b + 1)z - (\alpha + c) \quad \Rightarrow \quad -ab = (\alpha+1)(\alpha+a+b+1)$$

$$\text{Hence} \quad (\alpha + a + 1)(\alpha + b + 1) = 0 \quad \Rightarrow \quad \alpha = -a-1 \quad \text{or} \quad -b-1$$

Taking $\alpha = -a-1$ gives

$$f(z) = (a-c+1) - (a-b+1)z$$

$$\Rightarrow \quad \frac{\phi'(\zeta)}{\phi(\zeta)} = \frac{f(\zeta)}{r(\zeta)} = \frac{(a-c+1)-(a-b+1)\zeta}{\zeta(1-\zeta)} = \frac{a-c+1}{\zeta} + \frac{b-c}{1-\zeta}$$

$$\Rightarrow \quad \phi(\zeta) = \zeta^{a-c+1}(1-\zeta)^{c-b} \quad \text{and} \quad P(\zeta) = \zeta^{a-c}(1-\zeta)^{c-b-1}$$

without loss of generality, using (7.55). Hence solutions of (7.36) are given by $w = K\int_{\mathscr{C}} (\zeta - z)^{-a} \zeta^{a-c} (1-\zeta)^{c-b-1} d\zeta$ where $\mathscr{C}$ is any contour such that $\left[(\zeta - z)^{-a-1} \zeta^{a-c+1} (1-\zeta)^{c-b}\right]_{\mathscr{C}} = 0$.

In particular, suppose that $c > b > 0$. Then for $t \in \mathbb{R}$, $z \notin \mathbb{R}$, $(t-z)^{-a-1}\phi(t) = 0$ when $t = 1$, is non-singular for $t > 1$ and

$$(t-z)^{-a-1}\phi(t) = \frac{(1/t - 1)^{c-b}}{(1-z/t)^{a+1} t^b} \to 0 \quad \text{as } t \to \infty$$

so that a particular choice of $\mathscr{C}$ is $[1, \infty)$. Then

$$w = K\int_1^\infty t^{a-c}(1-t)^{c-b-1}(t-z)^{-a} dt$$

$$\Rightarrow \quad w = K' \int_0^1 u^{b-1}(1-u)^{c-b-1}(1-uz)^{-a} du$$

letting $t = 1/u$. The choice $K' = \Gamma(c)/\Gamma(b)\Gamma(c-b)$ gives **Euler's integral** representation of the hypergeometric function $F(a, b; c; z)$.

Exercise

7.4.1 Find all solutions to the equation

$$zw'' + cw' - w = 0 \qquad (c \in \mathbb{R})$$

in the form $w = \int_{\mathscr{C}} e^{z\zeta} P(\zeta) d\zeta$.

Exercise

7.4.2 Find all solutions to the equation

$$zw'' + (m - n - z)w' - mw = 0 \qquad (m,\, n \in \mathbb{N})$$

of the form $w = \int_{\mathscr{C}} e^{z\zeta} P(\zeta)d\zeta$. If $\operatorname{Re} z > 0$, show that one particular solution is $w = \int_0^\infty e^{-zt} t^{m-1}(1+t)^{-n-1}dt,\ \ t \in \mathbb{R}$. Show that a second, linearly independent solution in this case is $w = \int_{\mathscr{C}_3} e^{-z\zeta}\zeta^{m-1}(1+\zeta)^{-n-1}d\zeta$, where $\mathscr{C}_3$ is the contour shown in Fig. 7.1 with $\alpha = -1$.

Exercise

7.4.3 Find the solutions of the **confluent hypergeometric equation**

$$zw'' + (c - z)w' - aw = 0 \qquad (a,\, c \in \mathbb{R})$$

in the form $w = \int_{\mathscr{C}} e^{z\zeta} P(\zeta)d\zeta$. Show that in the case $c > a > 0$, if K is any constant, particular solutions are given by

$$w = K\int_0^1 e^{zt} t^{a-1}(1-t)^{c-a-1}dt \qquad (t \in \mathbb{R})$$

The **confluent hypergeometric function**, $F(a;\, c;\, z)$, is a solution of the given equation of this form satisfying $F(a;\, c;\, 0) = 1$. Use the properties of beta and gamma functions given in Chapter 6 to show that

$$F(a;\, c;\, z) = \frac{\Gamma(c)}{\Gamma(a)\Gamma(c-a)}\int_0^1 e^{zt} t^{a-1}(1-t)^{c-a-1}dt \qquad (c > a > 0)$$

Hence show that, as in Exercise 7.3.18,

$$F(a;\, c;\, z) = \frac{\Gamma(c)}{\Gamma(a)}\sum_{n=0}^{\infty}\frac{\Gamma(a+n)z^n}{\Gamma(c+n)n!}$$

Exercise

7.4.4 Find the solutions of **Hermite's equation**

$$w'' - 2zw' + 2nw = 0 \qquad (n \in \mathbb{N})$$

in the form $w = K\int_{\mathscr{C}} e^{z\zeta}P(\zeta)d\zeta$, where K is any constant.

The function $H_n(z)$ is the solution of Hermite's equation defined by

$$H_n(z) = \frac{n!}{2\pi i}\int_{\mathscr{C}}\frac{e^{2z\zeta-\zeta^2}}{\zeta^{n+1}}\,d\zeta$$

where $\mathscr{C}$ is any simple closed contour enclosing the origin. Use this definition to show that

(i) $\displaystyle\sum_{n=0}^{\infty}\frac{u^n H_n(z)}{n!} = e^{2zu-u^2},\ u \in \mathbb{C}$

(ii) $H_n(z) = (-1)^n e^{z^2}\dfrac{d^n}{dz^n}\left(e^{-z^2}\right)$ (Rodrigues' definition)

Note that it follows from (ii) that $H_n(z)$ is a polynomial of degree n, called the **Hermite polynomial of degree *n***. Use Rodrigues' definition to find $H_n(z)$ for $n = 1$, 2 and 3.

Exercise **7.4.5** Show that $w = K\int_{\mathscr{C}} e^{iz\zeta}(1-\zeta^2)^{\nu-1/2}d\zeta$ is a solution to

$$zw'' + (2\nu+1)w' + zw = 0 \qquad (\nu \in \mathbb{R})$$

provided that $\left[e^{iz\zeta}(1-\zeta^2)^{\nu+1/2}\right]_{\mathscr{C}} = 0$. Deduce that if $\nu > -1/2$ a solution is

$$w = 2K\int_0^1 (\cos zt)(1-t^2)^{\nu-1/2}dt \qquad (t \in \mathbb{R})$$

Hence use the Maclaurin series expansion of $\cos zt$ to show that

$$J_\nu(z) = \frac{2(z/2)^\nu}{\sqrt{\pi}\,\Gamma(\nu+1/2)}\int_0^1 (\cos zt)(1-t^2)^{\nu-1/2}dt \qquad (\nu > -1/2)$$

Deduce that

$$J_\nu(z) = \frac{(z/2)^\nu}{\sqrt{\pi}\,\Gamma(\nu+1/2)}\int_0^\pi \cos(z\sin\theta)\cos^{2\nu}\theta\, d\theta \qquad (\nu > -1/2)$$

This is known as **Poisson's integral.**

Exercise **7.4.6** Find all solutions of the equation

$$zw'' + (2\nu+1)w' + zw = 0 \qquad (\nu \in \mathbb{R})$$

of the form $w = K\int_{\mathscr{C}} e^{-z^2/4\zeta}P(\zeta)d\zeta$. If $\nu = m \in \mathbb{N}$ show that particular solutions are

$$w = K\int_{\mathscr{C}} \frac{e^{\zeta - z^2/4\zeta}}{\zeta^{m+1}}\,d\zeta$$

where $\mathscr{C}$ is any simple closed contour enclosing the origin. Hence show that

$$J_m(z) = \frac{(z/2)^m}{2\pi i}\int_{\mathscr{C}} \frac{e^{\zeta - z^2/4\zeta}}{\zeta^{m+1}}\,d\zeta \qquad \text{(Schläfli's integral)}$$

Exercise ***7.4.7** Find all solutions of Legendre's equation

$$(1-z^2)w'' - 2zw' + n(n+1)w = 0 \qquad (n \in \mathbb{N})$$

in the form $w = \int_{\mathscr{C}} (\zeta - z)^{\alpha+1}P(\zeta)d\zeta$.

Show that one particular solution is

$$w = \frac{1}{2^{n+1}\pi i}\int_{\mathscr{C}} \frac{(\zeta^2-1)^n}{(\zeta-z)^{n+1}}\,d\zeta$$

where $\mathscr{C}$ is any simple closed contour enclosing $\zeta = z$. This is **Schläfli's integral form** of $P_n(z)$.

Show that a second, linearly independent solution is given by

$$w = Q_n(z) = \frac{1}{2^{n+1}}\int_{-1}^1 \frac{(1-t^2)^n}{(z-t)^{n+1}}\,dt \qquad (t \in \mathbb{R},\ z \notin [-1, 1])$$

Use integration by parts to show that

$$Q_n(z) = \frac{1}{2}\int_{-1}^1 \frac{P_n(t)}{z-t}\,dt$$

8 Fourier and Laplace Transforms

Integral Transforms

An integral transform is a particular type of integral operator.

Definition

Let $A = \mathbb{C}$ or $\mathbb{R}$ and $x,\ a,\ b \in \mathbb{R},\ s \in A,\ f : \mathbb{R} \to \mathbb{C}$ and $K : A \times \mathbb{R} \to \mathbb{C}$. An **integral transform**, $\mathscr{T}$, of f, takes the general form

$$\mathscr{T}(f(x)) = g(s) = \int_a^b K(s,x) f(x) dx \tag{8.1}$$

where $K(s, x)$ is the **kernel** of the transform.

Note that, in general, a given transform of a particular function need not exist. Note also that the open interval (a, b) in (8.1) may be finite or infinite.

Integral transforms provide a powerful method for solving certain linear differential and integral equations, and can be used for evaluating certain definite integrals. This chapter deals exclusively with Laplace and Fourier transforms. We concentrate on those aspects of the theory which are directly related to complex analysis, so this chapter is certainly not intended to be a comprehensive account of such integral transforms. In addition, it is shown how certain definite integrals, which can be evaluated using residue theory, can be evaluated more simply using integral transforms. Finally, it is indicated how integral transforms can be used to solve certain linear ordinary differential equations (o.d.e.'s) and linear integral equations. Applications to linear partial differential equations (p.d.e.'s), are given in the next chapter.

Fourier Transforms and Their Applications

The Fourier transform and its variants are integral transforms closely related to Fourier series. It is a special case of (8.1), with $(a, b) = \mathbb{R}$ and $K(s, x) = e^{-isx}/\sqrt{2\pi}$. In this section, we show how these transforms may be used to obtain certain integral results, otherwise obtainable by applying residue theory, and how they can be used to solve certain integral equations.

Definition

Let $x, s \in \mathbb{R}$ and $f: \mathbb{R} \to \mathbb{C}$. Then the **Fourier transform** of f is denoted and defined by

$$\mathscr{F}(f(x)) = \hat{f}(s) = \frac{1}{\sqrt{2\pi}} \int_{-\infty}^{\infty} e^{-isx} f(x)dx$$

Notes

(i) Unfortunately, this definition is not universally accepted. Some books have e^{isx} as the kernel and some omit the factor $1/\sqrt{2\pi}$.

(ii) The form of this integral transform is a special case of the contour integral solution assumed when solving Laplace's linear o.d.e. in Chapter 7.

(iii) The integral in the definition is well defined, from Chapter 3, but note that the Fourier transform of a large class of functions will not exist. However, $\hat{f}(s)$ will exist, for instance, if $\int_{-\infty}^{\infty} |f(x)|dx$ exists.

We shall not dwell on the elementary properties of the Fourier transform, but note the following results, which follow directly from the definition. The proofs of these and other elementary results are included in Exercises 8.1.

Lemma 8.1. Elementary Properties of the Fourier Transform

Provided all the given transforms exist,

(i) $\mathscr{F}(af(x) + bg(x)) = a\mathscr{F}(f(x)) + b\mathscr{F}(g(x))$ for $a,\ b \in \mathbb{C}$; that is, the Fourier transform is a linear operator

(ii) $\mathscr{F}(e^{-ixa}f(x)) = \hat{f}(s+a)$ for $a \in \mathbb{R}$

(iii) $\mathscr{F}(f'(x)) = is\hat{f}(s)$

□

Example 8.1

(i) Let $f(x) = x^2$ for $|x| \leqslant 1$ and $f(x) = 0$ for $|x| > 1$. Then integration by parts gives

$$\mathscr{F}(f(x)) = \frac{1}{\sqrt{2\pi}} \int_{-1}^{1} x^2 e^{-isx} dx = \frac{1}{\sqrt{2\pi}} \left(\frac{2\sin s}{s} - \frac{2i}{s} \int_{-1}^{1} xe^{-isx} dx \right)$$

$$\Rightarrow \quad \mathscr{F}(f(x)) = \sqrt{\frac{2}{\pi}} \left(\frac{\sin s}{s} + \frac{2\cos s}{s^2} - \frac{2\sin s}{s^3} \right)$$

(ii) Clearly, residue theory is often useful when finding Fourier transforms. For example, using the residue theorem gives

$$\mathscr{F}\left(\frac{1}{1+x^2}\right) = \frac{1}{\sqrt{2\pi}}\int_{-\infty}^{\infty}\frac{e^{-isx}}{1+x^2}dx = \frac{2\pi i}{\sqrt{2\pi}}\operatorname*{Res}_{z=i}\left(\frac{e^{-isz}}{1+z^2}\right) = \sqrt{\frac{\pi}{2}}e^{s}$$

$$(s \leqslant 0)$$

using the standard technique, given in Chapter 5. Now let $x = -y$ in the above integral so that

$$\mathscr{F}\left(\frac{1}{1+x^2}\right) = \frac{1}{\sqrt{2\pi}}\int_{-\infty}^{\infty}\frac{e^{isy}}{1+y^2}dy = \sqrt{\frac{\pi}{2}}e^{-s} \qquad (s > 0)$$

by the above. Hence altogether

$$\mathscr{F}\left(\frac{1}{1+x^2}\right) = \sqrt{\frac{\pi}{2}}e^{-|s|}$$

(iii) Suppose we wish to find $\mathscr{F}(e^{-x^2/2})$ using residue theory. Consider first of all $\int_{\mathscr{C}} e^{-z^2/2}dz$ where $\mathscr{C}$ is the rectangular contour, shown in Fig. 8.1 for the case $s > 0$. It follows by Cauchy's theorem that

$$\int_{\mathscr{C}} e^{-z^2/2}dz = \int_{-R}^{R} e^{-x^2/2}dx + \int_0^s e^{-(R+iy)^2/2}i\,dy$$

$$-\int_{-R}^{R} e^{-(x+is)^2/2}dx - \int_0^s e^{-(-R+iy)^2/2}i\,dy = 0 \qquad (8.2)$$

Also $\left|\int_0^s e^{-(R+iy)^2/2}i\,dy\right| \leqslant e^{-R^2/2}\int_0^{|s|} e^{y^2/2}dy \to 0 \quad \text{as } R \to \infty$

and similarly for the fourth integral in (8.2). Hence, letting $R \to \infty$ in (8.2) gives

$$\int_{-\infty}^{\infty} e^{-x^2/2}dx = \sqrt{2\pi} = e^{s^2/2}\int_{-\infty}^{\infty} e^{-x^2/2-isx}dx$$

Hence $\mathscr{F}(e^{-x^2/2}) = e^{-s^2/2}$ and so this function is its own Fourier transform.

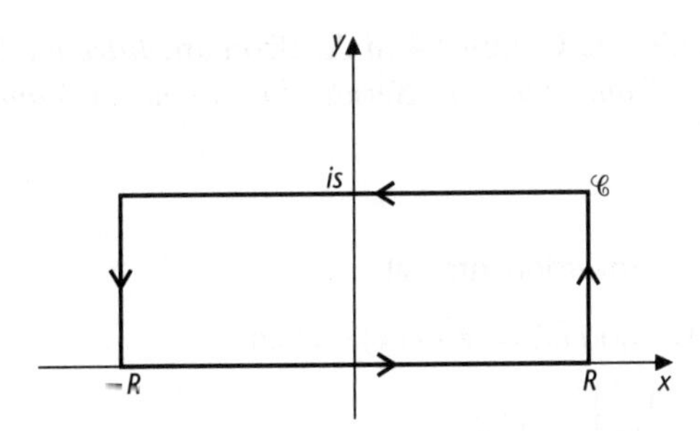

Figure 8.1

This result can also be derived without using residue theory. Let $I = \mathscr{F}(e^{-x^2/2})$. It can be shown that differentiation under the integral sign with respect to s is valid, so that

$$I'(s) = \frac{-1}{\sqrt{2\pi}} \int_{-\infty}^{\infty} ixe^{-isx - x^2/2} dx = \frac{i}{\sqrt{2\pi}} \left[e^{-isx - x^2/2} \right]_{-\infty}^{\infty} - sI = -sI$$

integrating by parts and integrating $xe^{-x^2/2}$ first. Also, when $s = 0$, $I = 1$ so that $I = e^{-s^2/2}$.

Definition

Let $\mathscr{F}(f(x)) = \hat{f}(s)$. Then f is an **inverse Fourier transform** of $\hat{f}$ and we write $f(x) = \mathscr{F}^{-1}(\hat{f}(s))$.

The Fourier inversion integral recovers $f(x)$ from $\hat{f}(s)$ and this result follows easily from an important result in real analysis, which we do not prove.

Definition

Function $f: \mathbb{R} \to \mathbb{C}$ belongs to **class D** if it is absolutely integrable on $\mathbb{R}$, piecewise continuously differentiable on each finite closed interval of $\mathbb{R}$ and

$$f(x) = \tfrac{1}{2} \lim_{\varepsilon \to 0} (f(x + \varepsilon) + f(x - \varepsilon)) \qquad (\text{for all } x \in \mathbb{R})$$

Note

This class of functions is named after Dirichlet.

Theorem 8.2. The Fourier Integral Theorem

Suppose that $f \in$ class D. Then

$$f(x) = \frac{1}{\pi} \int_0^{\infty} \int_{-\infty}^{\infty} f(u) \cos(s(x - u))\, du\, ds \qquad (x \in \mathbb{R})$$

□

For a proof of this result, see Chapter 4 in A. Broman, *Introduction to Partial Differential Equations from Fourier Series to Boundary-Value Problems*, Addison-Wesley, 1970.

Corollary 8.3. The Fourier Inversion Integral

Suppose that $f \in$ class D and $\hat{f}(s) = \mathscr{F}(f(x))$. Then

$$f(x) = \mathscr{F}^{-1}(\hat{f}(s)) = \frac{1}{\sqrt{2\pi}} \int_{-\infty}^{\infty} e^{isx} \hat{f}(s)\, ds$$

□

Proof

It is easily checked that the Fourier integral theorem can be written as

$$f(x) = \frac{1}{\sqrt{2\pi}} \int_{-\infty}^{\infty} \left(\frac{1}{\sqrt{2\pi}} \int_{-\infty}^{\infty} f(u) e^{-isu} du \right) e^{isx} ds$$

and the result follows by definition of $\hat{f}(s)$. ■

Since the Fourier inversion integral is the same as the Fourier transform apart from a sign change, it can be used to obtain Fourier transforms from known results.

Example 8.2

The result of Example 8.1(ii) and Corollary 8.3 give

$$\frac{1}{\sqrt{2\pi}} \int_{-\infty}^{\infty} e^{-|s|} e^{isx} ds = \sqrt{\frac{2}{\pi}} \frac{1}{1+x^2} \quad \Rightarrow \quad \mathscr{F}(e^{-|x|}) = \sqrt{\frac{2}{\pi}} \frac{1}{1+s^2}$$

Alternatively, finding $\mathscr{F}(e^{-|x|})$ directly gives $\mathscr{F}(1/(1+x^2))$ without the use of residue theory (see Exercises 8.1).

Related to the Fourier transform are the Fourier sine and cosine transforms.

Definitions

Let $x \in \mathbb{R}$ and $s \in \mathbb{R}$ with $s \geqslant 0$, and let $f : [0, \infty) \to \mathbb{C}$. The **Fourier sine transform** of f is denoted and defined by

$$\mathscr{F}_s(f(x)) = \hat{f}_s(s) = \sqrt{\frac{2}{\pi}} \int_0^{\infty} \sin(sx) f(x)\, dx$$

The **Fourier cosine transform** of f is denoted and defined by

$$\mathscr{F}_c(f(x)) = \hat{f}_c(s) = \sqrt{\frac{2}{\pi}} \int_0^{\infty} \cos(sx) f(x)\, dx$$

Notes

(i) If the domain of f is extended to $(-\infty, \infty)$ so that f is even then $\hat{f}_c(s) = \hat{f}(s)$, and if the domain of f is extended to $(-\infty, \infty)$ so that f is odd then $\hat{f}_s(s) = i\hat{f}(s)$.

(ii) For the Fourier sine or cosine transform of f to exist, it is sufficient for $\int_0^{\infty} |f(x)| dx$ to exist, but this is not necessary. For example, $\mathscr{F}_s(1/x) = \sqrt{\pi/2}$.

The inversion results for these transforms follow easily from Corollary 8.3.

Corollary 8.4 Inversion of Fourier Sine and Cosine Transforms

(i) If $\hat{f}_s(s) = \mathscr{F}_s(f(x))$ and f has an extension to $(-\infty, \infty)$ which is an odd function of class D, then $f(x) = \sqrt{2/\pi}\int_0^\infty \sin(sx)\hat{f}_s(s)\,ds$.

(ii) If $\hat{f}_c(s) = \mathscr{F}_c(f(x))$ and f has an extension to $(-\infty, \infty)$ which is an even function of class D, then $f(x) = \sqrt{2/\pi}\int_0^\infty \cos(sx)\hat{f}_c(s)\,ds$. □

In this sense, the Fourier sine and cosine transforms are self-reciprocal.

Proof

(i) Let the domain of f be extended to $\mathbb{R}$ so that f is odd and of class D. It follows by definition that

$$\hat{f}(s) = \frac{1}{\sqrt{2\pi}}\int_{-\infty}^{\infty}(\cos sx - i\sin sx)f(x)dx$$

$$= \frac{-2i}{\sqrt{2\pi}}\int_0^\infty \sin(sx)f(x)\,dx = i\hat{f}_s(s)$$

Hence, since $\hat{f}_s(s)$ is odd, it follows by 8.3 that

$$f(x) = \frac{1}{\sqrt{2\pi}}\int_{-\infty}^{\infty} -i\hat{f}_s(s)(\cos sx + i\sin sx)\,ds = \sqrt{\frac{2}{\pi}}\int_0^\infty \sin(sx)\hat{f}_s(s)\,ds$$

as required.

(ii) The proof is similar and is left as an exercise. ■

Example 8.3

Let $f(x) = e^{-ax}$ where a is a positive real constant.

$$\hat{f}_c(s) + i\hat{f}_s(s) = \sqrt{\frac{2}{\pi}}\int_0^\infty e^{(-a+is)x}dx = \sqrt{\frac{2}{\pi}}\frac{1}{a-is} = \sqrt{\frac{2}{\pi}}\frac{a+is}{a^2+s^2}$$

Hence $\hat{f}_c(s) = \sqrt{\frac{2}{\pi}}\frac{a}{a^2+s^2}$ and $\hat{f}_s(s) = \sqrt{\frac{2}{\pi}}\frac{s}{a^2+s^2}$

Then by Corollary 8.4 it follows that

$$\int_0^\infty \frac{a\cos sx}{s^2+a^2}ds = \int_0^\infty \frac{s\sin sx}{s^2+a^2}ds = \frac{\pi}{2}e^{-ax} \quad (x > 0)$$

a result which can be obtained using residue theory, as in Chapter 5.

If f is not of class D, there is no guarantee that the inversion formulae hold. For example, 8.4(i) does not hold for $f(x) = 1/x$ even though its Fourier sine transform exists.

The following result concerning the Fourier transform is useful in applications. Since it is essentially a result from real analysis, we shall not prove it in any detail.

Definition

Let f and g be absolutely integrable on $\mathbb{R}$. Their **Fourier convolution** is denoted and defined by

$$f(x) \hat{} g(x) = \frac{1}{\sqrt{2\pi}} \int_{-\infty}^{\infty} f(y) g(x-y) dy$$

Theorem 8.5. The Fourier Convolution Result

Let $\mathcal{F}(f(x)) = \hat{f}(s)$ and $\mathcal{F}(g(x)) = \hat{g}(s)$. Then, provided the transform exists,

$$\mathcal{F}(f(x) \hat{} g(x)) = \hat{f}(s)\hat{g}(s)$$ □

Note

It can be shown that, since f and g are absolutely integrable on $\mathbb{R}$, $f \hat{} g$ is also absolutely integrable on $\mathbb{R}$, so its Fourier transform exists. See Chapter 8 in J. W. Dettman, *Applied Complex Variables*, Macmillan, 1965.

Part Proof

From the definitions it follows that

$$\mathcal{F}(f(x) \hat{} g(x)) = \frac{1}{2\pi} \int_{-\infty}^{\infty} \int_{-\infty}^{\infty} e^{-isx} f(y)\, g(x-y)\, dy\, dx$$

It is possible to prove that the order of integration may be changed (see Dettman again). Then letting $u = x - y$ (y fixed) gives

$$\mathcal{F}(f(x) \hat{} g(x)) = \frac{1}{\sqrt{2\pi}} \int_{-\infty}^{\infty} e^{-isy} f(y) \left(\frac{1}{\sqrt{2\pi}} \int_{-\infty}^{\infty} e^{-isu} g(u) du \right) dy = \hat{f}(s)\hat{g}(s)$$ ■

It follows from 8.3 and 8.5 that if f and g are of class D then

$$\int_{-\infty}^{\infty} e^{isx} \hat{f}(s)\hat{g}(s)\, ds = \int_{-\infty}^{\infty} f(y)\, g(x-y)\, dy \tag{8.3}$$

Letting $x = 0$ and $g(-y) = \overline{f(y)}$ in (8.3), noting that in this case $\hat{g}(s) = \overline{\hat{f}(s)}$, gives the following corollary.

Corollary 8.6. The Parseval Identity

Let $f \in$ class D. Then

$$\int_{-\infty}^{\infty} |f(x)|^2 dx = \int_{-\infty}^{\infty} |\hat{f}(s)|^2 ds \qquad \square$$

The corresponding convolution results for the Fourier sine and cosine transforms are more complicated, but in applications very often it is only a Parseval relation that is required.

Corollary 8.7. Parseval Identities for Fourier Cosine and Sine Transforms

Let f and g have the appropriate even and odd extensions of class D. Then

$$\int_0^{\infty} \hat{f}_c(s)\hat{g}_c(s) ds = \int_0^{\infty} \hat{f}_s(s)\hat{g}_s(s)\, ds = \int_0^{\infty} f(x) g(x)\, dx \qquad \square$$

Proof

Let f and g be even functions in (8.3). Then $\hat{f}(s) = \hat{f}_c(s)$ where $\hat{f}_c(s)$ is even and similarly for g, and the result for Fourier cosine transforms follows immediately. Now let f and g be odd functions in (8.3), so that $\hat{f}(s) = -i\hat{f}_s(s)$, with $\hat{f}_s(s)$ odd and similarly for g. Then the second relation follows. ■

Example 8.4

(i) Let $f(x) = \begin{cases} 1 & \text{for } |x| \leqslant a \\ 0 & \text{for } |x| > a \end{cases}$

It is easily checked that $\mathscr{F}(f(x)) = (2/\sqrt{2\pi s}) \sin as$ (see Exercises 8.1). Then by the Parseval identity, Corollary 8.6, it follows that

$$\frac{2}{\pi} \int_{-\infty}^{\infty} \frac{\sin^2 as}{s^2} ds = \int_{-\infty}^{\infty} f^2(x) dx = 2a \quad \Rightarrow \quad \int_0^{\infty} \frac{\sin^2 as}{s^2} ds = \frac{a\pi}{2}$$

(ii) It follows from Example 8.1(iii) that $\mathscr{F}_c(e^{-x^2/2}) = e^{-s^2/2}$. Also, from Example 8.3,

$$\mathscr{F}_c(e^{-ax}) = \frac{\sqrt{2}a}{\sqrt{\pi}(s^2 + a^2)}$$

Then by 8.7,

$$\int_0^{\infty} \frac{ae^{-s^2/2}}{s^2 + a^2} ds = \frac{\sqrt{\pi} e^{a^2/2}}{\sqrt{2}} \int_0^{\infty} e^{-(x+a)^2/2} dx$$

Letting $\sqrt{2}u = x + a$ in the second integral gives

$$\int_0^\infty \frac{e^{-s^2/2}}{s^2 + a^2} ds = \frac{\sqrt{\pi}}{a} e^{a^2/2} \int_{a/\sqrt{2}}^\infty e^{-u^2} du = \frac{\pi}{2a} e^{a^2/2} \text{erfc } (a/\sqrt{2})$$

where erfc denotes the **complementary error function**, defined by $\text{erfc}(x) = (2/\sqrt{\pi}) \int_x^\infty e^{-x^2} dx$. This result is harder to obtain using residue theory.

Fourier transforms can be used to solve certain types of linear integral equations. For example, consider any equation of the form

$$\alpha f(x) = g(x) + \int_{-\infty}^\infty K(x-u) f(u) du \tag{8.4}$$

where f is an unknown function of a real variable, g and K are known functions of a real variable and $\alpha = 0$ or 1, without loss of generality. Taking the Fourier transform of (8.4), assuming that the transforms of K, f and g exist and that the sufficient conditions for inversion are met, it follows from the Fourier convolution and inversion results that

$$\alpha \hat{f}(s) = \hat{g}(s) + \sqrt{2\pi} \hat{K}(s) \hat{f}(s) \quad \Rightarrow \quad \hat{f}(s) = \frac{\hat{g}(s)}{\alpha - \sqrt{2\pi} \hat{K}(s)}$$

$$\Rightarrow \quad f(x) = \frac{1}{\sqrt{2\pi}} \int_{-\infty}^\infty \frac{e^{isx} \hat{g}(s)}{\alpha - \sqrt{2\pi} \hat{K}(s)} ds$$

Alternatively, in some cases, known results or the convolution result can be used to find $f(x)$ from $\hat{f}(s)$.

Example 8.5

Suppose we wish to solve the equation

$$f(x) = e^{-|x|} + \frac{1}{2}(1-k^2) \int_{-\infty}^\infty e^{-|x-u|} f(u)\, du$$

where k is a real positive constant. From Example 8.2 it follows that $\mathscr{F}(e^{-k|x|}/k) = \sqrt{2/\pi}(s^2 + k^2)^{-1}$. Then taking the Fourier transform of the given equation,

$$\hat{f}(s) = \frac{\sqrt{2}}{\sqrt{\pi}(s^2+1)} + \frac{1}{2}(1-k^2) \frac{2}{s^2+1} \hat{f}(s)$$

$$\Rightarrow \quad \hat{f}(s) = \frac{\sqrt{2}}{\sqrt{\pi}(s^2+k^2)} \quad \Rightarrow \quad f(x) = \frac{e^{-k|x|}}{k}$$

Exercise **8.1.1** Assuming that the Fourier transforms exist, prove that

(i) $\mathscr{F}(af(x)+bg(x)) = a\hat{f}(s)+b\hat{g}(s), a, b \in \mathbb{C}$

(ii) $\mathscr{F}(e^{-iax}f(x)) = \hat{f}(s+a), a \in \mathbb{R}$

(iii) $\mathscr{F}(f'(x)) = is\hat{f}(s)$

(iv) $\mathscr{F}(f(x/a)) = a\hat{f}(sa), a \in \mathbb{R}$

(v) $\mathscr{F}(xf(x)) = i\hat{f}'(s)$

In part (v) assume conditions on f sufficient to ensure that differentiation under the integral sign is valid. Use part (v) and the result of Example 8.1(ii) to find $\mathscr{F}(x/(x^2+1))$.

Exercise **8.1.2** Find the Fourier transform of f defined by

$$f(x) = \begin{cases} 1 & |x| \leqslant a \\ 0 & |x| > a \end{cases}$$

where a is a positive constant. Hence use the inversion theorem to evaluate

$$\int_{-\infty}^{\infty} \frac{\sin(ax)\cos(ax/2)}{x}\,dx$$

Exercise **8.1.3** Find $\mathscr{F}(e^{-|x|})$ directly using the definition. Use this result and the Fourier inversion integral to find

(i) $\displaystyle\int_0^{\infty} \frac{\cos x}{1+x^2}\,dx$

(ii) $\mathscr{F}\left(\dfrac{1}{1+x^2}\right)$

Exercise **8.1.4** Use residue theory to find $\mathscr{F}(1/x)$ and state why the inversion integral is not applicable in this case.

Exercise **8.1.5** Use residue theory to find

(i) $\mathscr{F}\left(\dfrac{x}{x^2+a^2}\right), a \in \mathbb{R}$

(ii) $\mathscr{F}\left(\dfrac{1}{(1+x^2)^2}\right)$

(Compare (i) with Exercise 8.1.1.)

Exercise **8.1.6** Prove Corollary 8.4(ii).

Exercise 8.1.7

$$\text{Let} \quad f(x) = \begin{cases} \sin x & 0 \leqslant x \leqslant \pi/2 \\ 0 & x > \pi/2 \end{cases}$$

Find the Fourier sine transform of f and hence use the inversion theorem to evaluate

$$\int_0^\infty \frac{x \sin(\pi x/4)\cos(\pi x/2)}{1 - x^2}\,dx$$

Exercise 8.1.8 Find the Fourier cosine transform of $f(x) = e^{-x}\cos x$. Hence evaluate

$$\int_0^\infty \frac{(2 + x^2)\cos ax}{4 + x^4}\,dx \qquad (a \in \mathbb{R}^+)$$

Exercise 8.1.9 Find $\mathscr{F}_s(xe^{-x^2/2})$ and hence evaluate $\int_0^\infty xe^{-x^2/2}\sin 4x\,dx$.

Exercise 8.1.10 Use residue theory and the contour shown in Fig. 8.2 to find $\mathscr{F}_s(x^{-1/2})$ and $\mathscr{F}_c(x^{-1/2})$.

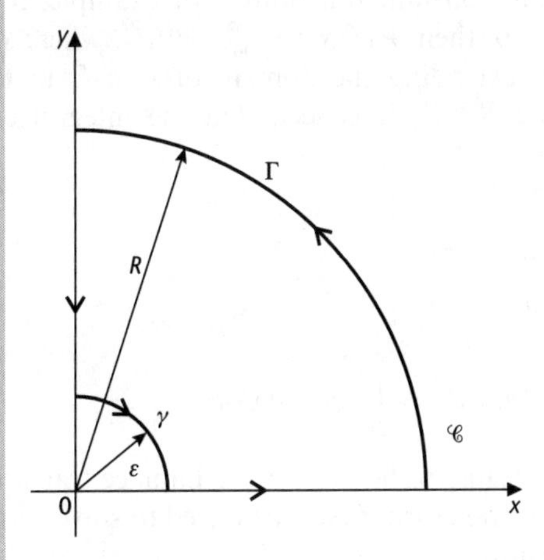

Figure 8.2

Exercise 8.1.11

$$\text{Let} \quad f(x) = \begin{cases} x & \text{for } |x| \leqslant 1 \\ 0 & \text{for } |x| > 1 \end{cases}$$

Find $\mathscr{F}(f(x))$ and hence deduce that $\int_{-\infty}^{\infty} x^{-4}(x\cos x - \sin x)^2 dx = \pi/3$ by using the Parseval identity.

Exercise **8.1.12** Let $f(x) = e^{-ax}$ and

$$g(x) = \begin{cases} a & \text{for } 0 < x < b \\ 0 & \text{for } x > b \end{cases}$$

where a and b are positive constants. Find the Fourier cosine transform of g. Use the Parseval identities and the result of Example 8.3 to evaluate

$$\int_0^\infty \frac{x^2}{(x^2+a^2)(x^2+b^2)}\,dx \quad \text{and} \quad \int_0^\infty \frac{\sin bx}{x(x^2+a^2)}\,dx$$

Exercise **8.1.13** Show that $\mathscr{F}(f''(x)) = -s^2\hat{f}(s)$, assuming that the transform exists. Given that $\mathscr{F}(e^{-x^2/2}) = e^{-s^2/2}$, solve the integral equation

$$e^{-x^2/2} = \int_{-\infty}^{\infty} e^{-|x-u|} f(u)\,du$$

Laplace Transforms

Fourier transform methods cannot in general be applied to functions which are not absolutely integrable on $\mathbb{R}$. Hence they cannot be applied to a large class of functions, including constant functions. For example, if $f(x) = 1$ for $x \geqslant 0$ and $f(x) = 0$ for $x < 0$, then $\mathscr{F}(f(x)) = \int_0^\infty e^{-isx}dx$, where $s \in \mathbb{R}$, so it is not convergent. However, extending the domain of s to $\mathbb{C}$ in this case and putting $s = \alpha + i\beta$, where $\alpha, \beta \in \mathbb{R}$, it is seen that the integral converges for $\beta < 0$.

In general, let

$$f(x) = \begin{cases} e^{-ax}g(x) & x \geqslant 0, a \in \mathbb{R} \\ 0 & x < 0 \end{cases}$$

$$\text{Then} \quad \hat{f}(s) = \int_0^\infty e^{-isx-ax}g(x)dx = \int_0^\infty e^{-\alpha x}g(x)\,dx$$

where $\alpha = a + is \in \mathbb{C}$. This leads to the idea of the Laplace transform, which is a special case of the Fourier transform first introduced to solve certain ordinary and partial differential equations.

Definition

Let $f\colon [0,\infty) \to \mathbb{C}$ and $s \in \mathbb{C}$. Then the **Laplace transform** of f is denoted and defined by

$$\mathscr{L}(f(t)) = \bar{f}(s) = \int_0^\infty e^{-st}f(t)\,dt$$

Note

For sufficiently 'well-behaved' functions, $\bar{f}(s)$ will be defined for certain values of s but not others. However, for certain functions, the integral does not converge for any values of s, so the the Laplace transform does not exist. For example, $\mathscr{L}(e^{t^2})$ does not exist. For historical reasons, t is used in the definition to denote a time variable.

Historical Note

Euler applied the Laplace transform to solve certain differential equations in 1737. Later and independently, Laplace used it to solve differential and difference equations in his book *Théorie Analytique des Probabilités* in 1812. Laplace's greatest achievements were in the field of analysis and he applied his methods to celestial mechanics, potential theory and probability.

In general, the following result gives sufficient (but not necessary) conditions for the Laplace transform of a given function to exist.

Definitions

The function $f: [0, \infty) \to \mathbb{C}$ is of **exponential order** if there exist positive real constants M, T and a such that $|f(t)| \leqslant Me^{at}$ for $t \geqslant T$. We shall say that f belongs to **class E** if f is integrable on $[0, b]$ for each $b > 0$ and of exponential order.

Lemma 8.8. Existence of a Laplace Transform

The Laplace transform of $f: [0, \infty) \to \mathbb{C}$ exists if $f \in$ class E. Specifically, $\bar{f}(s)$ is defined if and only if $\operatorname{Re} s > a$. □

Proof

It follows from the definition and by hypothesis that

$$|\mathscr{L}(f(t)| \leqslant \int_0^T |e^{-st}f(t)|\,dt + M\int_T^\infty |e^{-(s-a)t}|\,dt \tag{8.5}$$

The first integral in (8.5) is clearly finite and the second integral reduces to $\int_T^\infty e^{-(\operatorname{Re} s - a)t}\,dt$, which clearly converges if and only if $\operatorname{Re} s > a$. ■

Note

It follows by (8.5) that if $f \in$ class E, then $\lim_{\operatorname{Re} s \to \infty} \bar{f}(s) = 0$.

The following result is a list of the elementary Laplace transforms, together with the values of s for which they are valid.

Lemma 8.9. Elementary Laplace Transforms

Let $\alpha \in \mathbb{C}$. Then

(i) $\mathscr{L}(e^{\alpha t}) = \dfrac{1}{s-\alpha}$ for $\operatorname{Re} s > \operatorname{Re} \alpha$

(ii) $\mathscr{L}(\sin \alpha t) = \dfrac{\alpha}{s^2+\alpha^2}$; $\mathscr{L}(\cos \alpha t) = \dfrac{s}{s^2+\alpha^2}$ for $\operatorname{Re} s > -\operatorname{Im} \alpha$

(iii) $\mathscr{L}(t^n) = \dfrac{n!}{s^{n+1}}, n \in \mathbb{Z}_{\geqslant 0}; \mathscr{L}(t^a) = \dfrac{\Gamma(a+1)}{s^{a+1}}, a \in \mathbb{R}_{>-1}$, for $\operatorname{Re} s > 0$ □

Part Proof

All the results follow from the definition. Part (i) is left as an exercise and part (ii) follows from (i) by taking imaginary and real parts of $\mathscr{L}(e^{i\alpha t})$. Now for part (iii).

Let $s \in \mathbb{R}$. Then $\mathscr{L}(t^{a-1}) = \int_0^\infty e^{-st} t^{a-1} dt = s^{-a} \int_0^\infty e^{-u} u^{a-1} du$, letting $u = st$. Hence $\mathscr{L}(t^{a-1}) = s^{-a}\Gamma(a)$, for $a > 0$ and $s > 0$, from the definition of the gamma function given in Chapter 6. Now let $s \in \mathbb{C}$. Then the above result holds for $\operatorname{Re} s > 0$ since both sides of the equation are analytic for $\operatorname{Re} s > 0$ and agree along the positive real axis (see Theorem 10.1). For $n \in \mathbb{Z}_{\geqslant 0}$, $\Gamma(n) = (n-1)!$ by 6.12(i), as required. The result for $n \in \mathbb{Z}_{\geqslant 0}$ can also be obtained directly by integration by parts and induction (see Exercises 8.2). ■

Note

Since $\alpha \in \mathbb{C}$, $\mathscr{L}(\sinh at)$ and $\mathscr{L}(\cosh at)$ are easily obtained from (ii) by replacing α by $i\alpha$. By (iii), $\mathscr{L}(1) = 1/s$.

For convenience, we list below the standard elementary properties of Laplace transforms, most of which do not depend explicitly on any complex analysis.

Theorem 8.10. Elementary Properties of the Laplace Transform

Let α and $\beta \in \mathbb{C}$ and $n \in \mathbb{N}$ and suppose that $f, g \in$ class E.

(i) $\mathscr{L}(\alpha f(t) + \beta g(t)) = \alpha \bar{f}(s) + \beta \bar{g}(s)$

(ii) $\mathscr{L}(e^{\alpha t} f(t)) = \bar{f}(s-\alpha)$

(iii) If $f^{(n)}$ is continuous on $(0, \infty)$ and $f, f^{(k)} \in$ class E for $k = 1, \ldots, n-1$ then

$$\mathscr{L}(f^{(n)}(t)) = s^n \bar{f}(s) - s^{n-1} f(0+) - s^{n-2} f'(0+) - \ldots - f^{(n-1)}(0+)$$

where $f^{(k)}(0+) = \lim_{t \to 0+} f^{(k)}(t)$

(iv) If f is continuous on $(0,\infty)$ then

$$\mathscr{L}\left(\int_0^t f(u)\,du\right) = \frac{\bar{f}(s)}{s}$$

(v) If f is continuous on $(0,\infty)$ then

$$\mathscr{L}(t^n f(t)) = (-1)^n \bar{f}^{(n)}(s)$$

□

Part Proof

(i), (ii) These follow directly from the definition and are left as an exercise.

(iii) By hypothesis, integration by parts is valid and gives

$$\mathscr{L}(f'(t)) = [e^{-st}f(t)]_0^\infty + s\int_0^\infty e^{-st}f(t)\,dt = s\bar{f}(s) - f(0+) \tag{8.6}$$

for values of s in the half-plane of convergence. Then the result for $n > 1$ follows directly by induction.

(iv) Let $g(t) = \int_0^t f(u)\,du$. Then $g'(t) = f(t)$ with $g(0) = 0$. It then follows from (8.6) that $\bar{f}(s) = s\bar{g}(s)$, as required.

(v) By hypothesis, $\bar{f}(s) = \int_0^\infty e^{-st}f(t)\,dt$ for $\operatorname{Re} s > a$ say. It can be shown that differentiation under the integral sign with respect to s is valid for $\operatorname{Re} s > a$. This follows by Theorems 10.15 and 10.16, and we will return to this result again in Chapter 10. Differentiating under the integral sign with respect to s gives $\bar{f}'(s) = -\mathscr{L}(tf(t))$ so that the result is true for $n = 1$. Suppose that $\bar{f}^{(k)}(s) = (-1)^k\mathscr{L}(t^k f(t))$. Then differeniating again with respect to s gives $\bar{f}^{(k+1)}(s) = -(-1)^k\mathscr{L}(t^{k+1}f(t))$. Hence the result follows by induction. ■

Convention

For transforms in the examples, we omit the range of values of s over which they are valid.

Example 8.6

(i) $\mathscr{L}(\cos 2t) = s/(s^2+4)$ by 8.9, so that
$\mathscr{L}(e^{-t}\cos 2t) = (s+1)/\left((s+1)^2+4\right)$ by 8.10(ii). Then

$$\mathscr{L}(t^2e^{-t}\cos 2t) = \frac{d^2}{ds^2}\left(\frac{s+1}{(s+1)^2+4}\right) = \frac{2(s+1)(s^2+2s-11)}{((s+1)^2+4)^3}$$

by 8.10(v). Putting $s = 3$ gives

$$\int_0^\infty e^{-4t}t^2\cos 2t\,dt = \frac{8\cdot 4}{20^3} = \frac{1}{250}$$

(ii) $\mathscr{L}(\sin t) = 1/(1+s^2)$ by 8.9(iii). Let $g(t) = (\sin t)/t$, so that by 8.10(v), $\mathscr{L}(tg(t)) = -\bar{g}'(s) = 1/(1+s^2)$. Hence $\bar{g}(s) = k - \tan^{-1} s$. Letting $s \in \mathbb{R}$ and noting that $\lim_{s\to\infty} \bar{g}(s) = 0$, $k = \pi/2$. Then letting $s = 0$ gives $\int_0^\infty ((\sin t)/t)dt = \pi/2$. (Compare this with the residue theory method used to obtain the same result in Chapter 5.)

The function Si is defined by $\mathrm{Si}(t) = \int_0^t ((\sin u)/u)du$. Then by the above result and 8.10(iv) it follows that

$$\mathscr{L}(\mathrm{Si}(t)) = \frac{\pi}{2s} - \frac{\tan^{-1} s}{s} = \frac{\tan^{-1}(1/s)}{s}$$

Applications of other parts of Theorem 8.10 will be given later in the chapter.

The Laplace transform of a piecewise continuous function exists as long as the function is of class E. This leads to a practical advantage in using Laplace transforms over other methods for solving certain differential equations involving piecewise continuous functions. This will be demonstrated in the next section. A large number of piecewise continuous functions which occur in practical applications are defined in terms of the Heaviside unit function.

Definition

The **Heaviside unit function** is denoted and defined by

$$H(t) = \begin{cases} 0 & \text{for } t < 0 \\ 1 & \text{for } t \geqslant 0 \end{cases}$$

Lemma 8.11. The Shift Theorem

Let $f \in$ class E and $a \in \mathbb{R}^+$. Then $\mathscr{L}(H(t-a)f(t-a)) = e^{-as}\bar{f}(s)$. □

Proof

$\mathscr{L}(H(t-a)f(t-a)) = \int_a^\infty e^{-st}f(t-a)\,dt$. Let $u = t-a$ to obtain the result. ■

Note

In particular, $\mathscr{L}(H(t-a)) = e^{-as}/s$.

Example 8.7

The **square wave function** is defined by

$$f(t) = \begin{cases} 1 & 2na \leqslant t < (2n+1)a \\ -1 & (2n+1)a \leqslant t < (2n+2)a \end{cases}$$

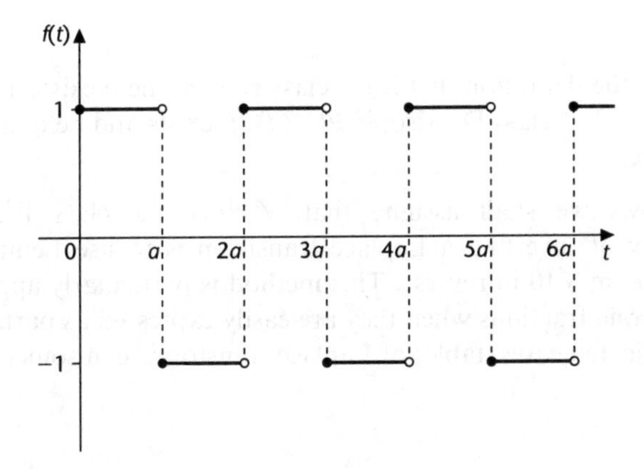

Figure 8.3

for all non-negative integers n (Fig. 8.3).

Then $f(t) = H(t) + 2\sum_{n=1}^{\infty}(-1)^n H(t - na)$

$$\Rightarrow \quad \bar{f}(s) = \frac{1}{s}\left(1 + 2\sum_{n=1}^{\infty}(-1)^n e^{-nas}\right) = \frac{1}{s}\left(1 - \frac{2e^{-as}}{1 + e^{-as}}\right) = \frac{1}{s}\tanh\left(\frac{as}{2}\right)$$

(for values of s for which the geometric series converges).

We now consider the standard methods employed for recovering a function of class E from its Laplace transform.

Definition

Let $\mathscr{L}(f(t)) = \bar{f}(s)$. Then f is an **inverse Laplace transform** of $\bar{f}$. We write $f(t) = \mathscr{L}^{-1}(\bar{f}(s))$.

Note that given f, $\mathscr{L}^{-1}(\bar{f}(s))$ will not be unique, if it exists, so it is convenient to introduce the following definition.

Definition

Function $f : \mathbb{R} \to \mathbb{C}$ belongs to **class E′** if its restriction to $[0, \infty)$ belongs to class E, $f(t) = 0$ for $t < 0$, and f is piecewise continuously differentiable on each finite closed interval with

$$f(t) = \frac{1}{2}\lim_{\varepsilon \to 0}(f(t + \varepsilon) + f(t - \varepsilon)) \quad \text{for all } t \in (0, \infty)$$

Note

It follows from the definition that if $f \in$ class E$'$ then there exists a real number b such that $f(t)e^{-bt} \in$ class D. Also, if $\mathscr{L}^{-1}(\bar{f}(s))$ exists and belongs to class E$'$, then it is unique.

In what follows we shall assume that $\mathscr{L}^{-1}(\bar{f}(s)) \in$ class E$'$. The most elementary way of inverting a Laplace transform is to use Lemmas 8.9 and 8.11, and Theorem 8.10 in reverse. This method is particularly appropriate for inverting algebraic fractions when they are easily expressed as partial fractions. In more complicated cases, tables of Laplace transforms can sometimes be used in reverse.

Example 8.8

(i) $\bar{f}(s) = \dfrac{1}{2s^2 - 7s - 4} = \dfrac{1}{(2s+1)(s-4)} = \dfrac{1}{9(s-4)} - \dfrac{1}{9(s+1/2)}$

$$\Rightarrow \quad f(t) = \frac{1}{9}e^{4t} - \frac{1}{9}e^{-t/2}$$

by 8.9(i) and 8.10(i).

(ii) $\bar{f}(s) = \dfrac{s-4}{s^2+6s+25} = \dfrac{s-4}{(s+3)^2+4^2} = \dfrac{s+3}{(s+3)^2+4^2} - \dfrac{7}{(s+3)^2+4^2}$

$$\Rightarrow \quad f(t) = e^{-3t}\cos 4t - \frac{7}{4}e^{-3t}\sin 4t$$

by 8.9(ii), 8.10(i), (ii).

(iii) $\bar{f}(s) = \dfrac{s}{(s^2+1)^2} = -\dfrac{1}{2}\dfrac{d}{ds}\left(\dfrac{1}{s^2+1}\right) \quad \Rightarrow \quad f(t) = \dfrac{t}{2}\sin t$

using 8.9(ii), 8.10(i), (v).

(iv) Let $\mathscr{L}(f(t)) = \bar{f}(s) = \mathrm{Log}(1 - a^2/s^2)$ where $a \in \mathbb{R}$. Then by 8.10(v),

$$\mathscr{L}(tf(t)) = -\bar{f}'(s) = \frac{-2a^2}{s(s^2-a^2)} = \frac{2}{s} - \frac{2s}{s^2-a^2}$$

$$\Rightarrow \quad f(t) = \frac{2(1-\cosh at)}{t}$$

using 8.9 and 8.10(i).

Another way of inverting Laplace transforms is via the Laplace inversion integral, and this is an area where complex analysis is directly applicable.

Theorem 8.12. The Laplace Inversion Integral

If $f \in$ class E$'$ and $\bar{f}(s) = \mathscr{L}(f(t))$ exists for Re $s > a$ then

$$f(t) = \mathscr{L}^{-1}(\bar{f}(s)) = \frac{1}{2\pi i}\int_{b-i\infty}^{b+i\infty} e^{st}\bar{f}(s)\,ds$$

for any real number $b > a$. □

If $\bar{f}$ has a finite number of singular points, then Theorem 8.12 can be justified in the following way. Construct a simple closed contour $\mathscr{C}$ in the z-plane comprising a semicircular arc Γ and a bounding diameter L (Fig. 8.4), such that all the singularities of $\bar{f}(z)$ lie to the left of L. Let $z = s$ be any point inside $\mathscr{C}$.

Since $\bar{f}$ is analytic inside and on $\mathscr{C}$, it follows by Cauchy's integral formula that

$$\bar{f}(s) = \frac{1}{2\pi i}\int_{\mathscr{C}} \frac{\bar{f}(z)}{z-s}\,dz$$

and by the ML lemma it follows that

$$\left|\int_{\Gamma} \frac{\bar{f}(z)}{z-s}\,dz\right| \leqslant \max_{\Gamma} \frac{|\bar{f}(z)|}{|z-s|}\pi|z-b|$$

where $\displaystyle\lim_{z\to\infty}\frac{|z-b|}{|z-s|} = 1$ and $\displaystyle\lim_{z\to\infty}|\bar{f}(z)| = 0$

(Recall that $\lim_{\text{Re}\, z\to\infty}\bar{f}(z) = 0$ since $f \in$ class E.) Hence

$$\bar{f}(s) = \frac{1}{2\pi i}\int_{L} \frac{\bar{f}(z)}{z-s}\,dz$$

$$\Rightarrow \quad f(t) = \mathscr{L}^{-1}(\bar{f}(s)) = \frac{1}{2\pi i}\mathscr{L}^{-1}\left(\int_L \frac{\bar{f}(z)}{z-s}\,dz\right) = \frac{1}{2\pi i}\int_L \bar{f}(z)\mathscr{L}^{-1}\left(\frac{1}{z-s}\right)dz$$

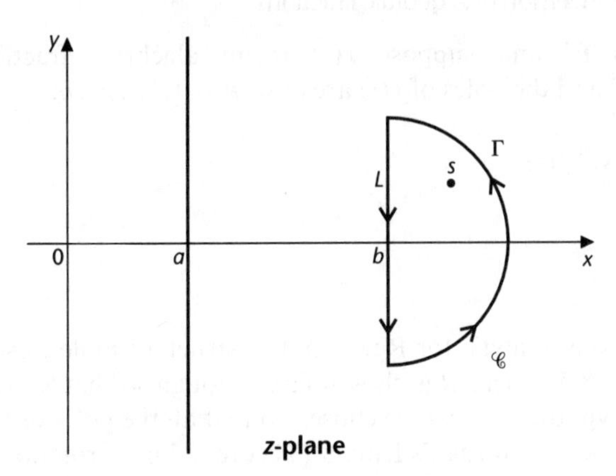

Figure 8.4

(The change in order of integration may be formally justified.)

$$\text{Hence} \quad f(t) = \frac{-1}{2\pi i}\int_L e^{zt}\bar{f}(z)dz$$

by 8.9, as stated in Theorem 8.12. The usual way of proving Theorem 8.12 is by invoking the Fourier integral formula as in the case of the Fourier inversion result.

Proof of 8.12

Recall that for $g \in$ class D, the Fourier integral formula can be written as

$$g(t) = \frac{1}{\sqrt{2\pi}}\int_{-\infty}^{\infty}\left(\frac{1}{\sqrt{2\pi}}\int_{-\infty}^{\infty} g(u)e^{-iuv}du\right)e^{ivt}dv \tag{8.7}$$

(see the proof of Corollary 8.3). Since $f \in$ class E$'$, $e^{-bt}f(t) \in$ class D for any real number $b > a$. Letting $g(t) = e^{-bt}f(t)$ and $s = b + iv$ gives

$$\int_{-\infty}^{\infty} g(u)e^{-iuv}du = \int_0^{\infty} e^{-su}e^{iuv}e^{-iuv}f(u)\,du = \bar{f}(s)$$

Hence (8.7) gives

$$e^{-bt}f(t) = \frac{1}{2\pi}\int_{-\infty}^{\infty} e^{ivt}\bar{f}(s)\,dv \quad \Rightarrow \quad f(t) = \frac{1}{2\pi i}\int_{b-i\infty}^{b+i\infty} e^{st}\bar{f}(s)\,ds$$

as required. ■

If $\bar{f}(s)$ is an algebraic fraction, Theorem 8.12 leads to a particularly simple result for inversion.

Corollary 8.13. Inversion of Algebraic Fractions

Let $f \in$ class E$'$ and suppose $\bar{f}(s)$ is an algebraic fraction. Then if $\lim_{s\to\infty}\bar{f}(s) = 0$ and the poles of $\bar{f}(s)$ are at s_k, $k = 1, 2, \ldots, n$,

$$f(t) = \sum_{k=1}^{n} \operatorname*{Res}_{s=s_k} e^{st}\bar{f}(s)$$ □

Proof

Suppose that $\bar{f}(s)$ is analytic for $\operatorname{Re} s > a$. Construct a simple closed contour $\mathscr{C}$ as shown in Fig. 8.5, where R is chosen large enough so that $\mathscr{C}$ encloses all the poles of $\bar{f}$. (By hypothesis, b can be chosen so that all the poles of f lie to the left of L.) Let $s = iz + b$ in Jordan's lemma (Theorem 5.6), a rotation through 90° followed by a translation, and take $m = t > 0$. Then by Jordan's lemma, since

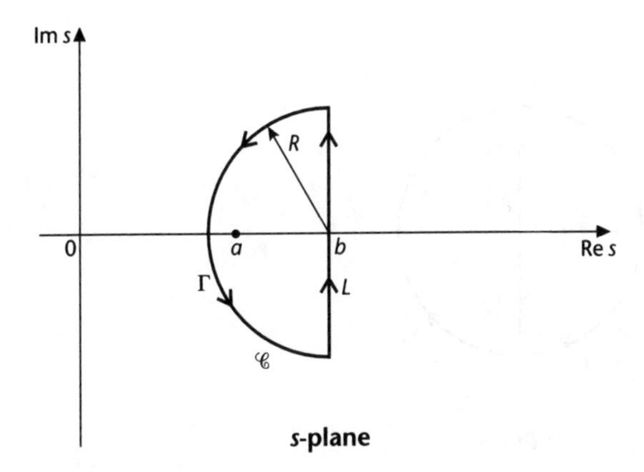

Figure 8.5

$\lim_{s\to\infty}\bar{f}(s) = 0$, it follows that $\lim_{R\to\infty}\int_\Gamma e^{st}\bar{f}(s)ds = 0$ and so by Cauchy's residue theorem and 8.12,

$$f(t) = \int_{\mathscr{C}} e^{st}\bar{f}(s)ds = \sum_{k=1}^{n} \operatorname*{Res}_{s=s_k} e^{st}\bar{f}(s)$$

as required. ■

Example 8.9

Let $\bar{f}(s) = e^{-s}/(s-1)^2$. Note that the only singular point of $\bar{f}$ is a double pole at $s = 1$. Construct a contour $\mathscr{C}_1$, enclosing $s = 1$, and construct also a contour $\mathscr{C}_2$ as shown in Fig. 8.6, where $b > 1$. Then by 8.12 and Jordan's lemma, with $s = -iz + b$ and $m = 1 - t > 0$,

$$f(t) = \frac{-1}{2\pi i}\int_{\mathscr{C}_2} \frac{e^{s(t-1)}}{(s-1)^2}ds = 0 \qquad (t < 1)$$

using Cauchy's theorem. Also, by Jordan's lemma with $s = iz + b$ and $m = t - 1 > 0$, as in the proof of 8.13,

$$f(t) = \frac{1}{2\pi i}\int_{\mathscr{C}_1} \frac{e^{s(t-1)}}{(s-1)^2}ds = \operatorname*{Res}_{s=1} e^{st}\bar{f}(s) = \operatorname*{Res}_{s=1} \frac{e^{s(t-1)}}{(s-1)^2} \qquad (t > 1)$$

by the residue theorem. Also,

$$\operatorname*{Res}_{s=1} e^{st}\bar{f}(s) = \frac{d}{ds}(e^{s(t-1)})|_{s=1} = (t-1)e^{t-1}$$

Hence, altogether, $f(t) = H(t-1)e^{t-1}(t-1)$.

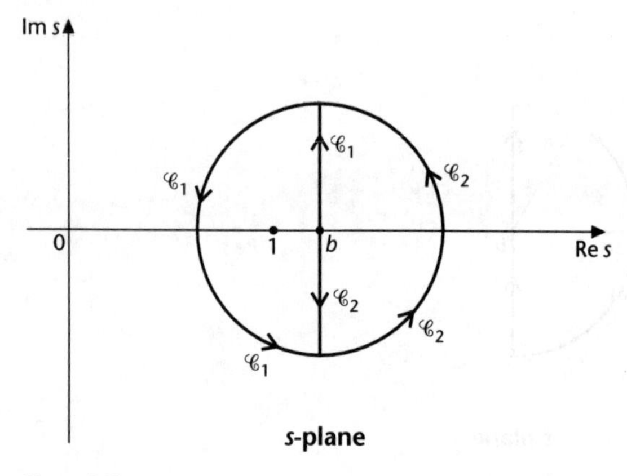

Figure 8.6

Example 8.10

(i) Let $\bar{f}(s) = 1/(s+a)^2$, $a \in \mathbb{C}$. Then $\bar{f}$ satisfies the conditions of Corollary 8.13 and has a double pole at $s = -a$. By the usual technique, $\text{Res}_{s=-a}\, e^{st}\bar{f}(s) = te^{-at}$ so that $\mathscr{L}^{-1}(\bar{f}(s)) = te^{-at}$ by 8.13.

(ii) Let $\bar{f}(s) = s/(s^2+a^2)$, $a \in \mathbb{C}$. Once again, $\bar{f}$ satisfies the conditions of Corollary 8.13 and has simple poles at $s = \pm ai$ with $\text{Res}_{s=\pm ai}\, e^{st}\bar{f}(s) = \frac{1}{2}e^{\pm ait}$. It follows by 8.13 that $\mathscr{L}^{-1}(\bar{f}(s)) = \cos at$.

(iii) Let $\bar{f}(s) = s^2/(s^3+1)$. In this case, $\bar{f}$ has simple poles at $s_k = e^{-i\pi/3 + 2ki\pi/3}$, $k = 0, 1, 2$, and

$$\operatorname*{Res}_{s=s_k} e^{st}\bar{f}(s) = \frac{e^{st}}{3}\bigg|_{s=s_k}$$

It follows by 8.13 that $f(t) = \frac{1}{3}(e^{-t} + 2e^{t/2}\cos\frac{1}{2}\sqrt{3}t)$

The inversion theorem is particularly useful when it comes to inverting functions with branch points, as the following example illustrates.

Example 8.11

Let $\bar{f}(s) = (1/s)e^{-as^{1/2}}$ where a is a positive real constant. From the Laplace inversion formula it follows that

$$\mathscr{L}^{-1}(\bar{f}(s)) = f(t) = \frac{1}{2\pi i}\int_{b-i\infty}^{b+i\infty} \frac{\exp(st - as^{1/2})}{s}\,ds$$

The integrand has only one singular point, a branch point, at $s = 0$, so consider the contour in the complex plane as shown in Fig. 8.7. By Cauchy's theorem it follows that

$$\int_L + \int_\Gamma + \int_{DC} + \int_\gamma + \int_{BA} e^{st}\bar{f}(s)ds = 0$$

Since $\operatorname{Re} s^{1/2} > 0$ on the arc Γ, $|\bar{f}(s)| \leqslant 1/|s|$ on Γ and so by Jordan's lemma, as in the proof of 8.13, $\lim_{R\to\infty} \int_\Gamma e^{st}\bar{f}(s)ds = 0$.

$$\text{Hence} \quad f(t) = \frac{1}{2\pi i} \lim_{R\to\infty} \lim_{\varepsilon\to 0} \left(\int_{CD} + \int_{AB} - \int_\gamma \frac{\exp(st - as^{1/2})}{s} ds \right)$$

On the arc γ, let $s = \varepsilon e^{i\theta}$, $-\pi \leqslant \theta < \pi$. Then

$$-\lim_{\varepsilon\to 0} \int_\gamma \frac{\exp(st - as^{1/2})}{s} ds = \lim_{\varepsilon\to 0} \int_{-\pi}^{\pi} i \exp(\varepsilon e^{i\theta} t - a\sqrt{\varepsilon} e^{i\theta/2}) d\theta = 2\pi i$$

In the limit as $\varepsilon \to 0$, $s = xe^{i\pi}$ on CD and $s = xe^{-i\pi}$ on BA. Hence

$$f(t) = \frac{1}{2\pi i} \left(\int_0^\infty \frac{e^{-xt - ai\sqrt{x}}}{x} dx - \int_0^\infty \frac{e^{-xt + ai\sqrt{x}}}{x} dx \right) + 1$$

$$\Rightarrow \quad f(t) = 1 - \frac{1}{\pi} \int_0^\infty \frac{e^{-xt} \sin(a\sqrt{x})}{x} dx = 1 - I(a)$$

$$\text{where} \quad I(a) = \frac{2}{\pi} \int_0^\infty \frac{e^{-u^2 t} \sin au}{u} du$$

letting $x = u^2$. It is possible to proceed further and simplify $f(t)$. Differentiation under the integral sign in $I(a)$ with respect to a is valid, using standard results

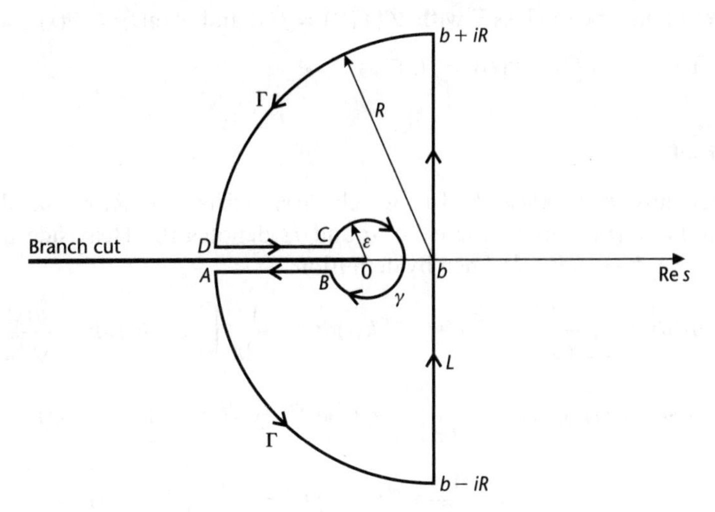

Figure 8.7

from real analysis. Differentiating $I(a)$ with respect to a twice and integrating by parts gives $I''(a) = -aI'(a)/2t$. Hence, since $I'(0) = 1/\sqrt{\pi t}$,

$$I'(a) = \frac{e^{-a^2/4t}}{\sqrt{\pi t}} \quad \Rightarrow \quad I = \frac{1}{\sqrt{\pi t}}\int_0^a e^{-u^2/4t}du = \frac{2}{\sqrt{\pi}}\int_0^{a/2\sqrt{t}} e^{-v^2}dv = \operatorname{erf}\left(\frac{a}{2\sqrt{t}}\right)$$

where erf denotes the **error function** defined by

$$\operatorname{erf}(x) = \frac{2}{\sqrt{\pi}}\int_0^x e^{-u^2}du$$

Then finally, $f(t) = \operatorname{erfc}(a/2\sqrt{t})$. This result can also be found indirectly without the use of the inversion formula (see Exercises 8.2).

As in the case of Fourier transforms, another very useful method for inverting Laplace transforms involves the convolution result.

Definition

Let f and g be two functions of class E. Their **Laplace convolution** is denoted and defined by

$$f(t) * g(t) = \int_0^t f(u)g(t-u)du$$

Theorem 8.14. The Laplace Convolution Result

Let f and g be of class E with $\mathcal{L}(f(t)) = \bar{f}(s)$ and $\mathcal{L}(g(t)) = \bar{g}(s)$.

Then $\mathcal{L}(f(t) * g(t)) = \bar{f}(s)\bar{g}(s)$ □

Proof

For any $h \in$ class E for which $\bar{h}(s)$ exists for $\operatorname{Re} s > a$, define h_1 by $h_1(t) = H(t)e^{-bt}h(t)$, where $b > a$ and H denotes the Heaviside unit function. Let $s = b + id$, $b \in \mathbb{R}$. Then by definition

$$\hat{h}_1(d) = \frac{1}{\sqrt{2\pi}}\int_{-\infty}^{\infty} e^{-idt}H(t)e^{-bt}h(t)dt = \frac{1}{\sqrt{2\pi}}\int_0^{\infty} e^{-st}h(t)dt = \frac{\bar{h}(s)}{\sqrt{2\pi}} \tag{8.8}$$

$$\text{Also} \quad f_1(t) \hat{*} g_1(t) = \frac{1}{\sqrt{2\pi}}\int_{-\infty}^{\infty} H(y)e^{-by}f(y)H(t-y)e^{-b(t-y)}g(t-y)dy$$

$$= \frac{1}{\sqrt{2\pi}}e^{-bt}f(t) * g(t) = \frac{1}{\sqrt{2\pi}}(f * g)_1(t)$$

(It can be shown that $f * g$ is of class E.) Then by the Fourier convolution result and (8.8) it follows that

$$\mathscr{F}(f_1(t)\cdot g_1(t)) = \mathscr{F}\left(\frac{1}{\sqrt{2\pi}}(f * g)_1(t)\right) = \hat{f}_1(d)\hat{g}_1(d) = \frac{1}{2\pi}\bar{f}(s)\bar{g}(s)$$

Also by (8.8),

$$\mathscr{F}\left(\frac{1}{\sqrt{2\pi}}(f * g)_1(t)\right) = \frac{1}{2\pi}\mathscr{L}(f(t) * g(t))$$

and the required result follows. ∎

Note

Putting $g(t) \equiv 1$ in 8.14 and noting that $\mathscr{F}(1) = 1/s$ gives 8.10(iv) as a special case.

Example 8.12

(i) From 8.9(ii) and 8.14 it follows that

$$\mathscr{L}^{-1}\left(\frac{1}{(s^2+1)^2}\right) = \int_0^t \sin u \sin(t-u)\, du = \frac{1}{2}\int_0^t (\cos(2u-t) - \cos t)\, du$$
$$= \frac{1}{2}(\sin t - t\cos t)$$

(Compare with the method of partial fractions in this case, given in Exercises 8.2.)

(ii) Let $f(t) = t^m$ and $g(t) = t^n$ where $m, n > -1$ but not necessarily integers. It then follows by 8.9(iii) that

$$\mathscr{L}^{-1}(\bar{f}(s)\bar{g}(s)) = \mathscr{L}^{-1}\left(\frac{\Gamma(m+1)\Gamma(n+1)}{s^{m+1}s^{n+1}}\right)$$
$$= \frac{\Gamma(m+1)\Gamma(n+1)t^{m+n+1}}{\Gamma(m+n+2)}$$

But also $\mathscr{L}^{-1}(\bar{f}(s)\bar{g}(s)) = \int_0^t u^m(t-u)^n du$, by Theorem 8.14. Hence putting $t = 1$ gives

$$B(m+1, n+1) = \int_0^1 u^m(1-u)^n du = \frac{\Gamma(m+1)\Gamma(n+1)}{\Gamma(m+n+2)}$$

This provides a proof of Theorem 6.12(iv).

(iii) Note that since $\Gamma(1/2) = \sqrt{\pi}$, it follows from 8.9(iii) that $\mathscr{L}(1/\sqrt{\pi t}) = 1/s^{1/2}$. Then using 8.9(i) and 8.14 it follows that

$$\mathscr{L}^{-1}\left(\frac{1}{s^{1/2}(s-1)}\right) = \frac{1}{\sqrt{\pi t}} * e^t = \int_0^t \frac{e^{t-u}}{\sqrt{\pi u}}\,du = \frac{2e^t}{\sqrt{\pi}}\int_0^{\sqrt{t}} e^{-v^2}\,dv$$

putting $u = v^2$.

Hence $\mathscr{L}^{-1}\left(\frac{1}{s^{1/2}(s-1)}\right) = e^t \operatorname{erf}(\sqrt{t})$

Then using 8.10(ii)

$$\mathscr{L}^{-1}\left(\frac{1}{s(s+1)^{1/2}}\right) = \operatorname{erf}(\sqrt{t})$$

Finally note that Laplace transforms of functions defined by power series can sometimes be found by termwise integration.

Example 8.13

By definition,

$$\mathscr{L}(J_0(2\sqrt{at})) = \int_0^\infty e^{-st}\sum_{n=0}^{\infty}\frac{(-1)^n(at)^n}{(n!)^2}\,dt$$

The power series can be integrated termwise by the results of Chapter 4, and the fact that the integral is uniformly convergent (see Theorem 10.16 and related results). Then

$$\mathscr{L}(J_0(2\sqrt{at})) = \sum_{n=0}^{\infty}\frac{(-a)^n}{(n!)^2}\mathscr{L}(t^n) = \sum_{n=0}^{\infty}\frac{(-a/s)^n}{n!s} = \frac{e^{-a/s}}{s}$$

within the domain of convergence.

Like Fourier transforms, Laplace transforms can be used to solve certain linear integral equations. Consider the equation

$$\alpha f(t) = g(t) + \int_0^t K(t-u)f(u)\,du \quad \text{for } t \in [0, \infty) \tag{8.9}$$

where f is an unknown function to be determined, g and K are known functions and $\alpha = 0$ or 1. Taking the Laplace transform of (8.9), assuming that the transforms of f, g and K exist and sufficient conditions for convolution and inversion are met, it follows from Theorems 8.12 and 8.14 that

$$\bar{f}(s) = \frac{\bar{g}(s)}{\alpha - \bar{K}(s)} \quad \Rightarrow \quad f(t) = \frac{1}{2\pi i}\int_{b-i\infty}^{b+i\infty} e^{st}\frac{\bar{g}(s)}{\alpha - \bar{K}(s)}\,ds$$

Tables or the convolution result may also be used to recover $f(t)$ from $\bar{f}(s)$. Alternatively in the case $\alpha = 1$,

$$\frac{\bar{g}(s)}{1-\bar{K}(s)} = \bar{g}(s) + \sum_{n=1}^{\infty} (\bar{K}(s))^n \bar{g}(s)$$

for $|\bar{K}(s)| < 1$ by the binomial series. By Theorem 8.14,

$$\mathcal{L}^{-1}(K(s))^n = \underbrace{K(t) * \ldots * K(t)}_{n} = K_n(t) \quad \text{say}$$

Then by 8.14 again,

$$f(t) = g(t) + \sum_{n=1}^{\infty} \int_0^t K_n(t-u) g(u)\, du$$

This is the **Liouville–Newman series** solution of (8.9).

Example 8.14

Consider the equation

$$f(t) = \sin t - \int_0^t \frac{f(u)}{\sqrt{\pi(t-u)}}\, du$$

Taking Laplace transforms and using 8.9 and 8.14 gives

$$\bar{f}(s) = \frac{1}{1+s^2} - \frac{\bar{f}(s)}{s^{1/2}} \quad \Rightarrow \quad \bar{f}(s) = \frac{s}{1+s^2} \frac{1}{s^{1/2}(s^{1/2}+1)}$$

From Exercises 8.2.10,

$$\mathcal{L}^{-1}\left(\frac{1}{s^{1/2}(s^{1/2}+1)}\right) = e^t \operatorname{erfc}(\sqrt{t})$$

so that

$$f(t) = \int_0^t e^u \operatorname{erfc}(\sqrt{u}) \cos(t-u)\, du$$

by 8.9 and 8.14. Alternatively,

$$\bar{f}(s) = \frac{1}{1+s^2}\left(1 + \sum_{n=1}^{\infty} \frac{(-1)^n}{s^{n/2}}\right) \qquad (\text{for } |s| > 1)$$

$$\text{Also} \quad \mathcal{L}^{-1}((-1)^n s^{-n/2}) = \frac{(-1)^n t^{n/2-1}}{\Gamma(n/2)}$$

$$\text{Then} \quad f(t) = \sin t + \sum_{n=1}^{\infty} \frac{(-1)^n}{\Gamma(n/2)} \int_0^t (t-u)^{n/2-1} \sin u\, du$$

Exercise 8.2.1

(a) Prove Lemma 8.9(i) and (ii).

(b) Prove by induction that $\mathscr{L}(t^n) = n!/s^{n+1}$ for $n \in \mathbb{Z}_{\geqslant 0}$, $\operatorname{Re} s > 0$.

Exercise 8.2.2

(i) Find $\mathscr{L}\left(\dfrac{e^{-t} - e^{-3t}}{t}\right)$

Hence evaluate $\displaystyle\int_0^\infty \frac{e^{-t} - e^{-3t}}{t}\,dt$

(ii) Find $\mathscr{L}\left(\dfrac{2(\cos at - \cos bt)}{t}\right)$ $\quad (a, b \in \mathbb{R})$

Exercise 8.2.3 Given that

$$\mathscr{L}\left(\frac{\cos(2\sqrt{at})}{\sqrt{\pi t}}\right) = \frac{e^{-a/s}}{s^{1/2}} \quad \text{find} \quad \mathscr{L}\left(\frac{\sin(2\sqrt{at})}{\sqrt{\pi a}}\right) \qquad (a \in \mathbb{R})$$

Exercise 8.2.4 Use the shift theorem to find the Laplace transform of the **half-rectified sine wave** function defined by

$$f(t) = \begin{cases} \sin(\pi t/a) & 2na \leqslant t < (2n+1)a \\ 0 & (2n+1)a \leqslant t < (2n+2)a \end{cases}$$

for all non-negative integers n.

Exercise 8.2.5 Use partial fractions to find the inverse Laplace transforms of the following functions:

(i) $\bar{f}(s) = \dfrac{1}{s^2(s^2+1)}$

(ii) $\bar{f}(s) = \dfrac{s}{s^4 + 4s^2 + 3}$

(iii) $\bar{f}(s) = \dfrac{1}{s^4 - 81}$

(iv) $\bar{f}(s) = \dfrac{1}{(s^2+1)^2}$

Exercise 8.2.6 Use Corollary 8.13 to find the inverse Laplace transforms of the following rational functions, where $a \in \mathbb{R}^+$:

(i) $\bar{f}(s) = \dfrac{1}{s^2 - a^2}$

(ii) $\bar{f}(s) = \dfrac{s^2}{(s^2 + a^2)^2}$

(iii) $\bar{f}(s) = \dfrac{1}{(s^2+1)^3}$

Exercise **8.2.7** Use the Laplace inversion result and the technique of Example 8.9 to find the inverse Laplace transforms of the following functions:

(i) $\bar{f}(s) = \dfrac{e^{-as}}{s}, a \in \mathbb{R}^+$

(ii) $\bar{f}(s) = \dfrac{e^{-s}}{(s+1)^3}$

Exercise **8.2.8** Use the Laplace inversion result and the technique of Example 8.11 to show that

$$\mathscr{L}^{-1}\left(\frac{e^{-as^{1/2}}}{s^{1/2}}\right) = \frac{1}{\pi}\int_0^\infty \frac{e^{-xt}\cos(a\sqrt{x})}{\sqrt{x}}\,dx = I(a)$$

say, where $a \in \mathbb{R}$. By differentiating under the integral sign with respect to a, deduce that $I'(a) = (-a/2t)\,I(a)$. Hence deduce that

$$\mathscr{L}^{-1}\left(\frac{e^{-as^{1/2}}}{s^{1/2}}\right) = \frac{e^{-a^2/4t}}{\sqrt{\pi t}}$$

Use this result to find $\mathscr{L}^{-1}((1/s)e^{-as^{1/2}})$ and $\mathscr{L}^{-1}(e^{-as^{1/2}})$.

Exercise **8.2.9** It can be shown using residue theory (see Chapter 5) that

$$\int_0^\infty \frac{dr}{r^\lambda(1+r)} = \frac{\pi}{\sin\lambda\pi} \qquad (0 < \lambda < 1)$$

Let $r = u/(t-u)$ in this result and use Theorem 8.14 to show that $\Gamma(\lambda)\Gamma(1-\lambda) = \pi/(\sin\lambda\pi)$. This gives a part proof of Theorem 6.12(ii).

Exercise **8.2.10** Use the Laplace convolution result to show that

$$\mathscr{L}^{-1}\left(\frac{1}{s^{1/2}(s-a)}\right) = \frac{e^{at}}{\sqrt{a}}\operatorname{erf}(\sqrt{at})$$

where a is a real constant. Hence find

$$\mathscr{L}^{-1}\left(\frac{1}{s^{1/2}(s^{1/2}+a)}\right) \quad \text{and} \quad \mathscr{L}^{-1}\left(\frac{1}{s^{1/2}+a}\right)$$

Exercise **8.2.11** Expand $f(t) = (1/\sqrt{\pi t})\cos 2(\sqrt{at})$ as a series about $t = 0$ to show that $\mathscr{L}(f(t)) = s^{-1/2}e^{-a/s}$.

Exercise **8.2.12** Let f be a function of class E for which $g(t) = \int_0^\infty f(u)J_0(2\sqrt{ut})\,du$ is well defined and of class E. Expand $J_0(2\sqrt{ut})$ as a Taylor series about $u = 0$ to prove that if $\mathscr{L}(f(t)) = \bar{f}(s)$ then

$$\mathscr{L}(g(t)) = \frac{\bar{f}(1/s)}{s}$$

Hence find $\int_0^\infty e^{-au}J_0(b\sqrt{u})\,du$ where $a, b \in \mathbb{R}^+$.

Exercise **8.2.13** Use Laplace transforms to solve the following integral equations for $t > 0$:

(i) $f(t) = e^t + \int_0^t f(u)\, du$

(ii) $f(t) = t^n + \int_0^t \sin(t-u) f(u)\, du,\ n \in \mathbb{N}$

Exercise **8.2.14** Use Laplace transforms to show that Abel's integral equation

$$g(t) = \int_0^t \frac{f(u)}{\sqrt{t-u}}\, du$$

has solution

$$f(t) = \frac{1}{\pi}\frac{d}{dt}\left(\int_0^t \frac{g(u)}{\sqrt{t-u}}\, du\right)$$

Find the explicit solution in the case $g(t) = t^2$.

Exercise **8.2.15** Find the Liouville–Newman series solution of

$$f(t) = g(t) + \int_0^t (t-u) f(u)\, du \qquad (t > 0)$$

Applications of Laplace Transforms to Ordinary Differential Equations

The Laplace transform can be applied to solve certain linear ordinary differential equations in which the domain of the independent variable is $[0, \infty)$, so that the equation describes a physical process which evolves with time. In many cases, a solution obtained using the Laplace transform may well be valid for a larger domain than $[0, \infty)$. The Laplace transform is particularly useful when a complete set of intial conditions are given, since the unique solution can then be obtained without having to introduce arbitrary constants. It is also a useful tool for solving certain linear equations with discontinuous right-hand sides. We give a brief treatment of these ideas in this section in order to illustrate an important use of Laplace transforms, although the material is not directly related to complex analysis. We do not deal with applications of Fourier transforms to ordinary differential equations explicitly, but a corresponding approach using such transforms can be made to equations in which the independent variable has domain $\mathbb{R}$.

We begin with an introductory example to indicate the technique.

Example 8.15

Consider the initial value problem

$$y''(t) + y(t) = 3\sin 2t \qquad t \in [0, \infty), y'(0) = 1, y'(0) = -2$$

Taking the Laplace transform of the equation using Theorems 8.9 and 8.10 gives

$$(s^2\bar{y}(s) - sy(0) - y'(0)) + \bar{y}(s) = \frac{6}{s^2+4}$$

$$\Rightarrow \quad \bar{y}(s) = \frac{6 + (s-2)(s^2+4)}{(s^2+1)(s^2+4)} = \frac{s}{s^2+1} - \frac{2}{s^2+4}$$

$$\Rightarrow \quad y(t) = \cos t - \sin 2t$$

Note that the solution is valid for all $t \in \mathbb{R}$ in this case.

In general, consider the nth order linear ordinary differential equation

$$a_n(t)y^{(n)}(t) + a_{n-1}(t)y^{(n-1)}(t) + \ldots + a_1(t)y'(t) + a_0(t)y(t) = f(t)$$

$$\text{for } t \in [0, \infty) \tag{8.10}$$

where the initial values $y^{(i)}(0), i = 1, \ldots, n-1$, are given.

Suppose, first of all, that the a_i are all constants. Then taking the Laplace transform of (8.10) gives an algebraic equation which can be solved for $\bar{y}(s)$. This can then theoretically be inverted to give the unique solution to the initial value problem. More precisely, assuming that the $y^{(i)}$ and f are of class E, taking the Laplace transform of (8.10) gives, using 8.10,

$$a_n(s^n\bar{y}(s) - s^{n-1}y(0) - s^{n-2}y'(0) - \ldots - y^{(n-1)}(0))$$
$$+ a_{n-1}(s^{n-1}\bar{y}(s) - s^{n-2}y(0) - \ldots - y^{(n-2)}(0))$$
$$+ \ldots + a_1(s\bar{y}(s) - y(0)) + a_0\bar{y}(s) = \bar{f}(s)$$

When initial values are substituted, this gives

$$P(s)\bar{y}(s) + Q(s) = \bar{f}(s) \quad \Rightarrow \quad y(t) = \mathcal{L}^{-1}\left(\frac{\bar{f}(s) - Q(s)}{P(s)}\right)$$

where P and Q are polynomials in s of degree n and of maximum degree $n - 1$ respectively. Inverting using tables, the inversion formula or the convolution result, gives $y(t)$. It is usually possible to find $y(t)$ explicitly if $\bar{f}(s)$ is an algebraic fraction. The method will work even if $f(t)$ is discontinuous, as long as $\bar{f}(s)$ exists. (In general, $y(t)$ is differentiable with continuous derivative as long as $f(t)$ is piecewise continuous.) It is not usually necessary to justify each step of the process – the fact that the formal solution obtained by the process is the unique solution of the problem is enough. Of course, the method will fail if the Laplace transform of $y^{(i)}(t)$ for some i, or $f(t)$ does not exist.

Example 8.16

Consider the initial value problem

$$y'''(t) - 3y''(t) + 3y'(t) - y(t) = e^{2t}$$
$$y(0) = 0 \qquad y'(0) = y''(0) = 1 \qquad t \in [0, \infty)$$

Taking the Laplace transform gives, by Theorems 8.9 and 8.10,

$$(s^3 - 3s^2 + 3s - 1)\bar{y}(s) + (2 - s) = \frac{1}{s-2}$$

$$\Rightarrow \quad \bar{y}(s) = \frac{s^2 - 4s + 5}{(s-1)^3(s-2)} = \frac{1}{s-2} - \frac{1}{s-1} - \frac{2}{(s-1)^3}$$

$$\Rightarrow \quad y(t) = e^{2t} - e^t - e^t t^2$$

Note that once again the solution is valid for $t \in \mathbb{R}$.

Example 8.17

Let $f(t) = n + 1$ for $n\pi \leqslant t < (n+1)\pi$, $n \in \mathbb{Z}_{\geqslant 0}$ and consider the initial value problem

$$y''(t) + y(t) = f(t) \qquad y(0) = y'(0) = 0$$

Taking the Laplace transform using 8.10 and 8.11 gives

$$(s^2 + 1)\bar{y}(s) = \mathscr{L}\left(\sum_{k=0}^{\infty} H(t - k\pi)\right) = \frac{1}{s}\sum_{k=0}^{\infty} e^{-ks\pi}$$

$$\Rightarrow \quad \bar{y}(s) = \left(\frac{1}{s} - \frac{s}{s^2+1}\right)\sum_{k=0}^{\infty} e^{-ks\pi} \quad \Rightarrow \quad y(t) = f(t) - g(t)$$

where $g(t) = \sum_{k=0}^{\infty} \cos(t - k\pi)H(t - k\pi)$, by Theorems 8.9 and 8.11. This solution is pictured in Fig. 8.8.

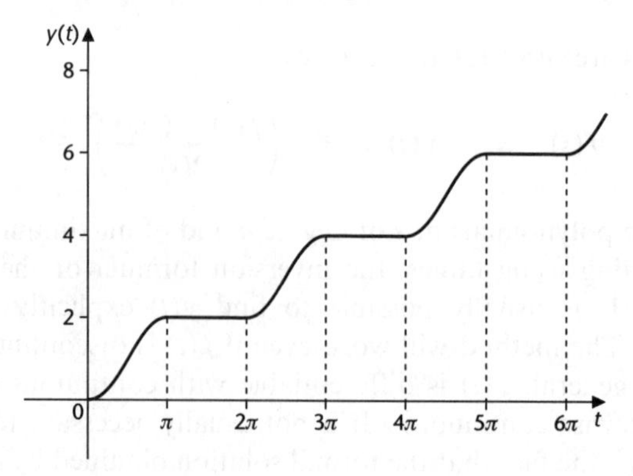

Figure 8.8

Now suppose that some of the coefficients $a_i(t)$ in (8.10) are polynomials in t. Taking the Laplace transform of (8.10), using Theorem 8.10, will clearly lead to another ordinary differential equation, this time in $\bar{y}(s)$. This differential equation will only have order less than n if the polynomials have degree at most $n-1$. Such an equation is a **Laplace linear equation**. The second-order case was first discussed in Chapter 7. If this is not the case, there is no advantage in taking Laplace transforms. The general Laplace linear equation of order 2 is

$$(a_2 + b_2 t)y''(t) + (a_1 + b_1 t)y'(t) + (a_0 + b_0 t)y(t) = f(t) \tag{8.11}$$

where the a_i and b_i are constants. Assuming that the necessary Laplace transforms and initial conditions exist, taking the Laplace transform of (8.11) using Theorem 8.10 gives

$$\left(a_2 - b_2\frac{d}{ds}\right)(s^2\bar{y}(s) - sy(0) - y'(0)) + \left(a_1 - b_1\frac{d}{ds}\right)(s\bar{y}(s) - y(0)) + \left(a_0 - b_0\frac{d}{ds}\right)\bar{y}(s) = \bar{f}(s)$$

After simplification this equation reduces to a linear first-order equation of the form

$$P(s)\bar{y}'(s) + Q(s)\bar{y}(s) = R(s) - \bar{f}(s)$$

where P and Q are polynomials of degree at most 2, and R is a polynomial of degree at most 1.

Example 8.18

Recall that the Bessel function $J_0(t)$ satisfies the Laplace linear equation

$$ty''(t) + y'(t) + ty(t) = 0 \quad \text{with } J_0(0) = 1$$

Taking the Laplace transform of this equation and simplifying,

$$(s^2 + 1)\bar{y}'(s) + s\bar{y}(s) = 0$$

In this case, the resulting linear equation is variables-separable and is easily integrated to give $\bar{y}(s) = k/(1 + s^2)^{1/2}$, where k is a constant. In general, $\mathscr{L}(f'(t)) = s\bar{f}(s) - f(0+)$ by 8.10, so it follows that $\lim_{s\to\infty} s\bar{f}(s) = f(0+)$. This gives $k = 1$ so that

$$\mathscr{L}(J_0(t)) = \frac{1}{(1 + s^2)^{1/2}}$$

Hence, for example, $\int_0^\infty e^{-t}J_0(t)\,dt = \frac{1}{2}\sqrt{2}$.

For a more detailed account of the use of Laplace transforms in solving ordinary differential equations, including their use in stability theory, see

Chapter 7 in A. D. Wunsch, *Complex Variables with Applications*, 2nd edn, Addison-Wesley, 1994.

Exercise **8.3.1** Using Laplace transforms, solve the following constant coefficient equations with the given initial conditions:

(i) $8(y''(t) + y(t)) = 3\sin 2t$, $y(0) = y'(0) = 0$

(ii) $y''(t) - 2y'(t) - 3y(t) = e^{-2t}, y(0) = -1, y'(0) = 3$

(iii) $y'''(t) + 2y''(t) + 2y'(t) = t^2, y(0) = y''(0) = 0, y'(0) = 2$

(iv) $y^{(4)}(t) + 8y''(t) + 16y(t) = t, y^{(i)}(0) = 0, i = 0,1,2,3$

Exercise **8.3.2** Solve these simultaneous differential equations by taking the Laplace transform of each one:

$$\begin{cases} x'(t) + 2x(t) - y'(t) = 3 \\ 3x'(t) + y'(t) - 2y(t) = 0 \end{cases} \qquad x(0) = y(0) = 0$$

Exercise **8.3.3** Use Laplace transforms to solve the following initial value problem.

$$y''(t) + 2y'(t) + y(t) = \begin{cases} e^t & t < 1 \\ 0 & t \geqslant 1 \end{cases}$$

$$y(0) = 0, y'(0) = 1$$

Exercise **8.3.4** Find the Laplace transform of the function f in Fig. 8.9. Hence solve the initial value problem

$$y''(t) - 3y'(t) + 2y(t) = f(t) \qquad y(0) = y'(0) = 0$$

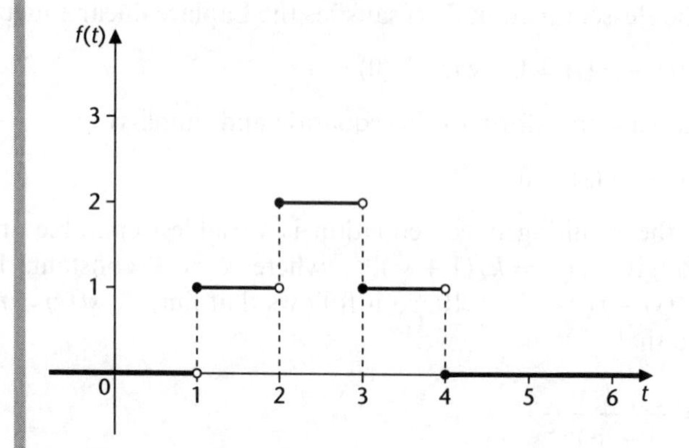

Figure 8.9

Exercise **8.3.5** Find the Laplace transform of the sawtooth wave function f in Fig. 8.10. Hence solve the initial value problem

$$y''(t) - y'(t) - 2y(t) = f(t) \qquad y(0) = 0, y'(0) = 1$$

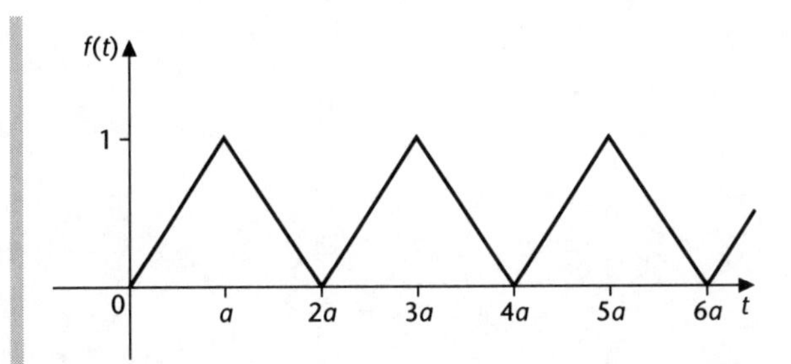

Figure 8.10

Exercise

8.3.6 Using Laplace transforms, solve the following Laplace linear equation with the given initial condition:

$$ty''(t) - (1 + 2t)y'(t) + (1 + t)y(t) = 0 \qquad y(0) = 0$$

Exercise

8.3.7 Use Laplace transforms to find the polynomial solution of Hermite's equation

$$y''(t) - 2ty'(t) + 4y(t) = 0$$

which satisfies $y(0) = -2, y'(0) = 0$.

Exercise

8.3.8 Let $n \in \mathbb{N}$ and let $y(t) = f_n(t)$ be a solution of Laguerre's differential equation

$$ty''(t) + (1 - t)y'(t) + ny(t) = 0$$

which satisfies $y(0) = 1$. Use Theorems 8.10 and 8.14 to show that

$$\int_0^t f_{m+n}(u)\,du = \int_0^t f_m(u)f_n(t-u)\,du$$

9 Laplace's Equation and Other Linear Partial Differential Equations

This chapter is particularly concerned with how complex variable techniques can be used to find specific solutions of Laplace's equation in two dimensions, and how these techniques can be applied in physical problems. We also demonstrate how Fourier and Laplace transforms can be used to solve Laplace's equation and certain other important partial differential equations in mathematical physics, calling upon some of the results of the previous chapter. We end by taking a brief look at the technique of separation of variables and demonstrate how some of the special functions introduced in Chapter 7 are of importance in practical applications. The emphasis in the later sections of this chapter is on methods and their applications.

Harmonic Functions

Recall, from Chapter 2, that a function $\phi(x, y)$ is **harmonic** in a region $\mathscr{R} \subseteq \mathbb{C}$ if it has continuous second-order partial derivatives and satisfies Laplace's equation,

$$\nabla^2\phi = \phi_{xx} + \phi_{yy} = 0 \tag{9.1}$$

in $\mathscr{R}$. Let $f: A \subseteq \mathbb{C} \to \mathbb{C}$ be analytic in $\mathscr{R}$ and let f be defined by

$$f(z) = f(x + iy) = u(x, y) + iv(x, y) \tag{9.2}$$

It follows from Theorem 2.4 that both $u(x, y)$ and $v(x, y)$ are harmonic in $\mathscr{R}$. The function v is a **harmonic conjugate** of u.

Laplace's equation can be expressed in terms of different coordinate systems by using the chain rule. In particular, in terms of plane polar coordinates (r, θ), equation (9.1) becomes

$$\nabla^2\phi = \phi_{rr} + \frac{1}{r}\phi_r + \frac{1}{r^2}\phi_{\theta\theta} = 0 \tag{9.3}$$

Then if f is analytic on $\mathscr{R}$ and is defined by

$$f(z) = f(re^{i\theta}) = u(r, \theta) + iv(r, \theta) \tag{9.4}$$

it is easily checked, using the Cauchy–Riemann equations in polar form, equation (2.6), that $u(r, \theta)$ and $v(r, \theta)$ both satisfy equation (9.3) on $\mathscr{R}$.

Example 9.1

(i) The analytic functions $f_0(z) = 1$, $f_1(z) = z$, $f_2(z) = z^2$, $f_3(z) = z^3$, $\ldots$, generate the harmonic functions $u_0(x, y) = 1$, $u_1(x, y) = x$, $v_1(x, y) = y$, $u_2(x, y) = x^2 - y^2$, $v_2(x, y) = 2xy$, $u_3(x, y) = x^3 - 3xy^2$, $v_3(x, y) = 3x^2y - y^3, \ldots$.

(ii) Let $f(z) = f(re^{i\theta}) = z^n = u(r, \theta) + iv(r, \theta)$, $n \in \mathbb{N}$. Then $u(r, \theta) = r^n \cos n\theta$ and $v(r, \theta) = r^n \sin n\theta$ are harmonic functions which satisfy (9.3).

(iii) Let $f(z) = \operatorname{Log} z$, $z \neq 0$ and $-\pi < \operatorname{Arg} z < \pi$. Then f is analytic and letting $f(re^{i\theta}) = u(r, \theta) + iv(r, \theta)$ gives the harmonic functions

$$u(r, \theta) = \operatorname{Log} r,\ r \neq 0 \quad \text{and} \quad v(r, \theta) = \theta \qquad (-\pi < \theta < \pi)$$

Clearly, given any two harmonic functions, $u(x, y)$ and $v(x, y)$, if v is not a harmonic conjugate of u, then f given by (9.2) is not analytic. However, the following result, which is of particular importance, shows the close connection between harmonic and analytic functions.

Theorem 9.1. Any Harmonic Function is the Real Part of an Analytic Function

Let $\mathscr{R} \subseteq \mathbb{C}$ be a simply connected region and let u be harmonic on $\mathscr{R}$. Then there exists a function $f: A \subseteq \mathbb{C} \to \mathbb{C}$, analytic on $\mathscr{R}$, such that $u = \operatorname{Re} f$. □

Proof

Let $u = u(x, y)$ and define g on $\mathscr{R}$ by

$$g(z) = g(x + iy) = U(x, y) + iV(x, y) = u_x - iu_y$$

The clue for this choice comes from Corollary 2.5.

Then since u is harmonic on $\mathscr{R}$,

$$U_x = u_{xx} = -u_{yy} = V_y \qquad U_y = u_{xy} = u_{yx} = -V_x$$

so that U and V satisfy the Cauchy–Riemann equations on $\mathscr{R}$ and have continuous first-order partial derivatives on $\mathscr{R}$. Hence by Theorem 2.6, g is analytic on $\mathscr{R}$. Then by 3.9 and 3.5 there exists an analytic function h on $\mathscr{R}$ such that $h' = g$. It then follows by 2.5 that

$$h'(z) = g(z) = (\operatorname{Re} h)_x - i(\operatorname{Re} h)_y \quad \Rightarrow \quad (u - \operatorname{Re} h)_x = (u - \operatorname{Re} h)_y = 0$$

so that $u - \operatorname{Re} h = k$, where k is a real constant. Letting $f(z) = h(z) + k$ gives the result. ■

Note

We can use the same technique to show that if v is harmonic on $\mathscr{R}$, then there exists an analytic function f on $\mathscr{R}$ such that $v = \operatorname{Im} f$. This is left as an exercise.

Example 9.2

Let $u(x, y) = \sin x \cosh y$. It is easily checked that u is harmonic on $\mathbb{C}$. Following the constructive proof of Theorem 9.1, let

$$g(z) = u_x - iu_y = \cos x \cosh y - i \sin x \sinh y = \cos(x + iy).$$

(The fact that $g(z) = \cos(x + iy)$ is suggested by putting $y = 0$.) Then $g(z) = \cos z$ and so

$$u = \operatorname{Re} f \quad \text{where} \quad f(z) = \sin z + ik \qquad (k \in \mathbb{R})$$

It follows that $v(x, y) = \cos x \sinh y$ is a complex conjugate of u.

Many problems in mathematical physics reduce to a **Dirichlet problem**, which involves finding a function that is harmonic on a region, given its values on the boundary of the region. In simple cases, it is often possible to see a solution directly.

Example 9.3

(i) Clearly, a function which is harmonic inside the unit circle $|z| = 1$ and takes the value 1 on $|z| = 1$ is $\phi(x, y) = 1$.

(ii) A function $\phi(x, y)$ which is harmonic on the strip $0 < x < 1$ with $\phi(0, y) = 0$ and $\phi(1, y) = 1$ is $\phi(x, y) = x$ (Fig. 9.1).

(iii) Suppose we wish to find a function $\phi(x, y)$, which is harmonic on $\operatorname{Im} z > 0$, such that $\phi(x, y) = 0$ on the positive x-axis and $\phi(x, y) = 1$ on the negative x-axis, as shown in Fig. 9.1. Since $v(r, \theta) = \theta$ is harmonic on $\operatorname{Im} z > 0$, it is clear that we should begin with the function

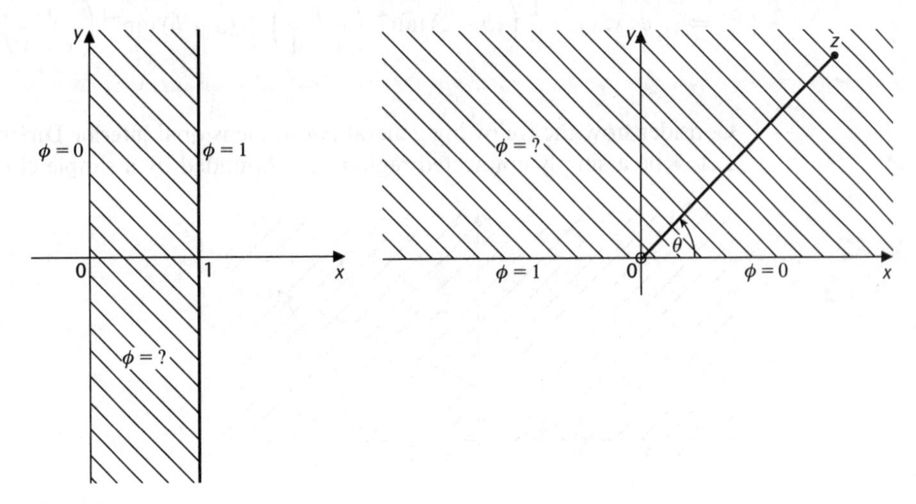

Figure 9.1

$\phi(r, \theta) = a\theta + b$, where a and b are constants. Then $\phi(r, 0) = 0 \Rightarrow b = 0$ and $\phi(r, \pi) = 1 \Rightarrow a = 1/\pi$. Hence $\phi(r, \theta) = \theta/\pi$, so that

$$\phi(x, y) = \frac{1}{\pi}\tan^{-1}(y/x) \qquad (0 \leqslant \tan^{-1}(y/x) \leqslant \pi)$$

It can be shown that the functions derived in Example 9.3 are the unique solutions to the given problems. The following example, which builds upon the idea of Example 9.3(iii), will prove very useful when it comes to applications.

Example 9.4

Find a function $\phi(x, y)$, which is harmonic on $\operatorname{Im} z > 0$, such that $\phi(x, 0) = a$ for $x < -1$, $\phi(x, 0) = b$ for $-1 < x < 1$ and $\phi(x, 0) = c$ for $x > 1$, as shown in Fig. 9.2.

Solution

Following the approach of Example 9.3(iii), begin with the function $\phi(x, y) = A\theta_1 + B\theta_2 + C$, where the angles θ_1 and θ_2 are indicated in Fig. 9.2 and A, B and C are constants. Note that ϕ is harmonic on $\operatorname{Im} z > 0$, since it is the imaginary part of the function $f(z) = A\operatorname{Log}(z-1) + B\operatorname{Log}(z+1) + C$, which is analytic on $\operatorname{Im} z > 0$. When $\theta_1 = \theta_2 = 0$, $\phi = c$, so that $C = c$. When $\theta_1 = \pi$ and $\theta_2 = 0$, $\phi = b$, so that $A\pi + c = b$. When $\theta_1 = \theta_2 = \pi$, $\phi = a$, so $(A+B)\pi + c = a$. Then the required function is

$$\phi(x, y) = \frac{1}{\pi}((b-c)\theta_1 + (a-b)\theta_2) + c$$

$$\Rightarrow \quad \phi(x, y) = \frac{1}{\pi}\left((b-c)\tan^{-1}\left(\frac{y}{x-1}\right) + (a-b)\tan^{-1}\left(\frac{y}{x+1}\right)\right) + c$$

Consider now the following general two-dimensional **interior Dirichlet problem**. Let $\mathscr{R}$ be a simply connected region of $\mathbb{C}$ bounded by a simple closed curve $\mathscr{C}$,

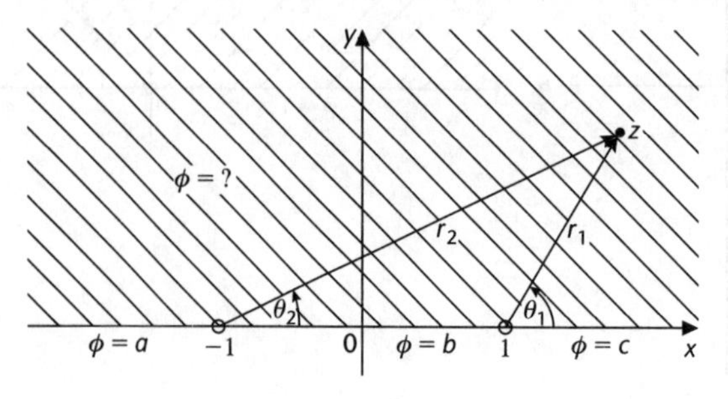

Figure 9.2

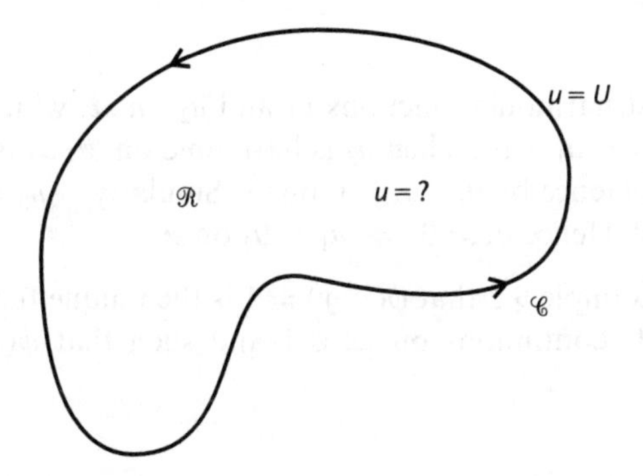

Figure 9.3

as shown in Fig. 9.3. Let $U = U(x, y)$ be a continuous function defined on $\mathscr{C}$. The question is, Can we always find a function $u = u(x, y)$, which is harmonic on $\mathscr{R}$, continuous on $\overline{\mathscr{R}} = \mathscr{R} \cup \mathscr{C}$, and such that $u = U$ on $\mathscr{C}$? It is by no means obvious that such a function exists. We examine the problem of existence shortly. However, if a solution to such a problem does exist, it is a simple matter to show that the solution is unique. This is a nice application of the maximum modulus theorem, 3.18. We begin with the following corollary.

Corollary 9.2 The Maximum Principle for Harmonic Functions

Let $\mathscr{R}$ be a simply connected region bounded by a simple closed curve $\mathscr{C}$. Let u be harmonic on $\mathscr{R}$ and continuous on $\overline{\mathscr{R}}$. Suppose that $u \leqslant M$ on $\mathscr{C}$, where M is a real constant. Then $u \leqslant M$ on $\overline{\mathscr{R}}$, i.e., u attains its maximum value on the boundary of $\mathscr{R}$. □

Note

Since u is continuous on $\mathscr{C}$, it has a maximum value on $\mathscr{C}$, by Theorem 4.5.

Proof

By 9.1, there exists a function f, analytic on $\mathscr{R}$ and continuous on $\overline{\mathscr{R}}$, such that $u = \operatorname{Re} f$. Let $g(z) = e^{f(z)}$ on $\mathscr{R}$. Then g is analytic on $\mathscr{R}$, continuous on $\overline{\mathscr{R}}$ and is never zero. Then $|g(z)| = e^u$ attains its maximum value on $\mathscr{C}$ by 3.18. Since e^u is a monotonic function of u, it follows that u attains its maximum value on $\mathscr{C}$. ■

Theorem 9.3 Uniqueness of Solution to the Interior Dirichlet Problem

Let $\mathscr{R}$ be a simply connected region bounded by $\mathscr{C}$. If there exists a harmonic function u on $\mathscr{R}$, which is continuous on $\overline{\mathscr{R}}$ and such that $u = U$ on $\mathscr{C}$, then u is unique. □

Proof

Suppose that there exist harmonic functions u_1 and u_2 on $\mathscr{R}$, which satisfy the given conditions. Let $u_3 = u_1 - u_2$. Then u_3 is harmonic on $\mathscr{R}$, u_3 is continuous on $\overline{\mathscr{R}}$ and $u_3 = 0$ on $\mathscr{C}$. Hence by 9.2, $u_3 \leqslant 0$ on $\mathscr{R}$. Similarly, $-u_3 \leqslant 0$ on $\mathscr{R}$ by 9.2, so that $u_3 \geqslant 0$ on $\mathscr{R}$. Hence $u_3 = 0 \Rightarrow u_1 = u_2$ on $\mathscr{R}$. ■

It follows by 9.3 and Example 9.3 that $\phi(x, y) = 1$ is the unique function which is harmonic for $|z| < 1$, continuous on $|z| \leqslant 1$ and such that $\phi(x, y) = 1$ on $|z| = 1$.

Note

The uniqueness result can easily be generalised to the case where U is piecewise continuous on $\mathscr{C}$.

Another general problem often encountered in applications is the two-dimensional **interior Neumann problem**. Once again, let $\mathscr{R}$ be a simply connected region of $\mathbb{C}$ bounded by a simple closed curve $\mathscr{C}$. The question is, Can we find a function $u = u(x, y)$, which is harmonic on $\mathscr{R}$ such that du/dn is continuous on $\mathscr{R}$ with $du/dn = U(x, y)$ on $\mathscr{C}$, where $U(x, y)$ is given on $\mathscr{C}$? Here, du/dn is the normal derivative of u, i.e. $du/dn = \nabla u \cdot \boldsymbol{n}$, where $\boldsymbol{n}$ is a unit vector normal to $\mathscr{C}$, as shown in Fig. 9.4.

The following result is analogous to Theorem 9.3 and is easily proved using the same technique. Its proof is left as an exercise.

Theorem 9.4 Uniqueness of Solution to the Neumann Problem

If there exists a harmonic function u on $\mathscr{R}$, such that du/dn is continuous on $\overline{\mathscr{R}}$, with $du/dn = U$ on $\mathscr{C}$, then u is unique up to an arbitrary constant. □

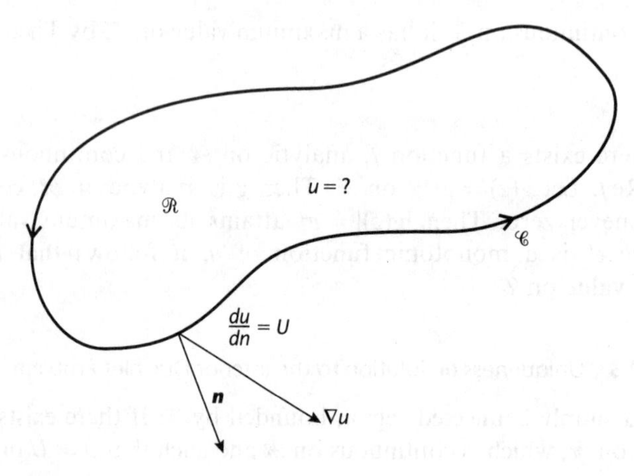

Figure 9.4

The Dirichlet Problem for the Unit Circle

The question of whether or not there exists a solution to the general interior Dirichlet problem is difficult to tackle directly. Instead, we study a particular Dirichlet problem, that for the unit circle. We shall show later how this result can help us to answer the more general question.

The clue to the existence of a harmonic function in the region $|z| < 1$, with given values on $|z| = 1$, lies in the following result.

Theorem 9.5 Poisson's Integral Formula for the Unit Circle

Let $f: A \subseteq \mathbb{C} \to \mathbb{C}$ be analytic inside and on the circle $\mathscr{C}$ given by $|z| = 1$. Then if $z = re^{i\theta}$ is any point inside $\mathscr{C}$,

$$f(re^{i\theta}) = \frac{1}{2\pi}\int_0^{2\pi} \frac{(1-r^2)f(e^{i\phi})}{r^2 - 2r\cos(\theta-\phi)+1}\,d\phi \qquad \square$$

Proof

If z is any point inside the circle $\mathscr{C}$, then by Cauchy's integral formula, $f(z) = (1/2\pi i)\int_{\mathscr{C}}(f(\zeta)/\zeta - z))\,d\zeta$. Also, since z lies inside $\mathscr{C}$, $|z| < 1 \Rightarrow |1/\bar{z}| > 1$, so that the point $1/\bar{z}$ lies outside $\mathscr{C}$. Then by Cauchy's theorem, $\int_{\mathscr{C}}(f(\zeta)/(\zeta - 1/\bar{z}))\,d\zeta = 0$, hence

$$f(z) = \frac{1}{2\pi i}\int_{\mathscr{C}} f(\zeta)\left(\frac{1}{\zeta - z} - \frac{1}{\zeta - 1/\bar{z}}\right)d\zeta = \frac{1}{2\pi i}\int_{\mathscr{C}} f(\zeta)\frac{z - 1/\bar{z}}{(\zeta - z)(\zeta - 1/\bar{z})}\,d\zeta$$

Then letting $z = re^{i\theta}$ and $\zeta = e^{i\phi}$, $0 \leqslant \phi \leqslant 2\pi$ on $\mathscr{C}$ gives

$$f(re^{i\theta}) = \frac{1}{2\pi}\int_0^{2\pi} e^{i(\theta+\phi)}\frac{f(e^{i\phi})(1-r^2)}{e^{i(\phi+\theta)} - r(e^{2i\theta}+e^{2i\phi}) + r^2e^{i(\phi+\theta)}}\,d\phi$$

as required. ■

The following corollary of 9.5 gives the values of a harmonic function inside the unit circle in terms of its values on the circle.

Corollary 9.6 Harmonic Functions Inside and on the Unit Circle

Let $u(r, \theta)$ be harmonic inside and on the circle $\mathscr{C}$, given by $|z| = 1$. Then if (r, θ) is any point inside $\mathscr{C}$,

$$u(r, \theta) = \frac{1}{2\pi}\int_0^{2\pi} \frac{(1-r^2)u(1,\phi)}{r^2 - 2r\cos(\phi-\theta)+1}\,d\phi \qquad \square$$

Proof

Since u is harmonic inside and on $\mathscr{C}$, it follows by 9.1 that there is a function f, analytic inside and on $\mathscr{C}$, such that $u = \operatorname{Re} f$. Then taking the real part of Poisson's integral formula gives the result. ■

Note

This result shows that there exists a solution to the Dirichlet problem for the unit circle, provided that the given function on the circle is harmonic on the circle. It suggests the more general result which follows. This more general result is necessary since in many applications, a given function is *not* harmonic on the boundary.

It is convenient to introduce the following notation for the **Poisson kernel** in Poisson's integral formula.

Notation

Let $P(r, \psi) = \dfrac{1 - r^2}{r^2 - 2r\cos\psi + 1}$

Theorem 9.7 The Solution of the Interior Dirichlet Problem for the Unit Circle

Let $\mathscr{C}$ be the circle $|z| = 1$ and let $U = U(\theta)$, $0 \leqslant \theta \leqslant 2\pi$ be a given continuous function on $\mathscr{C}$. Let $u(1, \theta) = U(\theta)$ and

$$u(r, \theta) = \frac{1}{2\pi}\int_0^{2\pi} P(r, \phi - \theta)U(\phi)\, d\phi \qquad (0 \leqslant r < 1)$$

Then u is harmonic on $|z| < 1$ and is continuous on $|z| \leqslant 1$. □

Proof

Step 1

It is easily checked that $P(r, \phi - \theta) = Re((\zeta + z)/(\zeta - z))$, where $\zeta = e^{i\phi}$ and $z = re^{i\theta}$. Then,

$$u(r, \theta) = \text{Re}\left(\frac{1}{2\pi i}\int_{\mathscr{C}} \frac{(\zeta + z)F(\zeta)}{\zeta(\zeta - z)}\, d\zeta\right) = \text{Re}\left(\frac{1}{\pi i}\int_{\mathscr{C}} \frac{F(\zeta)}{\zeta - z}\, d\zeta - \frac{1}{2\pi i}\int_{\mathscr{C}} \frac{F(\zeta)}{\zeta}\, d\zeta\right)$$

for $r < 1$, where $F(\zeta) = U(\phi)$. Note that the second integral is a multiple of $\int_0^{2\pi} U(\phi)\, d\phi$ and so is a constant, whereas the first integral can be shown to be analytic (this follows by Theorem 10.14). Hence u is the real part of an analytic function for $r < 1$ and so is harmonic inside $\mathscr{C}$.

Step 2

It remains to show that u is continuous on $|z| \leqslant 1$. Let $e^{i\alpha}$, $\alpha \in \mathbb{R}$, be any fixed point on $\mathscr{C}$. We need to show that $u(r, \theta) \to U(\alpha)$ as $r \to 1$ and $\theta \to \alpha$, independent of the direction of approach. Since 1 is harmonic everywhere, it follows by 9.5 that $\int_0^{2\pi} P(r, \phi - \phi)\, d\phi = 2\pi$. Then for $r < 1$,

$$u(r, \theta) - U(\alpha) = \frac{1}{2\pi} \int_0^{2\pi} P(r, \phi - \theta)(U(\phi) - U(\alpha))\, d\phi \tag{9.5}$$

Step 3

Now let $\varepsilon \in \mathbb{R}^+$ be given. Since U is continuous on $\mathscr{C}$, there exists $\beta \in \mathbb{R}^+$ such that $|U(\phi) - U(\alpha)| < \varepsilon/2$ whenever $|\phi - \alpha| < \beta$, i.e. for any ϕ such that $e^{i\phi}$ lies on the arc $\mathscr{C}_1$ as shown in Fig. 9.5. It follows from (9.5) that

$$u(r, \theta) - U(\alpha) = I_1 + I_2 \tag{9.6a}$$

$$\text{where} \quad 2\pi I_1 = \int_{\alpha - \beta}^{\alpha + \beta} P(r, \phi - \theta)(U(\phi) - U(\alpha))\, d\phi \tag{9.6b}$$

$$\text{and} \quad 2\pi I_2 = \int_{\alpha + \beta}^{\alpha - \beta + 2\pi} P(r, \phi - \theta)(U(\phi) - U(\alpha))\, d\phi \tag{9.6c}$$

By the above, since $P(r, \phi - \theta) \geqslant 0$ for $r \leqslant 1$,

$$|I_1| \leqslant \frac{1}{2\pi} \int_{\alpha - \beta}^{\alpha + \beta} P(r, \phi - \theta)|U(\phi) - U(\alpha)| d\phi < \frac{\varepsilon}{4\pi} \int_0^{2\pi} P(r, \phi - \theta)\, d\phi = \frac{\varepsilon}{2} \tag{9.7}$$

Step 4

Now let $\mathscr{C}_2$ be the arc complementary to $\mathscr{C}_1$, i.e. $\mathscr{C}_2 = \mathscr{C} \backslash \mathscr{C}_1$, as shown in Fig. 9.5. It is clear that there exists some $m \in \mathbb{R}^+$ such that for any $\zeta \in \mathscr{C}_2$ and any z inside the shaded sector, where $|\theta - \alpha| < \beta/2$ as shown, $|\zeta - z|^2 \geqslant m$.

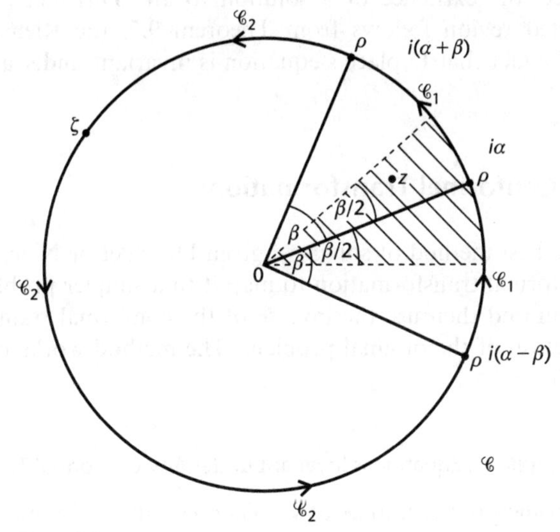

Figure 9.5

Then by (9.6c)

$$|I_2| \leqslant \frac{1}{2\pi}\int_{\alpha+\beta}^{\alpha-\beta+2\pi} \frac{(1-r^2)}{|\zeta - z|^2}|U(\phi) - U(\alpha)|\, d\phi < \frac{2\pi M(1-r^2)}{2\pi m}$$

where $|U(\phi) - U(\alpha)| \leqslant M$ on $\mathscr{C}_2$. Hence, as $1 + r < 2$ for $r < 1$,

$$|I_2| < \frac{2M(1-r)}{m} < \frac{2M\delta}{m} = \frac{\varepsilon}{2} \tag{9.8}$$

whenever $|r - 1| < \delta$, by choosing $\delta = m\varepsilon/4M$.

Finally, from (9.6), (9.7) and (9.8),

$$|u(r, \theta) - U(\alpha)| \leqslant |I_1| + |I_2| < \varepsilon$$

whenever $|r - 1| < \delta$ and $|\theta - \alpha| < \beta/2$, as required. ■

Note

The proof can be adapted to show that if U is piecewise continuous on $\mathscr{C}$, $u = U$ on $\mathscr{C}$ and u is given by the integral formula for $r < 1$, then u is harmonic in $|z| < 1$ and continuous on $|z| \leqslant 1$, except for a finite number of points on $|z| = 1$.

Historical Note

The classical Dirichlet problem in three dimensions was first solved by Green in 1828. A rigorous proof of an existence result was provided by Neumann in 1870.

As we shall see, the existence of a solution to any Dirichlet problem for a simply connected region follows from Theorem 9.7, the Riemann mapping theorem and the fact that Laplace's equation is invariant under any conformal mapping.

Harmonic Functions and Conformal Transformations

In practice, the best method of solving a given Dirichlet or Neumann problem is to use a conformal transformation to map it to a simpler problem, solve the simpler problem and then use the inverse of the conformal transformation to obtain the solution of the original problem. The method works because of the following result.

Theorem 9.8 Laplace's Equation is Invariant under Any Conformal Transformation

Let ϕ be harmonic on a region $\mathscr{R}$ of $\mathbb{C}$. Then ϕ remains harmonic under any conformal transformation. □

Note

We give a proof for the case when $\mathscr{R}$ is simply connected. It is possible to extend this result to any region. However, in the more general case, it is easier to prove the result directly by applying the chain rule to Laplace's equation. This method is given in the exercises.

Proof

Let $\mathscr{R}$ be simply connected and let F be a conformal transformation with $w = u + iv = F(z) = F(x + iy)$. By 9.1, there exists an analytic function f on $\mathscr{R}$ such that $\phi = \operatorname{Re} f$, so that $f(z) = \phi(x, y) + i\psi(x, y)$, where ψ is a harmonic conjugate of ϕ. Let $g(w) = f(z) = f(F^{-1}(w))$. Since F is conformal, g is analytic on $F(\mathscr{R})$. Let $g(w) = \Phi(u, v) + i\,\Psi(u, v)$. Then $g(w) = f(z) \Rightarrow \Phi(u, v) = \phi(x(u, v),\ y(u, v))$ and since $\Phi = \operatorname{Re} g$, Φ is harmonic on $F(\mathscr{R})$, as required. ■

Important Note

Recall that any simply connected region with boundary can be mapped to the unit circle and its inside by some conformal transformation by the Riemann mapping theorem. Hence by Theorems 9.7 and 9.8, the Dirichlet problem for any such region has a solution. However, in general, the Dirichlet problem for an arbitrary region may have no solution. It also follows from 9.8 and Exercise 9.1.12 that the Neumann problem for any simply connected region has a solution.

The solution to the interior Dirichlet problem for the circle $|w| = R$ is easily found from 9.7, using the conformal transformation $w = Rz$. The solution to the **exterior** Dirichlet problem for the circle $|w| = 1$ is found using the conformal transformation $w = 1/z$. These examples are left as exercises.

A specific case, worthy of separate study, is the Dirichlet problem for the half-plane.

Theorem 9.9 The Solution to the Dirichlet Problem for the Half-Plane

Let $U = U(x)$ be a given function, defined and continuous on the real axis, $\operatorname{Im} z = 0$. Let $u(x, 0) = U(x)$ on $\operatorname{Im} z = 0$ and

$$u(x, y) = \frac{y}{\pi}\int_{-\infty}^{\infty} \frac{U(t)}{(t - x)^2 + y^2}\, dt \qquad (y > 0)$$

Then u is harmonic on $\operatorname{Im} z > 0$ and continuous on $\operatorname{Im} z \geqslant 0$. □

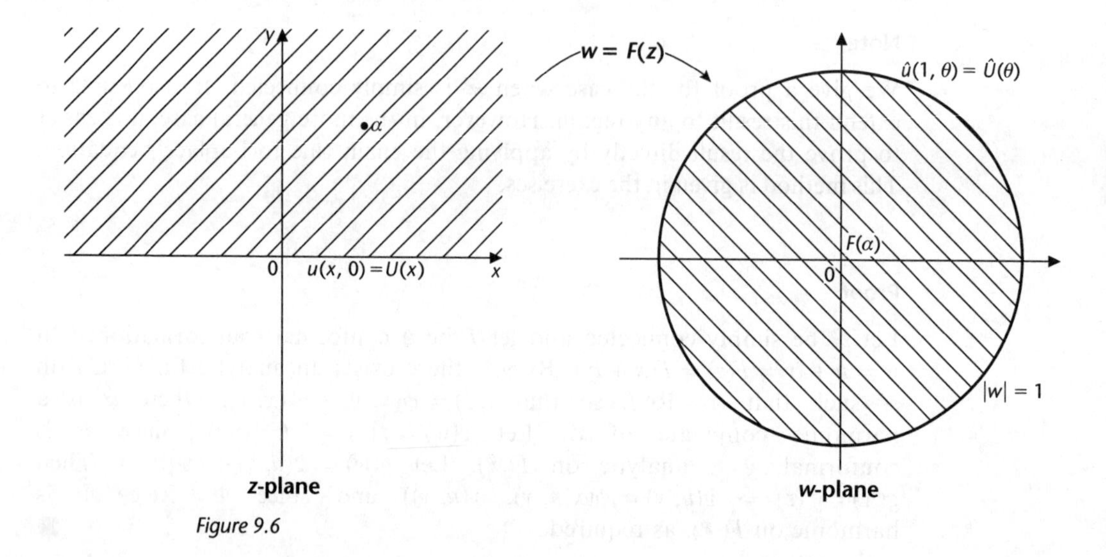

Figure 9.6

Part Proof

Let α be any fixed point in $\operatorname{Im} z > 0$, as shown in Fig. 9.6. The conformal transformation $w = F(z) = (z-\alpha)/(z-\overline{\alpha})$ maps $\operatorname{Im} z = 0$ to the circle $|w| = 1$ and $\operatorname{Im} z > 0$ to $|w| < 1$, as shown. This is because $F(\alpha) = 0$, $|F(0)| = 1$, $F(\infty) = 1$ and it is easily checked that $|F(1)| = 1$. Let $w = re^{i\theta}$ and $\hat{U}(\theta) = F(U(x))$. By 9.7, the function $\hat{u}(r, \theta)$, which is harmonic on $|w| < 1$, continuous on $|w| \leqslant 1$ and such that $\hat{u} = \hat{U}$ on $|w| = 1$ is given by $\hat{u}(r, \theta) = (1/2\pi)\int_0^{2\pi} P(r, \phi-\theta)\hat{U}(\phi)d\phi$ for $r < 1$. In particular, $\hat{u}(0) = (1/2\pi)\int_0^{2\pi} \hat{U}(\phi)\, d\phi$. By 9.8, $u = F^{-1} \circ \hat{u}$ is the solution to the given Dirichlet problem in the z-plane. In particular, $u(\alpha) = F^{-1}(\hat{u}(0))$. Now $F(t) = (t-\alpha)/(t-\overline{\alpha}) = e^{i\phi}$ for any $t \in \mathbb{R}$

$$\Rightarrow \quad ie^{i\phi}d\phi = \frac{\alpha-\overline{\alpha}}{(t-\overline{\alpha})^2}dt \quad \Rightarrow \quad d\phi = \frac{\alpha-\overline{\alpha}}{i(t-\alpha)(t-\overline{\alpha})}dt = \frac{2\operatorname{Im}\alpha}{|t-\alpha|^2}$$

$$\Rightarrow \quad u(\alpha) = F^{-1}(\hat{u}(0)) = \frac{1}{2\pi}\int_{-\infty}^{\infty}\frac{2\operatorname{Im}\alpha}{|t-\alpha|^2}U(t)\,dt$$

(The change of variable in the integral can be justified.) Then

$$u(x, y) = \frac{1}{\pi}\int_{-\infty}^{\infty}\frac{yU(t)}{|t-(x+iy)|^2}dt = \frac{y}{\pi}\int_{-\infty}^{\infty}\frac{U(t)}{(t-x)^2+y^2}dt \qquad \blacksquare$$

Note

The integral formula for $u(x, y)$ in 9.9 is often known as **Schwarz's integral formula**. Once again, 9.9 can be generalised to the case in which U is piecewise continuous on $\operatorname{Im} z = 0$.

Example 9.5

Suppose we wish to find the function $u(x, y)$ which is harmonic on $\operatorname{Im} z > 0$, continuous on $\operatorname{Im} z \geqslant 0$ except at 0, and such that $u(x, 0) = 1$ for $x > 0$ and $u(x, 0) = 0$ for $x < 0$. By Theorem 9.9, the required function is given by

$$u(x, y) = \frac{y}{\pi}\int_0^\infty \frac{dt}{(t-x)^2 + y^2} = \frac{y}{\pi}\int_{-x}^\infty \frac{dT}{T^2 + y^2} = \frac{1}{\pi}\left[\tan^{-1}(T/y)\right]_{-x}^\infty$$

letting $T = t - x$. Hence

$$u(x, y) = \frac{1}{\pi}\left(\frac{\pi}{2} + \tan^{-1}(x/y)\right) = \frac{1}{\pi}\tan^{-1}(-y/x) = 1 - \frac{1}{\pi}\tan^{-1}(y/x)$$

Note how this solution can be obtained directly using the technique of Example 9.3(iii).

It is clear that, in general, the boundary conditions in a given problem will change under a conformal transformation. However, there are two cases in which the boundary conditions remain invariant. One is a particular Dirichlet problem and one is a particular Neumann problem.

Theorem 9.10 Invariance of Particular Boundary Conditions

Let the transformation $w = F(z) = F(x + iy) = u(x, y) + iv(x, y)$ be conformal on a smooth arc $\mathscr{C}$ and let $\mathscr{C}' = F(\mathscr{C})$. If $\phi(x, y)$ satisfies

$$\phi(x, y) = k \quad \text{or} \quad \frac{d\phi}{dn} = 0 \quad \text{along } \mathscr{C}$$

where k is a constant, then $\Phi(u, v) = \phi(x(u, v), y(u, v))$ satisfies

$$\Phi(u, v) = k \quad \text{or} \quad \frac{d\Phi}{dn} = 0 \quad \text{along } \mathscr{C}'$$

□

Proof

If $\phi = k$ on $\mathscr{C}$ then $\Phi = k$ on $\mathscr{C}'$. Suppose now that $d\phi/dn = \nabla\phi \cdot \boldsymbol{n} = 0$ on $\mathscr{C}$, where $\nabla\phi = (\phi_x, \phi_y)$ and $\boldsymbol{n}$ is a unit vector normal to $\mathscr{C}$ at $P(x, y)$. Then either $\nabla\phi = \mathbf{0}$ or $\nabla\phi$ is orthogonal to $\boldsymbol{n}$. It is easily shown that if $\nabla\phi = \mathbf{0}$, then $\nabla\Phi = (\Phi_u, \Phi_v) = \mathbf{0}$, so that $d\Phi/dn = \mathbf{0}$, as required (see Exercises 9.1). Suppose now that $\nabla\phi \neq \mathbf{0}$, so that $\nabla\phi$ is orthogonal to $\boldsymbol{n}$ hence tangent to $\mathscr{C}$ at P (Fig. 9.7).

In general, $\nabla\phi$ is normal to any level curve $\phi(x, y) = a$, so that $\mathscr{C}$ is orthogonal to a level curve $\phi(x, y) = a$, passing through P as shown. The image of this curve under F is the level curve $\Phi(u, v) = a$ in the w-plane. Since F is conformal, angles between curves are preserved, so that $\mathscr{C}'$ is orthogonal to

$\Phi(u, v) = a$ at $P'(u, v)$ as shown in Fig. 9.7. Then $\nabla\Phi = (\Phi_u, \Phi_v)$ is tangent to $\mathscr{C}'$ at P'. Then if $\boldsymbol{n}$ is a unit normal vector to $\mathscr{C}'$ at P', $\nabla\Phi \cdot \boldsymbol{n} = 0 \Rightarrow d\Phi/dn = 0$ on $\mathscr{C}'$. ■

Example 9.6

Suppose we wish to find the function ϕ, which is harmonic on $|z| < 1$, continuous on $|z| \leqslant 1$, except at $\pm i$ and such that $\phi(1, \theta) = 1$ for $-\pi/2 < \theta < \pi/2$ and $\phi(1, \theta) = 0$ otherwise, as shown in Fig. 9.8. It is easy to

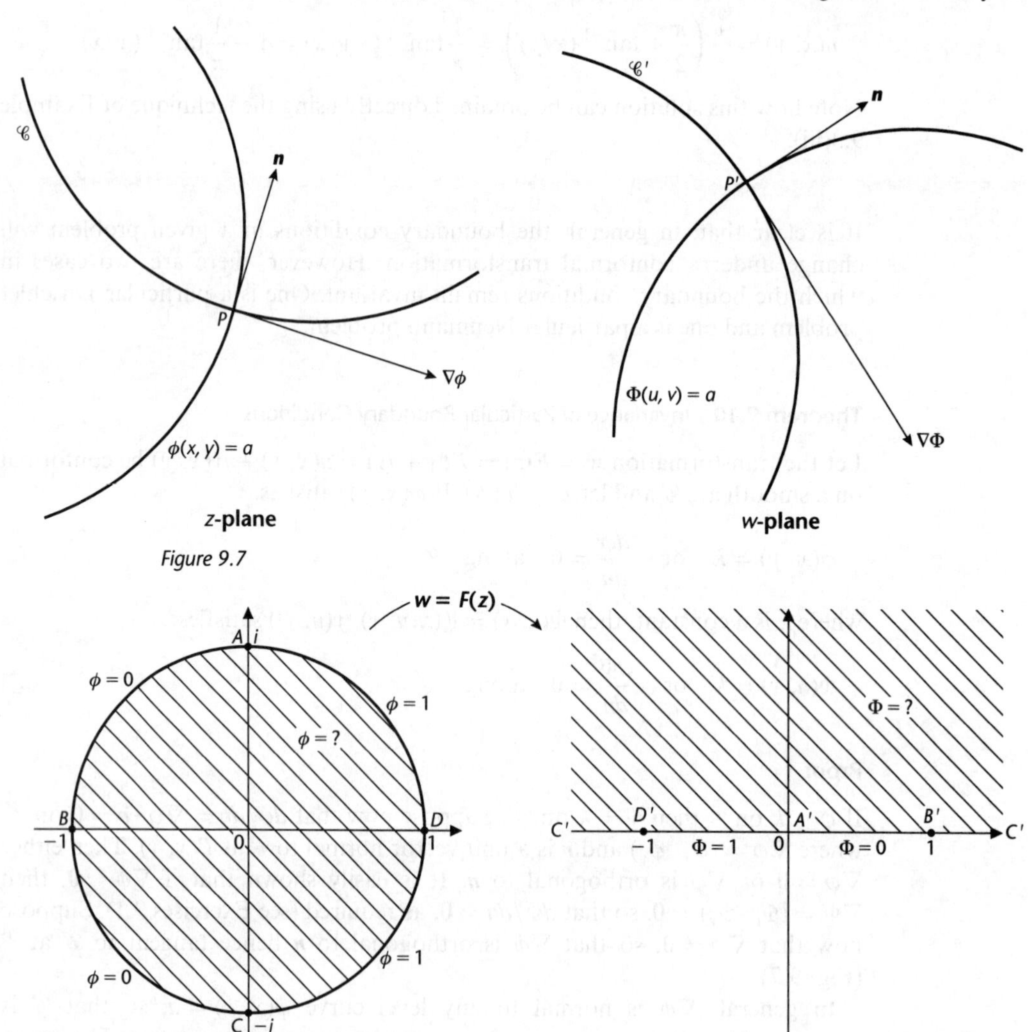

Figure 9.7

Figure 9.8

check that the bilinear transformation $w = F(z) = i(i - z)/(i + z)$ maps $|z| \leqslant 1$ to $\operatorname{Im} w \geqslant 0$ as shown. It follows by 9.8 and 9.10 that to solve the given problem, we just need to solve the problem in the w-plane. By Example 9.3(iii), the required harmonic function in the w-plane is

$$\Phi(u, v) = \theta/\pi = \frac{1}{\pi}\tan^{-1}(v/u)$$

$$\text{Also,} \quad z = x + iy \quad \Rightarrow \quad w = \frac{i((1 - x^2 - y^2) + 2ix)}{x^2 + (y+1)^2}$$

so that the desired solution to the original problem is

$$\phi(x, y) = \frac{1}{\pi}\tan^{-1}\left(\frac{x^2 + y^2 - 1}{2x}\right)$$

Compare this method of solution with the method of Exercises 9.1.10.

In the next section, we apply this fundamental technique to solve particular Dirichlet problems which occur in applications.

Exercise **9.1.1** Use the chain rule and equation (9.1) to verify that Laplace's equation in plane polar coordinates is equation (9.3). Show, using the Cauchy–Riemann equations in polar form, (2.6), that if f is analytic on a region $\mathscr{R}$ and is given by (9.4), then $u(r, \theta)$ and $v(r, \theta)$ satisfy equation (9.3).

Exercise **9.1.2** Show that $\phi(x, y) = \frac{x}{2}\operatorname{Log}(x^2 + y^2) - y\tan^{-1}(y/x)$ is harmonic on $\operatorname{Im} z > 0$ by

(i) verifying that ϕ is a solution of (9.1),

(ii) starting with a suitable analytic function on $\operatorname{Im} z > 0$

Exercise **9.1.3** Verify that

$$u(x, y) = \frac{x}{x^2 + y^2}$$

is harmonic on $\mathbb{C}\backslash\{0\}$. Use the constructive proof of Theorem 9.1 to find an analytic function f defined on $\mathbb{C}\backslash\{0\}$, such that $u = \operatorname{Re} f$.

Exercise **9.1.4** Use the technique of the proof of Theorem 9.1 to prove that if v is harmonic on a simply connected region $\mathscr{R}$, then there exists an analytic function f on $\mathscr{R}$ such that $v = \operatorname{Im} f$.

Exercise **9.1.5** For each of the cases (i) to (iv) shown in Fig. 9.9, write down a function which is harmonic in the given shaded region and which satisfies the given boundary conditions.

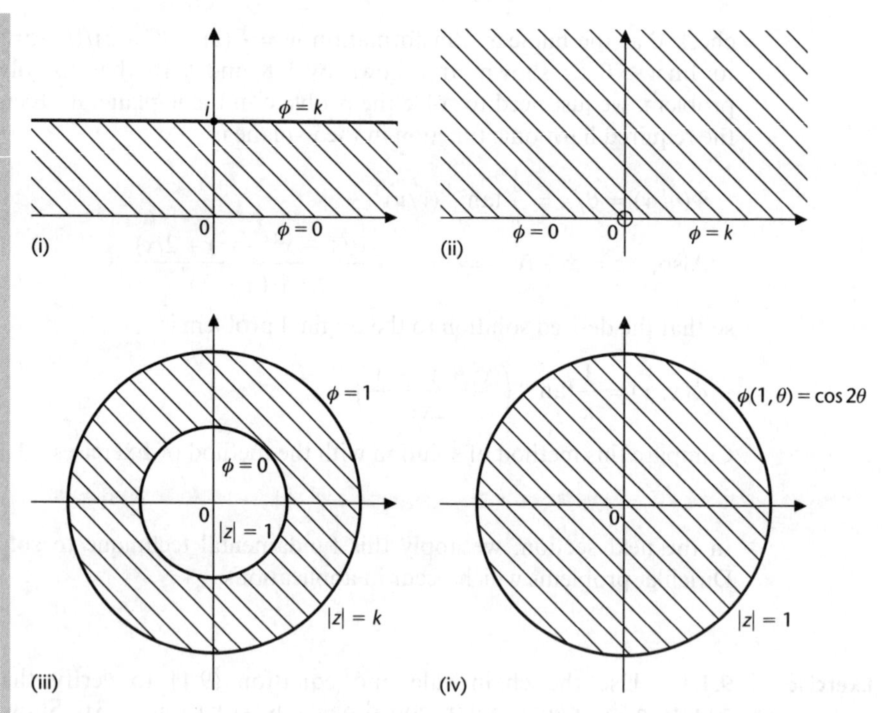

Figure 9.9

Exercise 9.1.6 Find a function $\phi(x, y)$, which is harmonic on $\operatorname{Im} z > 0$, such that $\phi(x, 0) = 0$ for $x < -1$, $\phi(x, 0) = 1$ for $-1 < x < 0$, $\phi(x, 0) = 0$ for $0 < x < 1$, and $\phi(x, 0) = 1$ for $x > 1$.

Exercise 9.1.7 Use Corollary 9.2 to prove Theorem 9.4.

Exercise 9.1.8 Use Corollary 9.2 to prove that if u and v are harmonic in a simply connected region $\mathscr{R}$, bounded by a simple closed contour $\mathscr{C}$, and are continuous on $\mathscr{R}$ $\overline{\mathscr{R}}$, with $u(x, y) \leqslant v(x, y)$ on $\mathscr{C}$, then $u(x, y) \leqslant v(x, y)$ on $\mathscr{R}$.

Exercise 9.1.9 Let $u(r, \theta)$ be harmonic on $|z| < 1$ and continuous on $|z| \leqslant 1$. Use Theorem 9.7 to show that $u(0) = (1/2\pi) \int_0^{2\pi} u(1, \phi)\, d\phi$. Deduce that if $u(1, \theta)$ is non-negative for all θ, then

$$\frac{1-r}{1+r} u(0) \leqslant u(r, \theta) \leqslant \frac{1+r}{1-r} u(0) \qquad \text{(Harnack's inequality)}$$

Exercise 9.1.10 Let $G(\psi) = 2 \tan^{-1}\left(\dfrac{1+r}{1-r} \tan(\psi/2)\right)$

Show that $G'(\psi) = P(r, \psi)$. Hence use Theorem 9.7 to find the function $u(r, \theta)$ which is harmonic on $|z| < 1$, continuous on $|z| \leqslant 1$, except at $\pm i$, and such that $u(1, \theta) = 1$ for $-\pi/2 < \theta < \pi/2$ and $u(1, \theta) = 0$ otherwise.

Exercise **9.1.11** Use Theorem 9.7 to find the function u which is harmonic inside the semicircle of unit radius and continuous inside and on the semicircle, such that $u = 0$ on the bounding diameter and $u(1, \theta) = U(\theta)$, where U is a given continuous function on the semicircular arc, as shown in Fig. 9.10. (*Hint*: Let $U(2\pi - \theta) = -U(\theta)$.)

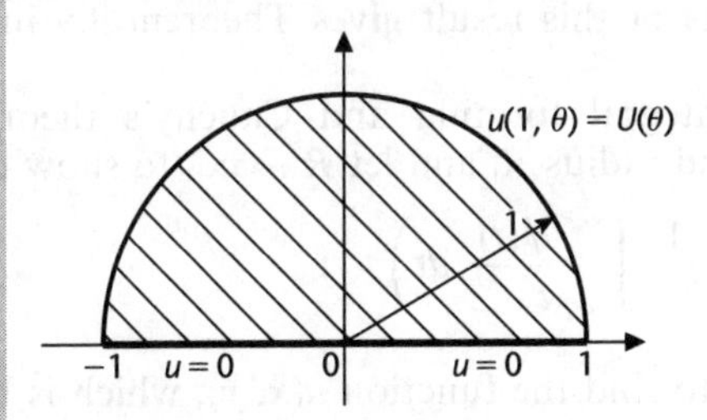

Figure 9.10

Exercise **9.1.12** ***The Neumann Problem for the Unit Circle***

(i) Let $N(r, \phi - \theta) = -\text{Log}\,(r^2 - 2r\cos(\phi - \theta) + 1)$. Show that N is the real part of $-2\,\text{Log}\,(z - \zeta)$, where $z = re^{i\theta}$ and $\zeta = e^{i\phi}$. Deduce that N is harmonic on $|z| < 1$.

(ii) Let $U = U(\theta)$ be a given continuous function on $|z| = 1$, with $\int_0^{2\pi} U(\phi)d\phi = 0$. Use Theorem 9.7 to show that if $u_r(r, \theta) = U(\theta)$ on $|z| = 1$ and

$$u(r, \theta) = \frac{1}{2\pi}\int_0^{2\pi} N(r, \phi - \theta)U(\phi)d\phi + K \qquad (0 \leqslant r < 1)$$

where K is any constant, then u is harmonic on $|z| < 1$ and continuous on $|z| \leqslant 1$.

Exercise **9.1.13** Show that $\phi(x, y) = e^{-y}\sin x$ is harmonic on $\mathbb{C}$. Show that ϕ remains harmonic on any region excluding 0, under the transformation $w = z^{1/2}$.

Exercise **9.1.14** Let f be given by $f(z) = u(x, y) + iv(x, y)$ and let f be analytic on a region $\mathscr{R}$. Use the chain rule to show that if $\phi = \phi(x, y)$ has continuous second partial derivatives on $\mathscr{R}$, then

$$\phi_{xx} + \phi_{yy} = |f'(z)|^2(\phi_{uu} + i\phi_{vv})$$

on $\mathscr{R}$. Deduce that Laplace's equation is invariant under any conformal transformation.

Exercise **9.1.15**

(i) Use Theorem 9.7 and the conformal transformation $w = Rz$ to find the solution to the interior Dirichlet problem for the circle $|w| = R$.

(ii) Use Theorem 9.7 and the transformation $w = 1/z$ to find the function $u(r, \theta)$, which is harmonic on $|w| > 1$, continuous on $|w| \geqslant 1$, and such that $u(1, \theta) = U(\theta)$ on $|w| = 1$.

Exercise **9.1.16** Let f be analytic on $\operatorname{Im} z \geqslant 0$ and such that $|z^a f(z)| < M$ for some a, $M \in \mathbb{R}^+$ on $\operatorname{Im} z \geqslant 0$. Prove that

$$f(z) = \frac{y}{\pi} \int_{-\infty}^{\infty} \frac{f(t)}{|t - z|^2} \, dt$$

Note that taking real parts of this result gives Theorem 9.9 in this special case.
(*Hint*: Apply Cauchy's integral formula and Cauchy's theorem using a semicircle, with centre 0 and radius R, and let $R \to \infty$, to show that

$$f(z) = \frac{1}{2\pi i} \int_{-\infty}^{\infty} \frac{f(t)}{t - z} \, dt - \frac{1}{2\pi i} \int_{-\infty}^{\infty} \frac{f(t)}{t - \bar{z}} \, dt \Big)$$

Exercise **9.1.17** Use Theorem 9.9 to find the function $u(x, y)$, which is harmonic on $\operatorname{Im} z > 0$, continuous on $\operatorname{Im} z \geqslant 0$, except at 1, and such that $u(x, 0) = 0$ for $x < 0$, $u(x, 0) = x$ for $0 \leqslant x < 1$ and $u(x, 0) = 0$ for $x \geqslant 1$.

Exercise **9.1.18** Let f given by $f(z) = u(x, y) + iv(x, y)$ be conformal on a region $\mathcal{R}$. Let $\phi(x, y) = \Phi(u(x, y), v(x, y))$ have continuous first derivatives on $\mathcal{R}$. Show that $|\nabla \phi| = |f'(z)||\nabla \Phi|$ on $\mathcal{R}$. Deduce that $\nabla \Phi = \mathbf{0}$ if and only if $\nabla \phi = \mathbf{0}$.

The Use of Conformal Mappings in Solving Laplace's Equation

We can often find the solution of Laplace's equation in two dimensions, subject to certain types of boundary conditions, by transforming the problem to a simpler one, such as a problem on a half-plane, strip or the inside of a circle, with the use of a conformal transformation. Once the simpler problem is solved, the solution to the original problem can be found by inverting the conformal transformation. In this section, we illustrate this technique with a number of different examples.

A standard problem is to find the steady temperature T at any point in a given region, for particular boundary conditions. If no thermal energy is created or destroyed within the region and T is unchanging with time, then T satisfies Laplace's equation. Curves along which T is constant are the **isotherms**.

Note

If part of the boundary of a given region is thermally insulated, then the heat flow across that part of the boundary is 0, so that $dT/dn = 0$ on that part of the boundary.

Example 9.6 can be interpreted as finding the steady temperature in a cylinder, where half its surface is kept at 0 °C and the other half is raised to 1 °C. Points of insulation occur at the top and bottom of the cylinder.

Convention

For simplicity, we shall always use the same letter to denote a function and its composition with a conformal transformation.

Example 9.7

Consider the steady temperature in a particular semi-infinite wall. Any cross-section of this wall is bounded by the lines $x = \pm\pi/2$ and $y = 0$ (Fig. 9.11). The temperature $T(x, y)$ satisfies Laplace's equation,

$$T_{xx} + T_{yy} = 0 \qquad (-\pi/2 < x < \pi/2, y > 0)$$

and is continuous on $-\pi/2 \leqslant x \leqslant \pi/2$ and $y \geqslant 0$. Suppose $T(\pm\pi/2, y) = 0$ and $T(x, 0) = 1$, as shown. Note that $T(x, y)$ must be bounded for all x and y, and that $T(x, y) \to 0$ as $y \to \infty$.

To solve the problem, we need to find the harmonic function $T(x, y)$, in the given region of the z-plane, satisfying the given boundary conditions. By 9.3, if this problem has a solution then it is unique. It is easily checked that the region in the z-plane can be mapped to the upper half w-plane by the transformation $w = f(z) = \sin z$, as shown in Fig. 9.11. Note that f is conformal except at the corner points B and C. By 9.8 and 9.10, f maps the given problem to the problem in the w-plane. We can easily find the harmonic function, $T(u, v)$, in the upper half w-plane, which satisfies the given boundary conditions. It follows by Example 9.4 that the required function is

$$T(u, v) = \frac{1}{\pi}(\theta_1 - \theta_2) = \frac{1}{\pi}\left(\tan^{-1}\left(\frac{v}{u-1}\right) - \tan^{-1}\left(\frac{v}{u+1}\right)\right)$$

where $0 < \theta_1, \theta_2 < \pi$. It follows from the addition formula for tan that, in general,

$$\tan^{-1} X + \tan^{-1} Y = \tan^{-1}\left(\frac{X+Y}{1-XY}\right) \tag{9.9}$$

$$\text{Hence} \quad T(u, v) = \frac{1}{\pi}\tan^{-1}\left(\frac{v/(u-1) - v/(u+1)}{1 + v^2/(u^2-1)}\right) = \frac{1}{\pi}\tan^{-1}\left(\frac{2v}{u^2+v^2-1}\right)$$

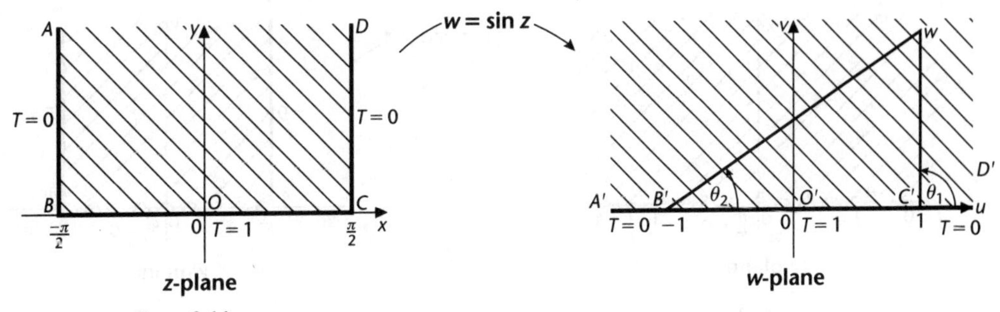

Figure 9.11

To obtain the solution of the original problem, we need $u(x, y)$ and $v(x, y)$. Now

$$w = u(x, y) + iv(x, y) = \sin(x + iy) = \sin x \cosh y + i \cos x \sinh y$$

$$\Rightarrow \quad u(x, y) = \sin x \cosh y, \; v(x, y) = \cos x \sinh y$$

$$\Rightarrow \quad u^2 + v^2 - 1 = \sinh^2 y - \cos^2 x$$

Then $$T(x, y) = \frac{1}{\pi}\tan^{-1}\left(\frac{2\cos x/\sinh y}{1 - (\cos^2 x/\sinh^2 y)}\right) = \frac{2}{\pi}\tan^{-1}\left(\frac{\cos x}{\sinh y}\right)$$

is the solution to the original problem (using (9.9) with $X = Y = \cos x/\sinh y$).

Example 9.8

Suppose we wish to determine the steady temperature $T(x, y)$, in the upper half-plane, caused by two metal plates, one kept at 0 °C and the other heated to K °C, separated by a strip of thermally insulated material (Fig. 9.12). (This can be considered as the cross-section of a three-dimensional problem.) To solve this problem, we need to find the function $T(x, y)$, which is harmonic on $\operatorname{Im} z > 0$, continuous on $\operatorname{Im} z \geqslant 0$ and which satisfies the given boundary conditions. In contrast to the last example, we cannot find the desired harmonic function in this plane directly. However, by 9.8 and 9.10, the inverse of the conformal mapping used in the previous problem, i.e. $w = f(z) = \sin^{-1} z$, maps the given problem to the one in the w-plane, as shown in Fig. 9.12. Clearly, the required harmonic function on the given region in the w-plane must be of the form $T(u, v) = au + b$, for some constants a and b. Since $T(-\pi/2, v) = 0$ and $T(\pi/2, v) = K$,

$$T(u, v) = K(u/\pi + 1/2)$$

We now need to find $u(x, y)$ and $v(x, y)$:

$$z = x + iy = \sin w = \sin(u + iv) = \sin u \cosh v + i \cos u \sinh v$$

$$\Rightarrow \quad x(u, v) = \sin u \cosh v \qquad y(u, v) = \cos u \sinh v$$

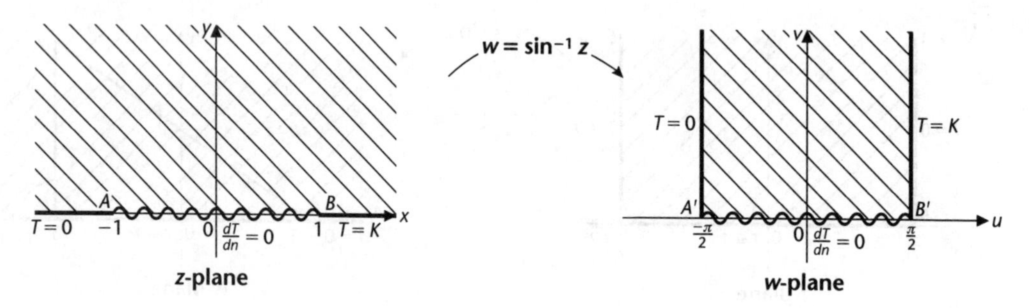

Figure 9.12

$$\Rightarrow \quad (x+1)^2 + y^2 = \sin^2 u \cosh^2 v + 2 \sin u \cosh v$$
$$+1 + (1 - \sin^2 u)(\cosh^2 v - 1)$$
$$= \cosh^2 v + \sin^2 u + 2 \sin u \cosh v = (\cosh v + \sin u)^2$$
$$\Rightarrow \quad (x-1)^2 + y^2 = (\cosh v - \sin u)^2$$
$$\Rightarrow \quad 2 \sin u = ((x+1)^2 + y^2)^{1/2} - ((x-1)^2 + y^2)^{1/2}$$

Then $T(x, y) = \dfrac{K}{2} + \dfrac{K}{\pi} \sin^{-1}\left(\dfrac{1}{2}\left(((x+1)^2 + y^2)^{1/2} - ((x-1)^2 + y^2)^{1/2}\right)\right)$

Another very important application of Laplace's equation lies in electrostatics. In general, the **intensity** $\boldsymbol{E}$ of an electric field at a point is the force which would be exerted on a unit positive charge placed at that point. The **electrostatic potential** V is a scalar function such that $\boldsymbol{E} = -\nabla V$. It follows by Gauss's flux theorem that V satisfies Laplace's equation. The **equipotential curves** are given by $V =$ constant and the electrostatic **flux lines**, which indicate the flow of electricity, are given by $U =$ constant, where U is a harmonic conjugate of V. All electric charge is located on the surface of a perfect **conductor**, so the surface of a conductor must be an equipotential curve. Along an **earthed** surface, $V = 0$.

Note

In general, if U is a harmonic conjugate of V, the curves $U =$ constant are orthogonal to the curves $V =$ constant (see Exercise 2.2.7). Thus, in this case, the flux lines are orthogonal to the equipotential lines.

As in the case of steady temperatures, many electrostatic problems, which involve finding the harmonic function $V(x, y)$, can be solved by using conformal transformations.

Note

The preceding examples can easily be reworded as problems in electrostatics. For instance, Example 9.6 can be interpreted as finding the electrostatic potential in a cylinder, where half the surface is charged and the other half is earthed. On the other hand, the following examples can easily be reworded as steady temperature problems. Both sets of examples and the exercises apply equally well in any number of physical problems which require finding a potential function.

Example 9.9

Consider two semi-infinite plane conductors, separated by a small gap, inclined at an angle α and charged to constant potentials V_0 and V_1. A plane section is shown in Fig. 9.13. To find the electrostatic potential in this case, we require

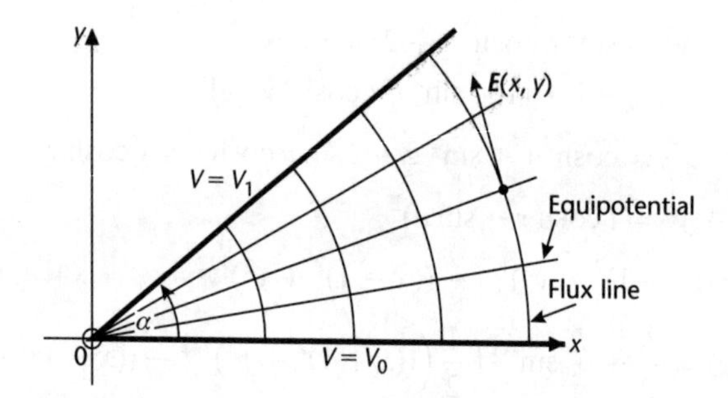

Figure 9.13

the function which is continuous except at the origin, is harmonic on the given region, and satisfies the given boundary conditions. This is clearly $V(x, y) = V(re^{i\theta}) = a\theta + b$ for some constants a and b. $V = V_0$ when $\theta = 0$ gives $b = V_0$. Then $V = V_1$ when $\theta = \alpha$ gives $a = (V_1 - V_0)/\alpha$. Hence the electrostatic potential is

$$V(x, y) = (V_1 - V_0)\theta/\alpha + V_0 = (V_1 - V_0)\frac{1}{\alpha}\tan^{-1}(y/x) + V_0$$

In this case, the electric field intensity is given by

$$\boldsymbol{E} = (-V_x, -V_y) = \frac{(V_1 - V_0)}{\alpha r^2}(y, -x)$$

so $\boldsymbol{E}$ is orthogonal to the position vector $\boldsymbol{r}$ and therefore acts in a direction between the conductors at any point, as shown in Fig. 9.13, with magnitude $|\boldsymbol{E}| = (V_1 - V_0)/\alpha r$, so that as $r \to \infty$, $|\boldsymbol{E}| \to 0$. The equipotential curves in this case are given by $\theta =$ constant, as shown. A complex conjugate of θ is $-\text{Log}\, r$, since these are the real and imaginary parts respectively of the analytic function $-i\,\text{Log}\, z$, for $-\pi < \theta < \pi$, $z \neq 0$. Hence the flux lines are given by $r =$ constant as shown.

Example 9.10

Consider a conductor whose surface consists of two plane plates together with a semicircular cylinder; a cross-section is shown in Fig. 9.14. Suppose that the cylindrical surface is charged to a constant potential $V = 1$ and the plates are earthed as shown in the figure. In order to find the electrostatic potential in the shaded region, outside the conductor, we use a conformal mapping to simplify the problem. It follows from Exercise 6.3.9 and Theorems 9.8 and 9.10 that $w = f(z) = (z + 1/z)/2$ maps the given problem to the problem in the w-plane. It then follows from Example 9.7 that the required harmonic function in the w-plane is

$$V(u, v) = \frac{1}{\pi}\tan^{-1}\left(\frac{2v}{u^2 + v^2 - 1}\right)$$

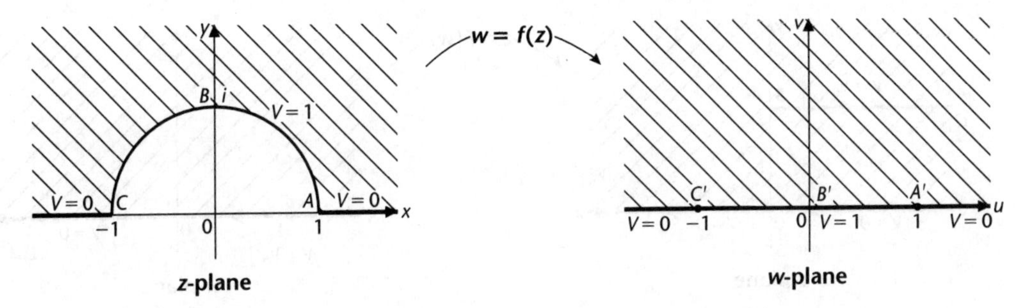

Figure 9.14

Now $w = u(x, y) + iv(x, y) = (z + 1/z)/2$ where $z = x + iy$ gives

$$u(x, y) = \frac{x(x^2 + y^2 + 1)}{2(x^2 + y^2)} \quad \text{and} \quad v(x, y) = \frac{y(x^2 + y^2 - 1)}{2(x^2 + y^2)}$$

$$\Rightarrow \quad V(x, y) = \frac{1}{\pi}\tan^{-1}\left(\frac{4y(x^2 + y^2 - 1)}{(x^2 + y^2)(x^2 + y^2 - 4) + (2x^2 - 2y^2 + 1)}\right)$$

Notice that the semicircular arc $x^2 + y^2 = 1$ and the line segments $y = 0$, $|x| > 1$, are equipotential lines.

In some examples, the Schwarz–Christoffel transformation can be used to determine a conformal transformation which maps a given problem to a simpler problem in the upper half-plane.

Example 9.11

Consider a semi-infinite plane conductor, charged to potential $V = V_0$, and a parallel infinite plane which is earthed and unit distance from the conductor (Fig. 9.15). Suppose we wish to determine the electrostatic potential in the upper half-plane. The two boundary lines and the upper half-plane can be treated as a degenerate polygon and its inside, so that by the Schwarz–Christoffel transformation, Theorems 6.9 and 6.10, the most general transformation which maps the real axis and the upper half-plane to the boundary lines and upper half-plane, with the correspondence of points as shown in Fig. 9.15, is

$$z = F(w) = K\int^{w} (\zeta + 1)^1 \zeta^{-1} d\zeta + k = K(w + \operatorname{Log} w) + k$$

Since $F(-1) = i$, we take $K = k = 1/\pi$. Then

$$z = F(w) = \frac{1}{\pi}(w + \operatorname{Log} w + 1)$$

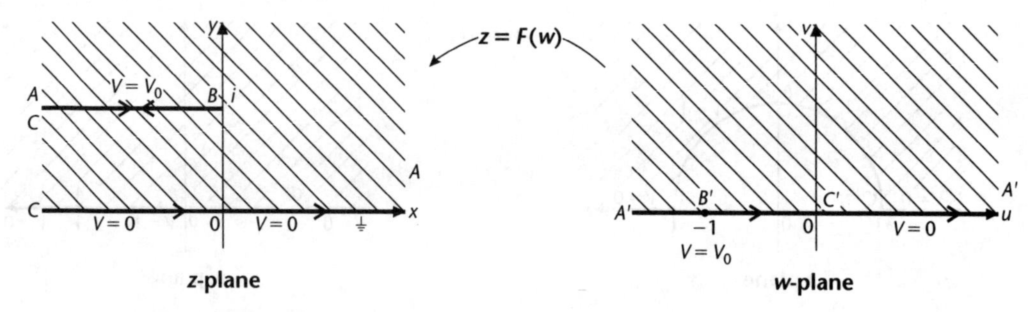

Figure 9.15

maps the problem in the upper half w-plane to the given problem. The solution to the problem in the w-plane is clearly

$$V(u, v) = \frac{V_0\theta}{\pi} = \frac{V_0}{\pi}\tan^{-1}(v/u)$$

$$\text{Now}\quad \pi z = \pi x + i\pi y = (u + iv) + 1 + \tfrac{1}{2}\operatorname{Log}(u^2 + v^2) + i\tan^{-1}(v/u)$$

$$\Rightarrow\quad 2\pi x(u, v) = 2u + 2 + \operatorname{Log}(u^2 + v^2) \qquad \pi y(u, v) = v + \tan^{-1}(v/u) \quad (9.10)$$

The solution to the original problem is then

$$V(x, y) = V_0 y - V_0\frac{v(x, y)}{\pi}$$

where $v(x, y)$ is given implicitly by (9.10).

In general, potential problems solved by the Schwarz–Christoffel transformation will have such an implicit solution.

Two-Dimensional Fluid Flow

A very important area in the application of conformal mappings lies in two-dimensional fluid flow. These applications follow from three-dimensional problems in which the basic flow pattern is essentially the same in any parallel plane. We only deal with the simplest case of perfect **steady** fluid flow, where there is no overall change to the flow pattern over time, and where the fluid is non-viscous. In addition, for initial simplicity, we assume there are no finite points in the plane at which fluid appears or disappears. Points where fluid appears are called **sources**; points where fluid disappears are called **sinks**.

It is convenient to study the motion of the fluid through the velocity of the flow at any point. Let us denote the velocity of the fluid at any point by

$$V(x, y) = (V_1(x, y), V_2(x, y))$$

Consider the fluid flow in a simply connected region $\mathscr{R} \subseteq \mathbb{C}$, bounded by a simple closed curve, $\mathscr{C}$.

Definition

The **circulation** of fluid in $\mathscr{R}$ is denoted and defined by

$$\Gamma = \int_{\mathscr{C}} \boldsymbol{V} \cdot d\boldsymbol{r} = \int_{\mathscr{C}} V_1\,dx + V_2\,dy \tag{9.11}$$

Physically, the circulation gives a measure of the amount of rotation about a central point in the region.

Definition

The fluid flow is **irrotational** if $\Gamma = 0$ in any simply connected region $\mathscr{R}$.

We shall only consider irrotational fluid flow. It then follows by (9.11) that there exists a function $\phi(x, y)$ such that

$$V_1\,dx + V_2\,dy = d\phi \Rightarrow V_1 = \phi_x \quad \text{and} \quad V_2 = \phi_y \Rightarrow \boldsymbol{V}(x, y) = \nabla\phi \tag{9.12}$$

We shall only consider **incompressible** fluids, i.e. fluids with constant density. Any incompressible fluid satisfies the **continuity condition**: the quantity of fluid contained within any simple closed curve is constant. This leads to the **equation of continuity**,

$$\frac{\partial V_1}{\partial x} + \frac{\partial V_2}{\partial y} = \nabla \cdot \boldsymbol{V} = 0 \tag{9.13}$$

It follows by (9.12) and (9.13) that $\nabla^2\phi = 0$, so $\phi(x, y)$ is harmonic in any simply connected region.

Definitions

A harmonic function $\phi(x, y)$ such that $\boldsymbol{V} = \nabla\phi$ is a **velocity potential** of the fluid flow. The curves given by $\phi(x, y) =$ constant are the **equipotentials**. A **stream function**, $\psi(x, y)$, for a fluid is a harmonic conjugate of $\phi(x, y)$. The **streamlines** are the curves $\psi(x, y) =$ constant.

Important Note

It follows by Exercise 2.2.7 that the streamlines are orthogonal to the equipotentials. Also, along the equipotentials, $d\phi/d\lambda = \nabla\phi \cdot d\boldsymbol{r}/d\lambda$ for any parameter λ, where $\boldsymbol{r}(\lambda)$ is the position vector of any point. Then $0 = \boldsymbol{V} \cdot d\boldsymbol{r}/d\lambda$, so that $\boldsymbol{V}$ is orthogonal to the tangent vector along the equipotentials. Hence $\boldsymbol{V}$ is tangent to the streamlines. In other words, the streamlines represent the actual physical direction of flow of the fluid. Notice that any rigid boundary must be a streamline.

Since $\phi(x, y)$ is harmonic and $\psi(x, y)$ is a harmonic conjugate of $\phi(x, y)$, ϕ and ψ are the real and imaginary parts of an analytic function defined on some simply connected region $\mathscr{R}$ of the plane.

Definition

The analytic function $F : \mathscr{R} \subseteq \mathbb{C} \to \mathbb{C}$ defined by

$$F(z) = \phi(x, y) + i\psi(x, y)$$

is the **complex potential** of the fluid flow.

Notice that the complex potential is unique up to an additive constant. It is convenient to treat the velocity of the fluid flow as a function mapping $\mathbb{C}$ to $\mathbb{C}$, so that

$$V(z) = V(x, y) = V_1(x, y) + iV_2(x, y)$$

Then

$$V = \nabla\phi \quad \Rightarrow \quad V(z) = \phi_x + i\phi_y$$

Since F is analytic in $\mathscr{R}$, it follows by Corollary 2.5 that

$$F'(z) = \phi_x + i\psi_x = \phi_x - i\phi_y \quad \Rightarrow \quad V(z) = \overline{F'(z)}$$

The speed of the fluid flow at any point is then given by $|V(z)| = |F'(z)|$, so that a **stagnation point**, i.e. a point at which the speed of fluid flow is zero, is given by $F'(z) = 0$.

Summary

For a steady, non-viscous, perfect, irrotational and incompressible fluid in two dimensions, the velocity of fluid flow at any point is given by $V(z) = \phi_x(x, y) + i\phi_y(x, y)$, where ϕ is the velocity potential, a two-dimensional harmonic function. Then $V(z) = \overline{F'(z)}$, where the complex potential F is the analytic function given by $F(z) = \phi(x, y) + i\psi(x, y)$, and where ψ is a harmonic conjugate of ϕ. The velocity $V(z)$ is tangential to the streamlines, $\psi(x, y) =$ constant, which thus represent the paths of fluid flow. The speed of the fluid is $|F'(z)|$, so that stagnation points are given by $F'(z) = 0$.

In studying fluid flow, we are once again concerned with finding a harmonic function, in this case the stream function $\psi(x, y)$ for a particular flow. Very often a problem may be simplified by use of a conformal mapping. The complex potential remains analytic under such a mapping and the velocity potential and stream function remain harmonic.

Example 9.12 Steady Fluid Flow Round a Bend

Consider the case of uniform steady horizontal fluid flow in the upper half-plane, with a rigid boundary along the real axis (Fig. 9.16). The velocity of fluid flow is constant and is tangential to the streamlines given by $y = k$ say, where k is any positive constant. Let $V(z) = a$ say, where $a \in \mathbb{R}^+$. Then the complex

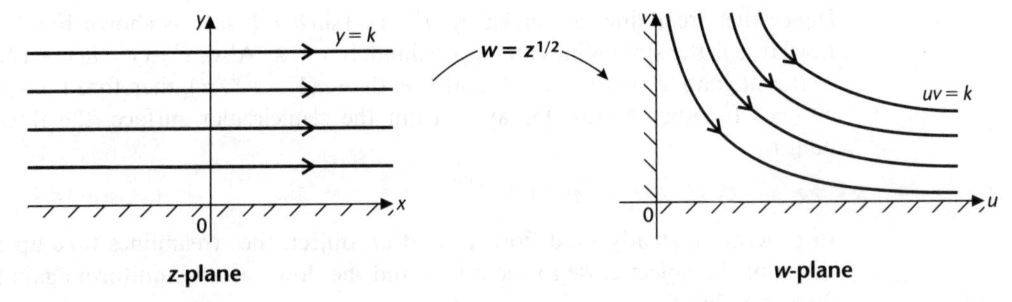

Figure 9.16

potential of the fluid flow is given by $F(z) = az$, without loss of generality, so $\phi(x, y) = ax$ and $\psi(x, y) = ay$. The fact that $a > 0$ indicates that the direction of flow is from left to right. Note that the rigid boundary is the streamline $y = 0$.

We can find the steady flow past a boundary of a different shape simply by using a conformal transformation to change the shape of the rigid boundary $y = 0$. For example, suppose we wish to find the steady flow round a right-angled bend, as shown in Fig. 9.16. The mapping $w = f(z) = z^{1/2}$ maps $y = 0$ to $u = 0$, $v \geqslant 0$ and $v = 0$, $u \geqslant 0$, and maps the upper half-plane to the first quadrant. Note that f is conformal except at the origin. Then the complex potential in the w-plane is given by

$$F(w) = aw^2 = a(u + iv)^2 = \phi(u, v) + i\psi(u, v)$$

where ϕ and ψ are the velocity potential and stream function in the w-plane. Hence $\phi(u, v) = a(u^2 - v^2)$ and $\psi(u, v) = 2auv$. It follows that the streamlines are given by $2auv = 2ak$ say, i.e. $uv = k$, where k is any constant such that the streamlines lie in the first quadrant. These curves are rectangular hyperbolae. The rigid boundary is the streamline given by $k = 0$, so that $u = 0$ or $v = 0$.

Lastly, $V(w) = \overline{F'(w)} \Rightarrow V(w) = 2a\bar{w}$. Hence the origin is the only stagnation point and, in general, the speed of the fluid is given by $|V(w)| = 2a(u^2 + v^2)^{1/2}$. The speed of the fluid is greatest along the line $u = v$, which makes physical sense.

Example 9.13 Steady Fluid Flow Past a Semicircular Cylinder

Consider the steady fluid flow around a semicircular cylinder (Fig. 9.17). As in Example 9.10, the transformation $z = f(w) = w + 1/w$ maps the given boundary to the x-axis in the z-plane and the region above the boundary to the upper half z-plane. It then follows by Example 9.12 that the complex potential for the flow in the w-plane is $F(w) = az = a(w + 1/w)$. Letting $w = re^{i\theta}$ gives

$$a(re^{i\theta} + e^{-i\theta}/r) = \phi(r, \theta) + i\psi(r, \theta)$$

$$\Rightarrow \quad \psi(r, \theta) = a\left(r - \frac{1}{r}\right)\sin\theta$$

Hence the streamlines are given by $(r - 1/r)\sin\theta = k$ say, as shown Fig. 9.17. For large r, the streamlines are approximately $y = k$. Also, $F'(w) = a(1 - 1/w^2)$ so the stagnation points are ± 1, and $V(r, \theta) = a(1 - e^{2i\theta}/r^2)$, therefore $V \to a$ as $r \to \infty$. In other words, far away from the semicircular surface, the flow is uniform.

In general, in steady fluid flow around an object, the streamlines take up the shape of the object close to the object and the flow becomes uniform again far from the object.

Example 9.14 Steady Fluid Flow Past a Flat Plate

For uniform fluid flow in the z-plane, inclined at angle α (Fig. 9.18), the complex potential is $F(z) = ae^{-i\alpha}z$, where $a \in \mathbb{R}^+$. (For then $V(z) = ae^{i\alpha}$ and the streamlines are $ay\cos\alpha - ax\sin\alpha = k$, i.e. $y = x\tan\alpha + K$.)

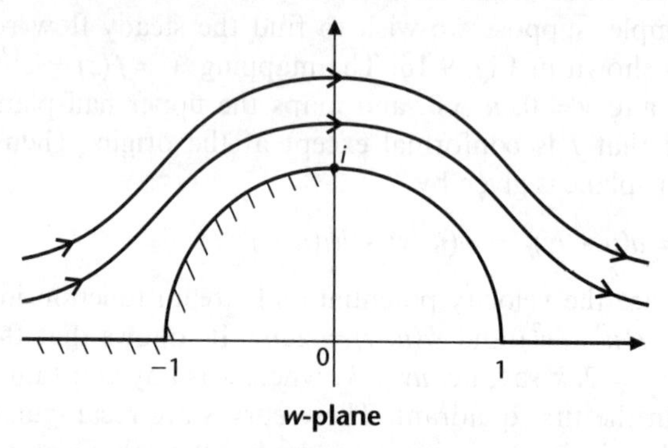

Figure 9.17

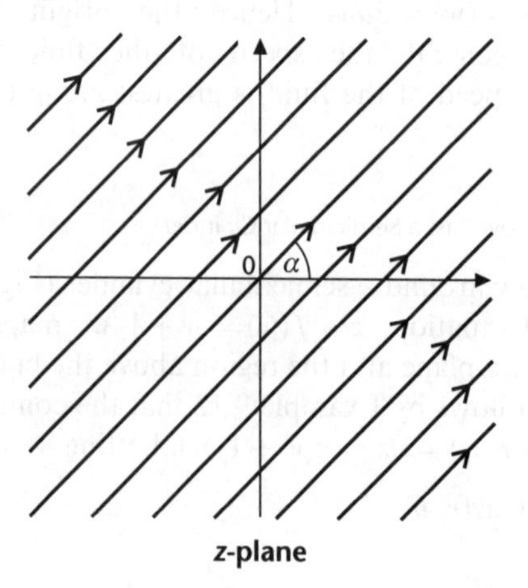

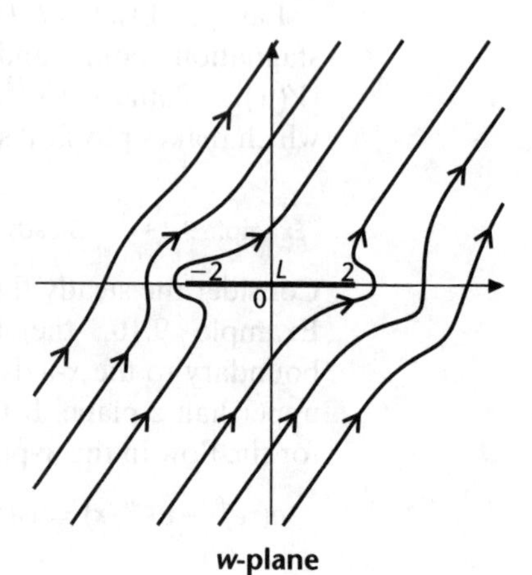

Figure 9.18

It follows from Exercise 9.2.13 that the mapping $w = \tau + 1/\tau$ maps the circle $|\tau| = 1$ to the line segment L in the w-plane, shown in Fig. 9.18, and $|\tau| > 1$ to the rest of the w-plane. It follows from Example 9.12 that the complex potential for steady fluid flow around this circle is $F(\tau) = a(\tau + 1/\tau)$, where $a \in \mathbb{R}^+$. Hence, by above, the complex potential for inclined steady fluid flow past the circle $|\tau| = 1$ is $F(\tau) = a(e^{-i\alpha}\tau + 1/(e^{-i\alpha}\tau))$. Now

$$w = \tau + 1/\tau \quad \Rightarrow \quad \tau = \frac{w + (w^2 - 4)^{1/2}}{2}$$

(the principal branch of the root being taken so that the transformation is injective). Hence, the complex potential for the inclined steady flow past the line segment L in the w-plane (which can be thought of as a cross-section of a thin plate) is

$$\begin{aligned} F(w) &= a\left(\frac{e^{-i\alpha}}{2}\left(w + (w^2 - 4)^{1/2}\right) + 2e^{i\alpha}\left(w + (w^2 - 4)^{1/2}\right)^{-1}\right) \\ &= \frac{a}{2}\left(e^{-i\alpha}\left(w + (w^2 - 4)^{1/2}\right) + e^{i\alpha}\left(w - (w^2 - 4)^{1/2}\right)\right) \\ \Rightarrow \quad F(w) &= a\left(w\cos\alpha - i\sin\alpha(w^2 - 4)^{1/2}\right) \end{aligned}$$

Notice that if $\alpha = 0$, $F(w) = aw$ and we are back to steady horizontal flow in the w-plane, as expected. A qualitative picture of the streamlines can be built up from $F(w)$ (see Exercises 9.2). Typical streamlines are shown in Fig. 9.18.

Point sources and sinks may be introduced into problems in steady fluid flow as follows. Suppose that fluid is emerging at a constant rate from a point *source* at $z = \alpha$, as shown in Fig. 9.19. It is clear that the streamlines are given by $\operatorname{Arg}(z - \alpha) = k$, where k is any constant, with $z \neq \alpha$. $\operatorname{Arg}(z - \alpha)$ is the imaginary part of the analytic function $\operatorname{Log}(z - \alpha)$. Hence the complex

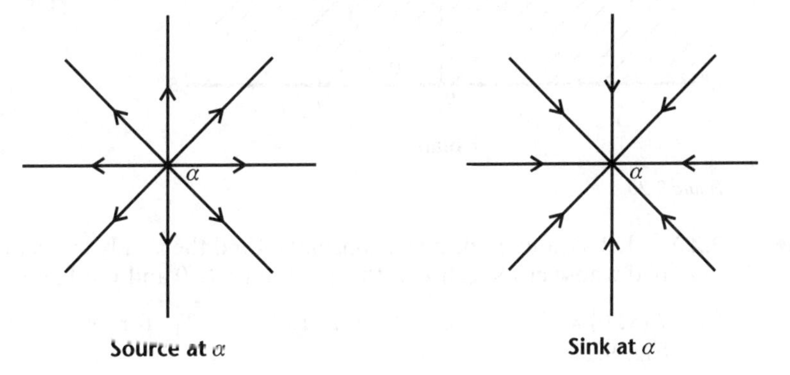

Figure 9.19

potential for fluid flow out of the source is $F(z) = a\,\mathrm{Log}\,(z - \alpha)$, where $a \in \mathbb{R}^+$, without loss of generality. Hence α is a singular point of the complex potential. Then $V(z) = a/(\overline{z - \alpha}) = (a/r)e^{i\theta}$, where $z - \alpha = re^{i\theta}$. The fact that $a > 0$ implies the direction of flow is away from the source. The constant a is the **strength** of the source. Similarly, a point *sink* at $z = \alpha$, as shown in Fig. 9.19, has complex potential $F(z) = -a\,\mathrm{Log}\,(z - \alpha)$, where $a \in \mathbb{R}^+$.

Note

If the nature of the fluid flow under the influence of two different effects is desired, the complex potentials for the flow determined by each effect are added to give the complex potential for the flow under both effects.

Exercise **9.2.1** Verify that in Example 9.7 $w = f(z) = \sin z$ maps the given points, boundary lines and region in the z-plane (Fig. 9.11) to the given points, u-axis and upper half w-plane. Show that

$$\tau = g(w) = \mathrm{Log}\left(\frac{w-1}{w+1}\right)$$

then maps the given points, u-axis and upper half w-plane to the points, lines and region in the τ-plane in Fig. 9.20.

Write down the function T, which is harmonic on the strip $0 \leqslant \mathrm{Im}\,\tau \leqslant \pi$ and which satisfies the boundary conditions. Then use g^{-1} to find $T(u, v)$. (This gives an alternative method for finding $T(u, v)$ in Example 9.7.)

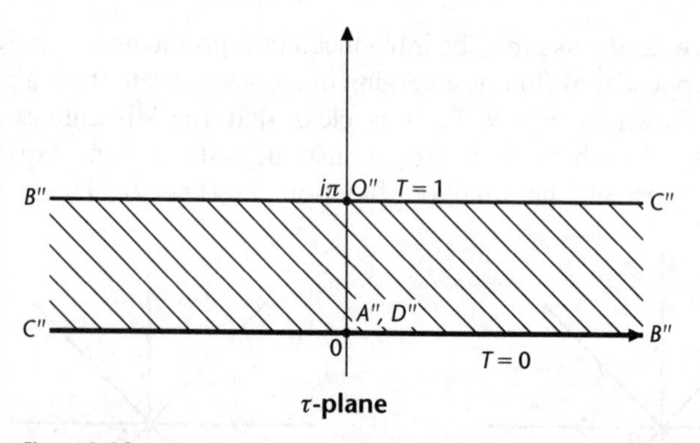

Figure 9.20

Exercise **9.2.2** Let T_0 and T_1 be given constants. Find the steady temperature, $T(x, y)$, in a solid whose cross section is the quadrant $x \geqslant 0$ and $y \geqslant 0$, given that

(i) $T(x, 0) = T_0$ for $x > 0$ and $T(0, y) = T_1$ for $y > 0$, as shown in Fig. 9.21(i)

(ii) $T(x, y)$ satisfies the boundary conditions as shown in Fig. 9.21(ii).

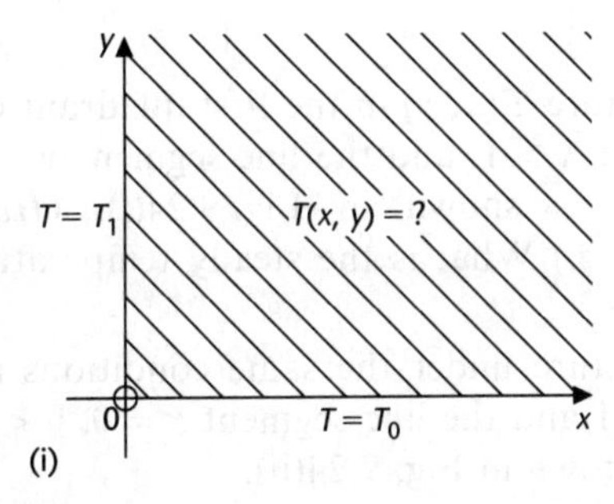

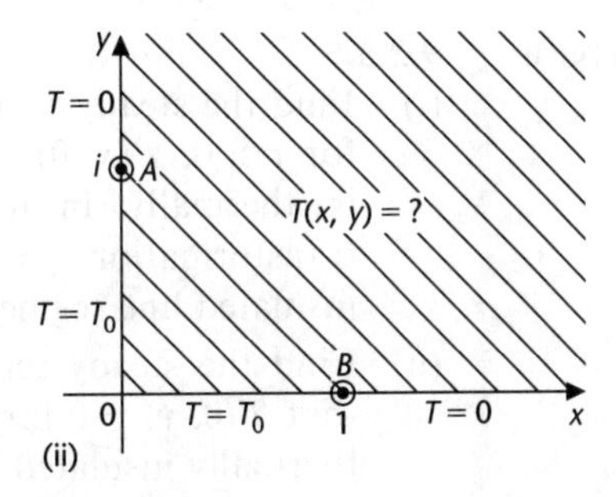

Figure 9.21

Exercise **9.2.3** Find the piecewise continuous function $T(x, y)$ that is harmonic in the semi-infinite region of the z-plane, shown in Fig. 9.22, if $T(0, y) = K$ for $y > 0$, $T(x, 0) = 0$ for $0 < x < a$, and $T(a, y) = 0$ for $y > 0$. (*Hint*: Use the mapping $w = f(z) = \sin(\pi z/2a)$ and the solution to Exercise 9.2.2(i).)

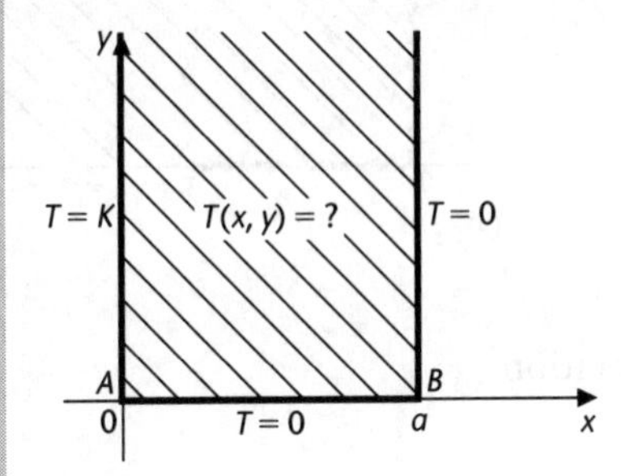

Figure 9.22

Exercise **9.2.4** Show that the bilinear transformation $w = f(z) = i(1 - z)/(1 + z)$ maps the given points, boundary and semicircular region of the z-plane shown in Fig. 9.23 to the given points, boundary and region of the w-plane. Hence find the steady temperature $T(x, y)$ inside a semicircular cylinder, given that $T(x, 0) = K$, $-1 < x < 1$ and that $T(x, y) = 0$ on its curved surface.

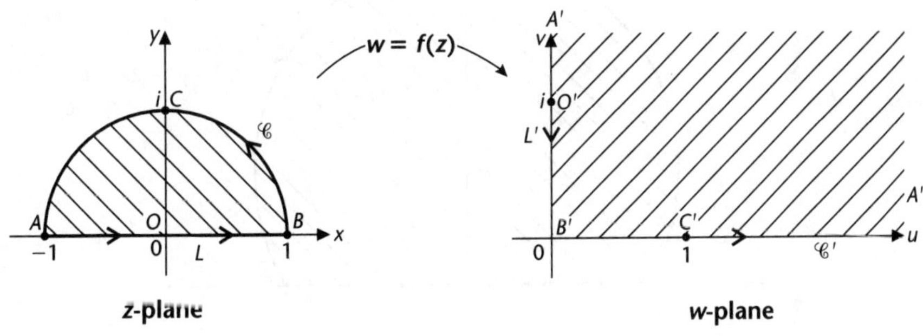

Figure 9.23

Exercise **9.2.5**

(i) Find the steady temperature $T(x, y)$ in the first quadrant if $T(0, y) = 0$ for $y > 0$, $T(x, 0) = 1$ for $x > 1$, and the line segment $y = 0$, $0 < x < 1$ is thermally insulated, as shown in Fig. 9.24(i). (*Hint*: Use the transformation $w = \sin^{-1} z$.) What is the steady temperature along the insulated line segment?

(ii) Find the steady temperature under the same conditions as (i), except that $T(0, y) = 0$ for $y > 1$ and the line segment $x = 0$, $0 < y < 1$ is also thermally insulated, as shown in Fig. 9.24(ii).

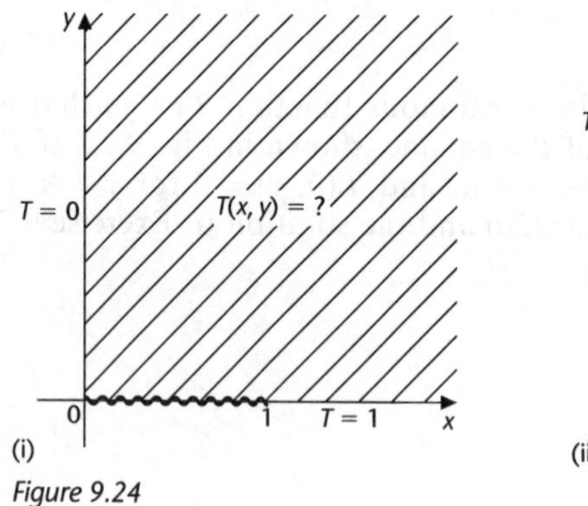

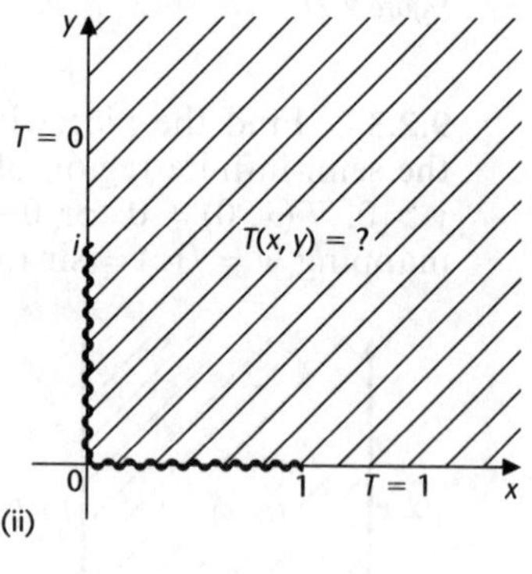

Figure 9.24

Exercise **9.2.6** Show that the tranformation

$$w = \text{Log}\left(\frac{z-1}{z+1}\right)$$

maps the region bounded by the segment of the circle, $x^2 + y^2 - 2y = 1$, $y \geqslant 0$, to an infinite strip of width $3\pi/4$ in the w-plane. Hence find the steady temperature $T(x, y)$ in the given region, such that $T(x, 0) = 0$ for $|x| < 1$, and $T(x, y) = K$ on the circular arc, as shown in Fig. 9.25.

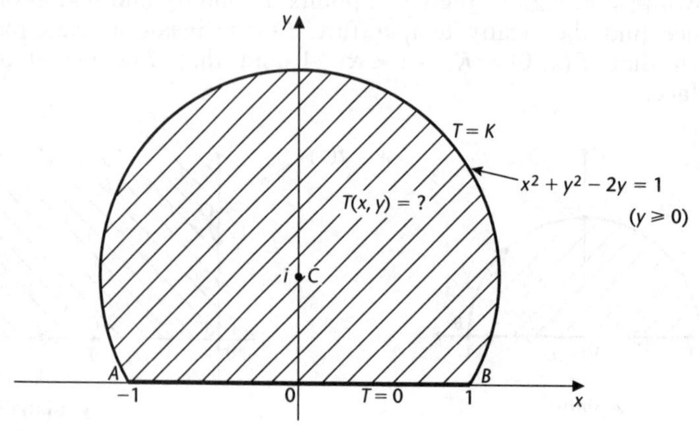

Figure 9.25

Exercise **9.2.7** A region is bounded by two concentric infinitely long circular cylindrical conductors of radii r_1 and r_2, with $r_2 > r_1$, which are charged to constant potentials of V_1 and V_2 respectively. Find the electrostatic potential within the region and show that the electric field intensity at any point is inversely proportional to the distance from the origin.

Exercise **9.2.8** A region is bounded by two infinitely long circular cylindrical conductors (Fig. 9.26). The inner cylinder is charged to a constant potential V_0 and the outer cylinder is earthed. Find the electrostatic potential within the region by first considering the mapping $w = f(z) = (1 - z)/z$. Hence find the equipotential lines and the flux lines in the region, and verify that the two families are orthogonal.

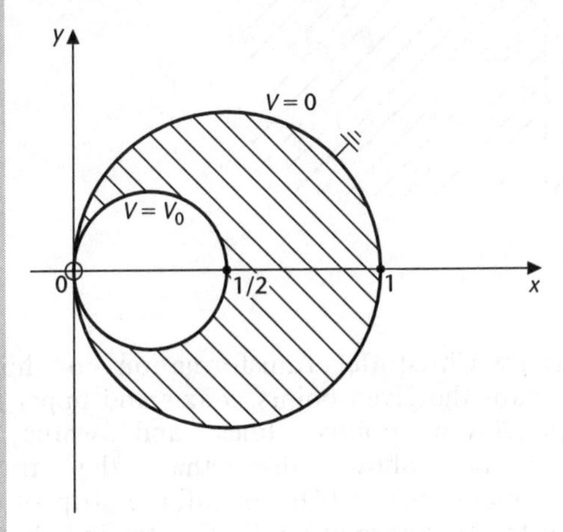

Figure 9.26

Exercise **9.2.9** Two parallel plane conductors are unit distance apart. Half of each conductor is charged to constant potential $V = 1$ and the other halves are earthed as shown in Fig. 9.27. Use the transformation $w = e^z$ to find the electrostatic potential at any point between the conductors. Find also the magnitude of the electric field intensity at any point.

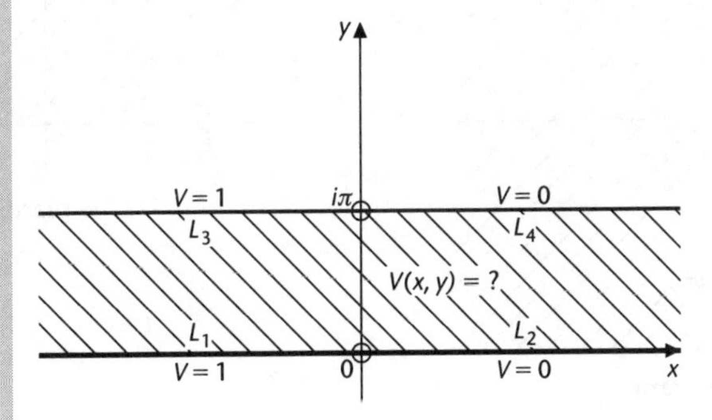

Figure 9.27

Exercise **9.2.10** Two plane conductors are distance 2 apart as shown in Fig. 9.28. One conductor is charged to constant potential $V = V_0$ and the other is earthed. Use the transformation $w = \cos^{-1} z$ to find the electrostatic potential surrounding the conductors.

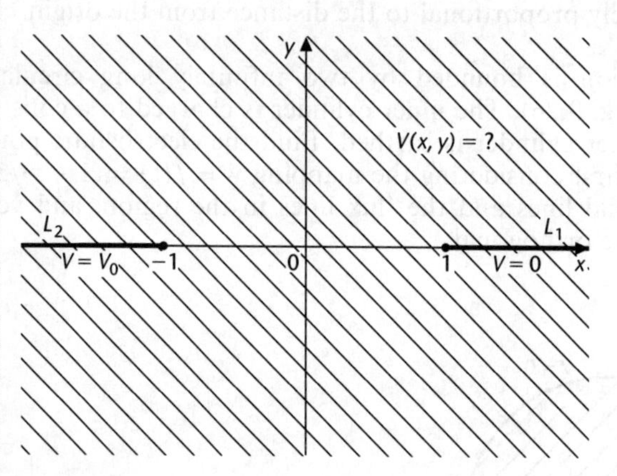

Figure 9.28

Exercise **9.2.11** Use the Schwarz–Christoffel transformation to show that the transformation which maps the given points, u-axis and upper half w-plane of Fig. 9.29 to the given points, lines and whole z-plane is $z = F(w) = 2\,\mathrm{Log}\,w - w^2 - i\pi$. Show also that the transformation $\tau = G(w) = 2\,\mathrm{Log}\,w - i\pi$ maps $\mathrm{Im}\,w \geqslant 0$ to an infinite strip of width 2π in the τ-plane. Hence use the transformation $F \circ G^{-1}$ to find the electrostatic potential in the z-plane if L_1 and L_2 are conductors with constant potentials $V = V_0$ and $V = 0$ respectively.

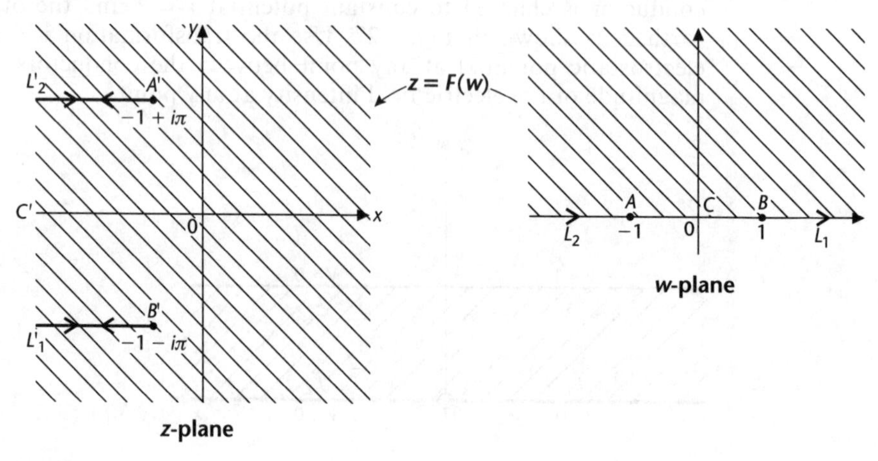

Figure 9.29

Exercise **9.2.12**

(i) Determine the complex potential of the steady fluid flow round the rigid boundary given in Fig. 9.30(i). Find the streamlines and give a qualitative picture of them. Find the speed of the flow at any point and show that the greatest speed occurs along the lines $v = Ku$, where $K^2 = 3 \pm 2\sqrt{2}$.

(ii) Find the complex potential of the steady fluid flow in the semi-infinite channel of width $2a$, as shown in Fig. 9.30(ii), by considering the transformation $z = \sin(\pi w/2a)$. Hence find the stagnation points and the general equation of the streamlines. Give a qualitative picture of the streamlines.

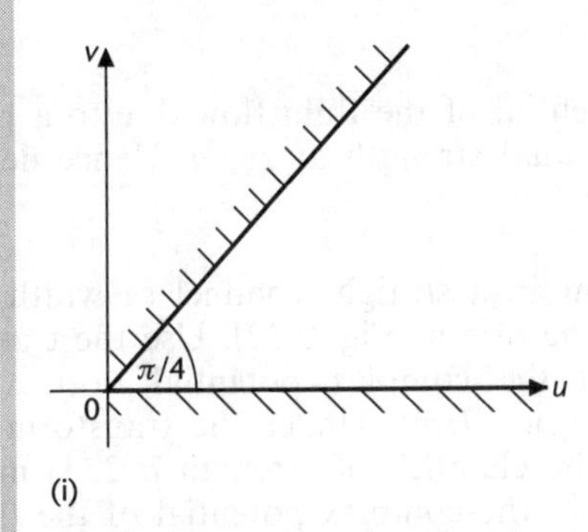

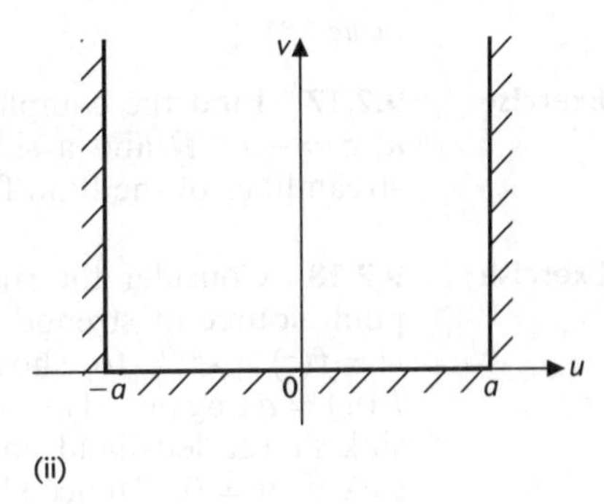

Figure 9.30

Exercise **9.2.13** Show that the transformation $z = f(w) = w + k^2/w$, $k \in \mathbb{R}^+$, maps the circle $|w| = k$ to the line segment joining $-2k$ to $2k$ and the exterior of the circle to the whole z-plane, excluding this line segment. Hence show that the complex potential for steady fluid flow around the circle $|w| = k$ is $F(w) = a(w + k^2/w)$, where $a \in \mathbb{R}^+$. (Note that for $k = 1$ this answer can be obtained from Example 9.13 and a symmetry argument.) Find the speed of the fluid at any point on the circle, hence show that the greatest speed occurs at the top and bottom.

Exercise **9.2.14** Find $F'(w)$ for the flat plate problem of Example 9.14. Investigate the behaviour of $F'(w)$ as $w \to 0$, $w \to \infty$ and $w \to \pm 2$, hence obtain the qualitative picture in Fig. 9.18.

Exercise **9.2.15** Use the result of Exercise 9.2.13 to find the complex potential of the steady fluid flow around a circle centred at $a \in \mathbb{R}^+$ and passing through -1. Hence determine the complex potential of the steady fluid flow around the aerofoil of Fig. 6.12. Find the stagnation point and investigate the qualitative behaviour of the streamlines.

Exercise **9.2.16** Use the Schwarz–Christoffel transformation to determine the transformation which maps $\operatorname{Im} z \geqslant 0$ to the given boundary lines and region in the w-plane; the correspondence between points is shown in Fig. 9.31. Hence find the complex potential $F(w)$ of the steady fluid flow past the 'dam' $A'B'$. Find any stagnation points and any singular points of $F'(w)$. Find also $\lim_{z\to\infty} F'(w)$ and so determine the qualitative behaviour of the streamlines.

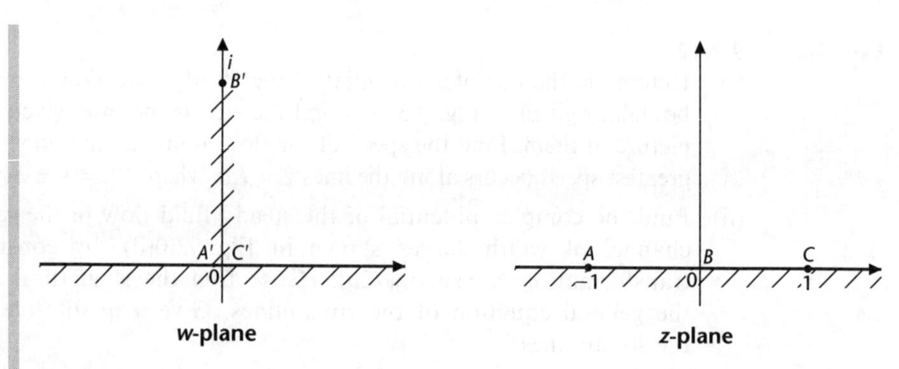

Figure 9.31

Exercise **9.2.17** Find the complex potential of the fluid flow due to a point source at $z = -a \in \mathbb{R}$ and a sink of equal strength at $z = a$. Hence determine the streamlines of the fluid flow.

Exercise **9.2.18** Consider the fluid flow in a straight channel of width h due to a point source of strength a at the origin (Fig. 9.32). Use the transformation $w = f(z) = e^{\pi z/h}$ to show that the complex potential in the w-plane is $F(w) = a\,\text{Log}\,(w-1) - (a/2)\,\text{Log}\,w$. (*Hint*: Under the transformation f, the sink at the left-hand end of the channel, of strength $a/2$, is mapped to a sink at $w = 0$.) Hence show that the complex potential of the fluid flow in the channel is $F(z) = a\,\text{Log}\,(\sinh(\pi z/2h))$ and determine the general equation of the streamlines.

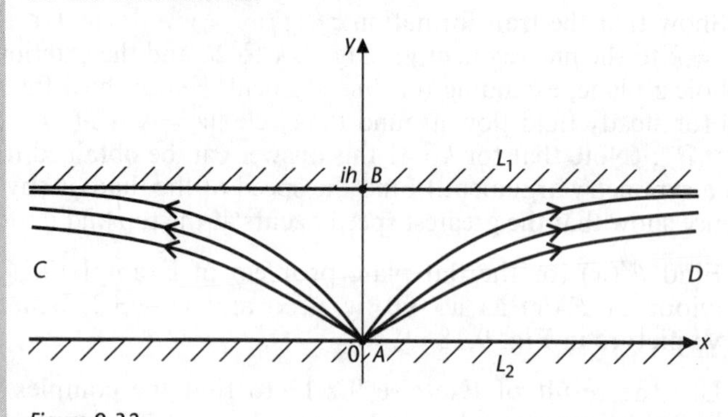

Figure 9.32

The Solution of Linear Partial Differential Equations Using Integral Transforms

Many physical problems require the solution of a linear partial differential equation (p.d.e.) subject to certain boundary or initial conditions. If we need to solve Laplace's equation with non-constant boundary conditions or a different p.d.e., then the methods given so far in this chapter are generally inappropriate. Many such problems can be solved by taking an integral transform of the equation. This technique is very general and widely applicable. We shall consider only problems in which the Fourier or Laplace transform is appropriate. Although the solution of some of these problems is not directly related to results in complex analysis, the technique enables us to use some of the properties and results concerning these transforms, derived in Chapter 8.

In general, the application of an integral transform reduces a linear p.d.e. in two independent variables to a linear ordinary differential equation (o.d.e.). The basic procedure consists of the following steps.

Step 1

Given a linear p.d.e. in a function of two independent variables, $\phi(x, y)$ say, choose one independent variable to be effectively eliminated. The choice will depend on the equation itself, the range of each of the independent variables and the given boundary and initial conditions.

Step 2

Take an appropriate integral transform, with respect to this independent variable, of the p.d.e., taking account of the given conditions. This choice will again depend on the equation and given conditions. Let $\mathscr{I}$ be the chosen transform and x the chosen independent variable; let $\mathscr{I}(\phi(x, y)) = \overline{\phi}(s, y)$. This step will produce a linear ordinary differential equation in $\overline{\phi}(s, y)$, in which s is treated as a constant.

Step 3

Solve the ordinary differential equation for $\overline{\phi}(s, y)$, taking account of the given conditions. It is usually necessary to take the transform of certain of the given conditions.

Step 4

Invert $\overline{\phi}(s, y)$ using tables, convolution or the appropriate inversion formula to obtain $\phi(x, y)$.

This general procedure is best illustrated by examples. We shall give a number of examples which arise from physical problems.

Important Note

When using this technique, it is not usual to justify the validity of each step. Instead, it is checked that the 'solution', once obtained, satisfies the given equation and conditions. It can be shown in each case that the problems we consider have *unique* solutions.

Example 9.15

Consider a very long metal bar, with initial temperature 0 °C everywhere. The left face is then raised to a constant temperature K°C. After this time, the temperature at any point on the bar depends only on the distance from the left face and the time elapsed. Let x denote the distance from the left face, t the time elapsed and $T(x, t)$ the temperature at any point of the bar. This is a problem in heat diffusion and it can be shown that $T(x, t)$ satisfies the **heat** or **diffusion equation**,

$$a^2 T_{xx} = T_t \tag{9.14}$$

for $x > 0$ and $t > 0$, where a_2 is a constant called the **diffusivity**, a physical property of the metal. For this problem we require

$$T(x, 0) = 0, T(0, t) = K \quad \text{and} \quad T(x, t) \to 0 \quad \text{as } x \to \infty \tag{9.15}$$

We can solve (9.14) subject to the conditions (9.15) by taking the Laplace transform. It is easily checked that, given (9.15), it is convenient to take the Laplace transform with respect to t. Following the notation of Chapter 8, let

$$\overline{T}(x, s) = \mathscr{L}(T(x, t)) = \int_0^{\infty} e^{-st} T(x, t)\, dt$$

Then taking the Laplace transform of (9.14) and supposing that differentiation under the integral sign is valid, we obtain

$$a^2 \frac{\partial^2}{\partial x^2}\left(\int_0^{\infty} e^{-st} T(x, t)\, dt\right) = \int_0^{\infty} e^{-st} T_t(x, t)\, dt = s\overline{T}(x, s) - T(x, 0)$$

by Theorem 8.10(iii), supposing that $\mathscr{L}(T_t(x, t))$ exists. Then using (9.15),

$$\overline{T}_{xx}(x, s) = \frac{s}{a^2}\overline{T}(x, s)$$

Treating s as a constant, this is a linear ordinary differential equation in $\overline{T}(x, s)$, which is easily solved to obtain

$$\overline{T}(x, s) = Ae^{-(\sqrt{s}/a)x} + Be^{(\sqrt{s}/a)x}$$

for some constants A and B, which in general are functions of s. Since $\overline{T}(x, s) \to 0$ as $x \to \infty$ from (9.15), we require s to be a real positive constant and $B = 0$. Also, since $\mathscr{L}(1) = 1/s$ and $T(0, t) = K$, $\overline{T}(0, s) = A = K/s$. Then

$$\overline{T}(x, s) = \frac{Ke^{-(\sqrt{s}/a)x}}{s} \quad \Rightarrow \quad T(x, t) = K\,\text{erfc}\left(\frac{x}{2a\sqrt{t}}\right)$$

from the result of Example 8.11. It is easily checked that this is the solution to the given problem.

Example 9.16

Suppose we wish to solve the heat equation (9.14), for $t > 0$ and $x \in \mathbb{R}$, subject to the conditions

$$T(x, 0) = f(x) \text{ with } T \text{ and } T_x \text{ bounded for all } x \text{ and } t \tag{9.16}$$

where f is some given continuous function. For example, we might be interested in finding the temperature at any point on a very long cable, with initial temperature $f(x)$ at any point. In this case, since $x \in \mathbb{R}$, it is convenient to take the Fourier transform with respect to x, which reduces (9.14) to a first-order linear o.d.e. Following the notation of Chapter 8, let

$$\hat{T}(s, t) = \mathscr{F}(T(x, t)) = \frac{1}{\sqrt{2\pi}} \int_{-\infty}^{\infty} e^{-isx} T(x, t)\, dx$$

For simplicity, let us take $a = 1$ say. We shall assume that the Fourier transforms of T, T_x and T_{xx} exist, and that differentiation under the integral sign is valid. Then taking the Fourier transform of (9.14) and integrating by parts gives

$$\hat{T}_t(s, t) = \frac{1}{\sqrt{2\pi}} \left(\lim_{a,b\to\infty} \left[e^{-isx} T_x(x, t)\right]_{x=-a}^{x=b} + is \int_{-\infty}^{\infty} e^{-isx} T_x(x, t)\, dx \right)$$

$$= \frac{is}{\sqrt{2\pi}} \left(\lim_{a,b\to\infty} \left[e^{-isx} T(x, t)\right]_{x=-a}^{x=b} + is \int_{-\infty}^{\infty} e^{-isx} T(x, t) \right)$$

$$\Rightarrow \quad \hat{T}_t = -s^2 \hat{T} \quad \Rightarrow \quad \hat{T}(s, t) = A(s) e^{-s^2 t}$$

for some $A(s)$. Now (9.16) gives $\hat{T}(s, 0) = A(s) = \hat{f}(s)$. Hence

$$\hat{T}(s, t) = \hat{f}(s) e^{-s^2 t} \tag{9.17}$$

Since f is not explicitly known, the Fourier convolution result is the most appropriate way to find $T(x, t)$ in this case. Recall that $\mathscr{F}(e^{-x^2/2}) = e^{-s^2/2}$, from Example 8.1(iii). It then follows that $\mathscr{F}(e^{-(bx)^2/2}) = (1/b)e^{-s^2/(2b^2)}$ for any constant b. Putting $b^2 = \frac{1}{2}k$ then gives $\mathscr{F}(e^{-x^2/(4k)}) = \sqrt{2k} e^{-ks^2}$. Then from (9.17) and the Fourier convolution result, Theorem 8.5, it follows that

$$T(x, t) = \frac{1}{2\sqrt{\pi t}} \int_{-\infty}^{\infty} f(y) e^{-(x-y)^2/4t} dy$$

Example 9.17

The Fourier transform can be used to give an alternative derivation of the solution to the Dirichlet problem for the half-plane. Suppose we wish to solve Laplace's equation for $\phi(x, y)$, i.e. equation (9.1), for $x \in \mathbb{R}$ and $y > 0$, subject to $\phi(x, 0) = f(x)$, with ϕ and ϕ_x bounded for all x, y. Taking the Fourier transform of (9.1) with respect to x, assuming as before that all the steps are valid, gives

$$\hat{\phi}_{yy} = s^2\hat{\phi} \quad \Rightarrow \quad \hat{\phi}(s, y) = A(s)e^{|s|y} + B(s)e^{-|s|y}$$

From the given conditions, it follows that $A(s) = 0$ and $\hat{\phi}(s, 0) = B(s) = \hat{f}(s)$. Hence

$$\hat{\phi}(s, y) = \hat{f}(s)e^{-|s|y} \tag{9.18}$$

It follows by Example 8.1(ii) that $\mathscr{F}(1/(1+x^2)) = \sqrt{\pi/2}e^{-|s|}$, so $\mathscr{F}^{-1}(e^{-a|s|}) = a\sqrt{2/\pi}/(x^2 + a^2)$, for any constant a. Then from (9.18) and the Fourier convolution result, Theorem 8.5,

$$\phi(x, y) = \frac{y}{\pi}\int_{-\infty}^{\infty} \frac{f(t)}{(t-x)^2 + y^2}\, dt$$

Compare this result with Theorem 9.9.

Example 9.18

Suppose a beam of length L has its left-hand end fixed and is initially at rest. A constant force F is then applied at the free end, in the direction along the length of the beam, which produces a longitudinal displacement. It can be shown that this displacement of the beam depends only on the distance x from the fixed end and the time elapsed t. Let $\phi(x, t)$ denote the longitudinal displacement at distance x from the free end and at time t. It can be shown that $\phi(x, t)$ satisfies the **wave equation**,

$$\phi_{tt} = c^2\phi_{xx} \tag{9.19}$$

for $0 < x < L$ and $t > 0$, where c is a positive constant depending on physical properties of the beam. From the given physical conditions, it follows that

$$\phi(x, 0) = \phi_t(x, 0) = 0 \qquad \phi(0, t) = 0 \qquad \phi_x(L, t) = K \tag{9.20}$$

where $K = F/E$, and E is Young's modulus, a constant depending on physical properties of the beam. Taking the Laplace transform of (9.19) with respect to t, using Theorem 8.10(iii), and taking account of the conditions (9.20) gives

$$c^2\overline{\phi}_{xx}(x, s) = s^2\overline{\phi}(x, s) - s\phi(x, 0) - \phi_t(x, 0) = s^2\overline{\phi}(x, s)$$

$$\Rightarrow \quad \overline{\phi}_{xx} = \frac{s^2}{c^2}\overline{\phi} \quad \Rightarrow \quad \overline{\phi}(x, s) = A(s)\cosh(sx/c) + B(s)\sinh(sx/c)$$

Notice that, with the given boundary conditions, it is more convenient to express $\overline{\phi}(x, s)$ in hyperbolic form than in exponential form. From (9.20), $\phi(0, t) = 0 \Rightarrow \overline{\phi}(0, s) = 0 \Rightarrow A(s) = 0$. Then $\phi_x(L, t) = K \Rightarrow \overline{\phi}_x(L, s) = K/s = (1/c)sB(s)\cosh(sL/c) \Rightarrow B(s) = Kc/(s^2\cosh(sL/c))$. Hence

$$\overline{\phi}(x, s) = \frac{cK\sinh(sx/c)}{s^2\cosh(sL/c)} \tag{9.21}$$

The appropriate way to determine $\phi(x, t)$ in this case is to use the Laplace inversion integral, and in particular, an extension of Corollary 8.13. We first need to identify the singular points of $\overline{\phi}(x, s)$, treated as a function of s, and to find the residue of $e^{st}\overline{\phi}(x, s)$ at each of these points. Clearly, ϕ has a singular point at $s = 0$ and at points given by $\cosh(sL/c) = 0$, i.e. $\cos(isL/c) = 0$, so that $s = (n + 1/2)i\pi c/L$, where $n \in \mathbb{Z}$.

Since $s = 0$ is a simple zero of $\sinh(sx/c)$, $s = 0$ is a simple pole of $\overline{\phi}(x, s)$ and by definition, using the Taylor series expansions of e^{st}, $\sinh(sx/c)$ and $\cosh(sL/c)$ in a neighbourhood of $s = 0$, we obtain

$$\operatorname*{Res}_{s=0} e^{st}\overline{\phi}(x, s) = cK(x/c) = Kx$$

Clearly, the points $s = \alpha_n = (n + \frac{1}{2})i\pi c/L$ are simple poles of $\overline{\phi}(x, s)$ and by Lemma 5.3,

$$\begin{aligned}\operatorname*{Res}_{s=\alpha_n} e^{st}\overline{\phi}(x, s) &= \frac{Kce^{\alpha_n t}\sinh(\alpha_n x/c)}{\alpha_n^2(L/c)\sinh(\alpha_n L/c)} = \frac{Kc^2 e^{\alpha_n t}\sin((n+\frac{1}{2})\pi x/L)}{L\alpha_n^2\sin((n+\frac{1}{2})\pi)} \\ &= \frac{-KL(-1)^n\exp((n+\frac{1}{2})i\pi ct/L)\sin((n+\frac{1}{2})\pi x/L)}{(n+\frac{1}{2})^2\pi^2}\end{aligned}$$

for any integer n. Since the only singular points of $\overline{\phi}(x, s)$ are poles, it can be shown, by using the technique of proof of Corollary 8.13 and a limiting argument, that

$$\phi(x, t) = \sum \text{Residues of } e^{st}\overline{\phi}(x, s)$$

Hence, by the above

$$\phi(x, t) = K\left(x - \frac{L}{\pi^2}\sum_{n=-\infty}^{\infty}\frac{(-1)^n\exp((n+\frac{1}{2})i\pi ct/L)\sin((n+\frac{1}{2})\pi x/L)}{(n+\frac{1}{2})^2}\right)$$

$$= K\left(x + \frac{L}{\pi^2}\sum_{n=1}^{\infty}\frac{(-1)^n\exp((n-\frac{1}{2})i\pi ct/L)\sin((n-\frac{1}{2})\pi x/L)}{(n-\frac{1}{2})^2}\right.$$

$$\left.-\frac{L}{\pi^2}\sum_{n=1}^{\infty}\frac{(-1)^n\exp((-n+\frac{1}{2})i\pi ct/L)\sin((-n+\frac{1}{2})\pi x/L)}{(-n+\frac{1}{2})^2}\right)$$

$$\Rightarrow \quad \phi(x, t) = K\left(x + \frac{2L}{\pi^2}\sum_{n=1}^{\infty}\frac{(-1)^n\cos((n-\frac{1}{2})\pi ct/L)\sin((n-\frac{1}{2})\pi x/L)}{(n-\frac{1}{2})^2}\right)$$

This infinite series solution converges for $0 \leqslant x \leqslant L$ and $t \geqslant 0$.

Exercise **9.3.1** Find the solution of

$$\phi_x(x, t) + \phi_t(x, t) = 0 \qquad (x > 0,\ t > 0)$$
$$\text{with} \quad \phi(x, 0) = e^{-x} \quad \text{and} \quad \phi(0, t) = e^t$$

by taking the Laplace transform with respect to t or x.

Exercise **9.3.2** An infinitely long elastic string is initially at rest on the x-axis with one end at $x = 0$. The end of the string at $x = 0$ then undergoes a continuous transverse displacement, given by $\phi(0, t) = f(t)$, as indicated in Fig. 9.33.

It can be shown that the transverse displacement $\phi(x, t)$ at any distance x from the fixed end and at any time t satisfies wave equation (9.19), for $x > 0$ and $t > 0$, subject to

$$\phi(x, 0) = \phi_t(x, 0) = 0, \quad \phi(0, t) = f(t) \quad \text{and} \quad \lim_{x\to\infty} \phi(x, t) = 0$$

(In this case, $c^2 = \rho/T$, where T is the constant tension of the string and ρ is its constant mass per unit length.)

Solve this problem by taking the Laplace transform with respect to t and using Lemma 8.11 to find $\phi(x, t)$ from $\bar{\phi}(x, s)$.

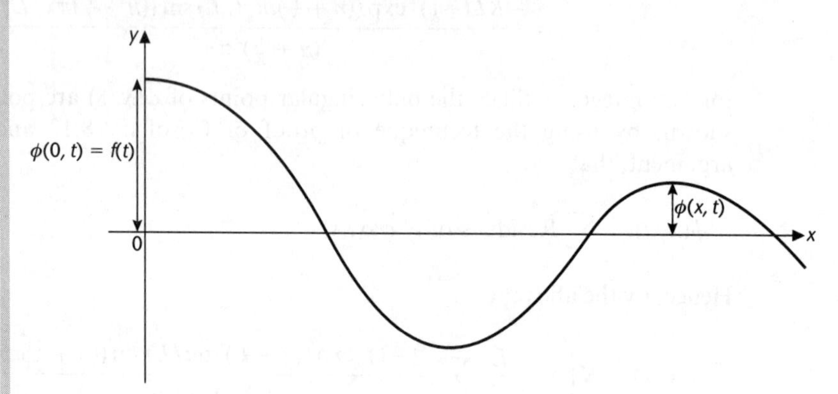

Figure 9.33

Exercise **9.3.3**

(i) Show that, as long as it exists, $\mathscr{F}\left(\int^x g(\tau)d\tau\right) = \hat{g}(s)/is$.

(ii) The transverse displacement $\phi(x, t)$ of an infinitely long elastic string that has an initial displacement $\phi(x, 0) = f(x)$, which is zero outside some finite interval, and an initial speed $\phi_t(x, 0) = g(x)$, also zero outside some finite interval, satisfies

$$c^2\phi_{xx} = \phi_{tt} \qquad (x \in \mathbb{R},\ t > 0)$$

$$\phi(x, 0) = f(x),\ \phi_t(x, 0) = g(x), \quad \lim_{x\to\pm\infty} \phi(x, t) = \lim_{x\to\pm\infty} \phi_x(x, t) = 0$$

Take the Fourier transform with respect to x, and use the Fourier inversion formula and the result of (i) to find $\phi(x, t)$ in terms of f and g.

Exercise 9.3.4 A metal bar of length L is initially heated to a constant temperature T_0, after which the right-hand end is kept at a constant temperature T_1. There is no change in temperature at the left-hand end. If $T(x, t)$ is the temperature at a point distance x along the bar from the left-hand end and t is the time elapsed, then $T(x, t)$ satisfies

$$T_{xx} = T_t \qquad (0 < x < L,\ t > 0)$$

$$T(x, 0) = T_0,\ T(L, t) = T_1 \qquad T_x(0, t) = 0$$

Take the Laplace transform with respect to t to show that, in the usual notation,

$$\overline{T}(x, s) = \frac{T_0}{s} + \frac{(T_1 - T_0)\left(e^{\sqrt{s}x} + e^{-\sqrt{s}x}\right)}{\left(e^{\sqrt{s}L} + e^{-\sqrt{s}L}\right)}$$

Use the binomial series to expand the denominator, and then the fact that $\mathscr{L}\left(\operatorname{erfc}\left(k/2\sqrt{t}\right)\right) = (1/s)e^{-k\sqrt{s}}$ to find a series solution for $T(x, t)$, valid for $t \in (0, a)$ for some $a > 0$.

Exercise 9.3.5 Find the solution of the non-homogeneous wave equation

$$\phi_{tt} = c^2\phi_{xx} + c^2\cos t \qquad (x > 0,\ t > 0)$$

$$\text{with} \quad \phi(x, 0) = 0,\ \phi_t(x, 0) = 0,\ \phi(0, t) = 0,\ \lim_{x\to\infty}\phi_x(x, t) = 0$$

by taking the Laplace transform with respect to t, and using using Lemma 8.11 to invert $\overline{\phi}(x, s)$.

Exercise 9.3.6 Consider an infinitely long beam, initially at rest. The beam is deformed in a direction perpendicular to its length and then released at time $t = 0$. Let $\phi(x, t)$ denote the transverse displacement of the beam, i.e. the displacement in a direction perpendicular to its length, at distance x from its midpoint, after any time t has elapsed. It can be shown that $\phi(x, t)$ satisfies

$$a^2\phi_{xxxx} + \phi_{tt} = 0 \qquad (x \in \mathbb{R},\ t > 0)$$

$$\text{with} \quad \phi(x, 0) = f(x),\ \phi_t(x, 0) = 0,\ \phi_{x^n}(x, t) \to 0 \quad \text{as} \quad x \to \pm\infty$$

where $n = 0, 1, 2, 3, 4$. Here a^2 is a constant which depends on physical properties of the beam, and the continuous function f represents the initial displacement at any point along the beam.

Assuming that all relevant transforms exist, take the Fourier transform with respect to x to show that

$$\phi(x, t) = \frac{1}{\sqrt{2\pi}}\int_{-\infty}^{\infty} e^{isx}\hat{f}(s)\cos(as^2t)\,ds$$

Exercise 9.3.7

(i) Show that if J_0 denotes the Bessel function of the first kind of order 0, defined in Chapter 7, and H denotes the Heaviside unit function, then

$$\int_0^{\infty} J_0(x)\cos(sx)\,dx = \frac{H(1 - s^2)}{\sqrt{1 - s^2}} \qquad (s \in \mathbb{R})$$

(*Hint*: Begin with the Laplace transform of $J_0(t)$ found in Example 8.18.)

(ii) By taking the Fourier transform, with respect to t, of the axially symmetric two-dimensional wave equation

$$\phi_{rr}(r, t) + \frac{1}{r}\phi_r(r, t) = \phi_{tt}(r, t) \qquad (r > 0 \text{ and } t \in \mathbb{R})$$

with $\phi(r, t)$ and $\phi_t(r, t) \to 0$ as $|t| \to \infty$ $\qquad \phi(0, t) = f(t)$

and $\phi(r, t)$ bounded for all $r \geqslant 0$ and $t \in \mathbb{R}$

show that

$$\hat{\phi}(r, s) = \hat{f}(s)J_0(rs)$$

Use the Fourier convolution result and (i) to deduce that

$$\phi(r, t) = \frac{1}{\pi}\int_{t-r}^{t+r} \frac{f(y)}{\sqrt{(r^2 - (t-y)^2)}}\, dy$$

Exercise **9.3.8** Let $T(x, t)$ be the solution of the problem in Exercise 9.3.4. Show that $\overline{T}(x, s)$ can be written as

$$\overline{T}(x, s) = \frac{T_0 \cosh(s^{1/2}L) + (T_1 - T_0)\cosh(s^{1/2}x)}{s\cosh(s^{1/2}L)}$$

Use the technique of Example 9.18 to find $T(x, t)$ for any $t > 0$ in the form of an infinite series.

Exercise **9.3.9** A uniform bar of unit cross-sectional area and of length L is initially at rest. A force is then applied to the right-hand end of the rod, at $x = L$, which induces a longitudinal displacement at this end of $K \sin t$, where K is a constant. The left-hand end, at $x = 0$, is kept fixed. If $\phi(x, t)$ denotes the longitudinal displacement at any point x after time t has elapsed, then $\phi(x, t)$ satisfies

$$\phi_{xx} = \phi_{tt} \qquad (0 < x < L,\ T > 0)$$

$$\phi(x, 0) = \phi_t(x, 0) = 0,\ \phi(0, t) = 0,\ \phi(L, t) = K\sin t$$

Solve this problem to find $\phi(x, t)$ as an infinite series, using the method of Example 9.18.

Exercise **9.3.10** The temperature $T(r, t)$ in a metal circular cylinder of unit radius satisfies the heat equation

$$T_{rr} + \frac{1}{r}T_r = \frac{1}{a}T_t \qquad (0 < r < 1,\ t > 0)$$

where a is a constant, r is the radial distance from the central axis and t is the time elapsed. Initially, the whole cylinder is kept at 0 °C. The surface is then raised to a constant temperature of K°C. Taking the Laplace transform with respect to t, show that

$$\overline{T}(r, s) = \frac{KJ_0(qr)}{sJ_0(q)} \qquad (\text{where } q^2 = -s/a)$$

Find $T(r, t)$ as an infinite series by using the technique of Example 9.18.

Separation of Variables

Although it is not directly related to complex variable results, we study briefly the technique of separation of variables for solving p.d.e.'s here for essentially two reasons. Firstly, it is often more appropriate to use this method than an integral transform. Secondly, the special functions introduced in Chapter 7 arise through solving important p.d.e.'s in mathematical physics by separating the variables. In this brief introduction, we are only concerned with the method itself, rather than the extensive theory of Fourier series behind it.

Example 9.19

Consider once again Dirichlet's problem for the circle $r = a$. In other words, we want to solve Laplace's equation in plane polar coordinates, i.e.,

$$u_{rr} + \frac{1}{r}u_r + \frac{1}{r^2}u_{\theta\theta} = 0 \qquad (0 < r \leqslant a,\ -\pi < \theta \leqslant \pi) \tag{9.22}$$

for $u(r, \theta)$, where $u(r, \theta)$ is bounded for all $r \leqslant a$ and $u(a, \theta) = f(\theta)$ for some given piecewise continuous function f. Recall that $U(r, \theta) = r^n \cos n\theta$ and $V(r, \theta) = r^n \sin n\theta$ are bounded solutions to (9.22), for any $n \in \mathbb{N}$, being the real and imaginary parts of the analytic function $F(z) = z^n$. These harmonic functions are **separable** in the sense that they can be expressed as the product of a function of r alone and a function of θ alone. These solutions are easily found by seeking solutions of (9.22) of the form $u(r, \theta) = R(r)T(\theta)$ to begin with. Substituting this *a priori* assumption into (9.22) gives

$$R''(r)T(\theta) + \frac{R'(r)T(\theta)}{r} + \frac{R(r)T''(\theta)}{r^2} = 0$$

$$\Rightarrow \quad r^2\frac{R''(r)}{R(r)} + r\frac{R'(r)}{R(r)} = -\frac{T''(\theta)}{T(\theta)} \tag{9.23}$$

multiplying by $r^2/(R(r)T(\theta))$. Since R is a function of r alone and T is a function of θ alone, the only way that (9.23) can be satisfied is if both sides are equal to a constant. Since T needs to be a periodic function of θ, this constant needs to be non-negative. Then (9.23) gives

$$r^2\frac{R''(r)}{R(r)} + r\frac{R'(r)}{R(r)} = -\frac{T''(\theta)}{T(\theta)} = m^2 \quad \text{say}$$

$$\Rightarrow \quad r^2R''(r) + rR'(r) - m^2R(r) = 0 \tag{9.24a}$$

$$\text{and} \quad T''(\theta) + m^2T(\theta) = 0 \tag{9.24b}$$

Equation (9.24a) is an Euler equation with general solution

$$R(r) = c(m)r^m + d(m)r^{-m} \quad (m \neq 0) \qquad R(r) = k_1 \operatorname{Log} r + k_2 \quad (m = 0)$$

Hence the only solutions which are bounded and analytic for $r \leqslant a$ are multiples of r^n, $n \in \mathbb{Z}_{\geqslant 0}$. The general solution of (9.24b) is

$$T(\theta) = a(m)\cos m\theta + b(m)\sin m\theta$$

Then the only bounded separable solutions of (9.22), which are analytic for $r \leqslant a$, take the form

$$u(r, \theta) = r^n(a_n \cos n\theta + b_n \sin n\theta) \qquad (n \in \mathbb{Z}_{\geqslant 0})$$

Any linear combination of these harmonic functions is a solution of (9.22), so that

$$u(r, \theta) = \sum_{n=0}^{\infty} (r/a)^n (a_n \cos n\theta + b_n \sin n\theta) \tag{9.25}$$

where the a_n and b_n are constants, is a solution of (9.22) and the unique solution to the given problem, by Theorem 9.3, as long as

$$u(a, \theta) = f(\theta) = \sum_{n=0}^{\infty} (a_n \cos n\theta + b_n \sin n\theta) \tag{9.26}$$

Assuming that the given trigonometric series converges uniformly to $f(\theta)$ for all θ, it is easily seen that termwise integration gives

$$a_0 = \frac{1}{2\pi}\int_{-\pi}^{\pi} f(\theta)\, d\theta \qquad a_n = \frac{1}{\pi}\int_{-\pi}^{\pi} f(\theta)\cos n\theta\, d\theta$$

$$b_n = \frac{1}{\pi}\int_{-\pi}^{\pi} f(\theta)\sin n\theta\, d\theta \qquad (n \geqslant 1) \tag{9.27}$$

Given a function f defined on $[-\pi, \pi]$, of period 2π, the trigonometric series defined in (9.26) with coefficients defined as in (9.27) is the **Fourier series** of f. The following result is a fundamental result from real analysis, which we shall not prove here.

Theorem 9.11. Convergence of a Fourier Series

Let $f: \mathbb{R} \to \mathbb{R}$ be periodic of period 2π and piecewise continuously differentiable on $(-\pi, \pi)$. Then the Fourier series of f converges to sum $g(\theta) = \frac{1}{2}(f(\theta-) + f(\theta+))$ for every $\theta \in \mathbb{R}$. □

Note

Since f is piecewise continuously differentable, it is integrable and so its Fourier series exists. At any point of continuity, $g(\theta) = f(\theta)$. In fact, the conditions of 9.11 can be weakened considerably.

Theorem 9.11 then allows us to solve the given Dirichlet problem using this approach. Suppose that the given function f satisfies the conditions of 9.11. Substituting (9.27) into (9.25) gives the solution

$$u(r, \theta) = \frac{1}{2\pi}\int_0^{2\pi} f(\phi)\, d\phi + \frac{1}{\pi}\sum_{n=1}^{\infty} (r/a)^n \int_0^{2\pi} f(\phi)\cos n(\phi - \theta)\, d\phi \tag{9.28}$$

Note that, since f is bounded, $|f(\theta)| \leqslant M$ say on $r = a$. Then $|(r/a)^n \int_0^{2\pi} f(\phi)\cos(\phi - \theta)\, d\phi| \leqslant 2\pi M(r/a)^n$ and $\sum_{n=1}^{\infty} (r/a)^n$ converges for $r < a$. Hence the series in (9.28) is uniformly convergent by Theorem 4.13. It then follows by (9.28) that

$$u(r, \theta) = \frac{1}{2\pi}\int_0^{2\pi} f(\phi)\left(1 + 2\sum_{n=1}^{\infty} (r/a)^n \cos n(\phi - \theta)\right) d\phi \tag{9.29}$$

Also,

$$1 + 2\sum_{n=1}^{\infty} t^n \cos n\psi = \mathrm{Re}\left(1 + 2\sum_{n=1}^{\infty} t^n e^{in\psi}\right) = \mathrm{Re}\left(1 + \frac{2te^{i\psi}}{1 - te^{i\psi}}\right)$$

by summing the geometric series. Hence

$$1 + 2\sum_{n=1}^{\infty} t^n \cos n\psi = \mathrm{Re}\left(\frac{1 + te^{i\psi}}{1 - te^{i\psi}}\right) = \frac{1 - t^2}{1 - 2t\cos\psi + t^2}$$

Then (9.29) gives

$$u(r, \theta) = \frac{a^2 - r^2}{2\pi}\int_0^{2\pi} \frac{f(\phi)}{a^2 - 2ar\cos(\phi - \theta) + r^2}\, d\phi$$

Notice that this is the result of Theorem 9.7 when $a = 1$.

In general, the technique of separation of variables looks for solutions of a particular type first. Given a linear partial differential equation in $u(x_1, x_2, \ldots, x_n)$, where the x_k are n independent variables, we assume there are solutions of the form $u(x_1, \ldots, x_n) = X_1(x_1)X_2(x_2)\ldots X_n(x_n)$. Substituting this assumption into the given equation very often results in a set of n linear ordinary differential equations for $X_1, \ldots, X_n$ which can be solved to obtain particular solutions of the given p.d.e. The most important p.d.e.'s of mathematical physics, relative to the standard coordinate systems can be separated into o.d.e.'s in this way. The resulting o.d.e.'s have solutions in terms of elementary functions or the special functions of Chapter 7 in these cases. It turns out that many functions can be expressed as Fourier series in terms of these special functions, so initial and boundary conditions can often be fitted to linear combinations of separable solutions to obtain the solution of a given problem.

Example 9.20

An elastic string of initial unit length has fixed endpoints at $x = 0$ and $x = 1$. The string is initially at rest and is then displaced in a transverse direction by an amount $f(x)$ at each point a distance x from the left-hand end. Let $\phi(x, t)$ denote the transverse displacement at distance x from the left-hand end after time t has elapsed. Then $\phi(x, t)$ satisfies the wave equation

$$c^2\phi_{xx} = \phi_{tt} \qquad (0 < x < 1, t > 0) \tag{9.19}$$

$$\text{with} \quad \phi(0, t) = \phi(1, t) = 0, \phi_t(x, 0) = 0, \phi(x, 0) = f(x) \tag{9.30}$$

We initially seek solutions of (9.19) in the form $\phi(x, t) = X(x)T(t)$. Substituting this assumption into (9.19) gives

$$c^2X''(x)T(t) = X(x)T''(t) \quad \Rightarrow \quad \frac{X''(x)}{X(x)} = \frac{T''(t)}{c^2T(t)} = -m^2$$

where m is a constant. Note that each side of this equation is equated to a negative constant so that ultimately $f(x)$ can be expressed as a Fourier series. Then

$$X''(x) + m^2X(x) = 0 \quad \text{and} \quad T''(t) + c^2m^2T(t) = 0$$

$$\Rightarrow X(x) = a(m)\cos mx + b(m)\sin mx$$

$$\text{and} \quad T(t) = A(m)\cos cmt + B(m)\sin cmt$$

where $a(m)$, $b(m)$, $A(m)$ and $B(m)$ are constants. We now use the conditions (9.30). Firstly, $\phi(0, t) = 0 \Rightarrow X(0) = 0 \Rightarrow a(m) = 0$ without loss of generality. Then $\phi(1, t) = 0 \Rightarrow X(1) = 0 \Rightarrow b(m)\sin m = 0 \Rightarrow m = n\pi$, $n \in \mathbb{Z}$, for non-trivial solutions. Also, $\phi_t(x, 0) = 0 \Rightarrow T'(0) = 0 \Rightarrow B(m) = 0$. Bearing in mind Theorem 9.11, a solution which satisfies these three boundary conditions is

$$\phi(x, t) = \sum_{n=1}^{\infty} b_n \sin(n\pi x)\cos(cn\pi t) \tag{9.31}$$

Finally, in order to satisfy $\phi(x, 0) = f(x)$, we need

$$\phi(x, 0) = f(x) = \sum_{n=1}^{\infty} b_n \sin n\pi x \quad \Rightarrow \quad f(\theta/\pi) = \sum_{n=1}^{\infty} b_n \sin n\theta$$

letting $\pi x = \theta$. We can extend the domain of f, which is initially [0, 1], to $\mathbb{R}$ by making f odd and periodic of period 2. Then as long as f is piecewise continuously differentiable, 9.11 gives

$$b_n = \frac{1}{\pi}\int_{-\pi}^{\pi} f(\theta/\pi)\sin n\theta\, d\theta = \int_{-1}^{1} f(x)\sin n\pi x\, dx = 2\int_0^1 f(x)\sin \pi x\, dx$$

Finally then, from (9.31), the required solution to the given problem, which can be shown to be unique, is

$$\phi(x, t) = \sum_{n=1}^{\infty} b_n \sin(n\pi x)\cos(cn\pi t) \quad \text{with} \quad b_n = 2\int_0^1 f(x)\sin \pi x\, dx$$

Most solutions found by separation of variables will take the form of an infinite series, as in this case.

Some of the special functions defined in Chapter 7 arise when solving linear p.d.e.'s by separating the variables. Consider the axially symmetric heat equation in cylindrical polar coordinates, i.e.

$$u_{rr} + \frac{1}{r}u_r = \frac{1}{k^2}u_t \tag{9.32}$$

where k is a constant. Letting $u(r, t) = R(r)T(t)$ in (9.32) gives

$$\frac{R''(r)}{R(r)} + \frac{R'(r)}{rR(r)} = \frac{T'(t)}{k^2T(t)} = \lambda$$

where λ is a real constant. Then $T(t) = a(\lambda)e^{\lambda k^2 t}$ and so for bounded solutions, $\lambda = -m^2$ say. Then

$$r^2R''(r) + rR'(r) + r^2m^2R(r) = 0 \tag{9.33}$$

Letting $mr = s$ in (9.33) gives

$$s^2R''(s) + sR'(s) + s^2R(s) = 0$$

which is Bessel's equation of order 0, (7.40). Hence the general solution to (9.33) is

$$R(r) = b(m)J_0(mr) + c(m)Y_0(mr)$$

Then separable solutions of (9.32), which are bounded for all t and for $0 \leqslant r \leqslant L$ say, take the form $u(r, t) = a(m)e^{-k^2m^2t}J_0(mr)$, for any constant m. Just as certain functions can be expanded in a Fourier series of sines and cosines, so certain functions can be expanded in a particular series of Bessel functions. Thus, in a large number of cases, these separable solutions of (9.32) can be used to solve a given initial-boundary value problem involving (9.32). For details of this technique, see H. F. Weinberger, *A First Course in Partial Differential Equations*, Blaisdell, 1965.

Now consider Laplace's equation in spherical polar coordinates, independent of the equatorial angle, i.e.

$$u_{rr} + \frac{2}{r}u_r + \frac{1}{r^2}u_{\theta\theta} + \frac{\cot\theta}{r^2}u_\theta = 0 \tag{9.34}$$

Letting $u(r, \theta) = R(r)T(\theta)$ in (9.34) gives

$$\frac{r^2R''(r)}{R(r)} + \frac{2rR'(r)}{R(r)} = \frac{-T''(\theta)}{T(\theta)} - \frac{\cot\theta\, T'(\theta)}{T(\theta)} = \lambda$$

say, where λ is a real constant. Then $T(\theta)$ satisfies

$$T''(\theta) + \cot\theta\, T'(\theta) + \lambda T(\theta) = 0 \tag{9.35}$$

Suppose that we require solutions of (9.34) which are bounded for all θ and for $0 \leqslant r \leqslant a$ say. It can be shown that (9.35) has bounded solutions for all θ if and only if $\lambda = n(n + 1)$, where $n \in \mathbb{N}$ or $n = 0$. Letting $x = \cos\theta$ in (9.35) gives, in this case,

$$(1 - x^2)T''(x) - 2xT'(x) + n(n + 1)T(x) = 0$$

which is Legendre's differential equation (7.37). It follows from Chapter 7 that the general solution to (9.35) is then

$$T(\theta) = a_n P_n(\cos\theta) + b_n Q_n(\cos\theta)$$

Since $Q_n(x)$ is not bounded at zero, $b_n = 0$. Now $R(r)$ satisfies

$$r^2 R''(r) + 2rR'(r) - n(n+1)R(r) = 0$$

which is an Euler equation with general solution

$$R(r) = c_n r^n + \frac{d_n}{r^{n+1}}$$

Since we require $u(r, \theta)$ to be bounded for all $r \leqslant a$, we have that $d_n = 0$. Then

$$u(r, \theta) = \sum_{n=0}^{\infty} a_n (r/a)^n P_n(\cos\theta) \tag{9.36}$$

is the most general bounded solution of (9.34) obtainable by this technique. A large number of functions can be expanded as a particular Fourier series in Legendre functions and so (9.36) can be used to solve particular boundary value problems involving (9.34).

Exercise 9.4.1 Show that if the series in (9.26) converges uniformly to $f(\theta)$ for $\theta \in [-\pi, \pi]$, then the coefficients are given by (9.27).

Exercise 9.4.2

(i) Give the solution of the elastic string problem of Example 9.20 when $f(x) = \sin \pi x$.

(ii) Show that if the string is initially plucked at its midpoint, with a transverse displacement of height h, then $f(x) = 2hx$, $0 \leqslant x \leqslant 1/2$ and $f(x) = 2h(1-x)$, $1/2 \leqslant x \leqslant 1$. Hence give the particular solution of the problem in this case.

Exercise 9.4.3 The temperature distribution, $u(x, t)$, in a rod of length L satisfies the heat equation

$$u_{xx} = \frac{1}{k} u_t \qquad (0 < x < L,\ t > 0)$$

where x is the distance from the left-hand end and t is the time elapsed. If the rod is insulated at each end and $u(x, 0) = Cx$, where C is a constant, find the subsequent temperature distribution in the rod by separating the variables.

Exercise 9.4.4 The steady temperature $T(x, y)$ in a semi-infinite plate satisfies Laplace's equation in the region bounded by $x = 0$, $y = 0$ and $y = b > 0$. The temperature satisfies the conditions $T(x, 0) = 0$, $T(x, b) = 0$, $T(0, y) = f(y)$, for some continuous function f, and $T(x, y) \to 0$ as $x \to \infty$. Solve this problem using separation of variables.

Exercise

9.4.5 The electric potential $V(x, y)$ in a metal plate satisfies Laplace's equation in the region bounded by $x = 0$, $y = 0$, $x = a > 0$ and $y = b > 0$. The potential satisfies $V(x, 0) = V(x, b) = 0$, $V(a, y) = 0$ and $V(0, y) = f(y)$, for some continuous function f. Solve this problem using separation of variables and so show that

$$V(x, y) = \sum_{n=1}^{\infty} \frac{b_n \sinh((a - x)n\pi/b)}{\sinh(n\pi a/b)} \sin(n\pi y/b) \quad b_n = \frac{2}{b}\int_0^b f(y)\sin(n\pi y/b)\,dy$$

Exercise

9.4.6 The function $\phi(r, t)$ satisfies the spherically symmetric wave equation,

$$\phi_{rr} + \frac{2}{r}\phi_r = \frac{1}{c^2}\phi_{tt} \qquad (r > a,\ t > 0)$$

where c is a constant, outside the sphere $r = a$. Solve this equation by separating the variables, given that $\phi(a, t) = \sin\omega t$, for all $t > 0$ and that $\phi(r, t) \to 0$ as $r \to \infty$. Hence show that

$$\phi(r, t) = \frac{\sqrt{a}\sin\omega t\left(AJ_{1/2}(\omega r/c) + BJ_{-1/2}(\omega r/c)\right)}{\sqrt{r}\left(AJ_{1/2}(\omega a/c) + BJ_{-1/2}(\omega a/c)\right)}$$

where A and B are arbitrary constants.

10 Analytic Functions

It has been shown in Chapter 4 that an analytic function can be represented as a Taylor series in a particular region of $\mathbb{C}$ and it may have a different Taylor series expansion in a different region. Sometimes it may be more convenient to represent an analytic function as an integral, as indicated in Chapter 7. In general, an analytic function may have many different representations, some of which may only be valid in certain regions. This chapter deals with infinite product and certain integral representations of analytic functions. It also investigates how the definition of an analytic function may be extended from one region to another, which is the idea of analytic continuation. The chapter also includes an introduction to asymptotic series representations of analytic functions.

Analytic Continuation

Given a function $f_1 : \mathscr{R}_1 \subset \mathbb{C} \to \mathbb{C}$, which is analytic on a region $\mathscr{R}_1$ of $\mathbb{C}$, it may be desirable to find a function f, analytic on a larger region $\mathscr{R}$, with $\mathscr{R}_1 \subset \mathscr{R}$, such that $f(z) = f_1(z)$ for all $z \in \mathscr{R}_1$. For example, it may be that $f_1(z)$, defined by a power series about 0, which converges for $|z| < r$ say, provides a solution to a linear differential equation. Then it is clearly desirable to extend the domain of this analytic function beyond the region $|z| < r$.

Definition

Let f_1 be analytic on the region $\mathscr{R}_1 \subset \mathbb{C}$. Let $\mathscr{R}_2$ be another region with $\mathscr{R}_1 \cap \mathscr{R}_2 = \mathscr{I} \neq \varnothing$. If there exists a function f_2, analytic on $\mathscr{R}_2$, with $f_1(z) = f_2(z)$ for all $z \in \mathscr{I}$, then f_2 is an **analytic continuation** of f_1 to $\mathscr{R}_2$.

Note that if f_2 is an analytic continuation of f_1, then the function f defined on $\mathscr{R}_1 \cup \mathscr{R}_2$ by $f(z) = f_1(z)$, $z \in \mathscr{R}_1$ and $f(z) = f_2(z)$, $z \in \mathscr{R}_2$, is clearly analytic on $\mathscr{R}_1 \cup \mathscr{R}_2$.

Example 10.1

Let f_1 be defined on the region $\operatorname{Re} z > 0$ by

$$f_1(z) = \int_0^\infty t e^{-zt}\, dt \qquad (t \subset \mathbb{R})$$

The given integral is convergent for $\operatorname{Re} z > 0$ since integration by parts gives

$$f_1(z) = \lim_{a\to\infty}\left(\left[\frac{-te^{-zt}}{z}\right]_0^a + \frac{1}{z}\int_0^a e^{-zt}\,dt\right) = \lim_{a\to\infty}\left[\frac{-te^{-zt}}{z} - \frac{e^{-zt}}{z^2}\right]_0^a = \frac{1}{z^2}$$

for $\operatorname{Re} z > 0$. The integral is clearly divergent for $\operatorname{Re} z \leqslant 0$ and so the given region is the largest domain for f_1 defined in this way. Now let $\mathscr{R}_2$ be the region $\mathbb{C}\backslash\{0\}$ and let $f_2(z) = 1/z^2$ on $\mathscr{R}_2$. Then f_2 is an analytic continuation of f_1 to $\mathscr{R}_2$.

It may happen that it is not possible to find any analytic continuation of f_1, defined on $\mathscr{R}_1$. In this case, the boundary of $\mathscr{R}_1$ is called a **natural boundary** to analytic continuation and $\mathscr{R}_1$ is then the largest possible domain of definition.

In order to show that the analytic continuation of an analytic function to another region is unique, we require the following result, which is of fundamental importance in its own right.

Theorem 10.1. The Uniqueness Theorem

Let f be analytic in a region $\mathscr{R}$ and let $f(z) = 0$ at an infinite number of distinct points in $\mathscr{R}$ with a limit point $\alpha \in \mathscr{R}$. Then $f(z) = 0$ for all $z \in \mathscr{R}$. □

Part Proof

Step 1

By hypothesis, any deleted neighbourhood of α contains points at which $f(z) = 0$. Since f is analytic at α, it follows by 4.18 that $f(z) = \sum_{n=0}^{\infty} a_n(z-\alpha)^n$ for $|z-\alpha| < r$ say, for some constants a_n. Suppose that at least one of these coefficients is non-zero, so let $a_k \neq 0$ with $a_n = 0$ for $n \leqslant k$. Therefore

$$f(z) = (z-\alpha)^k(a_k + a_{k+1}(z-\alpha) + a_{k+2}(z-\alpha)^2 + \ldots) = (z-\alpha)^k g(z)$$

say, where g is analytic at α with $g(\alpha) = a_k$. Then $g(z)$ is non-zero in some neighbourhood of α, so there is a deleted neighbourhood of α within which $f(z)$ is non-zero. This is a contradiction, so $a_n = 0$ for all n and $f(z) = 0$ for $|z-\alpha| < r$, i.e. $f(z) = 0$ throughout any neighbourhood of α contained in $\mathscr{R}$.

Step 2

Now let β be any point in $\mathscr{R}$ with $|\beta - \alpha| \geqslant r$, as otherwise, $f(\beta) = 0$ by step 1. Connect α and β by a curve $\mathscr{C}$ lying within $\mathscr{R}$, as shown in Fig. 10.1. It can be shown that there exists such a curve $\mathscr{C}$ and $\rho \in \mathbb{R}^+$ such that the distance from any point on $\mathscr{C}$ to the boundary of $\mathscr{R}$ is greater than ρ, i.e. $\mathscr{C}$ does not become 'arbitrarily close' to the boundary. Divide $\mathscr{C}$, in any manner, by means of points $\alpha = z_0, z_1, \ldots, z_m = \beta$, into a number of arcs whose lengths are all less than ρ, as shown in Fig. 10.1.

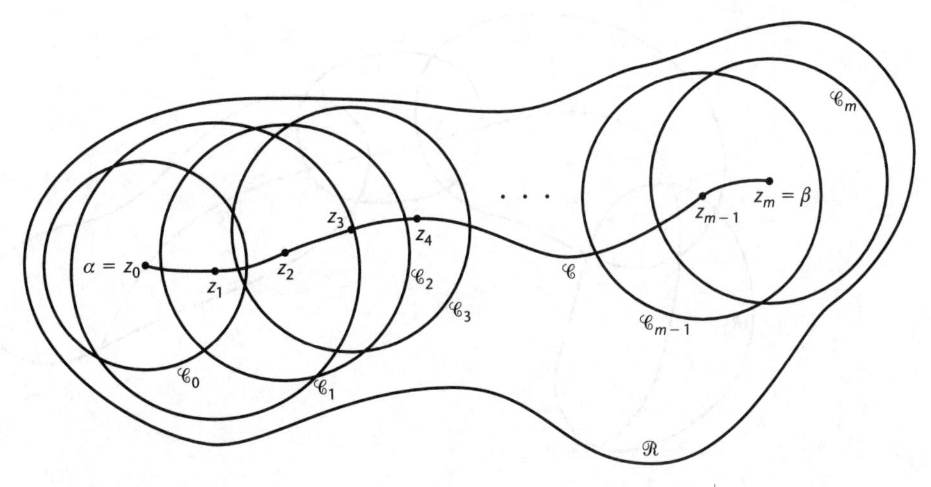

Figure 10.1

For each k, let $\mathscr{C}_k$ be a circle lying inside $\mathscr{R}$, with centre z_k, $0 \leqslant k \leqslant m$, with radius greater than ρ, as shown. Since the radii of these circles are all greater than ρ, each of the circles encloses the centre of the next. In this case, we say that the circles form a **circle chain**. By hypothesis, f can be expanded as a power series about any centre z_k and this series converges inside $\mathscr{C}_k$. Since $f(z) = 0$ inside $\mathscr{C}_0$ by step 1, $f(z) = 0$ along the arc joining z_0 to z_1 inside $\mathscr{C}_1$ and so by step 1, $f(z) = 0$ inside $\mathscr{C}_1$. Clearly, this argument can be repeated all along the circle chain so that $f(z) = 0$ inside $\mathscr{C}_m$ and, in particular, $f(\beta) = 0$. ■

Important Notes

(i) It follows by 10.1 that if $f(z) = 0$ in a neighbourhood of a point in $\mathscr{R}$ or if $f(z) = 0$ along an arc lying in $\mathscr{R}$, then $f(z) = 0$ for all $z \in \mathscr{R}$. Hence for example, if f is entire and $f(x) = 0$ for all $x \in \mathbb{R}$, then $f(z) = 0$ for all $z \in \mathbb{C}$. Very often if an identity holds for all real numbers, this result shows that it holds for all complex numbers. It also shows, for example, that there is a unique definition of $g(z) = e^z$ such that g is entire and g reduces to the real exponential function on the real axis.

(ii) Letting $f(z) = f_1(z) - f_2(z)$ for all $z \in \mathscr{R}$ in 10.1 shows that if $f_1(z) = f_2(z)$ at an infinite number of points in $\mathscr{R}$ with a limit point in $\mathscr{R}$, then $f_1(z) = f_2(z)$ in $\mathscr{R}$. Hence if $f_1(z) = f_2(z)$ in a subregion of $\mathscr{R}$, then $f_1(z) = f_2(z)$ throughout $\mathscr{R}$.

(iii) It follows from (ii) that if f_2 is the analytic continuation of f_1 from $\mathscr{R}_1$ to $\mathscr{R}_2$, then f_2 is unique. For if f_3 is analytic in $\mathscr{R}_2$ and $f_3(z) = f_1(z)$ in $\mathscr{R}_1 \cap \mathscr{R}_2$, then $f_3(z) = f_2(z)$ in $\mathscr{R}_1 \cap \mathscr{R}_2$ and so $f_3(z) = f_2(z)$ throughout $\mathscr{R}_2$.

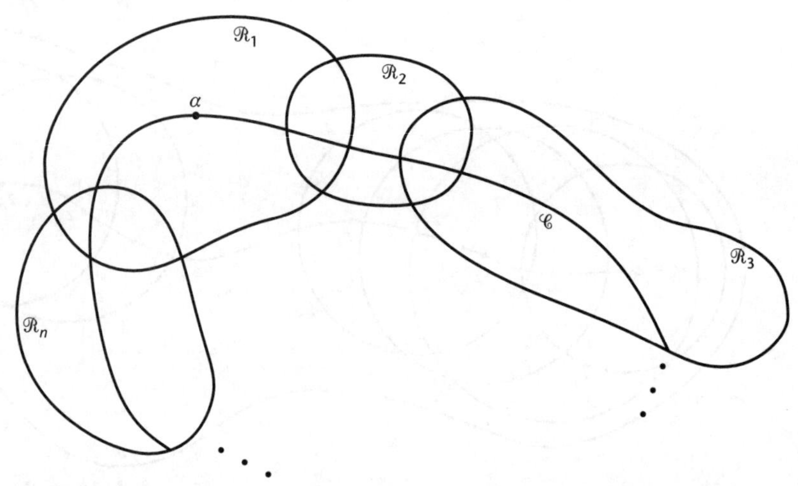

Figure 10.2

If analytic continuation of f_1 on $\mathcal{R}_1$ to f_2 on $\mathcal{R}_2$ is possible, it may be possible to analytically continue f_2 to f_3 on $\mathcal{R}_3$ and so on. In general, it may be possible to analytically continue f_1 to f_n defined on $\mathcal{R}_n$, through regions $\mathcal{R}_2, \ldots, \mathcal{R}_{n-1}$, as indicated in Fig. 10.2. In this case, suppose that $\mathcal{R}_1 \cap \mathcal{R}_n \neq \varnothing$. The obvious question that arises is when is the analytic continuation of f_n into $\mathcal{R}_1$ the original function f_1? If the intersection of $\mathcal{R}_1, \ldots, \mathcal{R}_n$ is non-empty, then $f_1(z) = f_2(z) = \ldots = f_n(z)$ in this intersection so that $f_n(z) = f_1(z)$ in $\mathcal{R}_1$ by 10.1. However, it is clear that if $\mathcal{R}_1, \ldots, \mathcal{R}_n$ have no region in common, the analytic continuation of f_n to $\mathcal{R}_1$ need not be f_1. For example, analytically continuing f_1 along a closed curve $\mathcal{C}$, which passes through the given regions from initial point α back to α, as shown in Fig. 10.2, will clearly not produce the same function if f_1 is a branch of a multifunction and $\mathcal{C}$ encloses a branch point of this multifunction. In this case, the analytic continuation of f_1 will lead to a different branch. The following definition makes the idea of analytic continuation along a curve more precise.

Definition

Let f be analytic at a point α in a region $\mathcal{R}$. Then f can be **analytically continued along a curve** $\mathcal{C}$ in $\mathcal{R}$, from α to β, if there are points $\alpha = z_0, z_1, \ldots, z_m = \beta$, on $\mathcal{C}$ and functions $f = f_0, f_1, \ldots, f_m$, such that each f_k is analytic inside any circle $\mathcal{C}_k$ with centre z_k lying inside $\mathcal{R}$, provided that the circles $\mathcal{C}_k$ form a **circle chain**, as defined in the proof of 10.1, and provided that $f_k(z) = f_{k+1}(z)$ in $\mathcal{C}_k \cap \mathcal{C}_{k+1}$, $0 \leqslant k \leqslant m-1$.

The following result shows that a function can be analytically continued along a closed curve to produce the original function as long as the curve does not enclose a branch point of an associated multifunction. The proof of this result is omitted, but see p. 105 in K. Knopp, *Theory of Functions, Part 1*, Dover, 1945.

Theorem 10.2. The Monodromy Theorem

Let f be analytic at a point α in a simply connected region $\mathcal{R}$. If f can be analytically continued from α along every curve within $\mathcal{R}$, the continuation gives rise to a unique function which is analytic in $\mathcal{R}$. □

Example 10.2

Let $f(z) = 1/z$, $g(z) = z^{1/2}$ and $h(z) = e^z$, for $|z - i| < 1$ say. All three functions are analytic in this region and, in particular, are analytic at i. Note that f can be analytically continued along any closed curve starting from i, which does not pass through 0, in the region $|z| < 2$. The process produces f again in a neighbourhood of i but f is clearly not analytic throughout $|z| < 2$. Similarly, g can be analytically continued along any closed curve starting from i, which does not enclose 0, in the region $|z| < 2$, but not along any closed curve enclosing 0, since any such curve will cross a branch cut of the multifunction $z^{1/2}$. Clearly, h can be analytically continued along every curve starting at i in $|z| < 2$ and the process produces a unique function $H(z) = e^z$ which is analytic in $|z| < 2$.

Definition

A **complete analytic (multi)function** f consists of a function f_1 analytic on a region $\mathcal{R}_1$, all its possible analytic continuations, all the possible analytic continuations of those continuations, and so on. Hence f is the analytic (multi)function with the largest possible domain such that $f(z) = f_1(z)$ on $\mathcal{R}_1$. The function f_1, with domain restricted to $\mathcal{R}_1$ is an **element** of f. For example, f defined by $f(z) = 1/z^2$ on $\mathbb{C}\backslash\{0\}$, is a complete analytic function and f_1, defined on $\operatorname{Re} z > 0$ in Example 10.1, is an element of f. On the other hand, g defined by $g(z) = z^{1/2}$ on $\mathbb{C}\backslash\{0\}$ is a complete analytic multifunction with an element $g_1(z) = z^{1/2}$, where $z^{1/2}$ denotes the principal value of $z^{1/2}$, on $|z - i| < 1$.

The construction of a complete analytic multifunction f on some region of $\mathbb{C}$, by analytic continuation, leads naturally to the construction of a Riemann surface on which f is a function. The regions of definition of elements of f can be considered as patches on the growing Riemann surface. If the continuation around a closed curve leads back to a different element at the starting point, then parts of two sheets of the Riemann surface are formed. If the continuation around every closed curve leads to the original element of f, the Riemann surface has one sheet and f is then a function.

Analytic Continuation by Means of Taylor Series

One of the simplest methods for analytically continuing a function is by means of its Taylor series expansions. Suppose that f is analytic in the region $\mathscr{R}$, so that by Theorem 4.18, f has a Taylor series expansion about any point in $\mathscr{R}$ which converges within any circle centred at this point and lying within $\mathscr{R}$. Hence, without loss of generality, we begin by supposing that a function f_1 is defined by the power series expansion

$$f_1(z) = \sum_{n=0}^{\infty} a_n(z - \alpha_1)n \qquad a_n = \frac{f_1{}^{(n)}(\alpha_1)}{n!}$$

within the region $\mathscr{R}_1$, given by $|z - \alpha_1| < r_1$, which is the largest domain of convergence. Note that at least one singular point of the complete analytic function f of which f_1 is an element must then lie on the circle $|z - \alpha_1| = r_1$, otherwise f would be analytic at every point on this circle and so there would be a larger circle with centre α_1 for which the power series converges. Then given any other point $\alpha_2 \in \mathscr{R}_1$, f_1 is analytic at α_2 and by Theorem 4.17,

$$\sum_{n=0}^{\infty} f_1{}^{(n)}(\alpha_2)\frac{(z - \alpha_2)^n}{n!} = f_2(z) \qquad (|z - \alpha_2| < r_2)$$

say, defines an analytic function f_2 in the region $\mathscr{R}_2$, given by $|z - \alpha_2| < r_2$. It follows by Theorem 4.18 that $f_2(z) = f_1(z)$ in the region $\mathscr{R}_1 \cap \mathscr{R}_2$. There are three possibilities, as indicated in Fig. 10.3.

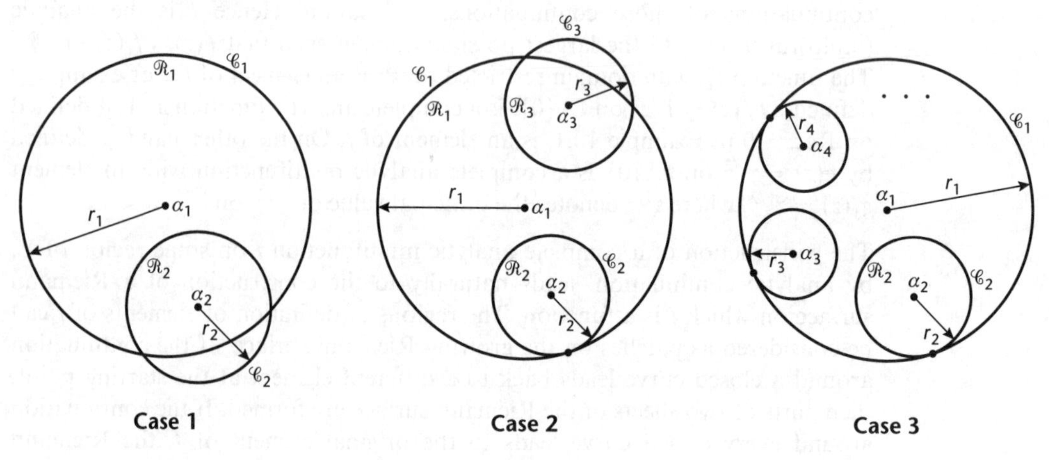

Figure 10.3

Case 1

The circle $\mathscr{C}_2$ given by $|z - \alpha_2| = r_2$ does not lie totally within the circle $\mathscr{C}_1$ given by $|z - \alpha_1| = r_1$, so that $\mathscr{R}_2$ is not a subset of $\mathscr{R}_1$. Then f_2 is the analytic continuation of f_1 to $\mathscr{R}_2$.

Case 2

The circle $\mathscr{C}_2$ lies within the circle $\mathscr{C}_1$, since a singular point of the complete analytic function of which f_1 is an element lies on $\mathscr{C}_1$. In this case, it may be possible to continue f_1 analytically to a different region $\mathscr{R}_3$ by constructing a Taylor series expansion about a different point $\alpha_3 \in \mathscr{R}_1$.

Case 3

The circle of convergence of the Taylor series expansion of f_1 about any point of $\mathscr{R}_1$ lies within the circle $\mathscr{C}_1$, since this circle is a 'barrier' of non-isolated singular points of the complete analytic function of which f_1 is an element. In this case, analytic continuation of f_1 beyond $\mathscr{R}_1$ is impossible and the circle $\mathscr{C}_1$ is a natural boundary. Such a function represented by a power series having a natural boundary is sometimes called a **lacunary function**.

Example 10.3

Noting that the radius of convergence of the given power series is 1, let f_1 be defined by

$$f_1(z) = \sum_{n=0}^{\infty} z^n \qquad (|z| < 1)$$

In this case, f_1 can be analytically continued beyond the region $\mathscr{R}_1$, given by $|z| < 1$. The given series is a convergent geometric series for $|z| < 1$ and $f_1(z) = 1/(1 - z)$ in this region. Then

$$f_1{}^{(n)}(z) = \frac{n!}{(1 - z)^{n+1}} \quad \text{for } |z| < 1, \text{ so that if } \alpha \in \mathscr{R}_1$$

$$\text{then } \sum_{n=0}^{\infty} \frac{(z - \alpha)^n}{(1 - \alpha)^{n+1}} = f_2(z) \quad \text{say, for } |z - \alpha| < |1 - \alpha|$$

by the ratio test. (Alternatively, f_2 can be obtained by noting that

$$f_1(z) = \frac{1}{(1 - \alpha)(1 - (z - \alpha)/(1 - \alpha))}$$

for any $\alpha \neq 1$). If $\alpha = a \in \mathbb{R}$ and $0 < a < 1$ then $|1 - a| < 1$, so the region of convergence of this new power series is contained within $\mathscr{R}_1$ and $f_2(z) = f_1(z)$ in this region. Then f_1 cannot be analytically continued beyond $\mathscr{R}_1$ using these points, since clearly 1 is a singular point of the function of which f_1 is an element. But if $\alpha = 3i/4$ say, the new power series converges in the region $\mathscr{R}_2$, given by $|z - 3i/4| < 5/4$ and $\mathscr{R}_2$ is not a subset of $\mathscr{R}_1$. Then f_2 is the analytic continuation of f_1 to $\mathscr{R}_2$. In general, if $\alpha \notin \mathbb{R}^+$ then $|1 - \alpha| > 1 - |\alpha|$, so the region $|z - \alpha| < |1 - \alpha|$ is not contained wholly within $\mathscr{R}_1$.

Example 10.4

Consider the function f given by

$$f(z) = 1 + \sum_{n=0}^{\infty} z^{3^n} \qquad (|z| < 1)$$

(It is easily shown that the given series converges for $|z| < 1$.) It follows from the definition of f that

$$f(z) = z + f(z^3) \quad f(z^3) = z^3 + f(z^9) \quad \ldots \quad f(z^{3^k}) = z^{3^k} + f(z^{3^{k+1}})$$

for any $k \in \mathbb{N}$. Letting $z \to 1$ and then $z^{3^k} \to 1$, $k = 1, 2, \ldots$, in the above set of equations shows that $f(z)$ is not finite at the points given by $z^{3^k} = 1$, for any $k \in \mathbb{Z}_{\geqslant 0}$. Hence the points $e^{2mi\pi/3^k}$, $m \in \mathbb{Z}$ and $k \in \mathbb{Z}_{\geqslant 0}$, are all singular points of f lying on $|z| = 1$. Since any arc of $|z| = 1$, no matter how small in length, contains at least one of these points, $|z| = 1$ is a natural boundary to analytic continuation of f, and f is a lacunary function.

Analytic Continuation Across a Boundary

Given a function f_1, which is analytic in a region $\mathscr{R}_1$, it is sometimes possible to analytically continue f_1 to another region if the regions themselves do not intersect, but the boundaries of the regions do. This is the content of the following result, which is of practical use in a large number of cases.

Theorem 10.3. Analytic Continuation Across a Boundary

Let $\mathscr{R}_1$ and $\mathscr{R}_2$ be two regions with $\mathscr{R}_1 \cap \mathscr{R}_2 = \emptyset$, while $\overline{\mathscr{R}_1} \cap \overline{\mathscr{R}_2} = \Gamma$, a simple contour, as shown in Fig. 10.4. Let f_1 be defined on $\overline{\mathscr{R}_1}$, analytic on $\mathscr{R}_1$ and continuous on $\overline{\mathscr{R}_1}$. Let there exist a function f_2, defined on $\overline{\mathscr{R}_2}$, analytic on $\mathscr{R}_2$ and continuous on $\overline{\mathscr{R}_2}$, such that $f_2(z) = f_1(z)$ for all $z \in \Gamma$. Then the function f defined by

$$f(z) = f_1(z) \quad \text{for } z \in \mathscr{R}_1 \cup \Gamma, \qquad f(z) = f_2(z) \quad \text{for } z \in \mathscr{R}_2 \cup \Gamma$$

is analytic on $\mathscr{R}_1 \cup \mathscr{R}_2 \cup \Gamma^*$, where Γ^* is Γ with the endpoints excluded. □

Part Proof

Note that f is well defined since $f_1(z) = f_2(z)$ on Γ. Let Γ_k, $k = 1, 2$, be any simple contour lying within $\mathscr{R}_k$, joining two distinct points on Γ, and let $\mathscr{C}_k$ be

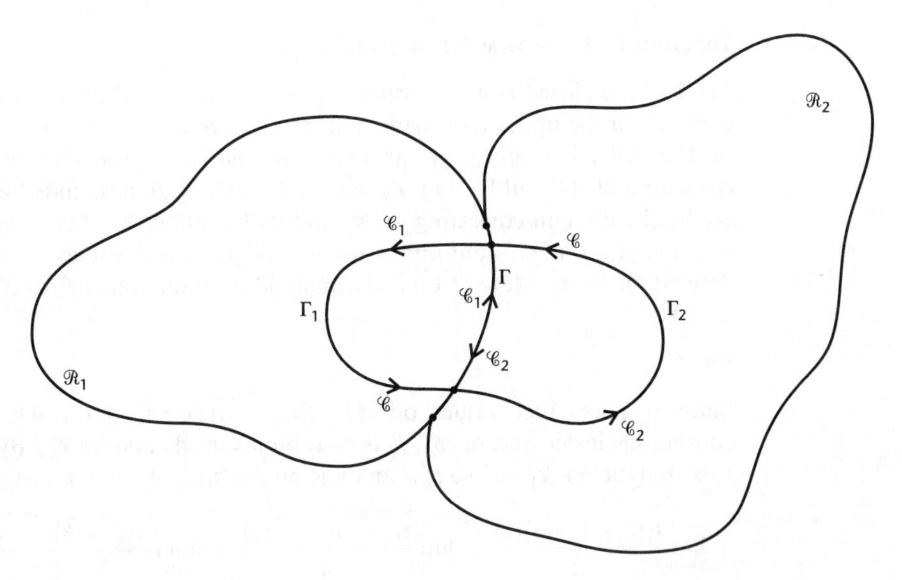

Figure 10.4

the simple closed curve consisting of Γ_k and part of Γ as shown in Fig. 10.4. Let $\mathscr{C} = \Gamma_1 \cup \Gamma_2$ and let F be defined by

$$F(z) = \frac{1}{2\pi i} \int_{\mathscr{C}} \frac{f(\zeta)}{\zeta - z} \, d\zeta$$

for all z inside $\mathscr{C}$. Since f is continuous on $\mathscr{C}$, it follows, using the technique used to prove Theorem 3.13, that F is analytic inside $\mathscr{C}$. This result is also a direct consequence of Theorem 10.14, proved later in this chapter. Also

$$F(z) = \frac{1}{2\pi i} \left(\int_{\mathscr{C}_1} \frac{f(\zeta)}{\zeta - z} \, d\zeta + \int_{\mathscr{C}_2} \frac{f(\zeta)}{\zeta - z} \, d\zeta \right)$$

Hence if z lies inside $\mathscr{C}_1$, then it follows by Cauchy's theorem and Cauchy's integral formula that

$$F(z) = \frac{1}{2\pi i} \left(\int_{\mathscr{C}_1} \frac{f_1(\zeta)}{\zeta - z} \, d\zeta + \int_{\mathscr{C}_2} \frac{f_2(\zeta)}{\zeta - z} \, d\zeta \right) = f_1(z) + 0 = f(z)$$

Similarly, if z lies inside $\mathscr{C}_2$, $F(z) = f(z)$. Since both F and f are continuous inside $\mathscr{C}$, if α is any point on Γ, then $F(\alpha) = \lim_{z \to \alpha} F(z) = \lim_{z \to \alpha} f(z) = f(\alpha)$, so that $F(z) = f(z)$ on Γ also. It follows that f is analytic inside $\mathscr{C}$, as required. ■

It is thus possible to generalise the definition of analytic continuation, based on the above result. Under the same conditions as Theorem 10.3, f_2 is the **analytic continuation** of f_1 to $\mathscr{R}_2$. A natural and very useful method of analytically continuing an analytic function across a boundary is by reflection.

Theorem 10.4. Schwarz's Reflection Principle

Let $\mathscr{C}_1$ be a closed contour consisting of a segment AB of the real axis and a curve Γ_1 in the upper half-plane, joining A to B, as shown in Fig. 10.5. Let Γ_2 be the reflection of Γ_1 in the real axis and let $\mathscr{C}_2$ be the closed contour consisting of AB and Γ_2. Let $\mathscr{R}_k$, $k = 1, 2$, be the region bounded by $\mathscr{C}_k$ and let $\mathscr{D}_k$ be the domain consisting of $\mathscr{R}_k$ and its boundary $\mathscr{C}_k$. Let f_1 be defined on $\mathscr{D}_1$, analytic on $\mathscr{R}_1$, continuous on $\mathscr{C}_1$ and take real values on AB. Then f_2 defined on $\mathscr{D}_2$ by $f_2(z) = \overline{f_1(\bar{z})}$ is the analytic continuation of f_1 to $\mathscr{R}_2$. □

Proof

Since f_1 takes real values on AB, $f_2(z) = \overline{f_1(\bar{z})} = f_1(z)$ on AB. Since f_1 is continuous inside and on $\mathscr{C}_1, f_2$ is continuous inside and on $\mathscr{C}_2$. By hypothesis, f_1 is analytic on $\mathscr{R}_1$ and so f_2 is analytic on $\mathscr{R}_2$, since if $z, z + h \in \mathscr{R}_2$,

$$\lim_{h\to 0} \frac{f_2(z+h) - f_2(z)}{h} = \lim_{h\to 0} \frac{\overline{f_1(\overline{z+h})} - \overline{f_1(\bar{z})}}{h} = \lim_{h\to 0} \left(\overline{\frac{f_1(\bar{z}+\bar{h}) - f_1(\bar{z})}{\bar{h}}} \right) = \overline{f_1'(\bar{z})}$$

i.e. $f_2'(z) = \overline{f_1'(\bar{z})}$. The result then follows by 10.3. ■

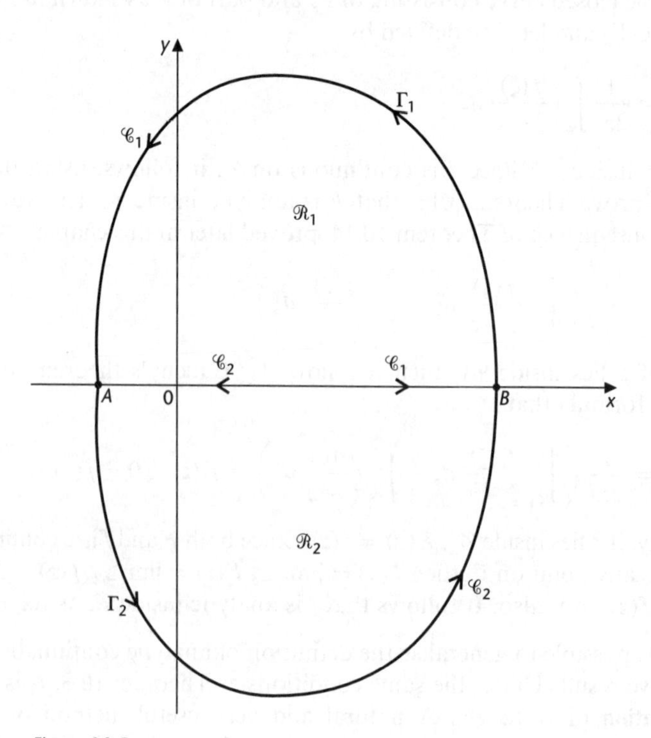

Figure 10.5

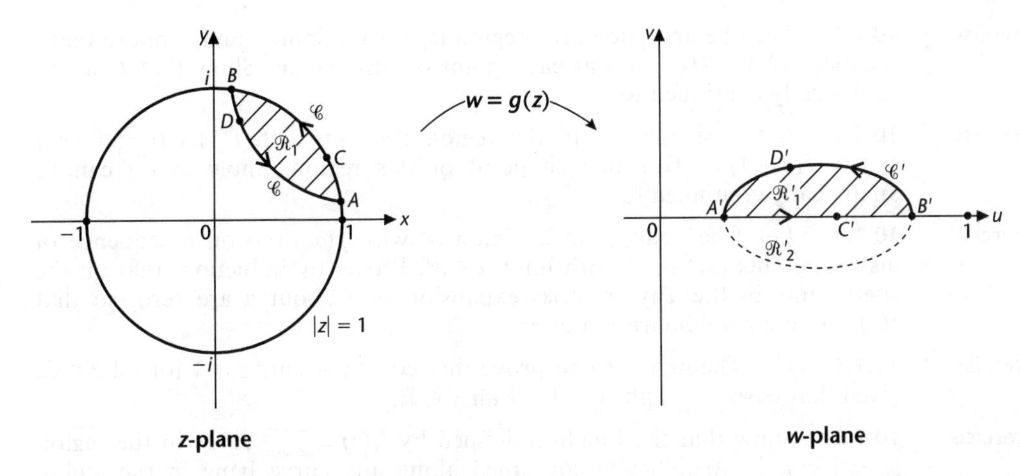

Figure 10.6

Extensions of this result can be obtained by conformally mapping a given region and boundary to a region of the form $\mathscr{R}_1$ given in Theorem 10.4, using the techniques of Chapter 6.

Example 10.5

Let $\mathscr{C}$ be a simple closed contour consisting of an arc ACB of the circle $|z| = 1$ and a curve ADB within the circle, joining A to B as shown in Fig. 10.6. Let a function f_1 be analytic inside $\mathscr{C}$, continuous on $\mathscr{C}$, and take real values on the arc ACB. It is easily seen that the bilinear transformation $w = g(z) = i(1 - z)/(1 + z)$ maps the arc ACB to a segment of the real axis, the curve ADB to a curve in the upper half-plane and the region $\mathscr{R}_1$, bounded by $\mathscr{C}$, to the region $\mathscr{R}'_1$. Let $f_1(z) = F_1(w)$. Then by Schwarz's reflection principle, F_1 can be analytically continued to the region $\mathscr{R}'_2$, by defining $F_2(w) = \overline{F_1(\overline{w})}$ for $w \in \mathscr{D}'_2$, consisting of $\mathscr{R}'_2$ and its boundary. Since $F_1(w) = f_1((i - w)/(i + w))$,

$$F_2(w) = \overline{f_1\left(\frac{i - \overline{w}}{i + \overline{w}}\right)} = \overline{f_1\left(\frac{i + i\overline{z} + i - i\overline{z}}{i + i\overline{z} - i + i\overline{z}}\right)} = \overline{f_1(1/\overline{z})}$$

Hence f_1 can be analytically continued across the boundary AB by defining $f_2(z) = \overline{f_1(1/\overline{z})}$ for $z \in g^{-1}(\mathscr{D}'_2)$.

Exercise

10.1.1 Show that the function f_1, defined by

$$f_1(z) = \int_0^\infty t^2 e^{-(z+1)t} dt \qquad (t \in \mathbb{R}, \quad \operatorname{Re}(z + 1) > 0)$$

is analytic in its domain. Find the analytic continuation of f_1 to $\operatorname{Re}(z + 1) \leqslant 0$ and find the value of this continuation at $i - 1$.

Exercise **10.1.2** Let f be analytic in the region $|z| < r$ for some r and suppose that f satisfies $f(2z) = 2f(z)f'(z)$ at each point of this region. Show that f can be analytically continued to $\mathbb{C}$.

Exercise **10.1.3** Let f be analytic in the region $\operatorname{Re} z > 0$ with $f(1) \neq 0$ and let f satisfy $f(z+1) = zf(z)$ at each point of this region. Show that f can be analytically continued to $\mathbb{C}\backslash\mathbb{Z}_{\leqslant 0}$.

Exercise **10.1.4** Let f be analytic in a region $\mathscr{R}$ with $f(\alpha_n) = 0$ on a sequence of distinct points (α_n) in $\mathscr{R}$ with limit $\alpha \in \mathscr{R}$. Prove, by induction, that all the coefficients in the Taylor series expansion of f about α are zero, so that $f(z) = 0$ in a neighbourhood of α.

Exercise **10.1.5** Use Theorem 10.1 to prove that $\cosh^2 z - \sinh^2 z = 1$ for all $z \in \mathbb{C}$, given that $\cosh^2 x - \sinh^2 x = 1$ for all $x \in \mathbb{R}$.

Exercise **10.1.6** Show that the function defined by $f_1(z) = \sum_{n=1}^{\infty} z^n/n$ in the region $|z| < 1$ can be analytically continued along any curve lying in the region $1 < |z-2| < 3$. Show, however, that this continuation does not give rise to an analytic function defined in the entire region. Why does this result not contradict the monodromy theorem?

Exercise **10.1.7** Let r_1 and r_2 be the respective radii of convergence for the series

$$\sum_{n=0}^{\infty} \frac{(n+1)z^n}{2^{n+2}} \quad \text{and} \quad \sum_{n=0}^{\infty} \frac{(n+1)(z+i)^n}{(2+i)^{n+2}}$$

Find r_1 and r_2 using the ratio test. Show that

$$f_1(z) = \sum_{n=0}^{\infty} \frac{(n+1)z^n}{2^{n+2}}, \ |z| < r_1 \qquad f_2(z) = \sum_{n=0}^{\infty} \frac{(n+1)(z+i)^n}{(2+i)^{n+2}}, \ |z+i| < r_2$$

are analytic continuations of each other. Sketch the two regions of convergence and their intersection. Find the complete analytic function f, of which f_1 and f_2 are elements. Show that

$$f_3(z) = \sum_{n=0}^{\infty} \frac{(-1)^n(1+n)(z-4)^n}{2^{n+2}}$$

is also an element of f but is not an analytic continuation of f_1.

Exercise **10.1.8** Let $f_1(z) = \sum_{n=0}^{\infty} z^{n+1}/3^n$ for $|z| < 3$. Show that f_1 can be analytically continued by constructing a Taylor series about any point α, $0 < |\alpha| < 3$, which is not real and positive. In the form of a power series find f_2, an analytic continuation of f_1 that is analytic at $2+3i$; find $f_2(2+3i)$.

Exercise **10.1.9** Show that $\sum_{n=1}^{\infty} (-1)^{n+1} z^n/n$ converges absolutely for $|z| < 1$ and that $\sum_{n=1}^{\infty} (-1)^{n+1}(z-i)^n/n(1+i)^n$ converges absolutely for $|z-i| < \sqrt{2}$. Show that

$$f_1(z) = \sum_{n=1}^{\infty} \frac{(-1)^{n+1} z^n}{n} \qquad (|z| < 1)$$

$$f_2(z) = \frac{\operatorname{Log} 2}{2} + \frac{i\pi}{4} + \sum_{n=1}^{\infty} \frac{(-1)^{n+1}(z-i)^n}{n(1+i)^n} \qquad (|z-i| < \sqrt{2})$$

are analytic continuations of each other.

Exercise **10.1.10** Show that $\sum_{n=0}^{\infty} z^{2^n}$ converges for $|z| < 1$. Show that the circle $|z| = 1$ is a natural boundary of analytic continuation of the function f defined by $f(z) = \sum_{n=0}^{\infty} z^{2^n}$, $|z| < 1$.

Exercise **10.1.11** Show that the function f defined by $f(z) = \sum_{n=1}^{\infty} z^{n!}$, $|z| < 1$, is a lacunary function by considering points on the circle $|z| = 1$ of the form $z = e^{2\pi ip/q}$ where $p \in \mathbb{Z}$ and $q \in \mathbb{N}$.

Exercise **10.1.12** Under the hypothesis of the Schwarz reflection principle, prove that f_2 is analytic in the region $\mathscr{R}_2$ by using the Cauchy–Riemann equations.

Exercise **10.1.13** Under the same hypothesis as Schwarz's reflection principle, except that $|f_1(z)| = 1$ on AB, show that f_1 can be analytically continued to $\mathscr{R}_2$ by defining $f_2(z) = 1/\overline{f_1(\bar{z})}$ on $\mathscr{R}_2$ and on $\mathscr{C}_2$.

Exercise **10.1.14** Let $\mathscr{C}_1$ be a closed contour consisting of a segment AB of the imaginary axis and a curve Γ_1 in the region $\operatorname{Re} z > 0$, joining A to B. Let Γ_2 be the reflection of Γ_1 in the imaginary axis and let $\mathscr{C}_2$ be the closed contour consisting of AB and Γ_2. Let $\mathscr{R}_k$, $k = 1, 2$, be the region bounded by $\mathscr{C}_k$. Let f_1 be analytic on $\mathscr{R}_1$, continuous on $\mathscr{C}_1$ and take imaginary values on AB. Prove that f_2, defined on $\mathscr{R}_2$ and its boundary by $f_2(z) = -\overline{f_1(-\bar{z})}$, is the analytic continuation of f_1 to $\mathscr{R}_2$.

Exercise **10.1.15** Let O and B be the points 0 and $a \in \mathbb{R}^+$ respectively in the complex plane. Let the point A represent a number in the first quadrant of the complex plane. Let the angle between OA and OB be ϕ and the angle between BO and BA be θ. Let f_1 be analytic inside the triangle AOB, continuous on the boundary lines and take real values on the boundary lines. Use Schwarz's reflection principle to show that

(i) f_1 can be analytically continued across OA to the reflection of AOB in OA by defining $f_2(z) = \overline{f_1(e^{2i\phi}\bar{z})}$

(ii) f_1 can be analytically continued across AB to the reflection of AOB in AB by defining $f_3(z) = \overline{f_1(e^{-2i\theta}(\bar{z} - a) + a)}$.

Notice that $e^{2i\phi}\bar{z}$ is the reflection of z in OA and $e^{-2i\theta}(\bar{z} - a) + a$ is the reflection of z in AB.

Exercise **10.1.16** Use the Schwarz–Christoffel transformation to determine an integral formula for the function f_1^{-1}, which is such that f_1 maps the triangle with vertices at $z = 0$, 2 and $1 + i$ to the real axis in the w-plane with $f_1(0) = -1$, $f_1(2) = 1$ and $f_1(1 + i) = \infty$, and which maps the inside of the triangle to the upper half-plane. Show that $f_1(1) = 0$.

If f is the complete analytic function of which f_1 is an element, indicate in a diagram, for $-2 \leqslant \operatorname{Re} z \leqslant 4$ and $-2 \leqslant \operatorname{Im} z \leqslant 2$, which triangular regions are mapped to the upper half w-plane by f. Indicate the poles and zeros of f in the diagram. (*Hint.* Use the results of Exercise 10.1.15.)

Exercise 10.1.17 Let A, O, B and C be the points $-1, 0, 1$ and i respectively in the z-plane. An analytic function f_1 maps the inside of the triangle ABC to the upper half w-plane in such a way that triangles BOC and AOC are mapped to the first and second quadrants; f_1 takes real values on the boundary lines; and A, O, B and C are mapped to $-1, 0, 1$ and ∞ respectively. If f is the complete analytic function of which f_1 is an element, show in a diagram, for $-1 \leqslant \operatorname{Re} z \leqslant 3$ and $-2 \leqslant \operatorname{Im} z \leqslant 2$, which triangular regions are mapped to which quadrant of the w-plane by f.

Infinite Products

As well as defining an infinite series of complex numbers, we can also define an infinite product of complex numbers. As in the case of an infinite series, an infinite product is just a special sequence; in this case, a sequence of partial products.

Definition

Let (z_n) be a sequence of complex numbers. Then the sequence (P_k), defined by $P_k = \prod_{n=1}^{k} z_n = z_1 z_2 \dots z_k$, is the sequence of **partial products** of the **infinite product** $\prod_{n=1}^{\infty} z_n$.

If the sequence (P_k) converges, it is desirable to exclude the possibility of the limit being 0. Otherwise, any infinite product with a single factor being 0 would converge to 0 and the convergence would not depend on the behaviour of any of the other terms. On the other hand, it is essential to consider infinite products with at least one factor being 0, for later results. Hence we make the following definitions.

Definitions

(i) Let $z_n \neq 0$ for any n. Then $\prod_{n=1}^{\infty} z_n$ **converges** to P if $P_k \to P \neq 0$ as $k \to \infty$. In this case, we write $\prod_{n=1}^{\infty} z_n = P$.

(ii) The infinite product $\prod_{n=1}^{\infty} z_n$ **converges to 0** if at most a finite number of factors are zero and if the sequence of partial products formed by the non-zero factors converges to a non-zero limit.

(iii) An infinite product **diverges** if it does not converge.

Note that, for example, according to the above definitions, $\prod_{n=1}^{\infty}(1 - 1/(n+1))$ diverges since the limit of the partial products is 0.

Important Note

It follows by the definitions above that when investigating convergence, we need only consider infinite products whose factors are all non-zero. It is easily seen that a necessary condition for convergence in this case is that $z_n \to 1$ as

$n \to \infty$. Hence conventionally, infinite products are written in the form $\prod_{n=1}^{\infty}(1+z_n)$.

Lemma 10.5. A Necessary Condition for Convergence

If the infinite product $\prod_{n=1}^{\infty}(1+z_n)$ converges to $P \neq 0$, then $z_n \to 0$ as $n \to \infty$.□

Proof

Let $P_k = \prod_{n=1}^{k}(1+z_n)$.

Then $P_k = P_{k-1}(1+z_k) \Rightarrow z_k = P_k/P_{k-1} - 1$ (since $P_k \neq 0$ for any k). By hypothesis, $P_k \to P \neq 0$ as $k \to \infty$, so that $P_{k-1} \to P \neq 0$ as $k \to \infty$. Then by the combination theorem, $z_k \to P/P - 1 = 0$ as $k \to \infty$. ■

Note

The condition $z_n \to 0$ as $n \to \infty$ is clearly not sufficient for convergence since $\prod_{n=1}^{\infty} z_n(1 - 1/(n+1))$ diverges.

Example 10.6

$$\text{Let} \quad P_k = \prod_{n=1}^{k}\left(1+\frac{1}{n(n+2)}\right) = \prod_{n=1}^{k}\frac{(n+1)^2}{n(n+2)}$$

Then $P_1 = (2\cdot 2)/3$, $P_2 = (2\cdot 3)/4, \ldots$ and by induction, $P_k = 2(k+1)/(k+2)$. Hence $P_k \to 2$ as $k \to \infty$ and so $\prod_{n=1}^{\infty}(1+1/n(n+2)) = 2$. Note that $1/n(n+2) \to 0$ as $n \to \infty$.

The following result gives a connection between the convergence of an infinite product and an infinite series.

Theorem 10.6. A Correspondence Between Infinite Products and Infinite Series

The infinite product $\prod_{n=1}^{\infty}(1+z_n)$ converges to $P \neq 0$ if and only if the infinite series $\sum_{n=1}^{\infty} \operatorname{Log}(1+z_n)$ converges. □

Proof

Step 1

Suppose that $\sum_{n=1}^{\infty} \operatorname{Log}(1+z_n)$ converges, so that $z_n \neq -1$ for any n and if $S_k = \sum_{n=1}^{k} \operatorname{Log}(1+z_n)$ then $S_k \to S$ say as $k \to \infty$. Let $P_k = \prod_{n=1}^{k}(1+z_n)$. Then $P_k = e^{S_k} \Rightarrow P_k \to e^S$ as $k \to \infty$ (by 2.2(i)), i.e. $\prod_{n=1}^{\infty}(1+z_n)$ converges to $e^S \neq 0$.

Step 2

Suppose that $\prod_{n=1}^{\infty}(1+z_n)$ converges to $P \neq 0$, so that $z_n \neq -1$ for any n and $P_k = \prod_{n=1}^{k}(1+z_n) \to P$ as $k \to \infty$. Let $S_k = \sum_{n=1}^{k} \operatorname{Log}(1+z_n)$ so that $S_k = \operatorname{Log} P_k + 2\pi i q_k$ for each k, where $q_k \in \mathbb{Z}$. Let $\theta_n = \operatorname{Arg}(1+z_n)$, $n \in \mathbb{N}$ and $\phi_k = \operatorname{Arg} P_k$, $k \in \mathbb{N}$. Then

$$\sum_{n=1}^{k} \theta_n = \phi_k + 2\pi q_k \quad \Rightarrow \quad \theta_k = (\phi_k - \phi_{k-1}) + 2\pi(q_k - q_{k-1}) \tag{10.1}$$

By hypothesis and 10.5, $1 + z_n \to 1$ as $n \to \infty$ so that $\theta_n \to 0$ as $n \to \infty$. Also, $\phi_k \to \operatorname{Arg} P$ as $k \to \infty \Rightarrow \phi_{k-1} \to \operatorname{Arg} P$ as $k \to \infty$. Hence by (10.1) and the combination theorem, $q_k - q_{k-1} \to 0$ as $k \to \infty$. Then as q_k is an integer, $q_k = q \in \mathbb{Z}$ for all $k > k_1$ say. Hence $S_k \to \operatorname{Log} P + 2\pi i q$ as $k \to \infty$, so that $\sum_{n=1}^{\infty} \operatorname{Log}(1+z_n)$ converges. ■

Example 10.7

Consider $\prod_{n=1}^{\infty} |1 + i/n| = \prod_{n=1}^{\infty}(1 + 1/n^2)^{1/2}$. Since $\sum_{n=1}^{\infty} 1/n^2$ converges and $\operatorname{Log}(1 + 1/n^2) < 1/n^2$, $\sum_{n=1}^{\infty} \operatorname{Log}(1 + 1/n^2)$ converges by the comparsion test. Hence the given product converges by 10.6. On the other hand, $\operatorname{Log}(1 + i/n) = \frac{1}{2}\operatorname{Log}(1 + 1/n^2) + i \tan^{-1}(1/n)$, where $0 < \tan^{-1}(1/n) < \pi/2$. For sufficiently large n, $|\tan^{-1}(1/n) - 1/n|$ is as small as we please and $\sum_{n=1}^{\infty} 1/n$ diverges. Hence $\sum_{n=1}^{\infty} \tan^{-1}(1/n)$ diverges so that $\sum_{n=1}^{\infty} \operatorname{Log}(1 + i/n)$ diverges. Then $\prod_{n=1}^{\infty}(1 + i/n)$ diverges by 10.6.

As in the case of infinite series, a useful concept for investigating convergence of infinite products is that of absolute convergence.

Definition

The infinite product $\prod_{n=1}^{\infty}(1+z_n)$ **converges absolutely** if $\prod_{n=1}^{\infty}(1+|z_n|)$ converges.

Note

At first sight, it may seem reasonable to say that the given infinite product converges absolutely if $\prod_{n=1}^{\infty}|1+z_n|$ converges. However, with this definition, every convergent product would be absolutely convergent; but the converse would not be true, as demonstrated in Example 10.7 and in contrast to the results for infinite series. It is easily shown that the adopted definition is equivalent to requiring $\sum_{n=1}^{\infty} \operatorname{Log}(1+z_n)$ to be absolutely convergent (see Exercises 10.2).

The following result gives a simpler correspondence than 10.6 in the case of absolutely convergent products.

Theorem 10.7. A Correspondence Between Absolutely Convergent Products and Series

The product $\prod_{n=1}^{\infty}(1+z_n)$ converges absolutely if and only if the series $\sum_{n=1}^{\infty} z_n$ converges absolutely. □

Proof

Let $|z_n| = a_n$. Then since $a_n \geqslant 0$, $1 + a_n \leqslant e^{a_n}$ and so

$$\sum_{n=1}^{k} a_n \leqslant \prod_{n=1}^{k}(1+a_n) \leqslant \exp\left(\sum_{n=1}^{k} a_n\right) \tag{10.2}$$

Let $P_k = \prod_{n=1}^{k}(1+a_n)$ and $S_k = \sum_{n=1}^{k} a_n$. If $\prod_{n=1}^{\infty}(1+a_n)$ converges to $P \neq 0$, since the sequence (P_k) is increasing, it is bounded above by P say. Then by (10.2), the sequence (S_k) is bounded above by P. It is also clearly increasing, and so converges by the monotonic-bounded principle, i.e. $\sum_{n=1}^{\infty} a_n$ converges. Similarly, if $\sum_{n=1}^{\infty} a_n$ converges to S, since the sequence (S_k) is increasing, it is bounded above by S. Then (P_k) is increasing and bounded above by e^S by (10.2), and so converges. ■

Notes

(i) The simple correspondence in 10.7 is only valid in the case of absolute convergence.

(ii) It follows by 10.7 and the fact that any rearrangement of an absolutely convergent series produces the same series, that any rearrangement of the factors in an absolutely convergent product produces the same product.

The following result is analogous to a result concerning infinite series.

Theorem 10.8. Absolutely Convergent Products Converge

Let $z_n \neq -1$ for any $n \in \mathbb{N}$. If the product $\prod_{n=1}^{\infty}(1+z_n)$ converges absolutely, then it converges to $P \neq 0$. □

Proof

Step 1

Let $P_k = \prod_{n=1}^{k}(1+z_n)$ and $P'_k = \prod_{n=1}^{k}(1+|z_n|)$. Then

$$P_k - P_{k-1} = \prod_{n=1}^{k-1}(1+z_n)z_k \quad \text{and} \quad P'_k - P'_{k-1} = \prod_{n=1}^{k-1}(1+|z_n|)|z_k|$$

$$\rightarrow \quad |P_k - P_{k-1}| \leqslant P'_k - P'_{k-1} \tag{10.3}$$

by the triangle inequality. Furthermore, defining $P_0 = P'_0 = 0$ gives $P_k = \sum_{n=1}^{k}(P_n - P_{n-1})$ and also $P'_k = \sum_{n=1}^{k}(P'_n - P'_{n-1})$. The series $\sum_{n=1}^{\infty}(P'_n - P'_{n-1})$ converges by hypothesis, so by (10.3) and the comparison test, $\sum_{n=1}^{\infty}(P_n - P_{n-1})$ converges absolutely and therefore converges. Hence $P_k \to P$ as $k \to \infty$ say.

Step 2

We now have to show that $P \neq 0$. Notice that

$$\prod_{n=1}^{\infty}\frac{1}{1+z_n} = \prod_{n=1}^{\infty}\left(1 - \frac{z_n}{1+z_n}\right)$$

and the partial products of this product are $1/P_k$. Hence if this product converges, $1/P_k \to 1/P$ as $k \to \infty \Rightarrow P \neq 0$. Also, by 10.7 and step 1, this product converges if $\sum_{n=1}^{\infty}|z_n/(1+z_n)|$ converges. It remains to show that this series converges. By hypothesis and 10.7, $\sum_{n=1}^{\infty}|z_n|$ converges, so that $z_n \to 0$ as $n \to \infty$. Hence there exists $N \in \mathbb{N}$ such that $|z_n| < 1/2$ say, for every $n > N$. Then since $|1+z_n| \geqslant 1 - |z_n|$, it follows that $|z_n/(1+z_n)| < 2|z_n|$ for every $n > N$. Then $\sum_{n=1}^{\infty}|z_n/(1+z_n)|$ converges by the comparison test, as required. ■

Example 10.8

(i) Since $\sum_{n=1}^{\infty}1/n^3$ converges, $\prod_{n=1}^{\infty}(1 + (-1)^n i/n^3)$ converges absolutely by 10.7 and so converges by 10.8.

(ii) The series $\sum_{n=1}^{\infty}1/\sqrt{n(n+1)}$ diverges by the comparison test and the fact that $\sum_{n=1}^{\infty}1/n$ diverges. Hence $\prod_{n=1}^{\infty}(1 + 1/\sqrt{n(n+1)})$ diverges by Theorem 10.7.

Infinite products whose factors involve functions of z are of particular importance. Given a sequence $(f_n(z))$ of functions defined on $A \subseteq \mathbb{C}$, previous results can be used to determine whether $\prod_{n=1}^{\infty}(1 + f_n(z))$ converges, converges absolutely, converges to 0, or diverges for a particular $z \in A$. Then the **product function** F is defined by $\prod_{n=1}^{\infty}(1 + f_n(z)) = F(z)$ on the subset of A on which the product converges or converges to 0.

Example 10.9

The product $\prod_{n=0}^{\infty}(1 + z^{2^n})$ converges absolutely for $|z| < 1$ by 10.7, since the series $\sum_{n=0}^{\infty}z^{2^n}$ converges absolutely for $|z| < 1$. Hence the given product converges for $|z| < 1$ by 10.8. (Notice that none of the factors are 0 for $|z| < 1$.) Let $P_k = \prod_{n=0}^{k}(1 + z^{2^n})$ for $|z| < 1$. Then $P_0 = 1 + z$, $P_1 = \sum_{m=0}^{3}z^m$ and in general, by induction, $P_k = \sum_{m=0}^{2^{k+1}-1}z^m$. Summing this finite geometric series gives $P_k = (1 - z^{2^{k+1}})/(1 - z)$ and so $P_k \to 1/(1-z)$ as $k \to \infty$ for $|z| < 1$.

Hence

$$\prod_{n=0}^{\infty}(1+z^{2^n}) = 1/(1-z) \text{ for } |z|<1.$$

As in the case of infinite series, the concept of uniform convergence plays a fundamental role in the study of infinite products of functions. Remember that we need only consider infinite products whose factors are all non-zero when investigating convergence.

Definition

Let $(f_n(z))$ be a sequence of functions such that $f_n(z) \neq -1$ for $z \in A \subseteq \mathbb{C}$. Then $\prod_{n=1}^{\infty}(1+f_n(z))$ **converges uniformly** to **product function** $F(z)$, on A if the sequence $\left(\prod_{n=1}^{k}(1+f_n(z))\right)$ converges uniformly to $F(z)$ on A, with $F(z) \neq 0$ for any $z \in A$.

The relevance of this concept is illustrated in the following results, which are analogous to the results for infinite series.

Theorem 10.9. Elementary Properties of Uniformly Convergent Products

(i) Let f_n be continuous on $A \subseteq \mathbb{C}$ for each $n \in \mathbb{N}$. If $\prod_{n=1}^{\infty}(1+f_n(z))$ converges uniformly to $F(z)$ on A then F is continuous on A.

(ii) Let f_n be analytic in a simply connected region $\mathscr{R}$ for each $n \in \mathbb{N}$. If $\prod_{n=1}^{\infty}(1+f_n(z))$ converges uniformly to $F(z)$ in every compact subset of $\mathscr{R}$, then F is analytic in $\mathscr{R}$. □

Proof

The results follow from the definition and Theorems 4.6 and 4.8. The proof of (i) is left as an exercise. By hypothesis, the sequence $\left(\prod_{n=1}^{k}(1+f_n(z))\right)$ converges uniformly to $F(z)$ in every compact subset of $\mathscr{R}$ and clearly, $\prod_{n=1}^{k}(1+f_n(z))$ is analytic in $\mathscr{R}$ for each k. Then F is analytic in $\mathscr{R}$ by 4.8, which proves (ii). ■

One of the most useful tests for uniform convergence is analogous to Weierstrass's test for series.

Theorem 10.10. Weierstrass's M Test

Let the sequence $(f_n(z))$ be defined on $A \subseteq \mathbb{C}$ with $f_n(z) \neq -1$ for $z \in A$. Let there exist constants $M_n \in \mathbb{R}^+$ such that $|f_n(z)| \leqslant M_n$ for all $z \in A$, for each $n \in \mathbb{N}$, where $\sum_{n=1}^{\infty} M_n$ converges. Then $\prod_{n=1}^{\infty}(1+f_n(z))$ converges uniformly on A. □

Proof

Let $P_k(z) = \prod_{n=1}^{k}(1 + f_n(z))$ and $Q_k = \prod_{n=1}^{k}(1 + M_n)$. Then by hypothesis, as in the proof of 10.8,

$$|P_k(z) - P_{k-1}(z)| \leqslant Q_k - Q_{k-1} \qquad \text{(for all } k \in \mathbb{N}) \tag{10.4}$$

Defining $P_0(z) = Q_0 = 0$ gives $P_k(z) = \sum_{n=1}^{k}(P_n(z) - P_{n-1}(z))$ and $Q_k = \sum_{n=1}^{k}(Q_n - Q_{n-1})$. Since $\sum_{n=1}^{\infty} M_n$ converges, $\prod_{n=1}^{\infty}(1 + M_n)$ converges by 10.7, so that $\sum_{n=1}^{\infty}(Q_n - Q_{n-1})$ converges. Then by (10.4) and Weierstrass's M test for series, Theorem 4.13, $\sum_{n=1}^{\infty}(P_n(z) - P_{n-1}(z))$ converges uniformly in A, i.e $(P_k(z))$ converges uniformly to $F(z)$, say, in A. It can be shown, as in the proof of 10.8, that $F(z) \neq 0$ for $z \in A$ (see Exercises 10.2). ■

Example 10.10

Prove that $\prod_{n=1}^{\infty}(1 + z^2/n^2)$ represents an entire function.

Solution

Note that before testing for uniform convergence, we need to extract any vanishing factors in the product. The infinite product has zeros at $z = \pm in$, $n \in \mathbb{N}$. Let $|z| \leqslant \rho$ say and choose $k \in \mathbb{N}$ such that $k > \rho$. This ensures that for $n \geqslant k$, $|z/n| < 1$, so that $1 + z^2/n^2 \neq 0$. Then $\prod_{n=1}^{\infty}(1 + z^2/n^2) = g(z)h(z)$ say, where $g(z) = \prod_{n=1}^{k-1}(1 + z^2/n^2)$ is clearly entire with zeros at $z = \pm in$, $n < k$, and $h(z) = \prod_{n=k}^{\infty}(1 + z^2/n^2)$, an infinite product with no zeros. Also $|z^2/n^2| \leqslant \rho^2/n^2$ and $\rho^2 \sum_{n=k}^{\infty} 1/n^2$ converges. Hence by Weierstrass's M test, this infinite product is uniformly convergent in $|z| \leqslant \rho$. Since ρ is as large as we please, the result follows by Theorem 10.9.

Although the product $\prod_{n=1}^{\infty}(1 - z/z_n)$, where $z_n \neq 0$ for any n, generally diverges, the technique used in Example 10.10 can be used to prove the following important theorem.

Theorem 10.11. Entire Infinite Products

Let z_n be a non-zero constant for each $n \in \mathbb{N}$, such that $|z_1| \leqslant |z_2| \leqslant \ldots$ with $\lim_{n\to\infty} z_n = \infty$. Then there exists a set of polynomials, $P_{s(n)}(z/z_n)$, such that $\prod_{n=1}^{\infty}(1 - z/z_n)\exp(P_{s(n)}(z/z_n))$ is an entire function with zeros at $z = z_n$. □

Proof

Step 1

Let $1 + f_n(z) = (1 - z/z_n)\exp(P_{s(n)}(z/z_n))$. To determine the uniform convergence of the product, we need to find an upper bound for $|f_n(z)|$. Note that

$$1+f_n(z)=\exp\left(P_{s(n)}(z/z_n)+\mathrm{Log}\,(1-z/z_n)\right)=\exp\left(R_{s(n)}(z/z_n)\right) \tag{10.5}$$

say. Now $\mathrm{Log}\,(1-z)=-\sum_{r=1}^{\infty}z^r/r$ for $|z|<1$. For each $n\in\mathbb{N}$, define

$$P_{s(n)}(z/z_n)=\sum_{r=1}^{s(n)}\frac{1}{r}\left(\frac{z}{z_n}\right)^r \tag{10.6}$$

where $s(n)$ has yet to be chosen, with $P_0(z/z_n)=1$. Then for $|z/z_n|\leqslant 1/2$ say,

$$R_{s(n)}(z/z_n)=-\sum_{r=s+1}^{\infty}\frac{1}{r}\left(\frac{z}{z_n}\right)^r=-\sum_{m=0}^{\infty}\frac{1}{m+s+1}\left(\frac{z}{z_n}\right)^{m+s+1}$$

$$\Rightarrow\quad |R_{s(n)}(z/z_n)|\leqslant\frac{1}{s+1}\left|\frac{z}{z_n}\right|^{s+1}\sum_{r=0}^{\infty}\frac{s+1}{r+s+1}\left|\frac{z}{z_n}\right|^r\leqslant\frac{1}{s+1}\left|\frac{z}{z_n}\right|^{s+1}\sum_{r=0}^{\infty}\frac{1}{2^r}$$

$$\Rightarrow |R_{s(n)}(z/z_n)|\leqslant\frac{2}{s(n)+1}\left|\frac{z}{z_n}\right|^{s(n)+1}\leqslant 1 \tag{10.7}$$

Then by (10.5) the series expansion of e^z, the triangle inequality and (10.7),

$$|f_n(z)|\leqslant\exp\left(|R_{s(n)}(z/z_n)|\right)-1\leqslant\exp\left(\frac{2}{s(n)+1}\left|\frac{z}{z_n}\right|^{s(n)+1}\right)-1 \tag{10.8}$$

Now $e^x-1<xe^x$ for all $x\in\mathbb{R}^+$ (there is equality when $x=0$ and $G(x)=e^x-1$, $H(x)=xe^x\ \Rightarrow\ G'(x)<H'(x)$ for $x>0$). Hence by (10.8) and (10.7),

$$|f_n(z)|<\frac{2e}{s(n)+1}\left|\frac{z}{z_n}\right|^{s(n)+1} \tag{10.9}$$

Step 2

Using the technique of Example 10.10, let $|z|\leqslant\rho$ and choose $k\in\mathbb{N}$ such that $|z_k|>2\rho$. This ensures that $|z/z_n|<1/2$ for $n\geqslant k$. Then $\prod_{n=1}^{\infty}(1-z/z_n)\exp(P_{s(n)}(z/z_n))=g(z)h(z)$ say, where

$$g(z)=\prod_{n=1}^{k-1}(1-z/z_n)(\exp P_{s(n)}(z/z_n))$$

and so is clearly entire with zeros at $z=z_n$, $n<k$, and $h(z)=\prod_{n=k}^{\infty}(1+f_n(z))$, with $1+f_n(z)$ defined in (10.5), has no zeros. Since $|f_n(z)|$ satisfies (10.9),

$$|f_n(z)|<\frac{2e\rho^{s(n)+1}}{(s(n)+1)|z_n|^{s(n)+1}}\qquad\text{(for each } n\in\mathbb{N})$$

It follows by 10.10 that $\prod_{n=k}^{\infty}(1+f_n(z))$ converges uniformly to $h(z)$ on $|z|\leqslant\rho$, provided that $s(n)$ can be chosen so that the series

$$\sum_{n=k}^{\infty}\frac{\rho^{s(n)+1}}{(s(n)+1)|z_n|^{s(n)+1}} \tag{*}$$

converges. Then by 10.9, h is entire, as required.

Step 3 (Choice of s(n))

Given the sequence (z_n), if there is a constant $p \geqslant 1$ so that $\sum_{n=1}^{\infty} 1/|z_n|^p$ converges, clearly we can just choose $s(n) = p - 1$ for every $n \geqslant k$ to ensure (*) converges. If there does not exist such a constant, we let $s(n) = n - 1$. Then since $|z_n| > 2\rho$ for each $n \geqslant k$,

$$\frac{\rho^{s(n)+1}}{(s(n)+1)|z_n|^{s(n)+1}} < \frac{\rho^n}{2^n n \rho^n} < \frac{1}{2^n}$$

and $\sum_{n=k}^{\infty} 2^{-n}$ is a convergent geometric series and so by the comparison test, series (*) converges. ■

Example 10.11

It follows by step 3 of the proof of Theorem 10.11 and 10.11 itself, that since $\sum_{n=1}^{\infty} 1/n^2$ converges, $\prod_{n=1}^{\infty} (1 - z/n^2)$ is entire with zeros at $z = n^2$, $n \in \mathbb{N}$. (This is so since $p = 1$ say in this case gives $s(n) = 0$ for all n.) Also, for example, $\prod_{n=1}^{\infty} (1 - z/n)e^{z/n}$ is entire with zeros at $z = n$, $n \in \mathbb{N}$ (since $p = 2$ say in this case gives $s(n) = 1$ for all n).

Weierstrass's Factor Theorem

Recall that any polynomial can be expressed as a product of factors involving its zeros. It turns out that any entire function with zeros can be expressed as a product of factors involving its zeros. In the case of an entire function with no zeros or a finite number of zeros, we have the following elementary result.

Theorem 10.12. Entire Functions with a Finite Number of Zeros

(i) If f is an entire function with no zeros, then there exists an entire function g such that $f(z) = e^{g(z)}$ for all $z \in \mathbb{C}$.

(ii) If f is an entire function with n zeros, $z_1, \ldots, z_n$ (counting multiplicities), then there exists an entire function g such that $f(z) = e^{g(z)} \prod_{k=1}^{n} (z - z_k)$. □

Proof

(i) Let f be entire with no zeros and let g be defined by $f(z) = e^{g(z)}$ for all $z \in \mathbb{C}$. Then $f'(z)/f(z) = g'(z)$ and since f has no zeros, g' is entire, hence g is entire.

(ii) Let F be defined by

$$F(z) = \frac{f(z)}{\prod_{k=1}^{n}(z - z_k)} \qquad (z \neq z_k,\ k = 1, \ldots, n)$$

and

$$F(z_k) = \lim_{z \to z_k} \frac{f(z)}{\prod_{k=1}^{n}(z - z_k)}$$

Then by hypothesis, F is entire with no zeros, and so by (i), there exists an entire function g such that $F(z) = e^{g(z)}$, as required. ■

Theorem 10.12 suggests that just as a meromorphic function with an infinite number of poles can be expressed as an infinite series involving its poles (see the results of Chapter 5), an entire function with an infinite number of zeros can be expressed as an infinite product involving its zeros. The following example indicates that, in certain cases, such an infinite product representation can be constructed by starting with the partial fraction expansion of an appropriate meromorphic function, found using Theorem 5.15.

Example 10.12

It follows by Theorem 5.15 that, for $z \neq \pm n\pi$, $n \in \mathbb{N}$

$$\cot z - \frac{1}{z} = \sum_{n=1}^{\infty}\left(\frac{1}{z - n\pi} + \frac{1}{n\pi}\right) + \sum_{n=1}^{\infty}\left(\frac{1}{z + n\pi} - \frac{1}{n\pi}\right)$$

(see Exercise 5.4.6(i)), where each series is uniformly convergent on any compact set not containing $\pm n\pi$. Hence integrating along a curve joining 0 to z, not passing through the poles at $\pm n\pi$ and taking exponentials gives

$$\frac{\sin z}{z} = \prod_{n=1}^{\infty}\left(1 - \frac{z}{n\pi}\right)e^{z/n\pi}\prod_{n=1}^{\infty}\left(1 + \frac{z}{n\pi}\right)e^{-z/n\pi}$$

$$\Rightarrow \quad \sin z = z\prod_{n=-\infty}^{\infty}{}'\left(1 - \frac{z}{n\pi}\right)e^{z/n\pi} = z\prod_{n=1}^{\infty}\left(1 - \frac{z^2}{n^2\pi^2}\right) \tag{10.10}$$

where $\prod'$ indicates that the term for $n = 0$ is omitted. The infinite product is uniformly convergent in any compact subset not containing $\pm n\pi$ by the proof of 10.11. Notice that $z = 0$ and $z = \pm n\pi$ are the zeros of the entire function $\sin z$.

The infinite product representation of $\sin z$ given in (10.10) is a special case of the following result, which can be thought of as an extension of Theorem 10.12(ii), modified so that the infinite product converges. It follows directly from Theorem 10.11.

Theorem 10.13. Weierstrass's Factor Theorem

Let f be an entire function with an infinite number of zeros $z_1, z_2, \ldots$ (counting multiplicities) such that $0 < |z_1| \leqslant |z_2| \leqslant \ldots$ with $z_n \to \infty$ as $n \to \infty$, in addition to a possible zero at 0. Then there exists an $m \in \mathbb{Z}_{\geqslant 0}$, an entire function g and a set of polynomials $P_{s(n)}(z/z_n)$ such that

$$f(z) = z^m e^{g(z)} \prod_{n=1}^{\infty} (1 - z/z_n) \exp\left(P_{s(n)}(z/z_n)\right)$$ □

Proof

Let f have a zero of order m at the origin, with $m = 0$ if f has no such zero. Then $f(z) = z^m h(z)$ where h is an entire function with an infinite number of zeros $z_1, z_2 \ldots$. It then follows by Theorem 10.11 that there exists a set of polynomials $P_{s(n)}(z/z_n)$ such that

$$\frac{h(z)}{\prod_{n=1}^{\infty} (1 - z/z_n) \exp\left(P_{s(n)}(z/z_n)\right)}$$

is an entire function without zeros. Then the result follows by 10.12(i). ■

Given a particular entire function f with an infinite number of zeros, the constant m in 10.13 is known and a suitable set of polynomials for the infinite product representation of $f(z)z^{-m}e^{-g(z)}$ can be determined using the proof of 10.11. However, with a particular choice of such polynomials, the entire function g in 10.13 cannot be determined without further information since different entire functions can have the same zeros.

Example 10.13

The entire function $\sin z$ has a simple zero at $z = 0$ and simple zeros at $z = \pm n\pi$, $n \in \mathbb{N}$. It follows by Theorem 10.11 that, although the infinite product $\prod_{n=-\infty}^{\infty}{}'(1 - z/n\pi)$ does not converge, the infinite product $\prod_{n=-\infty}^{\infty}{}'(1 - z/n\pi)e^{z/n\pi}$ represents an entire function with zeros at $z = \pm n\pi$. Hence by Theorem 10.13,

$$\sin z = ze^{g(z)} \prod_{n=-\infty}^{\infty}{}' \left(1 - \frac{z}{n\pi}\right) e^{z/n\pi} = ze^{g(z)} \prod_{n=1}^{\infty} \left(1 - \frac{z^2}{n^2\pi^2}\right) \tag{10.11}$$

for all $z \in \mathbb{C}$, for some entire function g. Equation (10.11) gives

$$\operatorname{Log}(\sin z/z) = g(z) + \sum_{n=1}^{\infty} \operatorname{Log}(1 - z^2/n^2\pi^2) \qquad (z \neq \pm n\pi)$$

Since this series converges uniformly on any compact set not containing $\pm n\pi$ by 10.6 and 10.10, differentiating gives

$$\cot z - 1/z = g'(z) + \sum_{n=1}^{\infty} \frac{2z}{z^2 - n^2\pi^2} \qquad (z \neq n\pi)$$

Since $\cot z - 1/z = \sum_{n=1}^{\infty} 2z/(z^2 - n^2\pi^2)$, $z \neq \pm n\pi$, by Exercises 5.4.6(i), it follows that $g(z)$ is a constant. Then letting $z \to 0$ in (10.11) gives $g(z) \equiv 0$, so that the result of Example 10.12 is obtained.

Exercise 10.2.1 Show that each of the following products converges and evaluate the product in each case by determining the sequence of partial products.

(i) $\displaystyle\prod_{n=1}^{\infty}\left(1 + \frac{1}{n(2n+3)}\right)$

(ii) $\displaystyle\prod_{n=2}^{\infty}\left(1 + \frac{2}{(n-1)(n+2)}\right)$

Exercise 10.2.2 Let $\prod_{n=1}^{\infty}(1+\alpha_n)$ and $\prod_{n=1}^{\infty}(1+\beta_n)$ converge, where α_n and β_n are complex constants for each n. Show that $\prod_{n=1}^{\infty}(1+\alpha_n)(1+\beta_n)$ converges and that

$$\prod_{n=1}^{\infty}(1+\alpha_n)(1+\beta_n) = \prod_{n=1}^{\infty}(1+\alpha_n)\prod_{n=1}^{\infty}(1+\beta_n)$$

Exercise 10.2.3 Let $z_n \neq -1$ for any $n \in \mathbb{N}$. Prove that $\sum_{n=1}^{\infty}|\operatorname{Log}(1+z_n)|$ converges if and only if $\sum_{n=1}^{\infty}|z_n|$ converges. Then by 10.7, $\prod_{n=1}^{\infty}(1+z_n)$ converges absolutely if and only if $\sum_{n=1}^{\infty}\operatorname{Log}(1+z_n)$ converges absolutely.

Exercise 10.2.4 Using appropriate results, determine whether or not each of the following products converges:

(i) $\displaystyle\prod_{n=1}^{\infty}(1 - i/n^4)$

(ii) $\displaystyle\prod_{n=1}^{\infty}\left(1 + \frac{n}{\sqrt{n^4+1}}\right)$

(iii) $\displaystyle\prod_{n=1}^{\infty}\left(1 - \frac{1}{\sqrt{n+1}}\right)$

Exercise **10.2.5** Show that $\prod_{n=1}^{\infty}(1+(-1)^n/(n+1))$ converges, by determining the limit of the sequence of its partial products. Show, however, that this product is not absolutely convergent.

Exercise **10.2.6** Determine the region of absolute convergence of

(i) $\displaystyle\prod_{n=1}^{\infty}(1-z^n)$

(ii) $\displaystyle\prod_{n=1}^{\infty}(1-n^{-z})$

(iii) $\displaystyle\prod_{n=1}^{\infty}(1+a_n z)$ where $\displaystyle\sum_{n=1}^{\infty} a_n$ is absolutely convergent

Exercise **10.2.7** Give the proof of Theorem 10.9(i).

Exercise **10.2.8** Complete the proof of Theorem 10.10, by showing that $F(z) \neq 0$.

Exercise **10.2.9** Use the technique of Example 10.10 to prove the following:

(i) $\displaystyle\prod_{n=1}^{\infty}(1+z^n)$ is analytic with no zeros for $|z|<1$

(ii) $\displaystyle\prod_{n=1}^{\infty}(1+e^{-nz}/n^2)$ is analytic with no zeros for $\operatorname{Re} z > 0$

(iii) $\displaystyle z\prod_{n=1}^{\infty}\left(1-\frac{z^2}{n^2\pi^2}\right)$ is entire, with zeros at 0 and $\pm n\pi$, $n \in \mathbb{N}$

Exercise **10.2.10** Show that, for each $n \in \mathbb{N}$, $\sin(z^2/n^2) = z^2 F(z)/n^2$ where F is entire. Hence use the technique of Example 10.10 to prove that $\prod_{n=1}^{\infty}(1-\sin(z^2/n^2))$ is entire.

Exercise **10.2.11** Use Theorem 10.11 to show that $\prod_{n=1}^{\infty}(1-z/\sqrt{n})\exp(z/\sqrt{n}+z^2/2n)$ is entire.

Exercise **10.2.12** Let f be an entire function with an infinite number of simple zeros $z_1, z_2, \ldots$ none of which are 0, such that $|z_1| \leqslant |z_2| \leqslant \ldots$ with $z_n \to \infty$ as $n \to \infty$. Show that the only singularities of f'/f are simple poles with residue 1 at z_n for each $n \in \mathbb{N}$. Let f'/f satisfy the other conditions of Theorem 5.15. Use the technique of Example 10.12 to show that

$$f(z) = f(0)\exp(zf'(0)/f(0))\prod_{n=1}^{\infty}(1-z/z_n)e^{z/z_n}$$

Exercise **10.2.13** Use Weierstrass's factor theorem to show that

$$\cos z = e^{g(z)}\prod_{n=1}^{\infty}\left(1-\frac{4z^2}{(2n-1)^2\pi^2}\right)$$

for some entire function g. Use the technique of Example 10.13 and the partial fraction expansion of $\tan z$ given in Example 5.14 to show that $g(z) \equiv 0$.

Exercise **10.2.14** Use Weierstrass's factor theorem to show that

$$e^z - 1 = ze^{g(z)} \prod_{n=1}^{\infty} \left(1 + \frac{z^2}{4n^2\pi^2}\right)$$

for some entire function g. Let $zf(z) = e^z - 1$. Assuming that f'/f satisfies the conditions of Theorem 5.15, use the result of Exercise 10.2.12 to show that $g(z) = e^{z/2}$.

Exercise **10.2.15** Determine the most general infinite product representation of an entire function with zeros of order 2 at all the integers, and no other zeros.

Functions Defined by Integrals

Many special functions can be conveniently defined by integrals involving z as a parameter. Clearly, it is important to know where such functions are differentiable, so we require results analogous to those for differentiating real integrals under the integral sign. The following result is the analogue of Leibniz's result for real integrals. In particular, this result shows that the formal differentiations carried out when seeking contour integral solutions of differential equations, as in Chapter 7, are valid. Recall that the result is also used in the proof of Theorem 10.3.

Theorem 10.14. Differentiation under the Integral Sign

Let $f(z, \zeta)$ be continuous for z in a simply connected region $\mathscr{R}$ and for ζ on a simple contour $\mathscr{C}$. Let f be analytic in $\mathscr{R}$ for each fixed ζ on $\mathscr{C}$. Then $F(z) = \int_{\mathscr{C}} f(z, \zeta)\, d\zeta$ is analytic in $\mathscr{R}$ and $F'(z) = \int_{\mathscr{C}} f_z(z, \zeta)\, d\zeta$. □

Note

It is possible to prove this result by elementary means, under the additional assumption that $f_z(z, \zeta)$ is continuous for z in $\mathscr{R}$ and for ζ on $\mathscr{C}$. This can be done by using the definition to express $F(z)$ in terms of two real integrals and then using Leibniz's result for real integrals. The details are left as an exercise. However, we shall prove 10.14 by using a uniform convergence result from Chapter 4.

Proof

The contour $\mathscr{C}$ has a parametrisation $\zeta = \zeta(t)$, $0 \leqslant t \leqslant 1$ say, where $\zeta'(t)$ is continuous without loss of generality. Divide $[0, 1]$ into n equal parts, $0 = t_0 < t_1 < t_2 \ldots < t_{n-1} < t_n = 1$ with $\Delta t_k = t_k - t_{k-1} = 1/n$, $k = 1, \ldots, n$. Define $F_n(z)$ by

$$F_n(z) = \sum_{k-1}^{n} f(z, \zeta(t_k))\zeta'(t_k)\Delta t_k \tag{10.12}$$

Then since $F(z) = \int_0^1 f(z, \zeta(t))\zeta'(t)\,dt$ and f is a continuous function of t, it follows that $\lim_{n\to\infty} F_n(z) = F(z)$.

Now $f(z, \zeta(t))\zeta'(t)$ is continuous for $t \in [0, 1]$ and for $z \in S$, where S is any compact subset of $\mathscr{R}$. Hence it follows by Exercises 10.3.1, that given any $\varepsilon > 0$, there exists $N(\varepsilon) \in \mathbb{N}$, independent of $z \in S$ and $t \in [0, 1]$, such that

$$|t^* - t| < \frac{1}{N} \quad \Rightarrow \quad |f(z, \zeta(t^*))\zeta'(t^*) - f(z, \zeta(t))\zeta'(t)| < \varepsilon$$

It then follows by (10.12) that

$$\begin{aligned}
|F_n(z) - F(z)| &= \left|\sum_{k=1}^{n} \int_{t_{k-1}}^{t_k} (f(z, \zeta(t_k))\zeta'(t_k) - f(z, \zeta(t))\zeta'(t))\,dt\right| \\
&\leqslant \sum_{k=1}^{n} \int_{t_{k-1}}^{t_k} |f(z, \zeta(t_k))\zeta'(t_k) - f(z, \zeta(t))\zeta'(t)|\,dt \\
&< \sum_{k=1}^{n} \int_{t_{k-1}}^{t_k} \varepsilon\,dt = \sum_{k=1}^{n} \varepsilon\Delta t_k = \varepsilon \quad \text{for every } n > N
\end{aligned}$$

where N is independent of z. Hence $(F_n(z))$ converges uniformly to $F(z)$ in S and $F'_n(z)$ clearly exists in S for each n. Then by Theorem 4.8, F is analytic for z in $\mathscr{R}$ and

$$F'(z) = \lim_{n\to\infty} F'_n(z) = \int_0^1 f_z(z, \zeta(t))\zeta'(t)\,dt = \int_{\mathscr{C}} f_z(z, \zeta)\,d\zeta \qquad ■$$

Example 10.14

Let f be analytic in a simply connected region $\mathscr{R}$ and let $\mathscr{C}$ be a simple closed contour in $\mathscr{R}$, with α any point inside $\mathscr{C}$. Then if $F(\alpha) = \int_{\mathscr{C}} (f(z)/(z-\alpha))\,dz$, the conditions of 10.14 are satisfied and so $F'(\alpha) = \int_{\mathscr{C}} (f(z)/(z-\alpha)^2)\,dz$. This result and Cauchy's integral theorem form the basis of an alternative proof to Theorem 3.13, Cauchy's integral theorem for derivatives.

We now investigate the possibility of differentiating certain improper integrals with respect to a parameter. Such integrals often define special functions, and also define certain integral transforms, as in Chapter 8.

Definitions

(i) An **infinite contour** $\mathscr{C}$ is a contour joining a point $\alpha \in \mathbb{C}$ to the point at infinity in $\tilde{\mathbb{C}}$; that is, $\mathscr{C}$ has a parametrisation $z = z(t)$, $t \geqslant a$, with $\alpha = z(a)$, such that for each $b > a$ the curve parametrised by $z = z(t)$, $a \leqslant t \leqslant b$, is a contour with finite length $L(b)$, and $\lim_{b\to\infty} L(b) = \infty$.

(ii) If $\mathscr{C}$ is such an infinite contour, then

$$\int_{\mathscr{C}} f(z)\,dz = \int_a^{\infty} f(z(t))z'(t)\,dt = \lim_{b\to\infty}\int_a^b f(z(t))z'(t)\,dt$$

provided that the associated real improper integrals exist. In this case, we say that $\int_{\mathscr{C}} f(z)\,dz$ **converges**.

(iii) Let $\mathscr{C}$ be an infinite contour parametrised by $\zeta = \zeta(t)$, $t \geqslant a$. Then $F(z) = \int_{\mathscr{C}} f(z, \zeta)d\zeta$ **converges uniformly** in $A \subseteq \mathbb{C}$ if $F_b(z) = \int_a^b f(z, \zeta(t))\zeta'(t)\,dt$ converges uniformly to $F(z)$ in A, i.e. given any $\varepsilon > 0$, there exists $B \in \mathbb{R}$, independent of $z \in A$, such that $|F_b(z) - F(z)| < \varepsilon$ for every $b > B$.

Important Note

If $F_b(z)$ converges uniformly to $F(z)$ in A, then the sequence $(F_n(z))$, $n \in \mathbb{N}$, converges uniformly to $F(z)$ in A. It then follows immediately by Theorem 4.6 that if $f(z, \zeta)$ is continuous for $z \in A$, then $F(z)$ is continuous in A.

The following result is a direct consequence of Theorem 10.14.

Theorem 10.15. Differentation of an Improper Integral with Respect to a Parameter

Let $\mathscr{C}$ be an infinite contour and let $f(z, \zeta)$ be continuous for z in a simply connected region $\mathscr{R}$ and for ζ on $\mathscr{C}$. Let $f(z, \zeta)$ be an analytic function of z in $\mathscr{R}$, for each ζ on $\mathscr{C}$, for any finite value of the parameter. Suppose that $F(z) = \int_{\mathscr{C}} f(z, \zeta)\,d\zeta$ is uniformly convergent in any compact subset of $\mathscr{R}$. Then F is analytic in $\mathscr{R}$ with $F'(z) = \int_{\mathscr{C}} fz(z, \zeta)\,d\zeta$. □

Proof

Let $\mathscr{C}$ have parametrisation $\zeta = \zeta(t)$, $t \geqslant a$ and let $F_n(z) = \int_a^n f(z, \zeta(t))\zeta'(t)\,dt$, $n \in \mathbb{N}$. Then by 10.14 and hypothesis $F'_n(z) = \int_a^n f_z(z, \zeta(t))\zeta'(t)\,dt$ for $z \in \mathscr{R}$. It then follows by hypothesis that $(F_n(z))$ is a sequence of analytic functions in $\mathscr{R}$, which converges uniformly to $F(z)$ on any compact subset of $\mathscr{R}$. Hence by 4.8, $F'(z) = \lim_{n\to\infty} F'_n(z)$ in $\mathscr{R}$ as required. ■

The simplest test for uniform convergence of improper integrals is the following, which is analogous to the corresponding real result. Uniform convergence of $\int_{\mathscr{C}} f(z, \zeta)\,d\zeta$ follows by this result with $g(z, t) = f(z, \zeta(t))\zeta'(t)$.

Theorem 10.16. Weierstrass's M Test

Let $g(z, t)$ be a continous function of t, $t \geqslant a$, for all z in a compact set $A \subseteq \mathbb{C}$, satisfying $|g(z, t)| \leqslant M(t)$ for all $z \in A$, where $M(t)$ is a positive function, independent of z. Then if $\int_a^\infty M(t)\,dt$ converges, $\int_a^\infty g(z, t)\,dt$ converges uniformly in A. □

Proof

Let $G(z) = \int_a^\infty g(z, t)\,dt$ and $G_b(z) = \int_a^b g(z, t)\,dt$. Since $\int_a^\infty M(t)\,dt$ converges, given any $\varepsilon > 0$, there exists $c \in \mathbb{R}$, independent of z, such that $\int_c^\infty M(t)\,dt < \varepsilon/2$. Then given this ε,

$$|G_b(z) - G_c(z)| = \left|\int_c^b g(z, t)\,dt\right| \leqslant \int_c^b M(t)\,dt < \int_c^\infty M(t)\,dt < \varepsilon/2$$

for every $b > c$. Hence $|G(z) - G_c(z)| < \varepsilon/2$ so that

$$|G_b(z) - G(z)| \leqslant |G_b(z) - G_c(z)| + |G(z) - G_c(z)| < \varepsilon$$

for every $b > c$, independent of z, as required. ■

Example 10.15

(i) Consider $\int_0^\infty te^{-zt}\,dt$. For $\operatorname{Re} z \geqslant \rho > 0$ and $t \geqslant 0$, $|te^{-zt}| \leqslant te^{-\rho t}$ and $\int_0^\infty te^{-\rho t}dt = 1/\rho^2$. Hence by 10.16, $\int_0^\infty te^{-zt}\,dt$ converges uniformly in any compact set of $\operatorname{Re} z > 0$ and so is analytic for $\operatorname{Re} z > 0$ by 10.15, as can be seen by direct evaluation.

(ii) Let f be continuous and of class E, so that there exist positive real constants M, T and a such that $|f(t)| \leqslant Me^{at}$ for $t \geqslant T$. Then by Lemma 8.8, the Laplace transform of f exists. Consider $\mathscr{L}(f(t)) = \bar{f}(z) = \int_0^\infty e^{-zt} f(t)\,dt$. For $\operatorname{Re} z \geqslant \rho > a$ and $t \geqslant T$, $|e^{-zt} f(t)| \leqslant Me^{(a-\rho)t}$ and $\int_T^\infty e^{(a-\rho)t}dt$ clearly converges for $\rho > a$. Hence by 10.16, $\int_0^\infty e^{-zt} f(t)\,dt$ converges uniformly in any compact subset of $\operatorname{Re} z > a$ and so is analytic for $\operatorname{Re} z > a$ by 10.15. Then $\bar{f}'(z) = \int_0^\infty -te^{-zt} f(t)\,dt = -\mathscr{L}(tf(t))$, a fact used in the proof of 8.10(v).

The Gamma Function

We now illustrate the use of some of the results from earlier in this chapter by investigating the gamma function, which occurs in many applications. The integral definition is a straightforward generalisation of the definition of the real gamma function, given in Chapter 6.

Definition

The **gamma function** is denoted and defined by

$$\Gamma(z) = \int_0^\infty e^{-t} t^{z-1} dt,\ t \in \mathbb{R},\ \text{for } z \in \mathbb{C} \text{ with } \operatorname{Re} z > 0 \tag{10.13}$$

Note

It is easily seen that the improper integral in (10.13) only converges for $\operatorname{Re} z > 0$. The next example shows that, in this case, the integral is uniformly convergent on any compact subset of $\operatorname{Re} z > 0$, so $\Gamma(z)$ is analytic on $\operatorname{Re} z > 0$. Note that $\Gamma(1) = 1$.

Example 10.16

The integrand of the improper integral in (10.13) has possible bad behaviour at $t = 0$, so let $\Gamma(z) = G(z) + H(z)$, where $G(z) = \int_0^1 e^{-t} t^{z-1} dt$ and $H(z) = \int_1^\infty e^{-t} t^{z-1} dt$.

Let $0 < \rho \leqslant \operatorname{Re} z$ and let $t = 1/s$ in $G(z)$, so that $G(z) = \int_1^\infty e^{-1/s} s^{-1-z} ds$. Since $|s^{-1-z} e^{-1/s}| \leqslant s^{-1-\rho} e^{-1/s} < s^{-1-\rho}$ for $s \geqslant 1$, and $\int_1^\infty s^{-1-\rho} ds = 1/\rho$, and so converges, $G(z)$ converges uniformly on any compact subset of Re $z > 0$ by 10.16. Hence $G(z)$ is analytic for $\operatorname{Re} z > 0$ by 10.15 since the integrand is analytic for $\operatorname{Re} z > 0$, for each $s \geqslant 1$.

Turning now to $H(z)$, for $\operatorname{Re} z \leqslant R$,

$$|t^{z-1} e^{-t}| \leqslant t^{R-1} e^{-t} \leqslant k e^{t/2} e^{-t} = k e^{-t/2} \quad \text{for some } k(R) \in \mathbb{R}$$

since $\lim_{t \to \infty} t^{R-1} e^{-t/2} = 0$. Since $\int_1^\infty e^{-t/2} dt$ converges, $H(z)$ converges uniformly on any compact subset of $\mathbb{C}$ by 10.16 and so is entire by 10.15.

Hence, altogether, $\Gamma(z)$ is analytic for Re $z > 0$. It follows by 10.15 that, for example, $\Gamma'(1) = \int_0^\infty e^{-t} \operatorname{Log} t \, dt$.

Note

It follows by the fundamental theorem of calculus and integration by parts that

$$\Gamma(z + 1) = z\Gamma(z) \tag{10.14}$$

Hence $\Gamma(n + 1) = n!$ for $n \in \mathbb{N}$ and also from (10.14),

$$\Gamma(z) = \frac{\Gamma(z + n)}{z(z + 1)(z + 2) \ldots (z + n - 1)} \qquad (n \in \mathbb{N})$$

This relation can be used to analytically continue the definition of Γ to $\operatorname{Re} z > -n$, excluding $0, -1, -2, \ldots, -n + 1$. Note that it shows that Γ has simple poles at $-n$, $n \in \mathbb{N}$ and at 0.

The gamma function has representations other than its integral definition, as indicated by the following result.

Theorem 10.17. Euler's Limit Formula for $\Gamma(z)$

$$\Gamma(z) = \lim_{n \to \infty} \frac{n^z n!}{z(z + 1)(z + 2) \ldots (z + n)} \qquad (\operatorname{Re} z > 0)$$

□

Note

It can be shown that the given sequence of functions converges uniformly on any compact set which does not contain $0, -1, -2, \ldots$. Hence this result once again enables us to analytically continue Γ to $\mathbb{C}\backslash\{0, -n : n \in \mathbb{N}\}$.

Historical Note

The problem of finding a function of a real variable x which is continuous when $x > 0$ and which reduces to $n!$ when $x = n$ was first solved by Euler in 1729. He announced his discovery that

$$\lim_{n\to\infty} \frac{n^x n!}{(x+1)(x+2)\ldots(x+n)}$$

had the desired properties in a letter to Goldbach. He was also responsible for the integral definition of $\Gamma(x)$, although not for the notation Γ.

Proof

Step 1

Let $\Gamma(z,n) = \int_0^n t^{z-1}(1 - t/n)^n dt$, $\operatorname{Re} z > 0$, for each $n \in \mathbb{N}$. Since $e^{-t} = \lim_{n\to\infty}(1 - t/n)^n$ for $t \in \mathbb{R}$, it seems reasonable that $\lim_{n\to\infty}\Gamma(z, n) = \Gamma(z)$. This can be shown rigorously as follows.

We first obtain an inequality involving the difference of the integrands. Let $f(t) = 1 - e^t(1 - t/n)^n$ for $0 \leqslant t \leqslant n$. Then $f'(t) = e^t(1 - t/n)^{n-1}t/n \geqslant 0$ for $0 \leqslant t \leqslant n$ so that $f(t)$ is monotonic increasing. Since $f(0) = 0$, $f(t) \geqslant 0$ for $0 \leqslant t \leqslant n$. Also

$$f(t) = \int_0^t f'(u)\,du \leqslant e^t \int_0^t \frac{u}{n}\,du = \frac{e^t t^2}{2n} \qquad (t \leqslant n)$$

Hence altogether

$$0 \leqslant e^{-t} - (1 - t/n)^n \leqslant \frac{t^2}{2n} \qquad (0 \leqslant t \leqslant n) \tag{10.15}$$

Let c be any positive real number. Then for $n \geqslant c$,

$$|\Gamma(z) - \Gamma(z, n)| = \left| \int_0^c t^{z-1}(e^{-t} - (1 - t/n)^n)\,dt + \int_c^\infty e^{-t}t^{z-1}dt - \int_c^n t^{z-1}(1 - t/n)^n dt \right|$$

$$\leqslant \int_0^c (e^{-t} - (1 - t/n)^n)t^{x-1}dt + \int_c^\infty e^{-t}t^{x-1}dt + \int_c^n (1 - t/n)^n t^{x-1}dt \tag{10.16}$$

since $|t^{z-1}| = t^{x-1}$, where $x = \operatorname{Re} z$. Then also by (10.15),

$$\int_c^n (1 - t/n)^n t^{x-1} dt \leqslant \int_c^n e^{-t} t^{x-1} dt \leqslant \int_c^\infty e^{-t} t^{x-1} dt$$

It then follows by (10.16) that

$$|\Gamma(z) - \Gamma(z, n)| \leqslant \int_0^c (e^{-t} - (1 - t/n)^n) t^{x-1} dt + 2\int_c^\infty e^{-t} t^{x-1} dt \tag{10.17}$$

Since $\Gamma(x)$ converges, given any $\varepsilon > 0$, c can be chosen so that $\int_c^\infty e^{-t} t^{x-1} dt < \varepsilon/4$. Then for $n \geqslant c^{x+2}/\varepsilon$, by (10.15),

$$\int_0^c (e^{-t} - (1 - t/n)^n) t^{x-1} dt \leqslant \int_0^c \frac{t^{x+1}}{2n} dt < \frac{c^{x+2}}{2n} \leqslant \varepsilon/2$$

Then by (10.17), $|\Gamma(z) - \Gamma(z, n)| < \varepsilon$ for $n \geqslant \max(c, c^{x+2}/\varepsilon)$, so that $\lim_{n\to\infty} \Gamma(z, n) = \Gamma(z)$, as required.

Step 2

Letting $u = t/n$ in $\Gamma(z, n)$ and integrating by parts gives

$$\Gamma(z, n) = n^z \int_0^1 u^{z-1}(1 - u)^n du = n^z \left(\frac{n}{z}\right) \int_0^1 u^z (1 - u)^{n-1} du$$

Integrating by parts another $n - 1$ times gives

$$\Gamma(z, n) = \frac{n^z n(n-1)(n-2)\ldots 1}{z(z+1)(z+2)\ldots(z+n-1)} \int_0^1 u^{z+n-1} du = \frac{n^z n!}{z(z+1)\ldots(z+n)}$$

The desired result then follows by step 1. ■

The following result follows easily from Theorem 10.17 and the infinite product expansion of $\sin \pi z$, which follows from Example 10.12. The proof is left as an exercise. It includes Theorem 6.12(ii) as a special case.

Theorem 10.18. The Reflection Formula

$$\Gamma(z)\Gamma(1 - z) = \frac{\pi}{\sin \pi z} \qquad (z \notin \mathbb{Z})$$ □

Euler's limit formula for Γ can also be used to derive the following infinite product expansion of $1/\Gamma(z)$.

Theorem 10.19. The Infinite Product Representation of $1/\Gamma(z)$

$$\frac{1}{\Gamma(z)} = z e^{\gamma z} \prod_{n=1}^\infty \left(1 + \frac{z}{n}\right) e^{-z/n} \qquad \text{(for all } z \in \mathbb{C}\text{)}$$

where $\gamma = \lim_{n\to\infty} \left(1 + \frac{1}{2} + \frac{1}{3} + \ldots + \frac{1}{n} - \operatorname{Log} n\right)$ □

Note

The sequence defining γ is easily shown to converge and $\gamma \approx 0.577\,2156$. The constant γ is known as **Euler's constant**.

Part Proof

It follows by Theorem 10.17 that

$$\frac{1}{\Gamma(z)} = z \lim_{n\to\infty} \frac{(1+z)(2+z)\ldots(n+z)}{n^z n!} \qquad \text{(for all } z \in \mathbb{C}\text{)}$$

$$\Rightarrow \quad \frac{1}{\Gamma(z)} = z \lim_{n\to\infty} e^{-z \operatorname{Log} n} \prod_{k=1}^{n} \left(1 + \frac{z}{k}\right) \tag{10.18}$$

It follows by Example 10.11 that although $\prod_{k=1}^{\infty}(1 + z/k)$ diverges, the infinite product $\prod_{k=1}^{\infty}(1 + z/k)e^{-z/k}$ converges. Hence we rewrite (10.18) in the form

$$\frac{1}{\Gamma(z)} = z \lim_{n\to\infty} \exp\left(z(1 + 1/2 + \ldots + 1/n - \operatorname{Log} n)\right) \prod_{k=1}^{n} \left(1 + \frac{z}{k}\right) e^{-z/k}$$

It is easily shown that $\gamma = \lim_{n\to\infty}(1 + 1/2 + \ldots + 1/n - \operatorname{Log} n)$ exists, so the result follows. ■

Note

It follows by this result and Weierstrass's factor theorem that $1/\Gamma(z)$ is an entire function with simple zeros at 0 and $-n$, $n \in \mathbb{N}$. Hence $\Gamma(z)$ is analytic on $\mathbb{C}\backslash\{0, -n : n \in \mathbb{N}\}$ and has no zeros.

Example 10.17

It follows by Theorem 10.19 that

$$\operatorname{Log}\Gamma(z) = -\operatorname{Log} z - \gamma z - \sum_{n=1}^{\infty}\left(\frac{-z}{n} + \operatorname{Log}\left(\frac{z+n}{n}\right)\right) \qquad (z \neq 0, -1, -2, \ldots)$$

$$\Rightarrow \quad \frac{\Gamma'(z)}{\Gamma(z)} = -\frac{1}{z} - \gamma + \sum_{n=1}^{\infty}\left(\frac{1}{n} - \frac{1}{z+n}\right)$$

Then $\Gamma'(1)/\Gamma(1) = \Gamma'(1) = -\gamma$. It follows from Example 10.16 that

$$\Gamma'(1) = -\gamma = \int_0^{\infty} e^{-t} \operatorname{Log} t \, dt$$

As another example of the use of 10.19, we prove the following theorem, of which 6.12(iii) is a special case.

Theorem 10.20. The Duplication Formula

$$\sqrt{\pi}\,\Gamma(2z) = 2^{2z-1}\Gamma(z)\Gamma(z+1/2) \qquad (z \neq 0, -n, n \in \mathbb{N})$$

□

Proof

Let $\psi(z) = \Gamma'(z)/\Gamma(z)$. It follows by Example 10.17 that

$$\psi(z) = -\frac{1}{z} - \gamma + \sum_{n=1}^{\infty}\left(\frac{1}{n} - \frac{1}{z+n}\right) \quad \Rightarrow \quad \psi'(z) = \sum_{n=0}^{\infty}\frac{1}{(z+n)^2}$$

$$\Rightarrow \quad \psi'(z) + \psi'(z+1/2) = 4\left(\sum_{n=0}^{\infty}\frac{1}{(2z+2n)^2} + \sum_{n=0}^{\infty}\frac{1}{(2z+2n+1)^2}\right)$$

$$= 4\sum_{m=0}^{\infty}\frac{1}{(2z+m)^2} = 2\psi'(2z)$$

Integrating this result twice and taking exponentials gives

$$\Gamma(z)\Gamma(z+1/2) = e^{az+b}\Gamma(2z)$$

for some constants a and b. Letting $z = 1/2$ and noting that $\Gamma(1/2) = \sqrt{\pi}$ by 10.18 gives $\sqrt{\pi} = e^{a/2+b}$. Letting $z = 1$ gives $\sqrt{\pi} = 2e^{a+b}$. Hence $e^{a/2} = 1/2 \Rightarrow e^{az} = 2^{-2z}$, and $e^b = 2\sqrt{\pi}$. ■

Finally, we give a representation of $1/\Gamma(z)$ in terms of a contour integral, which once again has the advantage over (10.13) of giving the largest domain of definition for $\Gamma(z)$. Since $e^z/z^n = \sum_{m=0}^{\infty} z^{m-n}/m!$ for all $z \in \mathbb{C}$, $n \in \mathbb{N}$,

$$\text{Res}_0\left(\frac{e^z}{z^n}\right) = \frac{1}{(n-1)!} = \frac{1}{\Gamma(n)}$$

It follows by the residue theorem that

$$\frac{1}{\Gamma(n)} = \frac{1}{2\pi i}\int_{\mathscr{C}}\frac{e^z}{z^n}\,dz \qquad (n \in \mathbb{N})$$

where $\mathscr{C}$ is any simple closed contour enclosing the origin. If $n \notin \mathbb{Z}$, the integrand has a branch point at 0 and $\mathscr{C}$ has to be amended to allow for a branch cut. This leads to the following result.

Theorem 10.21. A Contour Integral for $1/\Gamma(z)$

$$\frac{1}{\Gamma(z)} = \frac{1}{2\pi i}\int_{\mathscr{H}}\frac{e^\zeta}{\zeta^z}\,d\zeta \qquad (z \in \mathbb{C})$$

where $\mathscr{H}$ is the **Hankel contour** shown in Fig. 10.7. □

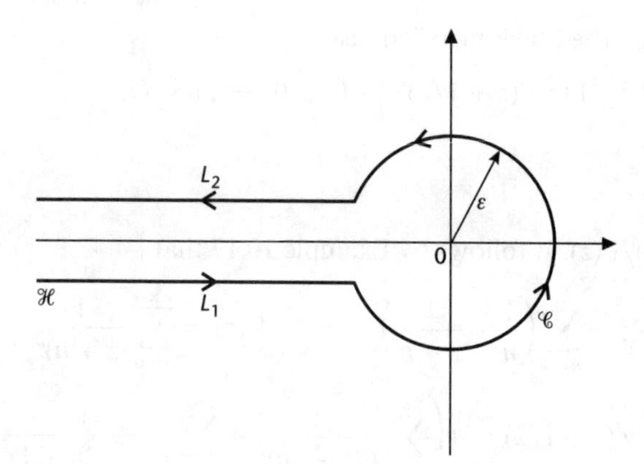

Figure 10.7

Note

It follows by the reflection formula and this result that

$$\Gamma(z) = \frac{\pi}{\sin \pi z\, \Gamma(1-z)} = \frac{1}{2i \sin \pi z} \int_{\mathscr{H}} e^{\zeta} \zeta^{z-1}\, d\zeta \qquad (z \notin \mathbb{Z}_{\leqslant 0})$$

Part Proof

Define f by $f(z) = (1/2\pi i) \int_{\mathscr{H}} e^{\zeta} \zeta^{-z}\, d\zeta$. Then by Theorems 10.14 to 10.16, f is an entire function. This is left as an exercise. Along the given arc of the circle $\mathscr{C}$, centre 0 and radius ε, $\zeta = \varepsilon e^{i\theta}$, $-\theta_1 \leqslant \theta \leqslant \theta_1$ say. Then using the definition and the ML lemma, it is straightforward to show that $\int_{L_i} e^{\zeta} \zeta^{-z}\, d\zeta \to 0$ as $\varepsilon \to 0$ provided that $\operatorname{Re} z < 1$. This is also left as an exercise.

Let $-L_1$ be the line segment given by $\zeta = -t - ic$, $a \leqslant t$, where $-a - ic = \varepsilon e^{-i\theta_1}$. Then $\int_{L_1} e^{\zeta} \zeta^{-z}\, d\zeta = e^{-ic} \int_a^{\infty} e^{-t} \exp(-z \operatorname{Log}(-t - ic))\, dt$. Similarly, if L_2 is the line segment given by $\zeta = -t + ic$, $a \leqslant t$, then $\int_{L_2} e^{\zeta} \zeta^{-z}\, d\zeta = -e^{ic} \int_a^{\infty} e^{-t} \exp(-z \operatorname{Log}(-t + ic))\, dt$. Hence, letting $\varepsilon \to 0$ and $\theta_1 \to \pi$ on the contour $\mathscr{H}$, gives, for $\operatorname{Re}(1-z) > 0$,

$$f(z) = \frac{1}{2\pi i}\left(\int_0^{\infty} e^{-t} e^{i\pi z} t^{-z}\, dt - \int_0^{\infty} e^{-t} e^{-i\pi z} t^{-z}\, dt \right) = \frac{\sin \pi z\, \Gamma(1-z)}{\pi}$$

$$\Rightarrow \quad f(z) = \frac{1}{\Gamma(z)} \text{ for } \operatorname{Re}(1-z) > 0 \text{ by Theorem 10.18.}$$

Since both $f(z)$ and $1/\Gamma(z)$ are entire, the result follows by Theorem 10.1. ■

Exercise

10.3.1 Let f be continuous on a compact subset S of $\mathbb{C}$. Show that given any $\varepsilon > 0$, there exists $\delta > 0$, independent of $z \in S$ such that $|z^* - z| < \delta \Rightarrow |f(z^*) - f(z)| < \varepsilon$. This fact is used in the proof of Theorem 10.14.

Exercise **10.3.2** Use Leibniz's theorem for differentiating a real integral with respect to a parameter and the definition of a complex contour integral to prove Theorem 10.14 under the additional assumption that $f_z(z, \zeta)$ is continuous for z in $\mathscr{R}$ and for ζ on $\mathscr{C}$.

Exercise **10.3.3** Let $F(z) = \int_0^\infty e^{-zt} \sin t\, dt$. Show that $F(z)$ is uniformly convergent on any compact subset of $\operatorname{Re} z > 0$ and hence analytic for $\operatorname{Re} z > 0$.

Exercise **10.3.4** Let $F(z) = \int_0^\infty e^{-zt^2}\, dt$. Use the fact that $\int_0^\infty e^{-x^2} = \sqrt{\pi}/2$ to show that $F(z)$ is uniformly convergent on any compact subset of $\operatorname{Re} z > 0$ and hence analytic for $\operatorname{Re} z > 0$. Hence show that $F(z) = \sqrt{\pi}/2\sqrt{z}$ for $\operatorname{Re} z > 0$; F can be analytically continued to $\mathbb{C}\backslash\{0\}$ by using this result as the definition for F.

Exercise **10.3.5** Use the integral definition of Γ to show that $\Gamma(1) = 1$ and that $\Gamma(z+1) = z\Gamma(z)$. This proves Theorem 6.12(i). Hence show that

$$\Gamma(z) = \frac{\Gamma(z+n)}{z(z+1)\ldots(z+n-1)} \qquad (\operatorname{Re} z > -n,\ n \in \mathbb{N})$$

Exercise **10.3.6** Use the integral definition of Γ to show that

$$\Gamma(z) = \sum_{n=0}^{\infty} \frac{(-1)^n}{n!(n+z)} + \int_1^\infty e^{-t} t^{z-1} dt \qquad (\operatorname{Re} z > 0)$$

Show that this result provides the analytic continuation of Γ to $\mathbb{C}\backslash\{0, -n : n \in \mathbb{N}\}$.

Exercise **10.3.7** Use Euler's limit definition for $\Gamma(z)$ to prove that

$$\Gamma(z)\Gamma(1-z) = \pi \csc \pi z \qquad (z \notin \mathbb{Z}_{\geqslant 0})$$

Exercise **10.3.8** Let $f(z) = 1/\Gamma(z+1)$. Assuming that f'/f obeys the conditions of Theorem 5.15, use the result of Question 12 in Exercise 10.2.12 to show that

$$\frac{1}{\Gamma(z)} = ze^{\gamma z} \prod_{n=1}^{\infty} \left(1 + \frac{z}{n}\right) e^{-z/n} \qquad (z \in \mathbb{C},\ \gamma = -\Gamma'(1))$$

Exercise **10.3.9** Use the monotonic-bounded principle to show that the sequence $(1 + 1/2 + 1/3 + \ldots + 1/n - \operatorname{Log} n)$ converges.

Exercise **10.3.10** Complete the proof of Theorem 10.21 by showing that

(i) $g(z) = \int_{\mathscr{H}} e^\zeta \zeta^{-z} d\zeta$ is entire

(ii) $\int_{\mathscr{C}} e^\zeta \zeta^{-z} d\zeta \to 0$ as $\varepsilon \to 0$ provided that $\operatorname{Re} z < 1$.

Exercise **10.3.11** Use the series definition of the Bessel function $J_\nu(z)$ and the Hankel contour representation of $1/\Gamma(z)$ to show that

$$J_\nu(z) = \frac{(z/2)^\nu}{2\pi i} \int_{\mathscr{H}} \frac{e^{\zeta - z^2/4\zeta}}{\zeta^{\nu+1}} d\zeta \qquad \text{(Schläfli's integral)}$$

(See Exercises 7.5.6)

Asymptotic Expansions

Very often in applications, given an analytic function $f: A \subseteq \mathbb{C} \to \mathbb{C}$, a good approximation of $f(z)$ for large $|z|$ is required so that the behaviour of $f(z)$ as $z \to \infty$ can be investigated. In many cases, we can use what is known as the asymptotic series expansion of f to do this.

Definition

The power series $\sum_{n=0}^{\infty} a_n z^{-n}$, where a_n is a complex constant for each n, which may either converge for large values of $|z|$ or diverge for all values of z, is an **asymptotic series** for $f(z)$, valid in a given range of values of $\operatorname{Arg} z$, if for every fixed $k \in \mathbb{Z}_{\geqslant 0}$, $\lim_{z\to\infty} z^k(f(z) - \sum_{n=0}^{k} a_n z^{-n}) = 0$, for $\operatorname{Arg} z$ in the given range. In this case, we write $f(z) \sim \sum_{n=0}^{\infty} a_n z^{-n}$. More generally, if $f(z)$ does not have such an asymptotic series, there may be a function $g(z)$ such that $f(z)/g(z) \sim \sum_{n=0}^{\infty} a_n z^{-n}$. We then write $f(z) \sim g(z)\sum_{n=0}^{\infty} a_n z^{-n}$.

Note that if $\sum_{n=0}^{\infty} a_n z^{-n}$ converges to $f(z)$ for $|z| > \rho$ say, then it follows easily from the definition that $f(z) \sim \sum_{n=0}^{\infty} a_n z^{-n}$. However, in general, an asymptotic series need not converge.

Historical Note

The definition of an asymptotic series was first proposed by Poincaré in 1886. It implies that the difference between $f(z)$ and the sum to k terms of the series is of the same order as the $(k+1)$th term when $|z|$ is large. Hence asymptotic series are very useful for numerical computations when $|z|$ is large, even though they may diverge. By taking the sum of successive terms of such a divergent series and stopping just before the terms increase again in modulus, a good approximation to $f(z)$ may be obtained for a particular large value of $|z|$.

Note

If a given function $f(z)$ has an asymptotic series, then the coefficients are clearly determined by the equations

$$a_0 = \lim_{z\to\infty} f(z) \qquad a_1 = \lim_{z\to\infty} z(f(z) - a_0) \qquad a_2 = \lim_{z\to\infty} z^2(f(z) - a_0 - a_1/z)$$

and so on.

Example 10.18

Recall that

$$e^{-1/z^2} = \sum_{n=0}^{\infty} \frac{(-1)^n}{n! z^{2n}} \qquad (z \neq 0)$$

Since this Laurent series converges for $|z| > 0$, it is the asymptotic expansion of e^{-1/z^2}. This follows because

$$\lim_{k\to\infty} z^{2k}\left(e^{-1/z^2} - \sum_{n=0}^{k} \frac{(-1)^n}{n!z^{2n}}\right) = \lim_{k\to\infty} z^{2k} \sum_{n=k+1}^{\infty} \frac{(-1)^n}{n!z^{2n}} = 0 \qquad (k \in \mathbb{Z}_{\geqslant 0})$$

A large number of special functions which occur in practice have integral representations. One of the easiest ways of obtaining the asymptotic expansion of a given special function is through such an integral representation.

Example 10.19

The **exponential integral** is denoted and defined by

$$\mathrm{Ei}(z) = \int_z^\infty \frac{e^{-\zeta}}{\zeta}\, d\zeta \qquad (\mathrm{Re}\, z > 0) \tag{10.19}$$

(It can be shown that the given improper integral converges.) Note that the integral is well defined since it is independent of the path of integration joining z to ∞ in $\tilde{\mathbb{C}}$. Letting $\tau = \zeta - z$ in (10.19) gives

$$\mathrm{Ei}(z) = e^{-z} \int_0^\infty \frac{e^{-\tau}}{\tau + z}\, d\tau = e^{-z} \int_0^\infty \frac{e^{-t}}{t + z}\, dt \qquad (t \in \mathbb{R}) \tag{10.20}$$

choosing the contour joining 0 to ∞ to be the positive real axis. Integrating by parts k times gives

$$\int_0^\infty \frac{e^{-t}}{t+z}\, dt = \frac{1}{z} - \frac{1}{z^2} + \frac{2}{z^3} - \ldots + \frac{(-1)^{k-1}(k-1)!}{z^k} + R_k(z)$$

$$\text{where} \quad R_k(z) = (-1)^k k! \int_0^\infty \frac{e^{-t}}{(t+z)^{k+1}}\, dt$$

$$\text{Hence} \quad |R_k(z)| \leqslant k! \int_0^\infty \frac{e^{-t}}{|t+z|^{k+1}}\, dt \leqslant \frac{k!}{|z|^{k+1}} \int_0^\infty e^{-t} dt = \frac{k!}{|z|^{k+1}}$$

so that $\lim_{z\to\infty} z^k R_k(z) = 0$ as $z \to \infty$ for any k. Then, by definition,

$$\int_0^\infty \frac{e^{-t}}{t+z}\, dt \sim \sum_{n=1}^{\infty} \frac{(-1)^{n-1}(n-1)!}{z^n} \quad \Rightarrow \quad \mathrm{Ei}(z) \sim e^{-z} \sum_{n=1}^{\infty} \frac{(-1)^{n-1}(n-1)!}{z^n}$$

from (10.20). It is easily seen that the asymptotic series diverges for all z by the ratio test, but it gives a good approximation of $\mathrm{Ei}(z)$ for large values of $|z|$, as long as it is terminated after the term of least modulus. For example,

$$\mathrm{Ei}(4i) \approx e^{-4} \sum_{n=1}^{5} \frac{(-1)^{n-1}(n-1)!}{(4i)^n} \qquad \mathrm{Ei}(10) \approx e^{-10} \sum_{n=1}^{11} \frac{(-1)^{n-1}(n-1)!}{10^n}$$

The larger the value of $|z|$, the closer the approximation will be. As $|z| \to \infty$, the dominant term in the asymptotic series for $\mathrm{Ei}(z)$ is e^{-z}/z.

Several elementary properties of asymptotic series, all of which follow from the definition, are given below.

Theorem 10.22. Elementary Properties of Asymptotic Series

(a) Let $f(z) \sim S(z) = \sum_{n=0}^{\infty} a_n z^{-n}$ and $g(z) \sim T(z) = \sum_{n=0}^{\infty} b_n z^{-n}$ for a common range of $\operatorname{Arg} z$. Then
  (i) $f(z) + g(z) \sim S(z) + T(z)$
  (ii) $f(z)g(z) \sim S(z)T(z)$.

(b) For a given range of $\operatorname{Arg} z$, a given function cannot have more than one asymptotic series. □

Proof

(a) (i) Let $P_k(z) = f(z) - \sum_{n=0}^{k} a_n z^{-n}$ and $Q_k(z) = g(z) - \sum_{n=0}^{k} b_n z^{-n}$. Then by hypothesis, $\lim_{z\to\infty} z^k P_k(z) = 0$ and $\lim_{z\to\infty} z^k Q_k(z) = 0$ for any $k \in \mathbb{Z}_{\geqslant 0}$. Hence $\lim_{z\to\infty} z^k(P_k(z) + Q_k(z)) = 0$ for any $k \in \mathbb{Z}_{\geqslant 0}$, as required

(ii) This is similar and is left as an exercise.

(b) Suppose that $f(z) \sim \sum_{n=0}^{\infty} a_n z^{-n}$ and $f(z) \sim \sum_{n=0}^{\infty} b_n z^{-n}$, where $a_n \neq b_n$ for at least one value of n. Then

$$\lim_{z\to\infty} z^k\Big(f(z) - \sum_{n=0}^{k} a_n/z^n\Big) = \lim_{z\to\infty} z^k\Big(f(z) - \sum_{n=0}^{k} b_n/z^n\Big) = 0 \qquad (k \in \mathbb{Z}_{\geqslant 0})$$

so that $\lim_{z\to\infty} z^k \sum_{n=0}^{k} (a_n - b_n)/z^n = 0$ for any $k \in \mathbb{Z}_{\geqslant 0}$. Hence

$$\lim_{z\to\infty} \big((a_0 - b_0)z^k + (a_1 - b_1)z^{k-1} + \ldots + (a_k - b_k)\big) = 0$$

for any $k \in \mathbb{Z}_{\geqslant 0}$ and so $a_n = b_n$ for all n, as required. ■

Note

Clearly, a given asymptotic series can be the asymptotic series for more than one function, since $f(z)$ and $g(z)$ may have the same asymptotic expansion if $\lim_{z\to\infty} z^k(f(z) - g(z)) = 0$ for each $k \in \mathbb{Z}_{\geqslant 0}$. For example, $f(x) = e^{1/x}$ and $g(x) = e^{-x} + e^{1/x}$, $x \in \mathbb{R}$, have the same asymptotic expansion. Hence a given asymptotic series does not determine a unique function.

There are several standard techniques for deriving the asymptotic expansion of a given function. We derive one very useful technique, which gives the asymptotic expansion of a function defined by a certain type of definite integral. The result is not so restrictive as it appears since many functions which occur in applications can be represented by an integral of the given form.

Suppose that $g(t) = \sum_{n=0}^{\infty} a_n t^{qn+p}$, $p, q, t \in \mathbb{R}$, for $|t| < \rho$ say. Then as long as the integrals exist, formally evaluating $\int_0^\infty e^{-zt} g(t)dt$ gives

$$\sum_{n=0}^{\infty} a_n \mathscr{L}(t^{qn+p}) = \sum_{n=0}^{\infty} a_n \frac{\Gamma(qn+p+1)}{z^{qn+p+1}}$$

This formal procedure is not valid in general since the resulting series may diverge. However, the following result shows that this procedure produces the asymptotic series for the given integral, under certain conditions.

Theorem 10.23. Watson's Lemma

Let $g(t) = t^p h(t^q)$, $t, p, q \in \mathbb{R}$, with $p > -1$, $q > 0$ and $q \geqslant -p$, where h is continuous for $t \geqslant 0$ and analytic in a neighbourhood of the origin, so that $g(t) = t^p \sum_{n=0}^{\infty} a_n t^{qn}$, for $0 < |t| < \rho$ say. Suppose that for $t \geqslant \rho > 0$, $|g(t)| < be^{ct}$ for some positive real constants b and c, independent of t. Then

$$f(z) = \int_0^\infty e^{-zt} g(t)dt \sim \sum_{n=0}^{\infty} a_n \frac{\Gamma(qn+p+1)}{z^{qn+p+1}}$$

for $|\text{Arg}\, z| \leqslant \pi/2 - \delta$, for any $0 < \delta < \pi/2$. □

Proof

Step 1

It follows by Theorems 10.15 and 10.16, as in Example 10.15(ii), that $f(z)$ exists and is analytic on $\text{Re}\, z > c$. Integrating termwise gives

$$f(z) = \sum_{n=0}^{k} a_n \int_0^\infty e^{-zt} t^{qn+p} dt + R_k(z) \tag{10.21}$$

$$\text{where} \quad R_k(z) = \int_0^\infty e^{-zt} \left(g(t) - \sum_{n=0}^{k} a_n t^{qn+p} \right) dt \tag{10.22}$$

Let $z = x \in \mathbb{R}^+$ and $u = xt$. Then since $p > -1$ and $q > 0$,

$$\int_0^\infty e^{-zt} t^{qn+p} dt = x^{-qn-p-1} \int_0^\infty e^{-u} u^{qn+p} du = \frac{\Gamma(qn+p+1)}{x^{qn+p+1}} \qquad (n \in \mathbb{Z}_{\geqslant 0})$$

Then by analytic continuation, using Theorem 10.1, it follows from (10.21) that for $\text{Re}\, z > c$,

$$f(z) = \sum_{n=0}^{k} a_n \frac{\Gamma(qn+p+1)}{z^{qn+p+1}} + R_k(z) \tag{10.23}$$

Step 2

In order to prove that the result follows from (10.23) we need to show that $\lim_{k\to\infty} z^{qk+p+1}R_k(z) = 0$ for each $k \in \mathbb{Z}_{\geqslant 0}$, by using (10.22). Notice that, by hypothesis, for any fixed $k \in \mathbb{Z}_{\geqslant 0}$,

$$\left|t^{-q(k+1)-p}\left(g(t) - \sum_{n=0}^{k} a_n t^{qn+p}\right)\right| = \left|\sum_{n=k+1}^{\infty} a_n t^{q(n-k-1)}\right| \qquad (0 \leqslant t < \rho)$$

and so is bounded for $0 \leqslant t < \rho$. Also by hypothesis,

$$\left|e^{-ct}t^{-q(k+1)-p}\left(g(t) - \sum_{n=0}^{k} a_n t^{qn+p}\right)\right| \leqslant bt^{-q(k+1)-p} + e^{-ct}\left|\sum_{n=0}^{k} a_n t^{q(n-k-1)}\right|$$

for $t \geqslant \rho$ and so is bounded for $t \geqslant \rho$. Hence, altogether, for any fixed $k \in \mathbb{Z}_{\geqslant 0}$, there exists a positive real number M such that

$$|g(t) - \sum_{n=0}^{k} a_n t^{qn+p}| \leqslant Me^{ct}t^{q(k+1)+p} \qquad (t \geqslant 0) \tag{10.24}$$

Step 3

It follows from (10.22) and (10.24) that

$$|R_k(z)| \leqslant M\int_0^\infty |e^{-zt}|e^{ct}t^{q(k+1)+p}dt = M\int_0^\infty e^{(c-x)t}t^{q(k+1)+p}dt$$

for fixed $k \in \mathbb{Z}_{\geqslant 0}$, where $x = \operatorname{Re} z$. Letting $u = (x - c)t$ gives

$$|R_k(z)| \leqslant M(x-c)^{-q(k+1)-p-1}\int_0^\infty e^{-u}u^{q(k+1)+p}du = \frac{M\Gamma(qk+q+p+1)}{(x-c)^{qk+q+p+1}}$$

for $x > c$. If $|\operatorname{Arg} z| \leqslant \pi/2 - \delta$, then $x > |z|\cos(\pi/2 - \delta)$ so that $x > |z|\sin\delta$. Then

$$0 \leqslant |z^{qk+p+1}R_k(z)| \leqslant \frac{M\Gamma(qk+q+p+1)}{|z|^q(\sin\delta - c/|z|)^{qk+q+p+1}} \to 0 \quad \text{as } z \to \infty$$

Hence $\lim_{z\to\infty} z^{qk+p+1}R_k(z) = 0$ for any $k \in \mathbb{Z}_{\geqslant 0}$, as required. ■

Example 10.20

The function $K_0(z)$, the **modified Bessel function of the second kind of order 0**, can be defined by

$$K_0(z) = \int_0^\infty e^{-z\cosh u}du \qquad (\operatorname{Re} z > 0)$$

In order to reduce the given integral to a type for which Watson's lemma applies, we let $\cosh u = t + 1$. Then

$$K_0(z) = e^{-z}\int_0^\infty e^{-zt}((t+1)^2 - 1)^{-1/2}dt$$

$$\Rightarrow \quad K_0(z) = \frac{e^{-z}}{\sqrt{2}}\int_0^\infty e^{-zt}t^{-1/2}(1 + t/2)^{-1/2}dt \tag{10.25}$$

Comparing (10.25) with the integral in Watson's lemma, in this case, $g(t) = t^{-1/2}(1+t/2)^{-1/2}$ so that by the binomial series,

$$g(t) = t^{-1/2}\left(1 + \sum_{n=1}^{\infty} \frac{(-1)^n(1/2)(3/2)(5/2)\ldots((2n-1)/2)}{n!}\left(\frac{t}{2}\right)^n\right)$$

for $|t| < 2$, and $|g(t)| < t^{-1/2} < 1 < e^{ct}$ for any $c > 0$, for $t \geqslant 2$. Then the conditions of Watson's lemma are satisfied, so

$$K_0(z) \sim \frac{e^{-z}}{\sqrt{2}}\left(\frac{\Gamma(1/2)}{z^{1/2}} + \sum_{n=1}^{\infty} \frac{(-1)^n 1.3.5\ldots(2n-1)}{2^{2n}n!} \cdot \frac{\Gamma(n+1/2)}{z^{n+1/2}}\right)$$

for $|\mathrm{Arg}\, z| < \pi/2$. From the duplication formula for the gamma function,

$$\Gamma(n+1/2)2^n = \frac{\sqrt{\pi}(2n-1)!}{2^{n-1}(n-1)!} = \sqrt{\pi}\,(1.3.5\ldots(2n-1)) \qquad (n \geqslant 1)$$

$$\Rightarrow \quad K_0(z) \sim \frac{e^{-z}}{\sqrt{2\pi z}} \sum_{n=0}^{\infty} \frac{(-1)^n(\Gamma(n+1/2))^2}{n!(2z)^n} \qquad (|\mathrm{Arg}\, z| < \pi/2)$$

Example 10.21

Watson's lemma can be used to find the asymptotic expansion of the gamma function. In this example, we give an idea of how this is done. Letting $z = x \in \mathbb{R}^+$ and substituting $t = ux$ in the integral representation of $\Gamma(z+1)$, given by (10.13),

$$\Gamma(x) = \frac{\Gamma(1+x)}{x} = \frac{1}{x}\int_0^{\infty} e^{-t}t^x\,dt = x^x e^{-x}\int_0^{\infty} (ue^{1-u})^x\,du$$

It then follows by Theorem 10.1 that in general

$$\Gamma(z) = z^z e^{-z}\left(\int_0^1 (ue^{1-u})^z\,du + \int_1^{\infty} (ve^{1-v})^z\,dv\right) \qquad (\mathrm{Re}\, z > 0)$$

Letting $e^{-t} = ue^{1-u}$, $0 \leqslant u \leqslant 1$, in the first integral, and $e^{-t} = ve^{1-v}$, $1 \leqslant v$, in the second gives

$$\Gamma(z) = z^z e^{-z}\int_0^{\infty} e^{-zt}(v'(t) - u'(t))\,dt \tag{10.26}$$

and this is an integral of the form where Watson's lemma may be applied. Now $u(t) = 1 + y_1(t)$ and $v(t) = 1 + y_2(t)$, where y_1 and y_2 are the two solutions to

$$t = y - \mathrm{Log}\,(1+y) \quad \Rightarrow \quad r^2 = 2\sum_{n=2}^{\infty} \frac{(-1)^n y^n}{n} \qquad (|y| < 1)$$

letting $r^2 = 2t$. Then supposing that $y = \sum_{n=0}^{\infty} a_n r^n$ and comparing coefficients of r^k, for $k \leqslant 6$ say, gives

$$y(r) = r + \frac{r^2}{3} + \frac{r^3}{36} - \frac{r^4}{270} + \frac{r^5}{4320} + \ldots$$

It can be shown that this series converges in some neighbourhood of the origin. Then $y_2(t) = y(\sqrt{2t})$ and $y_1(t) = y(-\sqrt{2t})$. Hence

$$v'(t) - u'(t) = y'(\sqrt{2t}) - y'(-\sqrt{2t})$$

$$\Rightarrow \quad v'(t) - u'(t) = \frac{1}{\sqrt{2t}}\left(2 + \frac{t}{3} + \frac{t^2}{108} + \ldots\right) \tag{10.27}$$

It can be shown that the conditions of Watson's lemma are satisfied, and so from (10.26) and (10.27), it follows that

$$\Gamma(z) \sim \frac{e^{-z}z^z}{\sqrt{2}}\left(\frac{2\Gamma(1/2)}{z^{1/2}} + \frac{\Gamma(3/2)}{3z^{3/2}} + \frac{\Gamma(5/2)}{108z^{5/2}} + \ldots\right)$$

$$\Rightarrow \quad \Gamma(z) \sim \frac{e^{-z}z^z\Gamma(1/2)}{\sqrt{2z}}\left(2 + \frac{1}{6z} + \frac{1}{144z^2} + \ldots\right)$$

$$\Rightarrow \quad \Gamma(z+1) \sim e^{-z}z^z\sqrt{2\pi z}\left(1 + \frac{1}{12z} + \frac{1}{288z^2} + \ldots\right) \qquad (|\operatorname{Arg} z| < \pi/2)$$

This result shows that for large $n \in \mathbb{N}$,

$$n! \approx e^{-n}n^n\sqrt{2\pi n}$$

which is **Stirling's asymptotic formula.**

Exercise **10.4.1** Use the definition to find the asymptotic expansions of $e^{1/x}$ and e^{-x}, $x \in \mathbb{R}$. Hence use Theorem 10.22 to show that $e^{1/x}$ and $e^{1/x} + e^{-x}$ have the same asymptotic expansions.

Exercise **10.4.2** Let $f(z) = \int_z^\infty e^{-\zeta}\zeta^{-m}d\zeta$, where $m \in \mathbb{N}$ and $\operatorname{Re} z > 0$. Show that

$$f(z) = e^{-z}\left(\frac{1}{z^m} + \sum_{n=1}^{k}\frac{(-1)^n m(m+1)\ldots(m+n-1)}{z^{m+n}} + R_k(z)\right)$$

where $R_k(z) = (-1)^{k+1}m(m+1)\ldots(m+k)e^z\int_z^\infty \frac{e^{-\zeta}}{\zeta^{m+k+1}}d\zeta \qquad (k \geqslant 0)$

Hence use the definition of an asymptotic series to show that

$$f(z) \sim e^{-z}\left(\frac{1}{z^m} + \sum_{n=1}^{\infty}\frac{(-1)^n m(m+1)\ldots(m+n-1)}{z^{m+n}}\right)$$

Show that the given series diverges for all z by the ratio test.

Exercise **10.4.3** Prove that if $f(z) \sim \sum_{n=0}^\infty a_n z^{-n}$ and $\alpha \in \mathbb{C}$ then $\alpha f(z) \sim \sum_{n=0}^\infty \alpha a_n z^{-n}$.

Exercise **10.4.4** Prove that if $f(z) \sim \sum_{n=0}^\infty a_n z^{-n}$ and $g(z) \sim \sum_{n=0}^\infty b_n z^{-n}$ in the same range of values of $\operatorname{Arg} z$, then $f(z)g(z) \sim \sum_{n=0}^\infty c_n z^{-n}$, where $c_n = a_0 b_n + a_1 b_{n-1} + \ldots + a_n b_0$.

Exercise **10.4.5** Prove that if for $x \in \mathbb{R}^+$, $f(x) \sim \sum_{n=2}^{\infty} a_n/x^n$, then

$$\int_x^{\infty} f(t)dt \sim \sum_{n=1}^{\infty} \frac{a_{n+1}}{nx^n}$$

Exercise **10.4.6** Find the asymptotic series for

$$f(z) = \int_0^{\infty} \frac{e^{-zt}}{1+t^2} dt \qquad (\operatorname{Re} z > 0)$$

by (i) integrating termwise and using the definition, (ii) using Watson's lemma.

Exercise **10.4.7**

(i) Use Watson's lemma to show that, for $\operatorname{Re} w > 0$,

$$f(w) = \int_0^{\infty} \frac{e^{-wt}}{1+t} dt \sim \sum_{n=0}^{\infty} \frac{(-1)^n n!}{w^{n+1}} \qquad (|\operatorname{Arg} w| < \pi/2)$$

(ii) Show that

$$\int_z^{\infty} \frac{e^{\pm i(\zeta - z)}}{\zeta} d\zeta = \int_0^{\infty} \frac{e^{\pm izt}}{1+t} dt$$

Hence use (i) to find the asymptotic expansions of

$$\int_z^{\infty} \frac{\cos(\zeta - z)}{\zeta} d\zeta \quad \text{and} \quad \int_z^{\infty} \frac{\sin(\zeta - z)}{\zeta} d\zeta$$

Exercise **10.4.8** $K_\nu(z)$, the **modified Bessel function of the second kind, of order** ν, can be defined for $\operatorname{Re} z > 0$ by

$$K_\nu(z) = \frac{\sqrt{\pi}(z/2)^\nu}{\Gamma(\nu + 1/2)} \int_1^{\infty} e^{-zu}(u^2 - 1)^{\nu - 1/2} du \qquad (\nu > -1/2)$$

Let $u = t + 1$ and apply Watson's lemma to obtain the asymptotic expansion of $K_\nu(z)$ in the form

$$K_\nu(z) \sim \left(\frac{\pi}{2z}\right)^{1/2} e^{-z} \sum_{n=0}^{\infty} \frac{\Gamma(\nu + 1/2 + n)}{\Gamma(\nu + 1/2 - n)n!(2z)^n} \qquad (|\operatorname{Arg} z| < \pi/2)$$

Exercise **10.4.9** The error function is defined and denoted by $\operatorname{erf}(z) = (2/\sqrt{\pi}) \int_0^z e^{-\zeta^2} d\zeta$. Show that

$$\operatorname{erf}(z) = 1 - \frac{2}{\sqrt{\pi}} \int_z^{\infty} e^{-\zeta^2} d\zeta = 1 - \frac{ze^{-z^2}}{\sqrt{\pi}} \int_0^{\infty} e^{-z^2 t}(1+t)^{-1/2} dt$$

Hence use Watson's lemma to show that

$$\operatorname{erf}(z) \sim 1 - \frac{e^{-z^2}}{\sqrt{\pi}} \left(\frac{1}{z} - \frac{1}{2z^3} + \frac{3}{4z^5} - \frac{15}{8z^7} + \ldots\right) \qquad (|\operatorname{Arg} z| < \pi/4)$$

Exercise **10.4.10**

If $\psi(z) = \dfrac{d}{dz}(\operatorname{Log}\Gamma(z))$ show that

$$\psi'(z+1) = \int_0^\infty e^{-zt}\left(\frac{t}{e^t - 1}\right) dt$$

for $\operatorname{Re} z > 0$. Hence use Watson's lemma to show that

$$\psi'(z+1) \sim \sum_{n=0}^{\infty} \frac{a_n}{z^{n+1}} \qquad (|\operatorname{Arg} z| < \pi/2)$$

where $$\frac{t}{e^t - 1} = \sum_{n=0}^{\infty} \frac{a_n t^n}{n!} \qquad (|t| < 2\pi)$$

Hence show that

$$\psi'(z+1) \sim \frac{1}{z} - \frac{1}{2z^2} + \frac{1}{6z^3} - \dots \qquad (|\operatorname{Arg} z| < \pi/2)$$

Exercise **10.4.11** In Example 10.21, let $y = \sum_{n=0}^{\infty} a_n r^n$ in $r^2 = 2\sum_{n=2}^{\infty}(-1)^n y^n/n$ and compare coefficients of r^k, $k \leqslant 6$, to verify that

$$y(r) = r + \frac{r^2}{3} + \frac{r^3}{36} - \frac{r^4}{270} + \frac{r^5}{4320} + \dots$$

11 Elliptic Functions

This chapter gives an introduction to elliptic functions; that is, meromorphic functions which are doubly periodic in the complex plane. Such functions possess many elegant properties which may be proved quite simply. An important special class are the Jacobian elliptic functions, which we introduce by means of a Schwarz–Christoffel transformation that maps the upper half-plane to a particular rectangular region. This approach provides the opportunity to apply some of the results of the previous chapter on analytic continuation and, in particular, the Schwarz reflection principle. These functions occur in many applications of differential equations; see Chapter 2 in P. G. Drazin and R. S. Johnson, *Solitons: An Introduction*, Cambridge University Press, 1989. Finally, we derive a number of results governing general elliptic functions and show how they can be applied in specific cases by giving a brief introduction to the Weierstrassian elliptic function.

Jacobian Elliptic Functions

Let $k \in \mathbb{R}$ with $0 < k < 1$ and consider the transformation F, defined on $\operatorname{Im} w \geqslant 0$, by

$$F(w) = \int_0^w \frac{d\zeta}{(1-\zeta^2)^{1/2}(1-k^2\zeta^2)^{1/2}} \tag{11.1}$$

Notice that $F(0) = 0$ and if $w \in \mathbb{R}$ with $w \in [-1, 1]$, then $F(w)$ is real and $F(-w) = F(w)$. Notice also that $F(1/k) - F(1)$ is purely imaginary. It follows by the Schwarz–Christoffel transformation, Theorems 6.9 to 6.11, that if $K = F(1)$ and $K + iK' = F(1/k)$, so that $K, K' \in \mathbb{R}$, then F maps $\operatorname{Im} w \geqslant 0$ and the points indicated in Fig. 11.1 to the given points and rectangular domain $\mathscr{D}$, consisting of the rectangle and its inside (see Exercises 6.4). Notice that, since $iK' = F(1/k) - F(1)$,

$$\begin{aligned} iK' &= F(1/k) - \int_0^1 \frac{d\zeta}{(1-\zeta^2)^{1/2}(1-k^2\zeta^2)^{1/2}} \\ &= F(1/k) + \int_{1/k}^{\infty} \frac{d\tau}{(1-\tau^2)^{1/2}(1-k^2\tau^2)^{1/2}} \end{aligned}$$

by letting $\tau = 1/k\zeta$. Hence $iK' = F(\infty)$ as shown.

Note that letting $k = 0$ in the integral (11.1) gives $F(w) = \sin^{-1} w$. Note also that F is an injection. This justifies the following notation and definition.

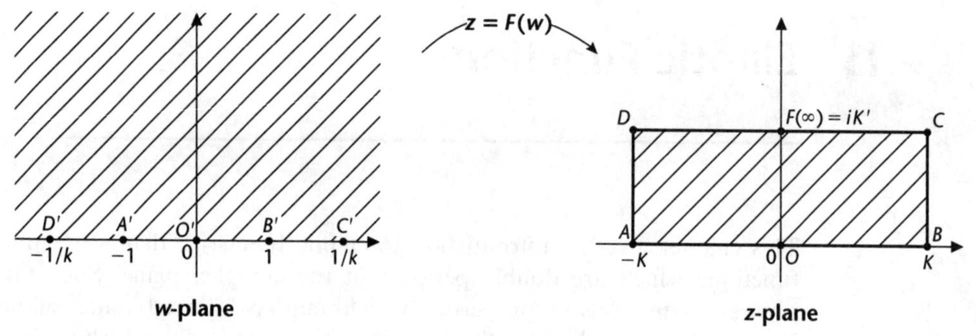

Figure 11.1

Definition

Let the domain $\mathscr{D}$ consist of the rectangle $ABCD$ and its inside (Fig. 11.1). The **Jacobian elliptic function** denoted by sn, is defined on $\mathscr{D}$ by $\text{sn} = F^{-1}$, i.e.

$$\text{sn}\,(z,k) = w \text{ if } z = F(w) \text{ for all } z \in \mathscr{D}$$

Notes

(i) It is usual to write $\text{sn}\, z$ for $\text{sn}\,(z,k)$ when there is no ambiguity. Generally $0 < k < 1$, but in the degenerate cases $k = 0$ or $k = 1$, $\text{sn}\,(z,0) = \sin z$ and $\text{sn}\,(z,1) = \tanh z$.

(ii) Function sn is analytic at all points of $\mathscr{D}$ except iK'.

Historical Note

(i) Real integrals of the type in (11.1) were first met in connection with finding the lengths of arcs of ellipses and are called **elliptic integrals of the first kind**.

(ii) Elliptic functions were first introduced by Jacobi in 1829. The notation sn was first used by Gudermann in 1838.

It follows by definition that sn maps the domain $\mathscr{D}$ to $\text{Im}\, w \geqslant 0$. In particular, we have the following values.

z	$-K + iK'$	$-K$	0	K	$K + iK'$	iK'
$\text{sn}\, z$	$-1/k$	-1	0	1	$1/k$	∞

Notation

It is usual to let $k' = \sqrt{1 - k^2}$.

This notation makes the next result, which shows the explicit relationship between K and K', particularly elegant.

Lemma 11.1. $K'(k) = K(k')$ □

Proof

$K = F(1)$ and $K + iK' = F(1/k)$, so that (11.1) gives

$$iK' = (K + iK') - K = \int_1^{1/k} \frac{d\zeta}{(1-\zeta^2)^{1/2}(1-k^2\zeta^2)^{1/2}}$$

Let $\zeta = (1 - k'^2\tau^2)^{-1/2}$. Then

$$K'(k) = \int_0^1 \frac{d\tau}{(1-\tau^2)^{1/2}(1-k'^2\tau^2)^{1/2}} = K(k').$$ ■

We can extend the domain of sn from $\mathscr{D}$ to the whole of $\mathbb{C}$ by analytic continuation using the Schwarz reflection principle. Let $\mathscr{D} = \mathscr{D}_1$ and $f_1(z) = \operatorname{sn} z$ on $\mathscr{D}_1$. Let the domain $\mathscr{D}_2$ consist of the rectangle with vertices at $\pm K$ and $-iK' \pm K$ and its inside, as shown in Fig. 11.2. Since f_1 takes real values on the boundary of $\mathscr{D}_1$ and is analytic in the region bounded by the rectangle, it follows by Theorem 10.4 that the analytic continuation of f_1 to $\mathscr{D}_2$ is given by $f_2(z) = \overline{f_1(\overline{z})}$ on $\mathscr{D}_2$. Then f_2 maps $\mathscr{D}_2$ to $\operatorname{Im} w \leqslant 0$.

Let the domain $\mathscr{D}_3$ consist of the rectangle with vertices at $iK' \pm K$ and $2iK' \pm K$, and its inside, as shown in Fig. 11.2. Let $\zeta = z - iK'$ and

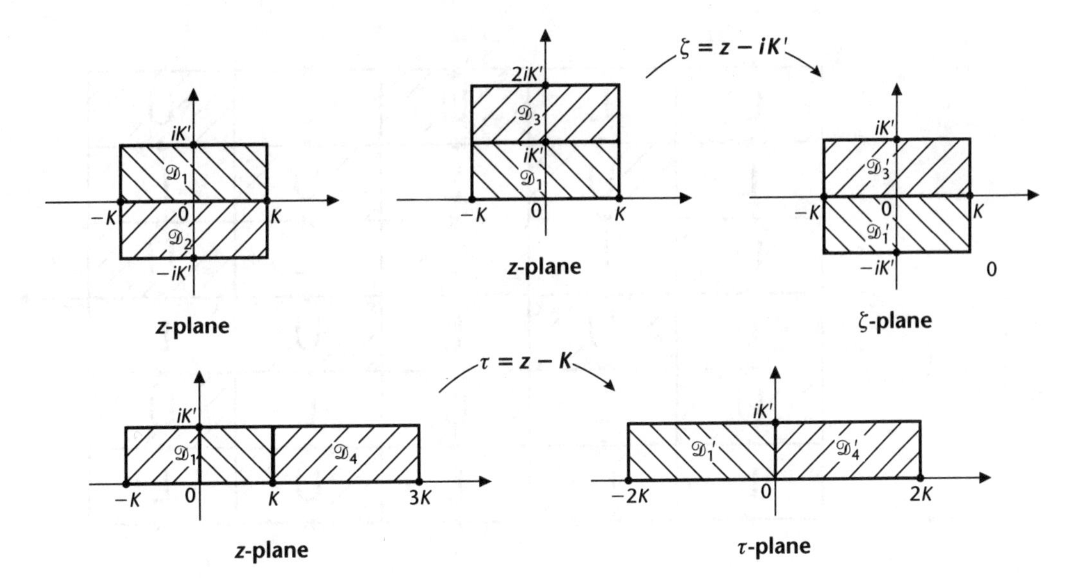

Figure 11.2

$f_1(z) = g_1(\zeta)$. Then $g_1(\zeta) = f_1(\zeta + iK')$. Since f_1 takes real values on the boundary of $\mathscr{D}_1$, it follows by 10.4 that g_1 can be analytically continued to $\mathscr{D}_3'$ in the ζ-plane, as shown, by $g_3(\zeta) = \overline{g_1(\bar{\zeta})}$ on $\mathscr{D}_3'$. Hence f_1 can be analytically continued to $\mathscr{D}_3$ in the z-plane by

$$f_3(z) = g_3(\zeta) = \overline{f_1(\bar{\zeta} + iK')} = \overline{f_1(\bar{z} + 2iK')} \quad \text{on } \mathscr{D}_3$$

and so f_3 maps $\mathscr{D}_3$ to $\operatorname{Im} w \leqslant 0$.

Now let $\mathscr{D}_4$ consist of the rectangle with vertices at K, $3K$, $K + iK'$ and $3K + iK'$ as shown in Fig. 11.2. Let $\tau = z - K$ and $h_1(\tau) = f_1(z)$ so that $h_1(\tau) = f_1(\tau + K)$. It follows by Theorem 10.4 and a rotation through $\pi/2$ radians that h_1 can be analytically continued to $\mathscr{D}_4'$ in the τ-plane by $h_4(\tau) = \overline{h_1(-\bar{\tau})}$. Hence f_1 can be continued to $\mathscr{D}_4'$ in the z-plane by

$$f_4(z) = h_4(\tau) = \overline{f_1(-\bar{\tau} + K)} = \overline{f_1(-\bar{z} + 2K)} \quad \text{on } \mathscr{D}_4.$$

Then f_4 maps $\mathscr{D}_4$ to $\operatorname{Im} w \leqslant 0$.

Continuing this process, the complete analytic function given by $f(z) = \operatorname{sn} z$ on $\mathbb{C}$, of which f_1 is an element, is easily constructed. Figure 11.3 shows some of the domain of f and indicates which rectangles are mapped to the upper half-plane (U) and which are mapped to the lower half-plane (L) by f. The zeros of f are indicated by a dot and the singular points of f are indicated by a cross.

Notice that for sn defined on $\mathbb{C}$,

$$\operatorname{sn}(z + 2iK') = \operatorname{sn}(z + 4K) = \operatorname{sn} z \tag{11.2}$$

so that sn is a doubly periodic function. Notice also that sn is an odd function.

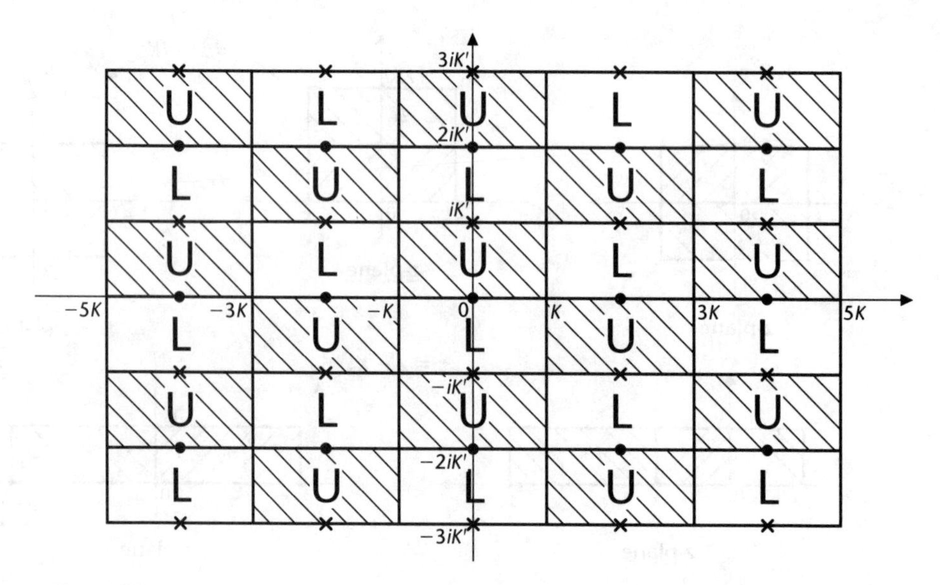

Figure 11.3

If $w = \operatorname{sn} z$ then $dw/dz = (1 - w^2)^{1/2}(1 - k^2w^2)^{1/2}$, hence if $w = 0$ then $dw/dz \neq 0$. Hence all the zeros of sn are simple. Note also that iK' is a singular point of sn which is not an essential singular point, nor a branch point since sn is an injection in some neighbourhood of iK'. It follows that iK' is a pole of sn. Notice that, by the above, if $w = \operatorname{sn} z$ and $\tau = 1/w$, then $d\tau/dz = -(\tau^2 - 1)^{1/2}(\tau^2 - k^2)^{1/2}$. Hence as $z \to iK'$, $\tau \to 0$ and $d\tau/dz$ remains finite and non-zero. Then iK' and hence all the singular points of sn are simple poles.

By partial analogy with the circular functions, we make the following definitions.

Definitions

The Jacobian elliptic functions cn and dn are defined by

$$\operatorname{cn} z = (1 - \operatorname{sn}^2 z)^{1/2} \qquad \operatorname{dn} z = (1 - k^2\operatorname{sn}^2 z)^{1/2}$$

for all $z \in \mathbb{C}$ with $\operatorname{cn} 0 = \operatorname{dn} 0 = 1$.

Note

Glaisher, in 1882, introduced a logical notation for reciprocals and quotients of Jacobian elliptic functions. For example, he defined $\operatorname{ns} z = 1/\operatorname{sn} z$, $\operatorname{sd} z = \operatorname{sn} z/\operatorname{dn} z$ and so on.

Note that cn and dn are even functions and must be doubly periodic. Figure 11.4 shows the image of the rectangular domain $\mathscr{D}$, under $w = \operatorname{dn} z$ and $w = \operatorname{cn} z$ respectively. The following example shows how the image of $\mathscr{D}$ under $w = \operatorname{dn} z$ is obtained. Its image under $w = \operatorname{cn} z$ is left as an exercise.

Example 11.1

Let $\zeta = \operatorname{sn} z$ so that $w = \operatorname{dn} z = g(\zeta) = (1 - k^2\zeta^2)^{1/2}$. By previous analysis, $\zeta = \operatorname{sn} z$ maps the rectangle $ABCD$ to the real axis and its inside to the upper half-plane, as shown in Fig. 11.5. Note that $g(0) = 1$, $g(\pm 1) = \sqrt{1 - k^2} = k'$ and $g(\pm 1/k) = 0$. Now $w = g(\zeta) = ik(\zeta + 1/k)^{1/2}(\zeta - 1/k)^{1/2}$. Let $\zeta - 1/k = r_1 e^{i\theta_1}$ and $\zeta + 1/k = r_2 e^{i\theta_2}$. Then $g(\zeta) = ik\sqrt{r_1 r_2}\exp(\frac{1}{2}i(\theta_1 + \theta_2) + in\pi)$, where $n = 0$ or 1. When $\zeta = 0$, $\theta_1 = \pi$, $\theta_2 = 0$ and $r_1 = r_2 = 1/k$, and since $g(0) = 1$, $n = 1$. Hence

$$w = g(\zeta) = -ik\sqrt{r_1 r_2}\exp(\tfrac{1}{2}i(\theta_1 + \theta_2)) \tag{11.3}$$

Suppose, first of all, that $\zeta \in \mathbb{R}$. For $\zeta < -1/k$, $\theta_1 = \theta_2 = \pi$ so that $w = ik\sqrt{r_1 r_2}$. Hence $w = iv$ say, where $v \in \mathbb{R}^+$. For $-1/k < \zeta < 1/k$, $\theta_2 = 0$ and $\theta_1 = \pi$ so $w = k\sqrt{r_1 r_2} = u$ say, where $u \in \mathbb{R}$ and $u \in (0, 1]$. For $\zeta > 1/k$,

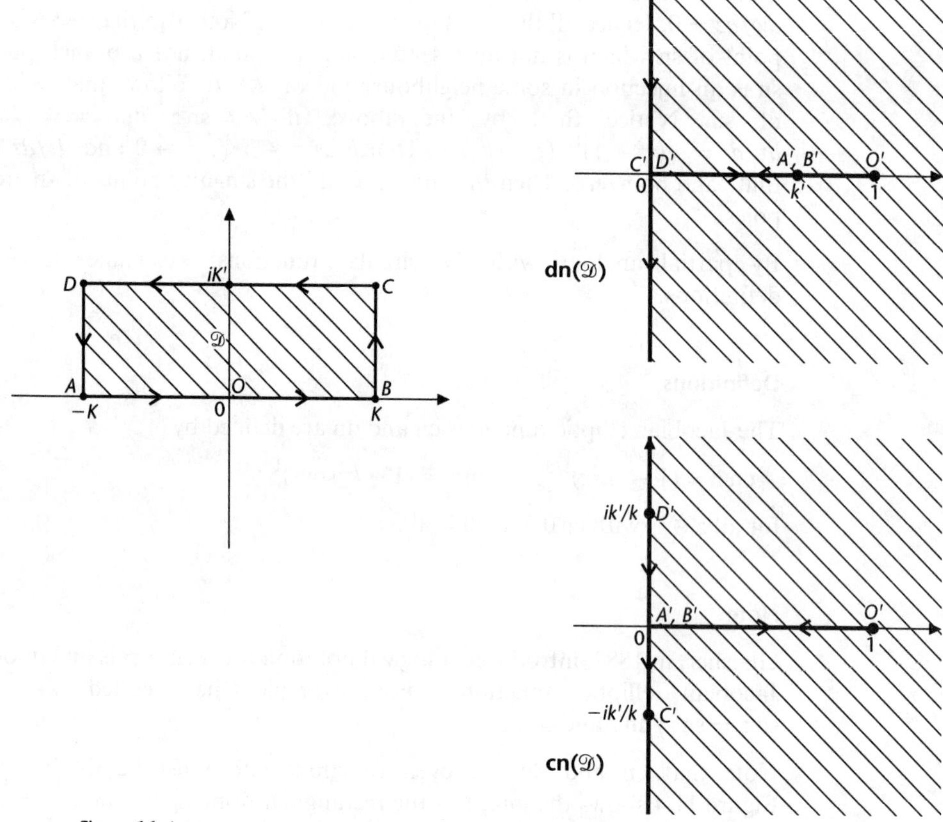

Figure 11.4

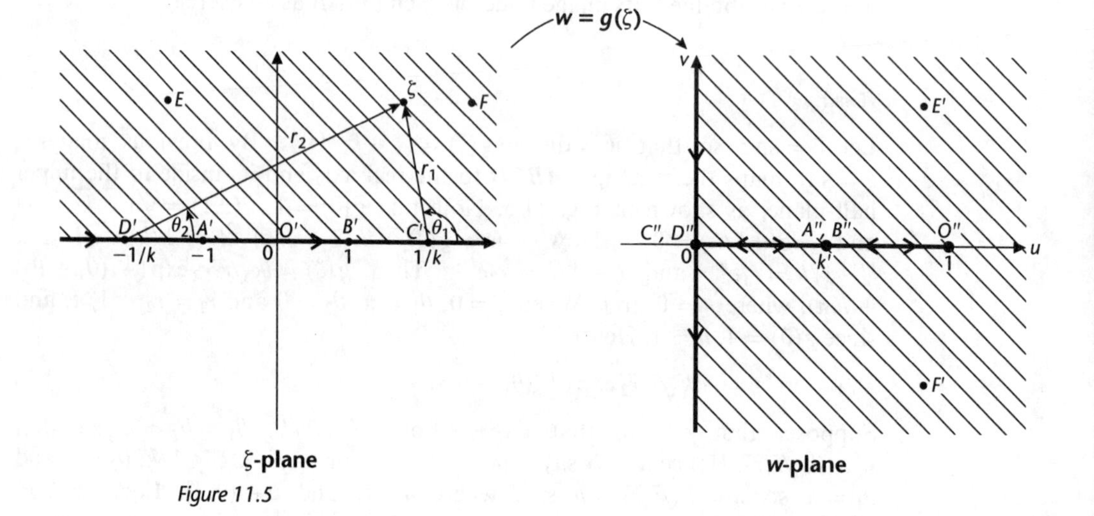

Figure 11.5

$\theta_1 = \theta_2 = 0$ and so $w = -ik\sqrt{r_1 r_2} = -iv$, where $v \in \mathbb{R}^+$, as shown. Now let $\zeta \in \mathbb{C}\backslash\mathbb{R}$ and lie in the first or second quadrant. It follows by (11.3) that

$$w = u + iv = g(\zeta) = k\sqrt{r_1 r_2}\left(-i\cos\left(\frac{\theta_1 + \theta_2}{2}\right) + \sin\left(\frac{\theta_1 + \theta_2}{2}\right)\right)$$

At any point E in the second quadrant, $\pi < \theta_1 + \theta_2 < 2\pi$, so that $u, v > 0$. At any point F in the first quadrant, $0 < \theta_1 + \theta_2 < \pi$, so that $u > 0$ and $v < 0$. Hence g maps the second quadrant to the first quadrant and the first quadrant to the fourth quadrant, as shown in Fig. 11.5.

The analytic continuation of cn and dn to $\mathbb{C}$ is obtained using Schwarz's reflection principle as in the case of sn. It is straightforward to show that under the same conditions as Theorem 10.4, except that f_1 takes purely imaginary values on AB, the analytic continuation of f_1 to $\mathcal{R}_2$ is $f_2(z) = \overline{-f_1(\bar{z})}$. This is left as an exercise. Also, under the same conditions as 10.4, except that AB lies on the imaginary axis and f_1 takes purely imaginary values on AB, the analytic continuation of f_1 to $\mathcal{R}_2$ is $f_2(z) = \overline{-f_1(-\bar{z})}$ (see Exercise 10.1.14). Notice that dn takes real values on the sides DA, AB, and BC of the rectangle $ABCD$ and purely imaginary values on CD; whereas cn takes real values on AB and purely imaginary values on DA, BC and CD.

Using the same approach as for sn, the complete analytic function dn, defined on $\mathbb{C}$, maps the square regions shown in Fig. 11.6 to the first, second, third or fourth quadrants, labelled 1, 2, 3 and 4, of the w-plane. Since iK' is a simple pole of sn, it is also a simple pole of dn. Once again, a dot denotes a zero and a cross denotes a pole.

Notice that dn is doubly periodic with

$$\text{dn}\,(z + 2K) = \text{dn}\,(z + 4ik') = \text{dn}\,z \tag{11.4}$$

In the same way, the complete analytic function cn maps the square regions shown in Fig. 11.7 to the first, second, third and fourth quadrants of the w-plane, as shown. Notice that iK' is a simple pole of cn and that cn is doubly periodic with

$$\text{cn}\,(z + 4K) = \text{cn}(z + 2K + 2iK') = \text{cn}\,z \tag{11.5}$$

The derivation of both these diagrams is left as an exercise.

Having defined sn, cn and dn on $\mathbb{C}$, we now derive some standard results involving these functions, starting with their derivatives.

Lemma 11.2 Differentiation Formulae

$$\frac{d}{dz}(\text{sn}\,z) = \text{cn}\,z\,\text{dn}\,z \qquad \frac{d}{dz}(\text{cn}\,z) = -\text{sn}\,z\,\text{dn}\,z \qquad \frac{d}{dz}(\text{dn}\,z) = -k^2\text{sn}\,z\,\text{cn}\,z$$

at all points where the derivatives exist. □

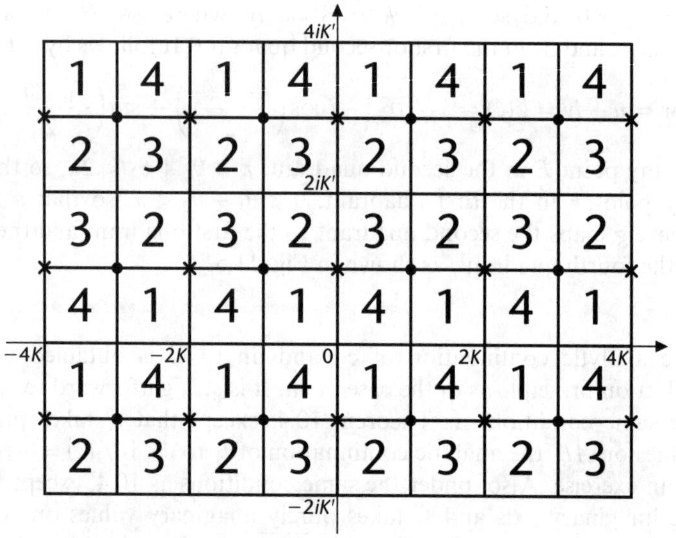

4iK'

1	4	1	4	1	4	1	4
2	3	2	3	2	3	2	3
3	2	3	2	3	2	3	2
4	1	4	1	4	1	4	1
1	4	1	4	1	4	1	4
2	3	2	3	2	3	2	3

2iK'
−4K
−2K
0
2K
4K
−2iK'

Figure 11.6

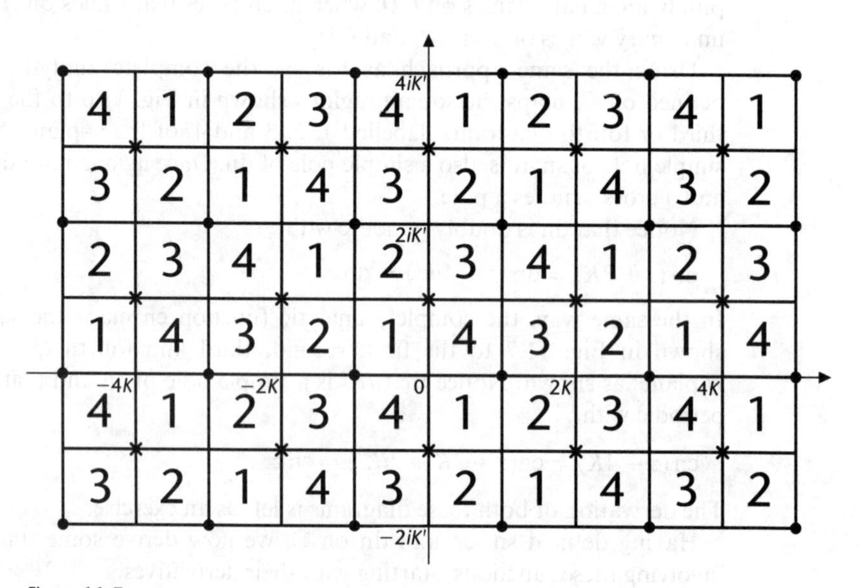

Figure 11.7

Proof

$$w = \operatorname{sn} z \text{ with } z \in \mathscr{D}_1 \quad \Rightarrow \quad z = \int_0^w \frac{d\zeta}{(1-\zeta^2)^{1/2}(1-k^2\zeta^2)^{1/2}}$$

$$\Rightarrow \quad \frac{dz}{dw} = \frac{1}{(1-w^2)^{1/2}(1-k^2w^2)^{1/2}}$$

$$\Rightarrow \quad \frac{dw}{dz} = (1-\operatorname{sn}^2 z)^{1/2}(1-k^2\operatorname{sn}^2 z)^{1/2} = \operatorname{cn} z \operatorname{dn} z$$

The other two results then follow from this result and the definitions (exercise). ■

Note

The differential equations satisfied by the Jacobian elliptic functions are easily constructed using 11.2. For instance, $w = \operatorname{cn} z$ is a solution to

$$\left(\frac{dw}{dz}\right)^2 = (1-w^2)(1-k^2(1-w^2))$$

These functions play an important role in the solution of first-order equations of the second degree and arise in many applications involving non-linear o.d.e.'s.

The Taylor series expansions of sn, cn and dn about any point at which they are analytic, are easily calculated using Lemma 11.2. In particular, these functions have Taylor series expansions about the origin.

Example 11.2

Let $f(z) = \operatorname{sn} z$, $z \in \mathbb{C}$. Recall that f has a simple pole at iK' and hence simple poles at $iK' + 2niK'$ and $iK' + 2nK$, $n \in \mathbb{Z}$. Hence the Taylor series expansion of f about 0 is valid for $|z| < K'$. It follows from 11.2 that

$$f'(z) = \operatorname{cn} z \operatorname{dn} z \quad \Rightarrow \quad f''(z) = -\operatorname{sn} z(\operatorname{dn}^2 z + k^2\operatorname{cn}^2 z) = -\operatorname{sn} z\, g(z)$$

say, where $g'(z) = -4k^2 \operatorname{sn} z \operatorname{cn} z \operatorname{dn} z$. Then

$$f'''(z) = \operatorname{cn} z \operatorname{dn} z(4k^2\operatorname{sn}^2 z - g(z)) = \operatorname{cn} z \operatorname{dn} z\, h(z)$$

say, where $h'(z) = 12k^2 \operatorname{sn} z \operatorname{cn} z \operatorname{dn} z$. Then

$$f^{(iv)}(z) = 12k^2 \operatorname{sn} z \operatorname{cn}^2 z \operatorname{dn}^2 z - \operatorname{sn} z\, g(z)h(z) = \operatorname{sn} z\, \phi(z) - \operatorname{sn} z\, g(z)\, h(z)$$

$$\Rightarrow \quad f^{(v)}(z) = \operatorname{sn} z(\phi'(z) - (g(z)h(z))') + \operatorname{cn} z \operatorname{dn} z(\phi(z) - g(z)\, h(z))$$

Hence, $f(0) = 0, f'(0) = 1, f''(0) = 0, f'''(0) = h(0) = -(1+k^2), f^{(iv)}(0) = 0$ and $f^{(v)}(0) = \phi(0) - g(0)h(0) = 12k^2 + (1+k^2)^2$. Then

$$\operatorname{sn} z = z - (1+k^2)\, z^3/3! + (1 + 14k^2 + k^4)\, z^5/5! + \ldots \qquad (|z| < K')$$

This reduces to the Maclaurin series expansion for $\sin z$ when $k = 0$ and the Maclaurin series for $\tanh z$ when $k = 1$.

As in the case of the circular and hyperbolic functions, it is possible to derive addition formulae for the Jacobian elliptic functions.

Theorem 11.3. Addition Formulae

(i) $$\operatorname{sn}(z_1+z_2)=\frac{\operatorname{sn} z_1 \operatorname{cn} z_2 \operatorname{dn} z_2+\operatorname{sn} z_2 \operatorname{cn} z_1 \operatorname{dn} z_1}{D(z_1,z_2)}$$

(ii) $$\operatorname{cn}(z_1+z_2)=\frac{\operatorname{cn} z_1 \operatorname{cn} z_2-\operatorname{sn} z_1 \operatorname{sn} z_2 \operatorname{dn} z_1 \operatorname{dn} z_2}{D(z_1,z_2)}$$

(iii) $$\operatorname{dn}(z_1+z_2)=\frac{\operatorname{dn} z_1 \operatorname{dn} z_2-k^2 \operatorname{sn} z_1 \operatorname{sn} z_2 \operatorname{cn} z_1 \operatorname{cn} z_2}{D(z_1,z_2)}$$

for any points z_1 and z_2 which are not poles, where $D(z_1,z_2)=1-k^2\operatorname{sn}^2 z_1 \operatorname{sn}^2 z_2$. □

Proof

To prove result (i), let $w_k=\operatorname{sn} z_k$, $k=1, 2$. For simplicity of notation let a prime denote differentiation with respect to z_1. Then

$$w_1'^2=(1-w_1{}^2)(1-k^2w_1{}^2) \quad\Rightarrow\quad w_1''=2k^2w_1{}^3-(1+k^2)w_1 \tag{11.6}$$

by Lemma 11.2, and similarly for $w_2(z_2)$. Now let $z_1+z_2=\alpha$, a constant. Then

$$w_2''=2k^2w_2{}^3-(1+k^2)w_2 \tag{11.7}$$

Multiplying (11.6) by w_2 and (11.7) by w_1 then subtracting gives

$$w_2w_1''-w_1w_2''=2k^2w_1w_2(w_1{}^2-w_2{}^2) \tag{11.8}$$

Also $w_2{}^2w_1'^2-w_1{}^2w_2'^2=(w_2{}^2-w_1{}^2)(1-k^2w_1{}^2w_2{}^2)$

$$\Rightarrow\quad w_2w_1'-w_1w_2'=\frac{(w_2{}^2-w_1{}^2)(1-k^2w_1{}^2w_2{}^2)}{w_2w_1'+w_1w_2'} \tag{11.9}$$

Dividing (11.8) by (11.9) gives

$$\frac{w_2w_1''-w_1w_2''}{w_2w_1'-w_1w_2'}=\frac{-2k^2w_1w_2(w_2w_1'+w_1w_2')}{1-k^2w_1{}^2w_2{}^2}$$

Notice that on each side of this equation, the numerator is the derivative of the denominator. It follows that

$$\operatorname{Log}(w_2w_1'-w_1w_2')=\operatorname{Log}(1-k^2w_1{}^2w_2{}^2)+\operatorname{Log} C$$

where C is a constant. Substituting for w_1 and w_2 gives

$$\operatorname{sn} z_2 \operatorname{cn} z_1 \operatorname{dn} z_1+\operatorname{sn} z_1 \operatorname{cn} z_2 \operatorname{dn} z_2=C(1-k^2\operatorname{sn} z_1^2 \operatorname{sn} z_2^2) \tag{11.10}$$

Since C and $z_1 + z_2$ are constants, $C = \phi(z_1 + z_2)$ for some choice of ϕ. When $z_2 = 0$, (11.10) reduces to $\operatorname{sn} z_1 = C = \phi(z_1)$ and so $\phi(z_1 + z_2) = \operatorname{sn}(z_1 + z_2)$ as required. This proof is constructive; the following alternative proof is more elegant, but non-constructive.

$$\text{Let}\quad F(z, w) = \frac{\operatorname{sn} z \operatorname{cn} w \operatorname{dn} w + \operatorname{sn} w \operatorname{cn} w \operatorname{dn} z}{D(z, w)}$$

It is straightforward to show that $\partial F/\partial z = \partial F/\partial w$ and this is left as an exercise. Then letting $G(z, w) = z + w$ gives

$$\begin{vmatrix} \dfrac{\partial F}{\partial z} & \dfrac{\partial F}{\partial w} \\ & \\ \dfrac{\partial G}{\partial z} & \dfrac{\partial G}{\partial w} \end{vmatrix} = 0$$

so that F and G are dependent and $F = H(G)$ say, so that $F(z, w) = H(z + w)$. When $w = 0$, $F(z) = \operatorname{sn} z$, so that $F(z, w) = \operatorname{sn}(z + w)$ as required.

Results (ii) and (iii) can be proved using similar techniques, or by using (i) and their definitions (exercise). ■

It is useful to introduce the following standard notation when using the addition formulae.

Notation

Let $s = \operatorname{sn} z$, $c = \operatorname{cn} z$ and $d = \operatorname{dn} z$. Let $s_k = \operatorname{sn} z_k$, $c_k = \operatorname{cn} z_k$ and $d_k = \operatorname{dn} z_k$ for any $k \in \mathbb{N}$.

Expressions are often simplified by reducing terms to those involving sn only.

Example 11.3

(i) It follows by 11.3 that

$$\operatorname{cn} 2z = \frac{c^2 - s^2 d^2}{1 - k^2 s^4} \quad \text{and} \quad \operatorname{dn} 2z = \frac{d^2 - k^2 s^2 c^2}{1 - k^2 s^4}$$

$$\Rightarrow \quad \frac{1 - \operatorname{cn} 2z}{1 + \operatorname{dn} 2z} = \frac{(1 - k^2 s^4) - (1 - s^2 - s^2(1 - k^2 s^2))}{(1 - k^2 s^4) + (1 - k^2 s^2 - k^2 s^2(1 - s^2))}$$

$$\Rightarrow \quad \frac{1 - \operatorname{cn} 2z}{1 + \operatorname{dn} 2z} = \frac{2s^2(1 - k^2 s^2)}{2(1 - k^2 s^2)} = s^2 = \operatorname{sn}^2 z$$

(ii) By the addition formulae,

$$\operatorname{sn}(z_1 + z_2) = \frac{s_1 c_2 d_2 + s_2 c_1 d_1}{D(z_1, z_2)} \quad \text{and} \quad \operatorname{sn}(z_1 - z_2) = \frac{s_1 c_2 d_2 - s_2 c_1 d_1}{D(z_1, z_2)}$$

$$\Rightarrow \quad \operatorname{sn}(z_1+z_2)\operatorname{sn}(z_1-z_2) = \frac{s_1^2c_2^2d_2^2 - s_2^2c_1^2d_1^2}{D^2(z_1,z_2)}$$

$$= \frac{s_1^2(1-s_2^2)(1-k^2s_2^2) - s_2^2(1-s_1^2)(1-k^2s_1^2)}{D^2(z_1,z_2)}$$

$$= \frac{D(z_1,z_2)(s_1^2-s_2^2)}{D^2(z_1,z_2)}$$

$$\Rightarrow \quad \operatorname{sn}(z_1+z_2)\operatorname{sn}(z_1-z_2) = \frac{\operatorname{sn}^2 z_1 - \operatorname{sn}^2 z_2}{1-k^2\operatorname{sn}^2 z_1\operatorname{sn}^2 z_2}$$

The following result, first discovered by Jacobi, is useful for finding certain values of the elliptic functions.

Theorem 11.4. Jacobi's Imaginary Transformation

(i) $\operatorname{sn}(iz,k) = \dfrac{i\operatorname{sn}(z,k')}{\operatorname{cn}(z,k')}$

(ii) $\operatorname{cn}(iz,k) = \dfrac{1}{\operatorname{cn}(z,k')}$

(iii) $\operatorname{dn}(iz,k) = \dfrac{\operatorname{dn}(z,k')}{\operatorname{cn}(z,k')}$ □

Proof

Let $w = f(z)$, where f is a function to be determined such that $f(0) = 0$ and

$$\operatorname{sn}(w,k) = \frac{i\operatorname{sn}(z,k')}{\operatorname{cn}(z,k')} \tag{11.11}$$

$$\text{Then} \quad \operatorname{cn}^2 w = \frac{\operatorname{cn}^2 z + \operatorname{sn}^2 z}{\operatorname{cn}^2 z} = \frac{1}{\operatorname{cn}^2 z}$$

and since $w = 0$ when $z = 0$,

$$\operatorname{cn}(w,k) = \frac{1}{\operatorname{cn}(z,k')} \tag{11.12}$$

$$\text{Also} \quad \operatorname{dn}^2 w = \frac{\operatorname{cn}^2 z + k^2\operatorname{sn}^2 z}{\operatorname{cn}^2 z} = \frac{1-k'^2\operatorname{sn}^2(z,k')}{\operatorname{cn}^2(z,k')} = \frac{\operatorname{dn}^2(z,k')}{\operatorname{cn}^2(z,k')}$$

$$\Rightarrow \quad \operatorname{dn}(w,k) = \frac{\operatorname{dn}(z,k')}{\operatorname{cn}(z,k')} \tag{11.13}$$

Differentiating (11.12) with respect to z gives

$$-f'(z)\,\mathrm{sn}\,(w,k)\,\mathrm{dn}\,(w,k) = \frac{\mathrm{sn}\,(z,k')\,\mathrm{dn}\,(z,k')}{\mathrm{cn}^2(z,k')}$$

$$\Rightarrow \quad -if'(z) = 1 \quad \Rightarrow \quad f(z) = iz$$

using (11.11) and (11.13) and the fact that $f(0) = 0$. Then $w = iz$ and the results follow from (11.11) to (11.13). ■

Example 11.4

It follows from the addition formulae that

$$\mathrm{sn}\,(z+K',k') = \frac{\mathrm{sn}\,(z,k')\mathrm{cn}\,(K',k')\mathrm{dn}\,(K',k') + \mathrm{sn}\,(K',k')\mathrm{cn}\,(z,k')\mathrm{dn}\,(z,k')}{1 - k'^2\,\mathrm{sn}^2(z,k')\,\mathrm{sn}^2(K',k')}$$

Now from Theorem 11.4, $\mathrm{cn}\,(K',k') = 0$, so that $\mathrm{sn}\,(K',k') = 1$ and $\mathrm{dn}\,(K',k') = k$. Hence $\mathrm{sn}\,(z+K',k') = \mathrm{cn}\,(z,k')/\mathrm{dn}\,(z,k')$ and, similarly,

$$\mathrm{cn}\,(z+K',k') = \frac{-k\,\mathrm{sn}\,(z,k')}{\mathrm{dn}\,(z,k')} \qquad \mathrm{dn}\,(z+K',k') = \frac{k}{\mathrm{dn}\,(z,k')}$$

Then using 11.4(i) twice,

$$\mathrm{sn}\,(i(z+K'),k) = \frac{i\,\mathrm{sn}\,(z+K',k')}{\mathrm{cn}\,(z+K',k')} = \frac{\mathrm{cn}\,(z,k')}{ik\,\mathrm{sn}\,(z,k')} = \frac{1}{k\,\mathrm{sn}\,(iz,k)}$$

Similarly,

$$\mathrm{cn}\,(i(z+K'),k) = \frac{-i\,\mathrm{dn}\,(iz,k)}{k\,\mathrm{sn}\,(iz,k)} \qquad \mathrm{dn}\,(i(z+K'),k) = \frac{-i\,\mathrm{cn}\,(iz,k)}{\mathrm{sn}\,(iz,k)}$$

Replacing iz by z gives

$$\mathrm{sn}\,(z+iK') = \frac{1}{k\,\mathrm{sn}\,z} \qquad \mathrm{cn}\,(z+iK') = \frac{-i\,\mathrm{dn}\,z}{k\,\mathrm{sn}\,z} \qquad \mathrm{dn}\,(z+iK') = \frac{-i\,\mathrm{cn}\,z}{\mathrm{sn}\,z}$$

The residue at any pole of sn, cn or dn can easily be calculated from these formulae, which cannot be derived from 11.3 directly.

Using the definitions, addition formulae, analytic continuation and Jacobi's imaginary transformation, it is possible to build up a list of standard values of sn, cn and dn. Table 11.1 summarises the important results.

Table 11.1

	0	$-z$	$z+K$	$z+2K$	$z+iK'$	$z+2iK'$	$z+K+iK'$	$z+2K+2iK'$
sn	0	$-\mathrm{sn}\,z$	$\dfrac{\mathrm{cn}\,z}{\mathrm{dn}\,z}$	$-\mathrm{sn}\,z$	$\dfrac{1}{k\,\mathrm{sn}\,z}$	$\mathrm{sn}\,z$	$\dfrac{\mathrm{dn}\,z}{k\,\mathrm{cn}\,z}$	$-\mathrm{sn}\,z$
cn	1	$\mathrm{cn}\,z$	$\dfrac{-k'\,\mathrm{sn}\,z}{\mathrm{dn}\,z}$	$-\mathrm{cn}\,z$	$\dfrac{-i\,\mathrm{dn}\,z}{k\,\mathrm{sn}\,z}$	$-\mathrm{cn}\,z$	$\dfrac{-ik'}{k\,\mathrm{cn}\,z}$	$\mathrm{cn}\,z$
dn	k'	$\mathrm{dn}\,z$	$\dfrac{k'}{\mathrm{dn}\,z}$	$\mathrm{dn}\,z$	$\dfrac{-i\,\mathrm{cn}\,z}{\mathrm{sn}\,z}$	$-\mathrm{dn}\,z$	$\dfrac{ik'\,\mathrm{sn}\,z}{\mathrm{cn}\,z}$	$-\mathrm{dn}\,z$

Example 11.5

$\mathrm{sn}\,0 = 0$ by definition and $\mathrm{sn}(-z) = -\mathrm{sn}\,z$ by definition and analytic continuation. It follows by 11.3 that

$$\mathrm{sn}(z+K) = \frac{\mathrm{sn}\,z\,\mathrm{cn}\,K\,\mathrm{dn}\,K + \mathrm{sn}\,K\,\mathrm{cn}\,z\,\mathrm{dn}\,z}{1 - k^2\,\mathrm{sn}^2 z\,\mathrm{sn}^2 K} = \frac{\mathrm{cn}\,z\,\mathrm{dn}\,z}{\mathrm{dn}^2 z} = \frac{\mathrm{cn}\,z}{\mathrm{dn}\,z}$$

Then $\mathrm{sn}\,2K = 0$ and similarly, $\mathrm{cn}\,2K = -1$ and $\mathrm{dn}\,2K = 1$. Using Theorem 11.3 again,

$$\mathrm{sn}(z+2K) = \frac{\mathrm{sn}\,z\,\mathrm{cn}\,2K\,\mathrm{dn}\,2K + \mathrm{sn}\,2K\,\mathrm{cn}\,z\,\mathrm{dn}\,z}{1 - k^2\,\mathrm{sn}^2 z\,\mathrm{sn}^2 2K} = -\mathrm{sn}\,z$$

$\mathrm{sn}(z+iK') = 1/k\,\mathrm{sn}\,z$ from Example 11.4. It follows by the double periodicity of sn, (11.2), that $\mathrm{sn}(z+2iK') = \mathrm{sn}\,z$. Also by 11.3

$$\begin{aligned}
&\mathrm{sn}(z+2K+2iK') \\
&= \frac{\mathrm{sn}(z+2iK')\mathrm{cn}\,2K\,\mathrm{dn}\,2K + \mathrm{sn}\,2K\,\mathrm{cn}(z+2iK')\mathrm{dn}(z+2iK')}{1 - k^2\,\mathrm{sn}^2(z+2iK')\,\mathrm{sn}^2 2K} \\
&= -\mathrm{sn}(z+2iK') = -\mathrm{sn}\,z
\end{aligned}$$

Lastly, $\mathrm{sn}(K+iK') = 1/k$ by definition, and $\mathrm{cn}(K+iK') = -ik'/k$ and $\mathrm{dn}(K+iK') = 0$ from Fig. 11.4, so that by 11.3 again,

$$\begin{aligned}
\mathrm{sn}(z+K+iK') &= \frac{\mathrm{sn}\,z\,\mathrm{cn}(K+iK')\mathrm{dn}(K+iK') + \mathrm{sn}(K+iK')\mathrm{cn}\,z\,\mathrm{dn}\,z}{1 - k^2\mathrm{sn}^2 z\,\mathrm{sn}^2(K+iK')} \\
&= \frac{\mathrm{cn}\,z\,\mathrm{dn}\,z}{k\,\mathrm{cn}^2 z} = \frac{\mathrm{dn}\,z}{k\,\mathrm{cn}\,z}
\end{aligned}$$

Similarly for the other entries in the table.

Exercise **11.1.1** Show that $w = \operatorname{cn} z$ maps the domain $\mathscr{D}$ to the points, boundary lines and region given in Fig. 11.4.

Exercise **11.1.2** Show that, under the same conditions as Schwarz's reflection principle, Theorem 10.4, except that f_1 takes purely imaginary values on AB, the analytic continuation of f_1 to $\mathscr{R}_2$ is given by $f_2(z) = -\overline{f_1(\bar{z})}$.

Exercise **11.1.3** Use Schwarz's reflection principle and its variations to extend the domain of definition of cn and dn to the whole of $\mathbb{C}$ and derive Figs 11.6 and 11.7.

Exercise **11.1.4** Show that $w = \operatorname{sn}^2 z$ maps the rectangle with vertices at $K \pm iK'$ and $\pm iK'$, and its inside, onto the real axis of the w-plane with $(0, 1)$ excluded, and the rest of the w-plane.

Exercise **11.1.5** Show that the transformation

$$w = \left(\frac{1 - \operatorname{cn} z}{1 + \operatorname{cn} z}\right)^{1/2}$$

maps the rectangle with vertices at $\pm K \pm iK'$ to the circle $|w| = 1$ and the inside of the rectangle to the inside of the circle.

Exercise **11.1.6** Find the image of the rectangle with vertices at 0, K, $K + iK'$ and iK', and its inside, under $w = \operatorname{cn} z/(1 + \operatorname{sn} z)$.

Exercise **11.1.7** Use the fact that

$$\frac{d}{dz}(\operatorname{sn} z) = \operatorname{cn} z \operatorname{dn} z$$

to prove that

$$\frac{d}{dz}(\operatorname{cn} z) = -\operatorname{sn} z \operatorname{dn} z \quad \text{and} \quad \frac{d}{dz}(\operatorname{dn} z) = -k^2 \operatorname{sn} z \operatorname{cn} z$$

Deduce that

$$\frac{d}{dz}\left(\frac{\operatorname{sn} z}{1 + \operatorname{cn} z}\right) = \frac{\operatorname{dn} z}{1 + \operatorname{cn} z}$$

Exercise **11.1.8** Use Taylor's theorem to show that

(i) $\operatorname{cn} z = 1 - \dfrac{z^2}{2!} + (1 + 4k^2)\dfrac{z^4}{4!} - \dots \quad \text{for } |z| < K'$

(ii) $\operatorname{dn} z = 1 - \dfrac{k^2 z^2}{2!} + k^2(4 + k^2)\dfrac{z^4}{4!} - \dots \quad \text{for } |z| < K'$

Hence determine the first three non-zero terms in the Taylor series expansion of $\operatorname{sech} z$ about the origin.

Exercise **11.1.9** Verify that

(i) $\displaystyle\int \operatorname{sn} z\,dz = \frac{1}{2k}\operatorname{Log}\left(\frac{\operatorname{dn} z - k\operatorname{cn} z}{\operatorname{dn} z + k\operatorname{cn} z}\right)$

(ii) $\displaystyle\int \operatorname{cn} z\,dz = \frac{1}{k}\tan^{-1}\left(\frac{k\operatorname{sn} z}{\operatorname{cn} z}\right)$

(iii) $\displaystyle\int \operatorname{dn} z\,dz = \sin^{-1}(\operatorname{sn} z)$

(iv) $\displaystyle\int \frac{\operatorname{cn} z}{\operatorname{sn} z}\,dz = \operatorname{Log}\left(\frac{1 - \operatorname{dn} z}{\operatorname{sn} z}\right)$

Exercise **11.1.10**

Let $\displaystyle F(z, w) = \frac{\operatorname{sn} z\operatorname{cn} w\operatorname{dn} w + \operatorname{sn} w\operatorname{cn} z\operatorname{dn} z}{1 - k^2\operatorname{sn}^2 z\operatorname{sn}^2 w}$

Show that $\displaystyle\frac{\partial F}{\partial z} = \frac{\partial F}{\partial w}$

Exercise **11.1.11** Use Theorem 11.3(i) and the definition of $\operatorname{cn} z$ to prove Theorem 11.3(ii).

Exercise **11.1.12** Use the addition formulae for the Jacobian elliptic functions to show that

(i) $\operatorname{cn}(z + 2K) = -\operatorname{cn} z$

(ii) $\operatorname{dn}(z + 2iK') = -\operatorname{dn} z$

(iii) $\displaystyle\operatorname{dn}^2 z = \frac{\operatorname{cn} 2z + \operatorname{dn} 2z}{1 + \operatorname{cn} 2z}$

(iv) $\displaystyle\operatorname{sn}^2 z = \frac{\operatorname{dn} 2z - \operatorname{cn} 2z}{k'^2 - k^2\operatorname{cn} 2z + \operatorname{dn} 2z}$

Exercise **11.1.13** Use the addition formulae to determine $\operatorname{cn}(z + \alpha)$ and $\operatorname{dn}(z + \alpha)$, for $\alpha = K$ and $\alpha = K + iK'$, in terms of $\operatorname{sn} z$, $\operatorname{cn} z$ and $\operatorname{dn} z$. Use these results, the expression for $\operatorname{sn}(z + K + iK')$ and the addition formulae to determine $\operatorname{cn}(z + iK')$ and $\operatorname{dn}(z + iK')$.

Exercise **11.1.14** Find the first two terms in the Taylor series expansions of $\operatorname{sn}(z + iK')$, $\operatorname{cn}(z + iK')$ and $\operatorname{dn}(z + iK')$ in a neighbourhood of $z = 0$. Hence find $\operatorname{Res}_{iK'}\operatorname{sn} z$, $\operatorname{Res}_{iK'}\operatorname{cn} z$ and $\operatorname{Res}_{iK'}\operatorname{dn} z$.

Exercise **11.1.15** Use the fact that $\operatorname{sn}^2(z/2) = (1 - \operatorname{cn} z)/(1 + \operatorname{dn} z)$ (Example 11.3) to show that

$$\operatorname{sn}(K/2) = \frac{1}{\sqrt{1 + k'}} \qquad \operatorname{cn}(K/2) = \frac{\sqrt{k'}}{\sqrt{1 + k'}} \qquad \operatorname{dn}(K/2) = \sqrt{k'}$$

Then use Jacobi's imaginary transformation to show that

$$\operatorname{sn}(iK'/2) = \frac{i}{\sqrt{k}} \qquad \operatorname{cn}(iK'/2) = \frac{\sqrt{1+k}}{\sqrt{k}} \qquad \operatorname{dn}(iK'/2) = \sqrt{1+k}$$

Hence use the addition formulae to deduce that

$$\operatorname{cn}((K+iK')/2) = \frac{(1-i)\sqrt{k'}}{\sqrt{2k}} \qquad (*)$$

Exercise **11.1.16** Show that

$$\operatorname{cn}^2(z/2) = \frac{\operatorname{cn} z + \operatorname{dn} z}{1 + \operatorname{dn} z}$$

using Theorem 11.3 and hence verify result (*) of Exercise 11.1.15.

Elliptic Functions in General

The study of Jacobian elliptic functions leads to the idea of elliptic functions in general; that is, doubly periodic functions which are meromorphic. A lot of useful general results follow directly from these properties, without reference to specific examples. We apply some of these results to Jacobian elliptic functions.

Definitions

Let ω_1 and ω_2 be any two complex numbers which satisfy $\operatorname{Im}(\omega_1/\omega_2) \neq 0$. A function $f: \mathbb{C} \to \mathbb{C}$ which satisfies

$$f(z+2\omega_1) = f(z+2\omega_2) = f(z) \qquad (\text{for all } z \in \mathbb{C})$$

is a **doubly periodic function** with **periods** $2\omega_1$ and $2\omega_2$. A doubly periodic function which has no singular points other than poles, i.e. a meromorphic function, is an **elliptic function** (e.f.). Periods $2\omega_1$ and $2\omega_2$ of an e.f. f are **primitive periods** of f if any period of f is of the form $2m\omega_1 + 2n\omega_2$ for some $m, n \in \mathbb{Z}$.

Notes

(i) The condition that $\operatorname{Im}(\omega_1/\omega_2) \neq 0$ ensures that the two complex numbers, as vectors, are linearly independent.

(ii) If $2\omega_1$ and $2\omega_2$ are two periods of a doubly periodic function, then so is $2m\omega_1 + 2n\omega_2$, for any integers m and n.

(iii) Notice that the derivative of an elliptic function is also an elliptic function.

Historical Note

It was Jacobi who first showed that if a meromorphic function, other than a constant, has two periods whose ratio is real, then each of these periods is a multiple of a single primitive period. He also proved that if a meromorphic function, other than a constant, has periods whose ratio is not real, it is necessarily doubly periodic.

Notation

Let f be an e.f. with primitive periods $2\omega_1$ and $2\omega_2$. Let Ω denote the **lattice** of points in $\mathbb{C}$ given by

$$\Omega = \Omega(\omega_1, \omega_2) = \{2m\omega_1 + 2n\omega_2 : m, n \in \mathbb{Z}\}$$

Hence Ω is the set of points in $\mathbb{C}$ which are the periods of f. By joining points in the lattice Ω by straight lines, we obtain a network of parallelograms which cover $\mathbb{C}$. Each parallelogram is called a **period parallelogram** or **mesh** of f.

Suppose that, once again, f is an e.f. with primitive periods $2\omega_1$ and $2\omega_2$, and without loss of generality, suppose that $\operatorname{Im}(\omega_2/\omega_1) > 0$. Then 0, $2\omega_1$, $2\omega_1 + 2\omega_2$ and $2\omega_2$ are the vertices of the parallelogram $\mathscr{P}$, shown in Fig. 11.8. This is a **primitive period parallelogram** (p.p.p.) for f.

Clearly, there is no point β inside or on the boundary of a p.p.p., except the vertices, such that $f(z+\beta) = f(z)$ for all z. Notice that any mesh of f is obtained from a p.p.p by a translation. Notice also that the points z and $z^* = z + 2m\omega_1 + 2n\omega_2$, m, $n \in \mathbb{Z}$ and not both zero, lie in different meshes but are a translation of the same point, γ say, in a p.p.p. Accordingly, z and z^* are said to be **congruent**. Since $f(z^*) = f(z) = f(\gamma)$, the properties of an e.f. are completely determined by its values in a p.p.p.

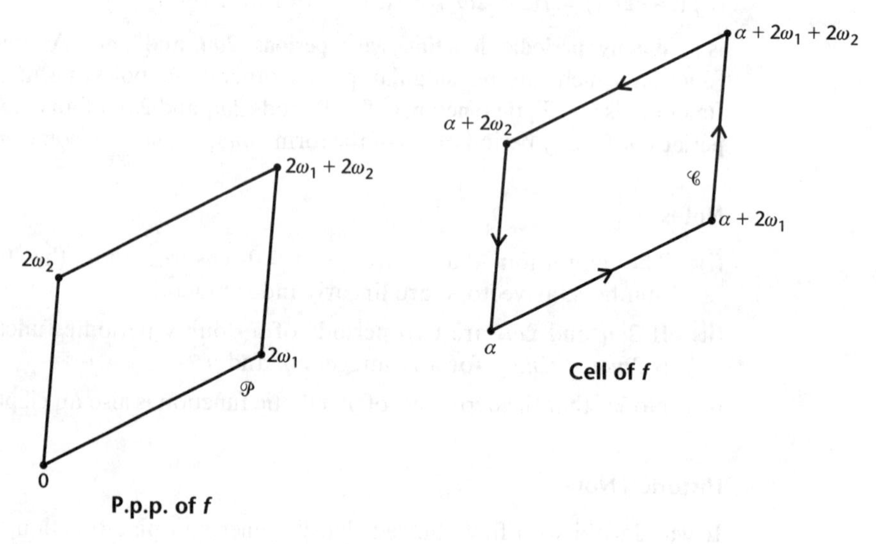

Figure 11.8

Example 11.6

The Jacobian elliptic functions sn, cn and dn, are clearly elliptic functions in the general sense. Two periods of cn, for example, are $4K$ and $4iK'$ (Fig. 11.7) but these are not primitive periods of cn since $2K+2iK'$ is not an integer linear combination of these. Two primitive periods of cn are $4K$ and $2K+2iK'$, and in this case the p.p.p. for cn has vertices at 0, $2K+2iK'$, $4K$ and $6K+2iK'$.

Important Note

If f is an e.f. then the number of poles of f in or on the boundary of any mesh is finite, by Theorem 4.24. Also, the number of zeros of f in or on the boundary of any mesh is finite. Otherwise, $1/f$ is an e.f. with an infinite number of poles in or on the boundary of any mesh.

Very often, as in the case of cn, poles or zeros occur on the boundary of a mesh of an elliptic function. It is sometimes convenient to consider a parallelogram obtained from a mesh by translation, which has no poles or zeros on its boundary. Such a parallelogram is called a **cell**. Note that it is always possible to construct a cell of an elliptic function since the number of poles and zeros in or on any mesh is finite. The Jacobian elliptic functions sn, cn and dn all have two zeros and two simple poles in any cell. Clearly, the properties of an e.f. are completely determined by its values in any cell.

We now derive a number of important properties of elliptic functions.

Lemma 11.5. A Non-constant Elliptic Function Has Poles

An elliptic function with no poles in any cell is a constant. □

Proof

Let an e.f. f have no poles inside any cell. Then it is analytic and bounded inside and on the boundary of the cell. Since f is doubly periodic, f is then entire and bounded on $\mathbb{C}$ and so is a constant by Liouville's theorem. ■

Example 11.7

It follows from the definition of sn on $\mathbb{C}$ that $\operatorname{sn}(z+iK')$ has a pole when $\operatorname{sn} z$ has a zero and conversely (Fig. 11.3). Since all the poles and zeros of sn are simple, $\operatorname{sn} z \operatorname{sn}(z+iK')$ is an e.f. with no poles and so is a constant by Lemma 11.5. When $z=K$, $\operatorname{sn} z \operatorname{sn}(z+iK') = \operatorname{sn}(K+iK') = 1/k$. Hence $\operatorname{sn}(z+iK') = 1/(k \operatorname{sn} z)$, a result derived earlier by using the addition formulae.

Theorem 11.6. The Integral Around a Cell

Let f be an elliptic function and let $\mathscr{C}$ be the boundary of any cell of f. Then $\int_{\mathscr{C}} f(z)dz = 0$. □

Proof

Let f have primitive periods $2\omega_1$ and $2\omega_2$. Then any cell of f has vertices shown in Fig. 11.8. Now

$$\int_{\mathscr{C}} f(z)\,dz = \left(\int_{\alpha}^{\alpha+2\omega_1} + \int_{\alpha+2\omega_1}^{\alpha+2\omega_1+2\omega_2} + \int_{\alpha+2\omega_1+2\omega_2}^{\alpha+2\omega_2} + \int_{\alpha+2\omega_2}^{\alpha}\right) f(z)\,dz$$

Let $z = \zeta + 2\omega_1$ in the second integral and $z = \tau + 2\omega_2$ in the third. Then

$$\int_{\mathscr{C}} f(z)\,dz = \int_{\alpha}^{\alpha+2\omega_1} f(z)\,dz + \int_{\alpha}^{\alpha+2\omega_2} f(\zeta + 2\omega_1)\,d\zeta$$

$$+ \int_{\alpha+2\omega_1}^{\alpha} f(\tau + 2\omega_2)d\tau + \int_{\alpha+2\omega_2}^{\alpha} f(z)\,dz$$

$$= \int_{\alpha}^{\alpha+2\omega_1} (f(z) - f(z+2\omega_2))\,dz + \int_{\alpha+2\omega_2}^{\alpha} (f(z) - f(z+2\omega_1))\,dz \tag{11.14}$$

The result then follows by the doubly periodic property of f. ■

Corollary 11.7. The Sum of the Residues of an Elliptic Function in a Cell

The sum of the residues of an elliptic function at its poles in any cell is zero. □

Proof

By the residue theorem, the sum of the residues of an e.f. f at its poles in any cell is $(1/2\pi i)\int_{\mathscr{C}} f(z)\,dz$, where $\mathscr{C}$ is the boundary of the cell, so the result follows by 11.6. ■

Theorem 11.8. The Number of Roots of the Equation $f(z) = \alpha$

Let α be any complex number and f an e.f. Then the number of roots of the equation $f(z) = \alpha$ in any cell of f is equal to the number of poles of f in that cell. □

Proof

Let $\mathscr{C}$ be the boundary of any chosen cell of f. Let Z be the number zeros of $f(z) - \alpha$ and P the number of poles of f in this cell. Then by Theorem 3.21,

$Z - P = (1/2\pi i)\int_{\mathscr{C}} (f'(z)/(f(z) - \alpha))\,dz$. Let $G(z) = f'(z)/(f(z) - c)$. Since f' is an e.f. with the same periods of f, it follows that G is an e.f. with the same periods as f. Hence by 11.6, $\int_{\mathscr{C}} G(z)\,dz = 0 \quad \Rightarrow \quad Z = P$, as required. ■

Note

This results shows that the number of roots of $f(z) = \alpha$ in any cell of f depends on f, but not on α.

Definition

The **order** of an e.f. f is the number of poles (counting multiplicities) of f in any cell.

Note

If an e.f. is not a constant, it must have at least two poles in any cell, since if there were only one pole, a simple pole, the residue at this pole would be zero by 11.7 and this gives a contradiction. It follows that the simplest e.f.'s have order 2. Clearly there are only two possible types of e.f.'s of order 2. Those of Jacobian type, already discussed, which have two simple poles in each cell, and those of **Weierstrassian** type, which have a double pole with residue zero in any cell. Weierstrassian e.f.'s will be discussed in the next section.

Theorem 11.9. Poles and Zeros of an Elliptic Function

Let $\mathscr{C}$ be the boundary of any cell of an e.f. f. Let α_k be a zero of order r_k of f, in this cell, and let β_k be a pole of order s_k of f in this cell, $k = 1, 2, \ldots, n$ say. Then

$$\frac{1}{2\pi i}\int_{\mathscr{C}} \frac{zf'(z)}{f(z)}\,dz = \sum_{k=1}^{n} r_k\alpha_k - \sum_{k=1}^{n} s_k\beta_k \qquad \square$$

Proof

Step 1

Let f have a zero of order r at $z = \alpha$. Then, by definition, $f(z) = (z - \alpha)^r\phi(z)$, where ϕ is analytic and non-zero at α. Then

$$f'(z) = r(z - \alpha)^{r-1}\phi(z) + (z - \alpha)^r\phi'(z)$$

$$\Rightarrow \quad \frac{zf'(z)}{f(z)} = (z - \alpha + \alpha)\left(\frac{r}{z - \alpha} + \psi(z)\right) = \frac{r\alpha}{z - \alpha} + r + z\psi(z)$$

in some neighbourhood of α, where $\psi - \phi'/\phi$ and so is analytic at α. Hence $\text{Res}_\alpha(zf'(z)/f(z)) = r\alpha$.

Step 2

Let f have a pole of order s at $z = \beta$. Then $f(z) = (z - \beta)^{-s}\phi(z)$, where ϕ is analytic and non-zero at β, and similar to step 1,

$$\frac{zf'(z)}{f(z)} = \frac{-s\beta}{z-\beta} - s + z\psi(z) \quad \Rightarrow \quad \text{Res}_\beta\left(\frac{zf'(z)}{f(z)}\right) = -s\beta$$

The required result then follows by the residue theorem. ■

Theorem 11.10. An Integral Related to a Period

Let f be an e.f. and $\mathscr{C}$ be the boundary of any chosen cell of f. Then $(1/2\pi i)\int_{\mathscr{C}}(zf'(z)/f(z))\,dz$ is equal to a particular period of f. □

Proof

Let the chosen cell of f be the cell in Fig. 11.8 and let $F(z) = zf'(z)/f(z)$. Then if f has fundamental periods $2\omega_1$ and $2\omega_2$, $F(z + 2\omega_1) = F(z) + 2\omega_1 f'(z)/f(z)$ and similarly for $2\omega_2$. It then follows by (11.14) in the proof of Theorem 11.6 that

$$\begin{aligned}
\int_{\mathscr{C}} F(z)\,dz &= -2\omega_2 \int_\alpha^{\alpha+2\omega_1} \frac{f'(z)}{f(z)}\,dz + 2\omega_1 \int_\alpha^{\alpha+2\omega_2} \frac{f'(z)}{f(z)}\,dz \\
&= -2\omega_2[\log f(z)]_\alpha^{\alpha+2\omega_1} + 2\omega_1[\log f(z)]_\alpha^{\alpha+2\omega_2} \\
&= -2\omega_2(-2n\pi i) + 2\omega_1(2m\pi i) \quad \text{for some } n, m \in \mathbb{Z}
\end{aligned}$$

$$\Rightarrow \quad \frac{1}{2\pi i}\int_{\mathscr{C}} F(z)dz = 2m\omega_1 + 2n\omega_2 \quad \text{as required}$$

■

Note

It follows from 11.9 and 11.10 that the sum of the affixes of the zeros minus the sum of the affixes of the poles of an e.f. in any cell is equal to some period of the e.f.

For example, consider any cell of sn containing the points iK', $2K + iK'$, 0 and $2K$. Recall that sn has a simple zero at 0 and $2K$, and a simple pole at iK' and $2K + iK'$. Then

$$(0 + 2K) - (iK' + 2K + iK') = -2iK'$$

and $-2iK'$ is a period of sn.

The following elementary but very powerful results imply that an e.f. is essentially determined by its periods, poles and zeros, or by its periods, poles and the principal part of its Laurent series expansion at each of its poles.

Theorem 11.11. Comparison of Elliptic Functions

Let f and g be two elliptic functions.

(i) If f and g have the same periods, the same poles and the same zeros (of the same order), then $f(z) = \gamma g(z)$ for all z, where γ is a non-zero constant.

(ii) If f and g have the same periods, the same poles and the same principal part in their Laurent series expansions at each of those poles, then $f(z) = \gamma + g(z)$ for all z, where γ is a constant. □

Proof

(i) Let $\phi(z) = f(z)/g(z)$ for all $z \in \mathbb{C}$. Then ϕ is doubly periodic (with the same periods of f and g), with possible singular points at the poles of f and the zeros of g. If f has a pole of order r at $z = \alpha$ say, then $f(z) = (z - \alpha)^{-r}h(z)$, where h is analytic and non-zero at α. Then by hypothesis, g also has a pole of order r at $z = \alpha$ and $g(z) = (z - \alpha)^{-r}H(z)$, where H is analytic and non-zero at α. Hence $\phi(\alpha) = h(\alpha)/H(\alpha)$ and so ϕ is non-singular at α. Similarly, if g has a zero of order s at $z = \beta$, ϕ is non-singular at β. Hence ϕ is doubly periodic and entire, and so is a constant by 11.5.

(ii) Let $\psi(z) = f(z) - g(z)$ so that ψ is doubly periodic and any singular points of ψ occur at the poles of f and g. Let f have a pole of order r at $z = \alpha$ say. Then by hypothesis

$$f(z) = \sum_{n=1}^{r} \frac{b_n}{(z-\alpha)^n} + F(z) \qquad g(z) = \sum_{n=1}^{r} \frac{b_n}{(z-\alpha)^n} + G(z)$$

in some neighbourhood of α, where F and G are analytic at α. Then $\psi(z) = F(z) - G(z)$ and so is analytic at α. Hence ψ is doubly periodic and entire and so is a constant by 11.5. ■

Theorem 11.11 can be used to derive the addition formulae for sn, cn and dn, as shown in the next example.

Example 11.8

Define f by $f(z) = \operatorname{sn} z \operatorname{sn}(z + \alpha)$, $z \in \mathbb{C}$ where α is a constant. Recall that sn has simple zeros at 0 and $2K$, simple poles at iK' and $2K + iK'$ and has primitive periods $2iK'$ and $4K$. Recall also that $\operatorname{sn}(z + 2K) = -\operatorname{sn} z$. It follows that f is an e.f. with primitive periods $2iK'$ and $2K$, simple poles at iK' and $iK' - \alpha$, and simple zeros at 0 and $-\alpha$.

Now define g by $g(z) = \operatorname{cn} z \operatorname{cn}(z + \alpha) - \operatorname{cn}\alpha$. Recall that cn has simple zeros at K and $3K$, simple poles at iK' and $2K + iK'$, and $\operatorname{cn}(z + 2K) = \operatorname{cn}(z + 2iK') = -\operatorname{cn} z$. It follows that g is an e.f. with the same primitive periods, zeros and poles as f. Hence by 11.11, $f(z) = \gamma g(z)$, where γ is a non-zero constant. Then $\gamma = \lim_{z\to 0} f(z)/g(z) = \lim_{z\to 0} f'(z)/g'(z)$ since 0 is a simple zero of f and g. Hence

$$\gamma = \lim_{z\to 0}\left(\frac{\operatorname{cn} z \operatorname{dn} z \operatorname{sn}(z+\alpha) + \operatorname{sn} z \operatorname{cn}(z+\alpha)\operatorname{dn}(z+\alpha)}{-\operatorname{sn} z \operatorname{dn} z \operatorname{cn}(z+\alpha) - \operatorname{cn} z \operatorname{sn}(z+\alpha)\operatorname{dn}(z+\alpha)}\right) = \frac{-1}{\operatorname{dn}\alpha}$$

$$\Rightarrow \quad \operatorname{cn} z \operatorname{cn}(z+\alpha) + \operatorname{sn} z \operatorname{sn}(z+\alpha)\operatorname{dn}\alpha = \operatorname{cn}\alpha \tag{11.15a}$$

Similarly, comparing f with $h(z) = \operatorname{dn} z \operatorname{dn}(z+\alpha) - \operatorname{dn}\alpha$ gives

$$\operatorname{dn} z \operatorname{dn}(z+\alpha) + k^2 \operatorname{sn} z \operatorname{sn}(z+\alpha)\operatorname{cn}\alpha = \operatorname{dn}\alpha \tag{11.15b}$$

This is left as an exercise. Letting $\alpha = -z - w$ in (11.15) gives

$$\operatorname{cn}(z+w) = \operatorname{cn} z \operatorname{cn} w - \operatorname{sn} z \operatorname{sn} w \operatorname{dn}(z+w)$$

$$\operatorname{dn}(z+w) = \operatorname{dn} z \operatorname{dn} w - k^2 \operatorname{sn} z \operatorname{sn} w \operatorname{cn}(z+w)$$

$$\Rightarrow \quad \operatorname{cn}(z+w) = \frac{\operatorname{cn} z \operatorname{cn} w - \operatorname{sn} z \operatorname{sn} w \operatorname{dn} z \operatorname{dn} w}{1 - k^2 \operatorname{sn}^2 z \operatorname{sn}^2 w}$$

$$\operatorname{dn}(z+w) = \frac{\operatorname{dn} z \operatorname{dn} w - k^2 \operatorname{sn} z \operatorname{sn} w \operatorname{cn} z \operatorname{cn} w}{1 - k^2 \operatorname{sn}^2 \operatorname{sn}^2 w}$$

Equation (11.15a) then gives the addition formula for sn.

Exercise **11.2.1** Let f be an elliptic function. Show that f' and $1/f$ are elliptic functions. Show also that if g is an elliptic function with the same periods as f, then $f \circ g$ is an elliptic function.

Exercise **11.2.2** Prove that the set of all elliptic functions with the same periods is a field under the the usual operations of addition and multiplication of functions.

Exercise **11.2.3** Let f be meromorphic and the derivative of f be an elliptic function. Show that f is not necessarily an elliptic function, but if f is even, then f is an elliptic function.

Exercise **11.2.4** Let f be an odd elliptic function with primitive periods $2\omega_1$ and $2\omega_2$. Show that ω_1 is either a zero or pole of f, and similarly for ω_2.

Exercise **11.2.5** Show that if $2\omega_1$ and $2\omega_2$ are two primitive periods of an elliptic function, then so are $2\omega_1'$ and $2\omega_2'$, given by

$$2\omega_1' = 2m\omega_1 + 2n\omega_2 \qquad 2\omega_2' = 2p\omega_1 + 2q\omega_2 \qquad (m,n,p,q \in \mathbb{Z})$$

if and only if $mq - np = \pm 1$.

Exercise **11.2.6** Illustrate a primitive period parallelogram for (i) sn, (ii) cn and (iii) dn. Indicate the position of the zeros and poles in each case. Illustrate a chosen cell of (i) sn, (ii) cn and (iii) dn in the same diagram in each case.

Exercise **11.2.7** Use the method of Example 11.7 to show that

(i) $\mathrm{cn}(z+K+iK') = \dfrac{-ik'}{k\,\mathrm{cn}\,z}$

(ii) $\mathrm{dn}(z+K) = \dfrac{k'}{\mathrm{dn}\,z}$

Exercise **11.2.8** Let f be an elliptic function and $\mathscr{C}$ the boundary of any cell of f. Let α_k be a zero of order r_k of f and β_k a pole of order s_k of f, inside $\mathscr{C}$, $k = 1, \ldots, n$ say. Let g be any function which is analytic inside and on $\mathscr{C}$. Prove that

$$\int_{\mathscr{C}} g(z)\frac{f'(z)}{f(z)}\,dz = \sum_{k=1}^{n} r_k g(\alpha_k) - \sum_{k=1}^{n} s_k g(\beta_k)$$

(Note that Theorem 11.9 is a special case of this more general result.)

Exercise **11.2.9**

(i) Evaluate $\displaystyle\int_{\mathscr{C}} \frac{z\,\mathrm{cn}\,z\,\mathrm{dn}\,z}{\mathrm{sn}\,z}\,dz$, where $\mathscr{C}$ is the boundary of any cell of sn.

(ii) Evaluate $\displaystyle\int_{\mathscr{C}} \frac{z\,\mathrm{cn}\,z\,\mathrm{sn}\,z}{\mathrm{dn}\,z}\,dz$ where $\mathscr{C}$ is the boundary of any cell of dn.

Exercise **11.2.10** Use Theorem 11.11, as in Example 11.8, to show that

$$\mathrm{dn}\,z\,\mathrm{dn}(z+\alpha) + k^2\mathrm{cn}\,\alpha\,\mathrm{sn}\,z\,\mathrm{sn}(z+\alpha) = \mathrm{dn}\,\alpha$$

Exercise **11.2.11** Use equation (11.15a) and the addition formula for cn to derive the addition formula for sn.

Introduction to the Weierstrassian Elliptic Function

The Weierstrassian elliptic function is, in some sense, the simplest elliptic function of order 2, possessing only one pole, a double pole, with residue zero, in any cell. The main reason for introducing it here is to show how some of the general properties of elliptic functions derived in the last section can be applied in a specific case.

Notation

Recall that if ω_1 and ω_2 are two complex numbers with $\mathrm{Im}\,(\omega_1/\omega_2) \neq 0$, then

$$\Omega = \Omega(\omega_1, \omega_2) = \{2m\omega_1 + 2n\omega_2 : m, n \in \mathbb{Z}\}$$

Let $\Omega' = \Omega\backslash\{0\}$.

Given ω_1 and ω_2, the Weierstrassian elliptic function is the e.f. having a double pole at each point $\omega \in \Omega$ and principal part $1/(z-\omega)^2$ in its Laurent series expansion about ω.

Definition

Let $\omega_1,\ \omega_2 \in \mathbb{C}$ with $\operatorname{Im}(\omega_1/\omega_2) \neq 0$. Then the **Weierstrassian elliptic function** (W.e.f.) is denoted and defined for all $z \in \mathbb{C}$ by

$$\wp(z\,|\,\omega_1, \omega_2) = \frac{1}{z^2} + \sum_{\omega \in \Omega'} \left(\frac{1}{(z-\omega)^2} - \frac{1}{\omega^2} \right) \tag{11.16}$$

Notes

(i) $\wp(z\,|\,\omega_1, \omega_2)$ is usually written simply as $\wp(z)$ when no ambiguity can occur.

(ii) It can be shown that the double series in the definition (11.16) converges uniformly and absolutely in every bounded closed set containing none of the points of Ω.

(iii) It follows by the definition (11.16) that

$$\wp(-z) = \frac{1}{z^2} + \sum_{\omega \in \Omega'} \left(\frac{1}{(z+\omega)^2} - \frac{1}{\omega^2} \right) = \wp(z)$$

since $\omega \in \Omega'$ if and only if $-\omega \in \Omega'$. Hence $\wp$ is an even function.

(iv) It can be shown that any elliptic function can be expressed as a rational function of $\wp$ and $\wp'$.

Historical Note

Weierstrass first introduced $\wp$ in 1895 and called it the **p-function.**

It follows immediately from its definition that $\wp$ is meromorphic with a double pole at 0 with $\operatorname{Res}_0 \wp(z) = 0$. To show $\wp$ is also doubly periodic, we examine its derivative. Since the series in (11.16) is uniformly convergent,

$$\wp'(z) = \frac{-2}{z^3} - 2\sum_{\omega \in \Omega'} \frac{1}{(z-\omega)^3} = -2\sum_{\omega \in \Omega'} \frac{1}{(z-\omega)^3}$$

$$\Rightarrow \quad \wp'(z + 2\omega_1) = -2\sum_{\omega \in \Omega} \frac{1}{(z-(\omega - 2\omega_1))^3} = \wp'(z)$$

and similarly, $\wp'(z + 2\omega_2) = \wp'(z)$. It follows that $\wp'$ is doubly periodic with a triple pole at each of the lattice points $\omega \in \Omega$. Clearly, $2\omega_1$ and $2\omega_2$ are a pair of primitive periods of $\wp'$, otherwise there would be a network of cells for $\wp'$, some of which would contain no singular points of $\wp'$. Now

$$\wp'(z + 2\omega_k) = \wp'(z) \quad \Rightarrow \quad \wp(z + 2\omega_k) = \wp(z) + \gamma \qquad (k = 1, 2)$$

where γ is a constant. Letting $z = -\omega_k$ gives $\wp(\omega_k) = \wp(\omega_k) + \gamma$ since $\wp$ is even, so that $\gamma = 0$. Hence $\wp$ is doubly periodic with primitive periods $2\omega_1$ and $2\omega_2$. In summary, we have the following results.

Theorem 11.12. $\wp$ and $\wp'$ are Elliptic Functions

(i) $\wp$ is an even elliptic function, with primitive periods $2\omega_1$ and $2\omega_2$, with a double pole at 0, and $\text{Res}_0 \wp(z) = 0$.

(ii) $\wp'$ is an odd elliptic function, with primitive periods $2\omega_1$ and $2\omega_2$, with a triple pole at 0, and $\text{Res}_0 \wp'(z) = 0$. □

We next derive a differential equation satisfied by $\wp(z)$. The following derivation illustrates the great power of the comparison theorem for elliptic functions. Without it, the differential equation would be very much more difficult to obtain from the definition.

Let $f(z) = \wp(z) - 1/z^2$. It follows from the definition of $\wp$ that f is analytic at 0 and is an even function since $\wp$ is even. Then $f(z) = \sum_{n=0}^{\infty} a_{2n} z^{2n}$ in some neighbourhood of the origin. Note also by (11.16) that $f(0) = 0$ and so $a_0 = 0$. Hence

$$\wp(z) = \frac{1}{z^2} + \sum_{n=1}^{\infty} a_{2n} z^{2n} = \frac{1}{z^2} + a_2 z^2 + a_4 z^4 + a_6 z^6 + \ldots \qquad (|z| < \rho)$$

say. The constants a_{2n} are easily determined from the definition (11.16) (see Exercises 11.3). Then

$$\wp'(z) = -2\left(\frac{1}{z^3} - a_2 z - 2a_4 z^3 - 3a_6 z^5 - \ldots\right)$$

$$\Rightarrow \quad \wp'^2(z) = 4\left(\frac{1}{z^6} - \frac{2a_2}{z^2} - 4a_4 + (a_2{}^2 - 6a_6)z^2 + \ldots\right)$$

for $|z| < \rho$. Then, clearly, the expansion of $\wp'^2(z)$ has the same principal part as the expansion of $4\wp^3(z) - \alpha\wp(z)$, for some suitably chosen constant α.

$$4\wp^3(z) - \alpha\wp(z) = 4\left(\frac{1}{z^6} + \frac{3a_2}{z^2} + 3a_4 + (3a_2{}^2 + 3a_6)z^2 + \ldots\right) - \alpha\left(\frac{1}{z^2} + a_2 z^2 + \ldots\right)$$

in some neighbourhood of the origin, and so the expansions of $\wp'^2(z)$ and $4\wp^3(z) - \alpha\wp(z)$ have the same principal parts as long as $-8a_2 = 12a_2 - \alpha$, i.e. $\alpha = 20a_2$. Then $\wp'^2(z)$ and $4\wp^3(z) - 20a_2\wp(z)$ both have a pole at 0, the same principal part in their Laurent series expansion about that pole, and clearly have the same periods. Hence by Theorem 11.11(ii),

$$\wp'^2(z) = 4\wp^3(z) - 20a_2\wp(z) + \gamma \qquad (11.17)$$

where γ is a constant. Comparing the constant terms in each expansion gives $\gamma = -28a_4$. The usual convention is to let $g_2 = 20a_2$ and $g_3 = 28a_4$. It is usually more convenient to express these constants in terms of values of $\wp$. Given primitive periods $2\omega_1$ and $2\omega_2$, let $\omega_3 = -(\omega_1 + \omega_2)$. Then $\wp'(z + 2\omega_k) = \wp'(z)$, $k = 1, 2, 3$. Letting $z = -\omega_k$ gives $\wp'(\omega_k) = \wp'(-\omega_k)$ and so, since $\wp'$ is odd,

$\wp'(\omega_k) = 0$. Then $\wp(\omega_k)$, $k = 1, 2, 3$, are the zeros of the right-hand side of (11.17). The following theorem summarises these results.

Theorem 11.13. A Differential Equation for $\wp(z)$

The w.e.f. $\wp(z|\omega_1, \omega_2)$ satisfies the differential equation

$$\wp'^2(z) = 4\wp^3(z) - g_2\wp(z) - g_3 = 4(\wp(z) - e_1)(\wp(z) - e_2)(\wp(z) - e_3)$$

where $e_k = \wp(\omega_k), k = 1, 2, 3$, and $\omega_3 = -(\omega_1 + \omega_2)$. □

We now derive an addition theorem for $\wp(z)$. Once again, the easiest proof uses general results for elliptic functions, rather than the explicit definition of $\wp(z)$.

Theorem 11.14. First Addition Formula for $\wp(z)$

$$\begin{vmatrix} \wp(u) & \wp(v) & \wp(u+v) \\ \wp'(u) & \wp'(v) & -\wp'(u+v) \\ 1 & 1 & 1 \end{vmatrix} = 0$$ □

Proof

Step 1

$$\text{Let } f(z) = \begin{vmatrix} \wp(u) & \wp(v) & \wp(z) \\ \wp'(u) & \wp'(v) & \wp'(z) \\ 1 & 1 & 1 \end{vmatrix}$$

where u and v are treated as constants. Then $f(z) = \alpha\wp(z) + \beta\wp'(z) + \gamma$ for some constants α, β and γ, so f is doubly periodic with the same periods as $\wp$. Since f is meromorphic, it is an elliptic function, and has a triple pole at 0 by 11.12. Then by 11.8, f has three zeros in any cell containing 0. Let u and v be in the same cell as 0 without loss of generality. Then, clearly, two of the zeros of f are u and v. Then there exists a third zero, ζ say, of f in this cell, and by 11.9 and 11.10, $u + v + \zeta$ is a period of f and hence a period of $\wp$.

Step 2

By step 1, letting $z = \zeta$ gives

$$\wp(z) = \wp(-u - v) = \wp(u + v) \quad \text{and} \quad \wp'(z) = \wp'(-u - v) = -\wp'(u + v)$$

and $f(z) = 0$, as required. ■

An alternative addition formula for $\wp(z)$ can be obtained by using Theorem 11.14 together with the differential equation for $\wp(z)$. Expanding $f(z) = 0$, where f is defined in the proof of 11.14 gives

$$\wp'(z)(\wp(u) - \wp(v)) = \wp(z)(\wp'(u) - \wp'(v)) + \wp(u)\wp'(v) - \wp(v)\wp'(u)$$

and so squaring and using (11.17) gives

$$(4\wp^3(z) - g_2\wp(z) - g_3)(\wp(u) - \wp(v))^2$$
$$= \wp^2(z)(\wp'(u) - \wp'(v))^2 + 2\wp(z)F(u, v) + G(u, v)$$

say. This is a cubic equation in $\wp(z)$ and so the sum of its roots is

$$\frac{(\wp'(u) - \wp'(v))^2}{4(\wp(u) - \wp(v))^2}$$

But by 11.14, the three roots are $\wp(u)$, $\wp(v)$ and $\wp(u+v)$. Hence the following result is obtained.

Theorem 11.15. Second Addition Formula for $\wp(z)$

$$\wp(u) + \wp(v) + \wp(u+v) = \frac{1}{4}\left(\frac{\wp'(u) - \wp'(v)}{\wp(u) - \wp(v)}\right)^2$$ □

Example 11.9

Show that $\wp(z - \alpha) - \wp(z + \alpha) = \dfrac{\wp'(z)\wp'(\alpha)}{(\wp(z) - \wp(\alpha))^2}$

Solution

It follows by the addition formula, 11.15, and the fact that $\wp$ is even and $\wp'$ is odd, that

$$\wp(z + \alpha) + \wp(z) + \wp(\alpha) = \frac{1}{4}\left(\frac{\wp'(z) - \wp'(\alpha)}{\wp(z) - \wp(\alpha)}\right)^2$$

$$\wp(z - \alpha) + \wp(z) + \wp(\alpha) = \frac{1}{4}\left(\frac{\wp'(z) + \wp'(\alpha)}{\wp(z) - \wp(\alpha)}\right)^2$$

Subtracting then gives

$$\wp(z - \alpha) - \wp(z + \alpha) = \frac{(\wp'(z) + \wp'(\alpha))^2 - (\wp'(z) - \wp'(\alpha))^2}{4(\wp(z) - \wp(\alpha))^2}$$

The numerator reduces to $4\wp'(z)\wp'(\alpha)$, as required.

The comparison theorem can be used to derive a number of identities involving $\wp$. An example is given in Exercises 11.3. It can also be used to express $\wp(z)$ in terms of the Jacobian elliptic functions. The derivation of this result and further properties of $\wp(z)$ may be found in Chapter 1 of E. G. Phillips, *Some Topics in Complex Analysis*, Pergamon, 1966.

Exercise **11.3.1** Let $\wp(z) = \wp(z \mid \alpha, i\beta)$, where α and β are real. Use the definition of $\wp(z)$ to show that $\wp(z)$ takes real values on the imaginary axis.

Exercise **11.3.2** Use the definition of $\wp(z)$ to show that if $\wp(z) - 1/z^2 = \sum_{n=1}^{\infty} a_{2n} z^{2n}$ in a neighbourhood of the origin, then $a_{2n} = (2n+1)\sum_{\omega\in\Omega'} \omega^{-2n-2}$.

Exercise **11.3.2** Use Theorem 11.13 to show that

$$\wp^{(iv)}(z) = 120\wp^3(z) - 18g_2\wp(z) - 12g_3$$

Exercise **11.3.4** Let $\wp(\omega_k) = e_k, k = 1, 2, 3$, where $\omega_3 = -(\omega_1 + \omega_2)$. Use Theorem 11.13 to show that $\wp''(\omega_1) = 2(e_1 - e_2)(e_1 - e_3)$, and find similar expressions for $\wp''(\omega_2)$ and $\wp''(\omega_3)$.

Exercise **11.3.5** Use the addition formula, Theorem 11.15, and the differential equation for $\wp$ to show that

$$\wp(z+\alpha) + \wp(z-\alpha) = \frac{(\wp(z) + \wp(\alpha))(2\wp(z)\wp(\alpha) - g_2/2) - g_3}{(\wp(z) - \wp(\alpha))^2}$$

Exercise **11.3.6** Use the second addition formula for $\wp(z)$ to deduce that

$$\wp(2z) + 2\wp(z) = \frac{1}{4}\left(\frac{\wp''(z)}{\wp'(z)}\right)^2 \qquad \text{(duplication formula)}$$

Use this result and the differential equation for $\wp$ to show that

$$\wp(2z) = \frac{\wp^4(z) + g_2\wp^2(z)/2 + 2g_3\wp(z) + g_2{}^2/16}{4\wp^3(z) - g_2\wp(z) - g_3}$$

Exercise **11.3.7** Prove, by induction, that $\wp(nz)$ is a rational function of $\wp(z)$ for $n \in \mathbb{N}$.

Exercise **11.3.8** Use the comparison theorem, 11.11(ii), to prove that

$$\wp(z) + \wp(z+\omega_1) + \wp(z+\omega_2) + \wp(z+\omega_1+\omega_2) = 4\wp(2z)$$

Exercise **11.3.9** Let $w = \wp(z)$. Show that

$$z = \int_{\infty}^{w} \frac{d\tau}{(4\tau^3 - g_2\tau - g_3)^{1/2}}$$

Let $\wp(z)$ have primitive periods $2\omega_1$ and $2\omega_2$ such that $\wp(\omega_1) = 1$, $\wp(\omega_2) = 0$ and $\wp(\omega_3) = \wp(\omega_1 + \omega_2) = -1$. Use the Schwarz–Christoffel transformation to show that $\wp$ maps the square with vertices 0, ω_1, $i\omega_1$ and $(1+i)\omega_1$ to the real line and the inside of the square to the lower half-plane. Show also that $\omega_1 = \int_0^1 (1-t^4)^{-1/2}\, dt$, $\omega_2 = -(1+i)\omega_1$ and $\omega_3 = i\omega_1$.

Exercise ***11.3.10** Prove that the double series

$$\sum_{\omega\in\Omega'} \left(\frac{1}{(z-\omega)^2} - \frac{1}{\omega^2}\right)$$

is absolutely convergent for any fixed $z \notin \Omega'$.

Bibliography

Ahlfors L. V. (1979). *Complex Analysis.* 3rd edn. London: McGraw-Hill.

Bak J. and Newman, D. J. (1982). *Complex Analysis.* New York: Springer-Verlag.

Brown J. W. and Churchill R. V. (1993). *Fourier Series and Boundary Value Problems.* 5th edn. London: McGraw-Hill.

Brown J. W. and Churchill R. V. (1996). *Complex Variables and Applications.* 6th edn. London: McGraw-Hill.

Conway J. B. (1993). *Functions of One Complex Variable.* 2nd edn. New York: Springer-Verlag.

Copson E. T. (1962). *Theory of Functions of a Complex Variable.* London: Oxford Univ. Press.

Derrick W. R. (1984). *Complex Analysis and Applications.* 2nd edn. Belmont CA: Wadsworth.

Dettman J. W. (1965). *Applied Complex Variables.* London: Collier-Macmillan.

Grove E. and Ladas G. (1974). *Introduction to Complex Variables.* Boston: Houghton-Mifflin.

Jameson G. J. O. (1970). *A First Course on Complex Functions.* London: Chapman and Hall.

Knopp K. (1945). *Theory of Functions.* New York: Dover.

Levinson N. and Redheffer R. M. (1988). *Complex Variables.* New York: McGraw-Hill.

Marsden J. E. and Hoffman M. J. (1987). *Basic Complex Analysis.* 2nd edn. New York: Freeman.

Mathews J. H. (1982). *Basic Complex Variables.* London: Allyn and Bacon.

Moretti G. (1964). *Functions of a Complex Variable.* Englewood Cliffs NJ: Prentice Hall.

Nevanlinna R. and Paatero V. (1964). *Introduction to Complex Analysis.* London: Addison-Wesley.

Phillips E. G. (1949). *Functions of a Complex Variable.* 6th edn. London: Oliver & Boyd.

Phillips E. G. (1966). *Some Topics in Complex Analysis.* Oxford: Pergamon.

Priestley H. A. (1990). *Introduction to Complex Analysis.* 2nd edn. Oxford: Clarendon.

Silverman R. A. (1967). *Introductory Complex Analysis.* Englewood Cliffs NJ: Prentice Hall.

Spiegel M. R. (1964). *Theory and Problems of Complex Variables.* New York: Schaum.

Titchmarsh E. C. (1939). *The Theory of Functions.* 2nd edn. London: Oxford Univ. Press.

Walker P. L. (1974). *An Introduction to Complex Analysis.* London: Adam Hilger.

Whittaker E. T. and Watson G. N. (1952). *A Course of Modern Analysis.* 4th edn. London: Cambridge Univ. Press.

Wunsch A. D. (1994). *Complex Variables with Applications.* 2nd edn. New York: Addison-Wesley.

Answers to Selected Exercises

Exercises

1.1.1 (ii) $12 + 13i$ (iii) $-117 - 44i$ (v) $110/97 - 5i/97$

1.1.3 (i) 0 (ii) 3/5

1.1.5 $a = 1$ and $b = 10$. Other two solutions are then -2 and $1 - 2i$.

1.1.7 $z = 1, -1/2 \pm i\sqrt{3}/2$

1.1.8 $\mathrm{Re}((2 + i)z) \leqslant 3/2$, etc.

Exercises

1.2.2 Use $|z|^2 = z\bar{z}$ to simplify the left-hand side, etc.

1.2.3 Letting $z = x + iy$ gives $x^2/27 + y^2/36 = 1$.

1.2.5 (ii) is closed and simply connected, (iii) is closed and bounded and so compact, (v) is connected and bounded.

1.2.7 The image is $2aX + 2bY = U(1 + \alpha) + (1 - \alpha)$, on the Riemann sphere, where $\alpha = r^2 - a^2 - b^2$, i.e. a circle, not passing through N.

Exercises

1.3.2 (i) 2^{500} (iii) $(\sqrt{3} + i)/4$ (vi) $\pm\sqrt[4]{2}(\alpha + i\beta)$ $\pm\sqrt[4]{2}(-\beta + i\alpha)$ where $\alpha = \cos(\pi/24)$ and $\beta = \sin(\pi/24)$, etc.

1.3.3 Roots are $z = -i, (-2 - 4i)/3$.

1.3.6 (i) Sixth roots of 1 are $\pm 1, \pm(1 + i\sqrt{3})/2, \pm(-1 + i\sqrt{3})/2$.

1.3.7 (i) Order is $2k/3$ (ii) order is $2k$

Exercises

1.4.1 (ii) $u = 2x^2 - 2y^2 - x$ and $v = 4xy - y$; f is a surjection but not an injection
(iv) $u = -(\tan^{-1}(y/x))^{-1}$, $-\pi < \tan^{-1}(y/x) \leqslant \pi$, and $v \equiv 0$; f is not a surjection and not an injection.

1.4.3 $u = \left(\left(x + \sqrt{x^2 + y^2}\right)/2\right)^{1/2}, v = \left(\left(-x + \sqrt{x^2 + y^2}\right)/2\right)^{1/2}$

1.4.5 (i) $\dfrac{\sin(n\theta/2)\sin((n + 1)\theta/2)}{\sin(\theta/2)}$ (ii) $\dfrac{\sin((n + 1)\phi/2)\cos(\theta + n\phi/2)}{\sin(\phi/2)}$

1.4.6 (ii) $z = (2n \pm 1/3)i\pi$, $n \in \mathbb{Z}$.

1.4.7 $z = (n + 1/2)\pi + i\tanh^{-1}(1/2)$

1.4.11 (iii) $((1 + i)^6)^{1/2} = 2(1 - i)$ whereas $(1 + i)^3 = -2(1 - i)$

Exercises

2.1.3 (i) $2i$ (ii) 0

2.1.4 (ii) Letting $z \to 0$ along the positive x-axis and along the line $y = x$ gives two different answers for the given limit.

2.1.8 The image of $|z| = a$ under $w = u + iv = \mathrm{Log}\, z$ is the line segment given by $u = \mathrm{Log}\, a$ and $v = \theta$, $-\pi < \theta \leqslant \pi$.

2.1.9 (i) ± 2 and $\pm 2i$ (ii) ± 1 (iii) ± 1

2.1.10 (b) Note that $g(z) = (z - 1)^{1/2}(z + 2)^{1/2}(z - i)^{1/2}$

2.1.11 $g(z)$ will only resume its initial value after 12 circuits.

Exercises

2.2.4 (ii) f is nowhere differentiable
(iii) f is entire with $f'(z) = \sinh z$
(vi) f is differentiable at $z = n\pi + \pi/2, n \in \mathbb{Z}$ only, with $f'(n\pi + \pi/2) = 0$.

2.2.5 $f(z) = aiz + b$, where a and b are real constants.

2.2.7 Since f is entire, the Cauchy–Riemann equations are satisfied everywhere. Hence, because the gradient of the second family of curves is $-v_x/v_y$, the product of the gradients is -1.

2.2.9 $R(x, y) = ke^{x^2-y^2}$, where k is a real constant, so $f(z) = ke^{z^2}$.

2.2.11 f is differentiable at all points except $z = 0$ and those for which $\operatorname{Arg} z = \pi$. Where it exists, $f'(z) = 1/2z^{1/2}$.

2.2.12 $f(z) = \tanh^{-1} z \quad \Rightarrow \quad f'(z) = 1/(1 - z^2), z \neq \pm 1$

2.2.13 (i) $v(x, y) = x^3 - 3xy^2 + k \quad \Rightarrow \quad f(z) = iz^3 + ik$

Exercises

2.3.1 (i) $11/7$ (ii) $\pi^2/2$ (iii) e

2.3.2 (ii) $\pm 3i$ are branch points of the associated multifunction and -1 is a pole of order 2
(iv) $(2n + 1)i\pi/2, n \in \mathbb{Z}$, are simple poles
(v) 0 is a removable singular point
(vi) 0 is an essential singular point

2.3.3 (ii) $in\pi, n \in \mathbb{Z}$, are zeros of order 2
(iii) 0 is a simple zero

Exercises

3.1.1 $-2/3$

3.1.2 (i) i (ii) $1 + i$ and $1 + i(1 - \pi/2)$

3.1.5 (i) $8i$ (ii) $9 + 24i$

3.1.8 $2\pi i$

3.1.12 (i) 0 and $-4iR^{3/2}/3$ (ii) 0 and $-2Ri\pi$

Exercises

3.2.2 (i) 0 (ii) -3π (iii) 3π

3.2.3 (second part) $(\sin a)/a, a \neq 0$

3.2.4 (ii) $\sqrt{\pi}e^{-b^2}/2$

3.2.7 $4\pi i$

3.2.9 $2\pi i, \pi/2$

3.2.11 $4\pi i, 2\pi$

Exercises

3.3.7 f has three zeros and two poles inside $|z| = 5$ and f'/f is the given integrand. Hence the given integral $= 2\pi i$.

3.3.9 The equation has one root on the negative real axis, two roots in each of the first and fourth quadrants, and one root in each of the second and third quadrants.

3.3.12 Let $f(z) = 16z^4$ and $g(z) = -8z + 3$ to show that the given equation has three zeros inside $|z| = 1$.

Exercises

4.1.2 If $z_n = i^n$ then (z_n) diverges, whereas $|z_n| = 1 \to 1$ as $n \to \infty$.

4.1.4 (i) $(-1 - i)/2$ (ii) i

Exercises 4.2.8 (i) The series is absolutely convergent for all $z \in \mathbb{C}$.
(ii) $1/2$ (iii) $1/\sqrt{3}$
(iv) The series converges only for $z = 0$.
(v) 1, using the nth root test
(vi) $1/2$, using the nth root test

Exercises 4.3.1 (ii) $1/z^2 = \sum_{n=0}^{\infty} \frac{(-1)^n(n+1)}{4}\left(\frac{z-2}{2}\right)^n$ for $|z-2| < 2$

4.3.2 (i) $(1+z)^{-1} = \sum_{n=0}^{\infty}(-1)^n \frac{(z-3)^n}{4^{n+1}}$ for $|z-3| < 4$

(ii) $(1+z)^{-2} = \sum_{n=1}^{\infty}(-1)^{n+1} n z^{n-1}$ for $|z| < 1$

4.3.5 (i) $S(z) = \sqrt{(2/\pi)} \sum_{n=0}^{\infty} \frac{(-1)^n z^{4n+3}}{(4n+3)(2n+1)!}$ for all $z \in \mathbb{C}$

(ii) $f(z) = \sum_{n=0}^{\infty} c_n z^n$, where $c_n = \sum_{r=0}^{n} a_r$ and a_r is the coefficient of z^r in the expansion of $\sin z$. Then $f^{(n)}(0) = n! c_n$.

4.3.6 By hypothesis, $f(z) = \sum_{n=0}^{\infty} a_n z^n$ for all z and $|a_n| \leqslant k r^{m-n}$ for any $r > 0$, etc.

Exercises 4.4.1 (i) $\csc z = \frac{1}{z} + \frac{z}{6} + \frac{7z^3}{360} + \frac{31z^5}{15\,120} + \dots, \; 0 < |z| < \pi$

4.4.2 Let $w = z - 1$. Then $z^2 \cos(z-1)^{-1} = (w^2 + 2w + 1) \sum_{n=0}^{\infty} \frac{(-1)^n}{w^{2n}(2n)!}$

4.4.3 (ii) Let $w = z - 1$. Then $f(z) = \frac{e \cdot e^w}{2w(1 + w/2)}$, so that

$$f(z) = \frac{e}{2w}\left(\sum_{n=0}^{\infty} \frac{w^n}{n!}\right)\left(\sum_{n=0}^{\infty}(-1)^n (w/2)^n\right) \qquad (w \neq 0)$$

where the first series converges for all w and the second series converges for $|w| < 2$, etc.

4.4.8 (ii) f has a removable singular point at 0 and $\mathrm{Res}_0 f(z) = 0$.
(iv) f has poles of order 3 at $ni\pi$, $n \in \mathbb{Z}$ and $\mathrm{Res}_{ni\pi} f(z) = (-1)^{3n+1}/2$
(vii) f has an essential singular point at 0 and $\mathrm{Res}_0 f(z) = 0$.

4.4.10 (ii), (iv) ∞ is an essential singular point of f
(vii) f is analytic at ∞.

Exercises 5.1.1 (i) $-\pi i$ (ii) $2\pi e i$

5.1.2 $2\pi i$

5.1.4 (ii) $\mathrm{Res}_{\pm 2i} f(z) = \mp i/8$ (iv) $\mathrm{Res}_{-1} f(z) = 0$, $\mathrm{Res}_{\pm i} f(z) = \pm 5/4i$

5.1.5 (ii) $\mathrm{Res}_{n\pi i} f(z) = n\pi i$ (iv) $\mathrm{Res}_{n\pi} f(z) = e^{n\pi}$

5.1.7 (iii) $\frac{-2\sinh \pi}{\pi}$

5.1.8 $\int_{\mathscr{C}} f(z)dz = \pi \mathrm{Log}\, 2 + i\pi^2/2$, so that $\int_0^{\infty} \frac{\mathrm{Log}(x^2+1)}{x^2+1} dx = \pi \mathrm{Log}\, 2$

Exercises

5.2.1 (iii) $5\pi/32$ (iv) $\pi a^2/(1-a^2)$ (v) $\pi\sqrt{2}/4$ (vii) $2\pi/ab$

5.2.3 (ii) $\pi/4a$ (iii) $\pi/6a^3$ (vi) $\pi\sqrt{3}/6$ (vii) $4\pi/3\sqrt{3}$

5.2.4 (i) $\pi(1+e^{-2a})/4a$ (iii) $\pi(\cos 1+\sin 1)/e$
(iv) $(m^2+3m+3)\pi/8e^m$

5.2.6 (ii) $\pi/6$ (iv) $3\pi/8$ (v) π

5.2.7 $\pi\sqrt{3}/3$

Exercises

5.3.1 (i) $\pi\sqrt{2}/4a^3$ (iii) $4\pi\sqrt{3}/27$

5.3.3 $\pi\sqrt{3}/9$

5.3.4 (ii) $-\pi/4$, using contour of Fig. 5.7(i) say
(v) $(\pi^2\cos\pi\lambda)/(\sin^2\pi\lambda), -1<\lambda<0$, using contour of Fig. 5.7(iv) say
(vi) $2\pi/3$, using contour of Fig. 5.7(iv) say

5.3.7 (i) (first part) $\dfrac{\pi}{\cos(\pi\lambda/2)}, \ -1<\lambda<1$ (iii) $\pi^2/4$

5.3.9 $1/2-\pi/(2\sinh\pi)$

Exercises

5.4.1 (i) $\pi^2\csc^2 a\pi$ (iii) $\dfrac{\alpha(\sin 2\alpha+\sinh 2\alpha)}{2a^4(\cosh 2\alpha-\cos 2\alpha)}-\dfrac{1}{2a^4}, \alpha=\pi a/\sqrt{2}$
(iv) $\pi^4/90$

5.4.3 (i) $\dfrac{\pi\operatorname{csch}\pi a}{2a}+\dfrac{1}{2a^2}$ (ii) $\pi^3/16$

5.4.4 $\pi/8$

Exercises

6.1.1 (b) $w=f(z)$ gives $(u-1)^2+v^2=e^{2k}$, a circle

6.1.3 $1, \pm i$ and $(1\pm\sqrt{2}i)/3$

6.1.4 Note that 1 is a critical point of f.

6.1.8 $F(z)=e^{iz}$, for example, maps the real axis to $|w|=1$.

Exercises

6.2.5 (i) $w=2z-(2+i)$ (ii) $w=\dfrac{(i-1)z}{z-2}$, say

6.2.6 $f(z)=\dfrac{i-z}{i+z}$

6.2.8 $f(z)=\dfrac{i(z+1)}{z-1}$

6.2.9 (iii) $f(z)=\dfrac{1-z}{z+2}$

6.2.11 Converse is true.

6.2.13 If z^* is the inverse point of z with respect to the circle with equation (6.6) and $z^*=f(z)$, then $f(z)=(-\overline{B}z-c)/a\overline{z}+B)$.

6.2.14 $w=\dfrac{e^{i\theta}(z-\mu)}{z-\overline{\mu}}, \theta\in\mathbb{R}$ and $\operatorname{Im}\mu>0$

Exercises

6.3.1 The half-plane $\operatorname{Re} w \leqslant 0$

6.3.2 $w = \dfrac{(iz+a)^2 - i(iz-a)^2}{(iz+a)^2 + i(iz-a)^2}$

6.3.7 The ellipse has equation $u^2/9 + v^2/5 = 1$. Hence a suitable choice is $w = az + b/z$, where $a = (3+\sqrt{5})/8, b = 2(3-\sqrt{5})$.

6.3.8 $|w| < 1$

6.3.15 The annular region $e^{-a} < |w| < e^a$.

6.3.17 The given region is mapped to $\operatorname{Im} w > 0$.

6.3.18 $w = (\sin z)^{1/2}$

Exercises

6.4.1 (i) $z = (1 - \cosh w)/2$

6.4.2 $w = f(z) = K\displaystyle\int_0^z \zeta^{-1/2}(1-\zeta)^\alpha d\zeta \qquad K = \frac{a(2\alpha+1)\Gamma(2\alpha)i}{\alpha 2^{2\alpha}\Gamma^2(\alpha)}$

using the beta function and the duplication formula.

6.4.3 (second part) $z = (w^2 + a^2)^{1/2}/a$

6.4.7 Side length $= \dfrac{2^{1/3}\Gamma^2(1/3)}{6\Gamma(2/3)}$

Exercises

7.1.1 (ii) $w = az + b(1-z^2)^{1/2}$, valid for all z, except along the branch cut joining -1 to 1.

7.1.2 (i) $w = a(1-2z^2) + b\left(z + \displaystyle\sum_{n=1}^{\infty} \frac{(4n-6)(4n-10)\dots(-2)z^{2n+1}}{(2n+1)!}\right) \forall z$

7.1.4 $P_0(z) = 1,\ P_1(z) = z,\ P_2(z) = (3z^2-1)/2$
$P_3(z) = (5z^3 - 3z)/2,\ P_4(z) = (35z^4 - 30z^2 + 3)/8$

7.1.5 $w = a\left(1 + \displaystyle\sum_{n=1}^{\infty}\frac{(-2)^n z^n}{(2n-1)(2n-3)\dots 1}\right) + bz^{1/2}\sum_{n=0}^{\infty}\frac{(-1)^n z^n}{n!}, \forall z$

except along the branch cut for $z^{1/2}$.

7.1.6 (ii) $w = az^{-1/2}\sin z + bz^{-1/2}\cos z,\ z \neq 0,\ z \notin \mathbb{R}^-$

(v) $w = \dfrac{z(a+bz)}{1-z^2}, z \neq \pm 1$

7.1.7 If $w(0) = 0$, then $w = az^{2/3}e^{z/3}$.

7.1.10 One solution is $w = w_1(z) = 1/z$. Then, by (7.23), a second solution is $w = w_2(z) = -e^{1/z}$.

Exercises

7.2.1 When $k = 1/2$, $w = a\cosh(2z^{1/2}) + b\sinh(2z^{1/2})$. When $k = 1$, $w = w_1(z) = \sum_{n=0}^{\infty} z^n/(n!)^2$ and $w = w_2(z) = w_1 \operatorname{Log} z - 2\sum_{n=0}^{\infty}\Phi(n)z^n/(n!)^2$ are two solutions.

7.2.3 One solution is $w = w_1(z) = e^z$. A second solution is then $w = w_2(z) = e^z \operatorname{Log} z - \sum_{n=0}^{\infty}\Phi(n)z^n/n!$.

7.2.4 $w = w_1(z) = z/(1-z)^2$ and $w = w_2(z) = w_1\operatorname{Log} z + 1/(1-z)$ are two linearly independent solutions.

Exercises **7.3.2** The required series solution is $w = a\sum_{n=0}^{\infty}\frac{(-1)^n}{(n!)^2 z^n}$.

7.3.4 Note that $a = -n$, $b = n + 1$ and $c = 0$ gives (7.37).

7.3.13 The general solution to $w'' + zw = 0$ is

$$w = az^{1/2}J_{1/3}(2z^{3/2}/3) + bz^{1/2}J_{-1/3}(2z^{3/2}/3)$$

Exercises **7.4.1** $w = \int_{\mathscr{C}} e^{z\zeta + 1/\zeta}\zeta^{c-2}d\zeta$, where $[e^{z\zeta+1/\zeta}\zeta^c]_{\mathscr{C}} = 0$

7.4.3 (second part) Note that for the given particular solution, $w(0) = K\Gamma(a)\Gamma(c-a)/\Gamma(c)$. Then using the Maclaurin series for e^{zt} and integrating termwise gives the final result.

7.4.4 (ii) Let $\zeta = z - \tau$ in the definition and use Cauchy's formula for derivatives. $H_1(z) = 2z$, $H_2(z) = 4z^2 - 2$ and $H_3(z) = 8z^3 - 12z$.

7.4.5 (second part) From the given solution, using the Maclaurin series for $\cos zt$, integrating termwise, letting $u = t^2$ and using the duplication formula for the gamma function, it follows that

$$w = \frac{K\sqrt{\pi}\Gamma(v + 1/2)}{(z/2)^v}J_v(z)$$

7.4.7 $w = K\int_{\mathscr{C}}\frac{(1-\zeta^2)^n}{(\zeta - z)^{n+1}}d\zeta$, where $[(1-\zeta^2)^{n+1}(\zeta - z)^{-n-2}]_{\mathscr{C}} = 0$, or

$$w = K\int_{\mathscr{C}}\frac{(\zeta - z)^n}{(1-\zeta^2)^{2n+1}}d\zeta, \text{ where } [(1-\zeta^2)^{-2n}(\zeta - z)^{n-1}]_{\mathscr{C}} = 0$$

The first class of solutions includes both the given particular solutions as special cases. Integrating by parts n times and using Rodrigues' definition of $P_n(z)$ gives the final result.

Exercises **8.1.2** (second part) π

8.1.3 (i) $\pi/2e$

8.1.4 $f(s) = -i\sqrt{\pi/2}\,\mathrm{sgn}(s)$

8.1.5 (i) $\hat{f}(s) = -i\sqrt{\pi/2}e^{-a|s|}\,\mathrm{sgn}(s)$

8.1.7 $\hat{f}_s(s) = \frac{\sqrt{2}s\cos(\pi s/2)}{1 - s^2}$; $\frac{\sqrt{2}\pi}{4}$

8.1.9 $\hat{f}_s(s) = se^{-s^2/2}$

8.1.10 $\hat{f}_c(s) = \hat{f}_s(s) = s^{-1/2}$

8.1.13 $\hat{f}(s) = \frac{1}{2}(1 + s^2)e^{-s^2/2} \quad \Rightarrow \quad f(x) = e^{-x^2/2}(1 - x^2/2)$

Exercises **8.2.2** (ii) $\bar{f}(s) = \mathrm{Log}\left(\frac{s^2 + b^2}{s^2 + a^2}\right)$

8.2.3 $\bar{f}(s) = \frac{e^{-a/s}}{s^{3/2}}$

8.2.4 $\bar{f}(s) = \frac{a\pi}{(a^2s^2 + \pi^2)(1 - e^{-as})}$

8.2.5 (ii) $f(t) = \frac{1}{2}\cos t - \frac{1}{2}\cos\sqrt{3}t$ (iii) $f(t) = \frac{1}{54}(\sinh 3t - \sin 3t)$

8.2.6 (ii) $f(t) = \dfrac{1}{2a}(at\cos at + \sin at)$

8.2.7 (ii) $f(t) = \dfrac{e^{-(t-1)}}{2}(t-1)^2 H(t-1)$

8.2.8 $\mathscr{L}^{-1}(e^{-as^{1/2}}) = \dfrac{ae^{-a^2/4t}}{2\sqrt{\pi}t^{3/2}}$

8.2.10 $\mathscr{L}^{-1}\left(\dfrac{1}{s^{1/2}(s^{1/2}+a)}\right) = e^{a^2 t}\operatorname{erfc}(a\sqrt{t})$

8.2.13 (i) $f(t) = e^t(1+t)$

8.2.15 $f(t) = g(t) + \displaystyle\sum_{n=1}^{\infty}\int_0^t g(u)\frac{(t-u)^{2n-1}}{(2n-1)!}du$

Exercises

8.3.1 (ii) $y(t) = -\dfrac{7}{4}e^{-t} + \dfrac{1}{5}e^{-2t} + \dfrac{11}{20}e^{3t}$

(iii) $y(t) = 2 + \dfrac{t}{2} - \dfrac{t^2}{2} + \dfrac{t^3}{6} - 2e^{-t}\cos t - \dfrac{1}{2}e^{-t}\sin t$

8.3.4
$$y = \sum_{k=1}^{2}\left(\frac{1}{2} - e^{t-k} + \frac{1}{2}e^{2(t-k)}\right)H(t-k)$$
$$-\sum_{k=3}^{4}\left(\frac{1}{2} - e^{t-k} + \frac{1}{2}e^{2(t-k)}\right)H(t-k)$$

8.3.6 $y(t) = -\dfrac{k}{2}e^t t^2$

8.3.7 $y(t) = 4t^2 - 2$

Exercises

9.1.2 (ii) $f(z) = z\operatorname{Log} z$ is analytic on $\operatorname{Im} z > 0$ and $\phi = \operatorname{Re} f$

9.1.3 $f(z) = -1/z$ say

9.1.5 (iii) $\phi(r) = (\operatorname{Log} r)/(\operatorname{Log} k)$ (iv) $\phi = r^2\cos 2\theta$

9.1.6 $\phi(x,y) = \dfrac{1}{\pi}\left(-\tan^{-1}\left(\dfrac{y}{x-1}\right) + \tan^{-1}\left(\dfrac{y}{x}\right) - \tan^{-1}\left(\dfrac{y}{x+1}\right)\right) + 1$

9.1.11 $u(r,\theta) = \dfrac{1}{2\pi}\displaystyle\int_0^{\pi}(P(r,\phi-\theta) - P(r,\phi+\theta))U(\phi)\,\mathrm{d}\phi$, where the domain of U is extended to $|z| = 1$ by $U(2\pi-\theta) = -U(\theta)$.

9.1.13 If $w = u + iv$ then $\phi(u,v) = e^{-2uv}\sin(u^2 - v^2)$.

9.1.15 (ii) $u(r,\theta) = \dfrac{1}{2\pi}\displaystyle\int_0^{2\pi}\frac{(r^2-1)U(\phi)}{r^2 - 2r\cos(\phi+\theta)+1}\mathrm{d}\phi, r > 1$

9.1.17 $u(x,y) = \dfrac{y}{2\pi}\operatorname{Log}\left(\dfrac{(1-x)^2+y^2}{x^2+y^2}\right) + \dfrac{x}{\pi}\tan^{-1}\left(\dfrac{y}{x^2+y^2-x}\right)$

Exercises

9.2.2 (ii) $T(x, y) = \dfrac{T_0}{\pi} \tan^{-1}\left(\dfrac{4xy}{(x^2+y^2)^2 - 1}\right)$

9.2.3 $T(x, y) = \dfrac{2K}{\pi} \tan^{-1}\left(\dfrac{\tanh(\pi y/2a)}{\tan(\pi x/2a)}\right)$

9.2.5 (i) $T(x, y) = \dfrac{2}{\pi} \sin^{-1}\left(\dfrac{1}{2}(((x+1)^2 + y^2)^{1/2} - ((x-1)^2 + y^2)^{1/2})\right)$

9.2.6 $T = \dfrac{4K}{3}\left(1 - \dfrac{1}{\pi}\tan^{-1}\left(\dfrac{2y}{x^2 + y^2)^2 - 1}\right)\right)$

9.2.8 $V(x, y) = V_0\left(\dfrac{x}{x^2+y^2} - 1\right)$

9.2.9 $V(x, y) = \dfrac{1}{\pi}\tan^{-1}\left(\dfrac{\sin y}{\sinh x}\right)$

9.2.11 $V(x, y) = \dfrac{V_0}{2\pi}(v + \pi)$, where $x = u + e^u \cos v$ and $y = v + e^u \sin v$

9.2.12 (ii) $F(w) = k \sin(\pi w/2a)$ and there are stagnation points at $\pm a$. The streamlines have equation $\sinh(\pi v/2a) = K \sec(\pi u/2a)$.

9.2.15 $F(w) = K((z-a) + (a+1)^2/(z-a))$, where $z = w + (w^2-1)^{1/2}$. The only stagnation point is $w = (2a^2 + 2a + 1)/(2a+1)$, i.e. C'.

9.2.16 $F(w) = a(1 + w^2)^{1/2}$

9.2.17 $F(z) = k\,\mathrm{Log}\left(\dfrac{z+a}{z-a}\right)$

Exercises

9.3.2 $\phi(x, t) = H(t - x/c)f(t - x/c)$, so that the displacement at $x = 0$ propagates down the string at speed c.

9.3.3 (ii) $\phi(x, t) = \dfrac{1}{2}(f(x+ct) + f(x-ct)) + \dfrac{1}{2c}\displaystyle\int_{x-ct}^{x+ct} g(\tau)\,d\tau$

9.3.5 $\phi(x, t) = c^2(1 - \cos t) + c^2 H(t - s/c)(1 - \cos(t - s/c))$

9.3.8
$$T(x, t) = T_1 + \frac{8(T_1 - T_0)}{\pi} \times \sum_{n=1}^{\infty} \frac{(-1)^n \exp(-(n-1/2)^2\pi^2 t/L^2)\cos((n-1/2)\pi X/L)}{2n-1}$$

9.3.10 $T(r, t) = K\left(1 - 2\displaystyle\sum_{n=1}^{\infty} \dfrac{e^{-a\alpha_n^2 t} J_0(\alpha_n r)}{\alpha_n J_1(\alpha_n)}\right)$

where α_n is the nth zero of $J_0(x)$.

Exercises

9.4.2 (ii) $\phi(x, t) = \dfrac{8h}{\pi^2}\displaystyle\sum_{n=1}^{\infty} \dfrac{(-1)^{n-1}\sin((2n-1)\pi x)\cos((2n-1)\pi ct)}{(2n-1)^2}$

9.4.3 $$u(x,t)=\frac{CL}{2}-\frac{4CL}{\pi^2}\sum_{n=1}^{\infty}\frac{\cos((2n-1)\pi x/L)\exp(-(2n-1)^2\pi^2kt/L^2)}{(2n-1)^2}$$

Exercises

10.1.1 $f_2(z)=2/(z+1)^3$

10.1.2 Use the given condition and induction to show that f is analytic for $|z|<2^n r$ for any $n\in\mathbb{N}$.

10.1.6 Note that $f_1(z)=-\text{Log}(1-z)$ for $|z|<1$, except along the branch cut of $-\log(1-z)$. The monodromy theorem is not contradicted since the region $1<|z-2|<3$ is not simply connected.

10.1.7 $r_1=2$ and $r_2=\sqrt{5}$. Then $f_1(z)=1/(2-z)^2$ for $|z|<2$ and $f_2(z)=f_1(z)$ in the non-empty intersection of their domains. Notice that the domains of f_1 and f_3 do not intersect.

10.1.9 Note that $f_1(z)=\text{Log}(1+z)$ for $|z|<1$ and expanding $\text{Log}(1+z)$ in a Taylor series about $z=i$ gives $f_2(z)$ for $|z-i|<\sqrt{2}$.

10.1.11 Let $z=e^{2\pi ip/q}$, $p,q\in\mathbb{N}$, so that if $n>q$, $z^{n!}=1$. It then follows by the series definition of f that these points are singular points of f which lie on $|z|=1$.

10.1.13 For any $x\in AB$, $f_2(x)=1/\overline{f_1(x)}=f_1(x)/|f_1(x)|^2=f_1(x)$. Then use the definition of $f_2'(z)$, as in the proof of 10.4, to show that f_2 is analytic on $\mathscr{R}_2$.

10.1.15 (i) The mapping $w=-e^{-i\phi}z$ rotates AOB through $\pi-\phi$ radians so that $O'A'$ lies on the real axis.
(ii) The mapping $w=-e^{i\theta}(z-a)$ maps B to the origin and then rotates AOB through $-(\pi-\theta)$ radians, so that $A'B'$ lies on the real axis.

10.1.16 $z=f_1^{-1}(w)=K\displaystyle\int_{-1}^{w}\frac{d\zeta}{(1-\zeta^2)^{3/4}}$, where $2=K\displaystyle\int_{-1}^{1}\frac{d\zeta}{(1-\zeta^2)^{3/4}}$

Exercises

10.2.1 (i) 3/2 (ii) 3

10.2.3 *Hint*: Use the Maclaurin series expansion of $\text{Log}(1+z)$ to show that $|z_n|/2<|\text{Log}(1+z_n)|<3|z_n|/2$ for $|z_n|<1/2$.

10.2.5 The product converges to 1/2 but is not absolutely convergent, by 10.7, since $\sum_{n=1}^{\infty}1/(n+1)$ diverges.

10.2.6 Using 10.7, (i) converges for $|z|<1$, (ii) converges for $\text{Re}\,z>1$ and (iii) converges for all z.

10.2.11 Note that since $\sum_{n=1}^{\infty}n^{-3/2}$ converges, a suitable choice for $s(n)$ in 10.11 is $s=2$ for all n.

10.2.12 Note that by hypothesis and 5.15,

$$\frac{f'(z)}{f(z)}=\frac{f'(0)}{f(0)}+\sum_{n=1}^{\infty}\left(\frac{1}{z-z_n}+\frac{1}{z_n}\right)$$

Integrating termwise and taking exponentials gives the result.

10.2.15 $f(z)=z^2e^{g(z)}\displaystyle\prod_{n=1}^{\infty}(1-z^2/n^2)^2$, where g is entire.

Exercises

10.3.4 Note $|e^{-zt^2}|\leqslant e^{-\rho t^2}$ for $\text{Re}\,z\geqslant\rho>0$ and $\int_0^{\infty}e^{-\rho t^2}\,dt$ converges. Then the results follow by 10.16 and 10.15. The final result follows by applying 10.1 to $G(z)=F(z)-\sqrt{\pi}/2z^{1/2}$.

10.3.6 From Example 10.16, $\Gamma(z) = G(z) + H(z)$, where $H(z) = \int_1^\infty e^{-t}t^{z-1}dt$ and is entire, and $G(z) = \int_0^1 e^{-t}t^{z-1}dt$, $\operatorname{Re} z > 0$. By Theorem 4.16 $G(z) = \sum_{n=0}^\infty (-1)^n/n!(n+z)$, and this series is uniformly convergent in any compact subset of $\mathbb{C}$ excluding 0, −1, −2, ...

10.3.7 Using Euler's limit definition,
$\Gamma(z)\Gamma(1-z) = (z\prod_{n=1}^\infty (1 - z^2/n^2))^{-1}$
and the result follows by Example 10.12.

10.3.11 $J_\nu(z) = \sum_{n=0}^{\infty} \frac{(-1)^n(z/2)^{2n+\nu}}{n!2\pi i} \int_{\mathscr{H}} e^\zeta \zeta^{-n-\nu-1} d\zeta$, etc.

Exercises 10.4.1 $e^{1/x} \sim \sum_{n=0}^{\infty} \frac{1}{x^n n!}$ and $e^{-x} \sim 0$

10.4.5 Let $S_k(t) = \sum_{n=2}^k a_n t^{-n}$. Then $\int_x^\infty S_k(t)dt = \sum_{n=1}^{k-1} a_{n+1}/nx^n$. By hypothesis, $\lim_{x\to\infty} x^k(f(x) - S_k(x)) = 0$ for $k \geqslant 2$. Use this to show that $\lim_{x\to\infty} x^{k-1}\int_x^\infty (f(t) - S_k(t))\,dt = 0$ for $k \geqslant 2$.

10.4.6 $f(z) \sim \sum_{n=0}^{\infty} (-1)^n \frac{(2n)!}{z^{2n+1}}$ for $|\operatorname{Arg} z| < \pi/2$

10.4.8 Letting $u = 1 + t$ in the given integral gives

$$K_\nu(z) = \frac{\sqrt{\pi} z^\nu e^{-z}}{\sqrt{2}\Gamma(\nu + 1/2)} \int_0^\infty e^{-zt} t^{\nu-1/2}(1 + t/2)^{\nu-1/2} dt$$

Then use Watson's lemma with $g(t) = t^{\nu-1/2}(1 + t/2)^{\nu-1/2}$.

10.4.10 $\psi'(z) = \sum_{n=0}^{\infty} \frac{1}{(z+n)^2} = \sum_{n=0}^{\infty} \int_0^\infty te^{-(z+n)t}dt$, $\operatorname{Re} z > 0$
(See the proof of Theorem 10.20.) Then

$$\psi'(z) = \int_0^\infty te^{-zt} \sum_{n=0}^{\infty} (e^{-t})^n dt, \text{ etc.}$$

Exercises 11.1.6 Note $w = \tau^{1/2}$, where $\tau = (1-\zeta)/(1+\zeta)$ and $\zeta = \operatorname{sn} z$. The required image is a quarter-circular domain in the fourth quadrant.

11.1.8 Letting $k = 1$ in (i) gives $\operatorname{sech} z = 1 - \frac{z^2}{2} + \frac{5z^4}{24} - \ldots$

11.1.9 (iv) Let $g(z) = \frac{1 - \operatorname{dn} z}{\operatorname{sn} z}$ then $g'(z) = \frac{\operatorname{cn} z(1 - \operatorname{dn} z)}{\operatorname{sn}^2 z}$

and $f(z) = \operatorname{Log}(g(z)) \Rightarrow f'(z) = \frac{g'(z)}{g(z)} = \frac{\operatorname{cn} z}{\operatorname{sn} z}$

11.1.12 (i) $\operatorname{cn} 2K = \dfrac{\operatorname{cn}^2 K - \operatorname{sn}^2 K \operatorname{dn}^2 K}{1 - k^2 \operatorname{sn}^4 K} = \dfrac{-k'^2}{1-k^2} = -1$

and since $\operatorname{sn} 2K = 0$

$$\operatorname{cn}(z + 2K) = \frac{\operatorname{cn} 2K \operatorname{cn} z - \operatorname{sn} 2K \operatorname{dn} 2K \operatorname{sn} z \operatorname{dn} z}{1 - k^2 \operatorname{sn}^2 2K \operatorname{sn}^2 z} = -\operatorname{cn} z$$

11.1.14 $\operatorname{sn}(z + iK') = 1/(k \operatorname{sn} z)$ and so in a neighbourhood of 0,

$$\operatorname{sn}(z + iK') = \frac{1}{kz} + \frac{(1+k^2)z}{6k} + \dots$$

using the Maclaurin series expansion of $\operatorname{sn} z$. So $\operatorname{Res}_{iK'} \operatorname{sn} z = 1/k$. Similarly, $\operatorname{Res}_{iK'} \operatorname{cn} z = -i/k$ and $\operatorname{Res}_{iK'} \operatorname{dn} z = -i$.

11.1.15 For example, $\operatorname{sn}(iK'/2) = \dfrac{i \operatorname{sn}(K'/2, k')}{\operatorname{cn}(K'/2, k'} = \dfrac{i\sqrt{1+k}}{\sqrt{1+k}\sqrt{k}} = \dfrac{i}{\sqrt{k}}$.

Exercises

11.2.4 Note that $f(\omega_k) = -f(\omega_k), k = 1, 2$

11.2.8 Use the same technique as for proving 11.9.

11.2.9 (i) Let $f(z) = \operatorname{sn} z$ and let $\mathscr{C}$ be any cell of f containing 0, $2K$, iK' and $2K + iK'$. Use Theorem 11.9 to obtain $4\pi K'$.

Exercises

11.3.1 Use the definition (11.16) to show $\overline{\wp(iy)} = \wp(iy)$.

11.3.7 Suppose $\wp(kz) = P(z)/Q(z)$, where P and Q are polynomials in $\wp(z)$. Then $\wp'(kz) = R(z)\wp'(z)/Q^2(z)$, where $R(z)$ is another polynomial in $\wp(z)$. Use Theorem 11.15 to show that $\wp((k+1)z) = S(z)\wp'^2(z) - P(z)/Q(z) - \wp(z)$, where $S(z)$ is another polynomial in $\wp(z)$.

11.3.8 Let $f(z)$ denote the left-hand side of the given identity and let $g(z) = 4\wp(2z)$. Show that f and g both have primitive periods ω_1 and ω_2, and a double pole at 0. Show also that both have principal part $1/z^2$ in their Laurent series about 0. Hence $f(z) = g(z) + \gamma$. Then let $z \to 0$ in this result.

Index